ENCYCLOPEDIA OF

FORENSIC SCIENCE

REVISED EDITION

ENCYCLOPEDIA OF FORENSIC SCIENCE

REVISED EDITION

SUZANNE BELL, PH.D.

BENNETT DEPARTMENT OF CHEMISTRY,
WEST VIRGINIA UNIVERSITY, MORGANTOWN

FOREWORD BY

BARRY A. J. FISHER

AMERICAN ACADEMY OF FORENSIC SCIENCES,
PAST PRESIDENT, 1998–1999
CRIME LABORATORY DIRECTOR,
LOS ANGELES COUNTY SHERIFF'S DEPARTMENT

PREFACE BY

ROBERT C. SHALER, PH.D.

DIRECTOR, FORENSIC SCIENCE PROGRAM,
PENNSYLVANIA STATE UNIVERSITY

Encyclopedia of Forensic Science, Revised Edition

Facts On File, Inc.
An imprint of Infobase Publishing
132 West 31st Street
New York NY 10001

Library of Congress Cataloging-in-Publication Data

Bell, Suzanne.
Encyclopedia of forensic science / Suzanne Bell; foreword by Barry A. J. Fisher; preface by Robert C. Shaler.—Rev. ed.
p. cm.
Includes bibliographical references and index.
ISBN-13: 978-0-8160-6799-2 (acid-free paper)
ISBN-10: 0-8160-6799-6 (acid-free paper) 1. Forensic sciences—Encyclopedias. I. Title.
HV8073.B425 2008
363.2503—dc22 2008005862

Facts On File books are available at special discounts when purchased in bulk quantities for businesses, associations, institutions, or sales promotions. Please call our Special Sales Department in New York at (212) 967-8800 or (800) 322-8755.

You can find Facts On File on the World Wide Web at http://www.factsonfile.com

Text design adapted by James Scotto-Lavino
Illustrations by Sholto Ainslie
Photo research by Suzanne M. Tibor

Printed in the United States of America

VB Hermitage 10 9 8 7 6 5 4 3 2 1

This book is printed on acid-free paper and contains 30 percent postconsumer recycled content.

To my husband, Mike, and to my departed mother—

thanks for all the years of faith and support.

Contents

FOREWORD

Hardly a week passes when I do not receive a phone call or e-mail from a college or high school student asking about careers in forensic science or wishing to interview someone for a class report they must do on forensic science. To be sure, much of this interest is a result of nearly a decade of media hype from such television programs as *CSI*, *Forensic Files*, and dating all the way back to the O. J. Simpson trial coverage. But for whatever reason, there is a genuine fascination about the subject. What is forensic science? What sort of academic preparation do I need to enter this field? Where do I turn for answers?

Crime laboratories and medical examiner/coroner's offices play a vital role in our nation's criminal justice system. Forensic science brings science, medicine, and technology to bear in criminal investigations and helps police and prosecutors do their jobs better. Those who work in this field have university degrees in chemistry, biology, and the like.

Suzanne Bell's *Encyclopedia of Forensic Science, Revised Edition,* an excellent reference, again addresses many of the questions people have who are considering careers in this field. The encyclopedia is also a wonderful resource for the general public who is simply interested in the subject matter. Arranged as a collection of literate entries enhanced by 14 relevant essays, Bell covers the expansive area of forensic science well. She provides readers interested in forensic science with a single source that is easy to understand, accessible, and very well written.

The book is organized in alphabetical order—A through Z—whereas entries and their cross-references that have to do with the subject are easily found and explained along with essays on a number of subject areas. She has also done a fine job in bringing the volume up to date with both new and revised entries to be as complete a reference as possible.

For those seriously considering forensic science as a career, I also recommend one of the many fine professional organizations in the area. Membership information is easily found on the Web. Three of the many fine organizations are the American Academy of Forensic Sciences, the International Association for Identification, and the National Association of Medical Examiners. These are but a few of the other resources available.

Professor Bell's *Encyclopedia of Forensic Science, Revised Edition* is an excellent start to begin your exploration of this intriguing field of study.

—Barry A. J. Fisher
American Academy of Forensic Sciences President, 1998–1999
Crime Laboratory Directory Director
Los Angeles County Sheriff's Department

Preface

Young students nationwide are increasingly attracted to what might arguably be the "new wave" profession: forensic science. Whether this interest stems from the plethora of television shows, such as the numerous documentaries and docudramas currently broadcast, or from an inherent interest in forensics, students are flocking to the profession in unprecedented numbers. Universities offering forensic science degrees are springing up throughout the country to meet the demand, and high school science teachers are adding forensic courses to their science curricula. Forensic science touches the everyday world in multiple ways, so it stimulates impressionable young minds and propels students toward careers in forensic-related professions: law enforcement, the law, medicine, nursing, and science.

Re-creations of the practice of forensic science on television programs such as *C.S.I.* and *Forensic Files* also affect events in the real world. For example, jurors are demanding more science in the courtroom, a phenomenon known as the C.S.I. effect, which forces prosecutors and defense attorneys to consider how the science will affect their cases when going to trial. Law-enforcement and legal officials working in the criminal justice system are subjected to forensic practice almost daily, and many struggle to understand the science of forensics without having a scientific background.

Thus, many individuals and organizations in our complex modern society, because of an interest in forensics, career aspirations, or professional practice need an authoritative source of forensic information. Certainly many forensic-science textbooks are readily available, but these are mostly written for practicing professionals or college-level students and do not consider the nonscientist. Such a resource should be written by a forensic scientist, an educator who understands the nuances of forensic practice, its legal implications, and relationship to our criminal justice system. The resource must be comprehensive and explain the science behind forensic practice accurately but at a level that a nonscientist easily comprehends.

Suzanne Bell's *Encyclopedia of Forensic Science, Revised Edition* is such a resource: It is a comprehensive, factual reference for anyone interested in learning about topics in forensic science from a lay perspective. Like any encyclopedia, the book lists subjects alphabetically, is extensively cross-referenced, and provides references for further reading. For example, Bell cross-references "shotguns" under "firearms," where she writes about various types of firearms and includes a history of firearms. Under "FBI," she gives a brief history of the agency. In another section she discusses how the scientific method relates to forensic scientific investigations.

The book also includes helpful "Feature Essays," sprinkled throughout the encyclopedia, that illustrate how contemporary topics—drugs, fingerprinting, sports, etc.—relate to forensic science. Dr. Bell pulls no punches when candidly discussing confounding issues in modern forensic science, such as ethics, bias, and individualization versus identity.

An interesting aspect of the encyclopedia is her treatment of historically famous cases. In the Lindbergh kidnapping and murder case of the 1930s, for example, she explains how questioned documents and toolmark examinations helped to prove Bruno Hauptmann's guilt. In the Sam Sheppard murder case of the 1950s, Bell explains why Dr. Paul Kirk believed Sheppard was innocent, based on bloodstain patterns he analyzed at the crime scene. She mentions the pioneers of forensic science in the encyclopedia and the role they played in shaping contemporary forensic science. She dives into the influence of recent court decisions, such as the Daubert decision in 1993 that defined the admissibility of scientific evidence in federal courts.

In an era where the word *forensic* seems ubiquitous and most every profession has or wants a "forensic" bent—forensic meteorologists, veterinarians, kinesiologists, accountants, and others—it is reassuring that the core disciplines employing the application of the basic sciences (chemistry, physics, and biology) to the analysis of physical evidence is still alive and covered extremely well by Bell in the *Encyclopedia of Forensic Science, Revised Edition.*

—Robert C. Shaler, Ph.D.
Pennsylvania State University
Professor, Biochemistry and Molecular Biology
Director, Forensic Science Program
Retired Director of the Forensic Biology Department
Office of Chief Medical Examiner, City of New York

ACKNOWLEDGMENTS

Forensic science is a field growing larger every day and no one person could ever hope to grasp it all. This book would have been impossible without the assistance of dedicated and generous forensic scientists who volunteered their time, advice, expertise, and library of photos to this effort. Aaron Brunedelle (Idaho State Patrol and later Washington State Patrol and Arizona agencies) supplied many fine images, particularly of firearms evidence and microscopic images. Chet Park, also of the Idaho State Patrol, also provided images. Art Craig of the New Mexico Department of Public Safety (NMDPS) took photos of presumptive chemical tests and shared good times with me while we worked together in Santa Fe in the 1980s. I owe a debt to many New Mexicans in the State Patrol—past and present—for hiring me there and letting me earn my forensic wings. Other colleagues who graciously supplied images are Heather Schafstall (Oklahoma State Bureau of Investigation) and Bill Schneck of the Washington State Patrol and Microvision. My photo researcher, Ms. Suzanne Tibor, located a number of commercial images, and I greatly appreciate her efforts.

Many members of the American Academy of Forensic Sciences (AAFS) provided hours of time and assistance. First and foremost, I am indebted to Dr. R. E. Gaensslen of the University of Illinois (Chicago) not only for reviewing my DNA entries, but also for starting me in this direction when I was a graduate student. Teacher, scholar, and gentleman—no one could dare hope to find a better mentor. I also wish to thank Jim Hurley and the Executive Council of the AAFS for assisting in finding reviewers and ensuring that this book was the best it could be. Barry Fisher, Max Houck, and Dr. Bruce McCord were gracious enough to review the text in the first edition on behalf of the Academy and to provide feedback and suggestions. Reviewers are often unsung heroes, and I extend a hearty thanks to all for their time and effort. A special thanks to Dr. Robert Shaler of Pennsylvania State University for career support and the preface.

Finally, I wish to thank the wonderful folks at Facts On File, particularly Frank K. Darmstadt, executive editor, who is a joy to work with and the walking definition of "professional." I am proud to be associated with a company that is dedicated to education, particularly of young people. I am grateful to have been given the opportunity to produce this book under their expert guidance.

INTRODUCTION

The two words *forensic* and *science* each relate to the common theme of truth, either spoken or seen. The word *forensic* can be traced to the Latin *forum*, or "in the public." The definition is roughly translated as "to speak the truth in public." In the modern world, this extends to speaking the truth in court, today's equivalent of the forum. Thus, the role of science is to help society define what is fact; the role of forensic science is to help the legal system define it.

As a recognized scientific field, forensic science is a relative newcomer, yet it represents a natural expansion of existing disciplines such as chemistry, biology, geology, medicine, anthropology, engineering, and many others. The scope and depth of forensic science grows daily as new technologies are discovered and as our society becomes more dependent on the judicial system to solve disputes. At the same time, the public's fascination with forensic science is surging as evidenced by the success of movies, novels, and television shows with forensic themes. Recently, the CBS television show *C.S.I.* has been among the highest rated dramas, resulting in a 2002 spin-off entitled *C.S.I.: Miami* and later, *C.S.I.: New York*. Novels such as those authored by Kathy Reichs and Patricia Cornwell top best-seller lists, and detective and police movies routinely integrate forensic science into their plot lines. Coupled to these fictional portrayals are real events such as the attacks of September 11, 2001, the anthrax mailings, the sniper attacks around Washington, D.C., in late 2002, terrorist attacks in Madrid and London, and the 2004 tsunami, in which forensic science was a highly publicized component.

With high levels of public exposure and awareness, particularly through entertainment, comes the need for public education. There is a danger that media coverage and fictional portrayals of forensic science can lead people to exaggerate and misunderstand the role, capabilities, and limitations of forensic science. This growing gap between forensic fiction and forensic science served as a primary motivation for this project, which grew from my experiences as a forensic laboratory scientist, researcher, and forensic science educator. *Encyclopedia of Forensic Science* represents my small contribution toward linking public perception of forensic science to its reality.

Encyclopedia of Forensic Science gathers the core topics of forensic science into one comprehensive volume and provides an overview of each. Interested readers will find resources in the references and appendixes for further investigation of topics of interest to them. These references range from textbooks to articles in professional journals, all of which are available at or through public or college libraries. The encyclopedia is richly illustrated with drawings and figures and supplemented with photographs—most supplied by working forensic scientists in many different organizations. A color insert containing selected photographs provides additional detail in areas such as firearms, toolmarks, and DNA analysis. Fourteen essays interspersed throughout the encyclopedia describe how forensic science relates to areas such as drug testing in sports, privacy concerns, and the interface of forensic engineering and forensic science. One essay, "The Top Ten Myths of Forensic Science," deals directly with misconceptions about forensic science. In a larger sense, the intent of the encyclopedia is to provide students, teachers, and the public with a comprehensive and reliable reference that shows forensic science as it really is, devoid of romance, glamour, and hype.

TARGET AUDIENCE AND READERS

Encyclopedia of Forensic Science is geared to reach a wide audience ranging from students to working professionals. At the high school and college level, the encyclopedia can serve as the starting point for class projects or career research. General readers can use the encyclopedia to learn about timely topics and issues relevant to forensic science. Working professionals and forensic laboratories will find the encyclopedia a useful addition to their libraries, a quick and easy source of information on more than 600 forensic science topics.

ORGANIZATION AND CONVENTIONS

The International System of Units (SI) conventions have been followed in this book, the only exceptions being in entries or topics in which forensic scientists conventionally use other units. A Periodic Table of the Elements is provided in Appendix III, and references to these elements are provided as both names and symbols (for example, *arsenic* and *As*). Appendix II comprises an extensive list of the common abbreviations used in forensic science. For alphabetical ordering, the different areas of forensic science are listed in the order of their subject area. For example, forensic anthropology can be found under "ANTHROPOLOGY, FORENSIC." Similarly, biographical or other named entries are ordered based on last name. Thus, a discussion of the O. J. Simpson case is listed under SIMPSON, O. J.

TERMINOLOGY IN AN EVOLVING WORLD

As this volume went to press, issues were being debated within the forensic science community and the courts concerning the significance and uniqueness of pattern evidence such as TOOLMARKS and BITE MARKS, FINGERPRINTS, and DNA evidence. For example, successful DNA TYPING of 13 CODIS loci can result in a probability of duplication that, while not zero, is astoundingly small. However, whether it is correct to say that such a combination of DNA types can be considered unique (INDIVIDUALIZED) has not been fully resolved. Similar debates about the uniqueness of fingerprints and impression evidence continue. An understanding or qualifying statement such as "unique to reasonable degree of scientific certainty" is often appropriate and should be understood by the reader even if not stated explicitly.

Finally, the reader should know that every reasonable effort was expended to find reliable, peer-reviewed sources and to present a technically correct and balanced treatment of controversial subjects and cases. Two distinguished members of the American Academy of Forensic Science have reviewed the encyclopedia, while many more assisted with individual entries. However, these entries are not intended as a primary reference, but rather as an overview and gateway to the wealth of information available on the fascinating field that is forensic science.

Entries A–Z

ABO blood group system The first human BLOOD GROUP SYSTEM discovered and the first used in forensic SEROLOGY. Although ABO typing was an indispensable tool in forensic serology for decades, DNA TYPING has largely replaced blood group typing.

The ABO system consists of antigens found on the surfaces of red blood cells (also called erythrocytes and commonly abbreviated RBCs) and corresponding ANTIBODIES in the serum. KARL LANDSTEINER discovered the ABO blood group system in 1900 and for a quarter of a century, it was the only one known. Confusion over naming conventions continued until 1941, when the U.S. military adopted the ABO standard. By the 1960s ABO typing of BLOODSTAIN and BODY FLUID evidence was commonplace in forensic laboratories.

The ABO system is a POLYMORPHIC blood group system, meaning that the antigens (and corresponding antibodies) have more than one observable variant (PHENOTYPES or "type"). These variants are summarized in the accompanying figure. In the serum portion of blood, a person will have the antibodies associated with the antigen *not* found on the RBC surface. For example, someone with Type B blood has B antigens on the surface of their RBCs and anti-A antigens in their serum. For blood transfusions, people with Type AB blood are considered to be universal recipients since their serum does not contain any anti-A or anti-B antibodies. Conversely, people who are Type O are universal donors since their RBCs have neither antigen. There is also an H antigen, and an anti-H antiserum that will cause O cells to agglutinate (see below), and so the notation ABH blood group system is sometimes used. There are also subgroups within Types A, B, and AB, but these are not routinely used in forensic work. In the U.S. population, the approximate frequencies of the types are

- Type A 42 percent
- Type O 43 percent
- Type B 12 percent
- Type AB 3 percent

In addition, a large percentage of people (~80 percent) are SECRETORS, meaning that the antigens present in their blood are also found in other body fluids such as saliva.

When antibodies react with corresponding antigens on the RBC surface, the cells clump together in a reaction called agglutination, which is illustrated in the figure. For example, if RBCs with A antigens on the surface (Type A blood) are mixed with serum containing anti-A antibodies (Type B blood), agglutination results. This was the process that, prior to Landsteiner's discovery, caused many people to die after receiving blood transfusions. In recognition of his life-saving discovery, he received the Nobel Prize in 1930.

Different tests were developed to type whole blood beginning in 1915 with the Lattes crust test. This test is applied to the serum and works by adding known blood cell types to the unknown serum and looking for agglutination. In 1923, Vittorio Siracusa developed the ABSORPTION-INHIBITION TEST, which detects the type of antigens on the RBC surface. ABSORPTION-ELUTION followed in 1930, and all of these tests,

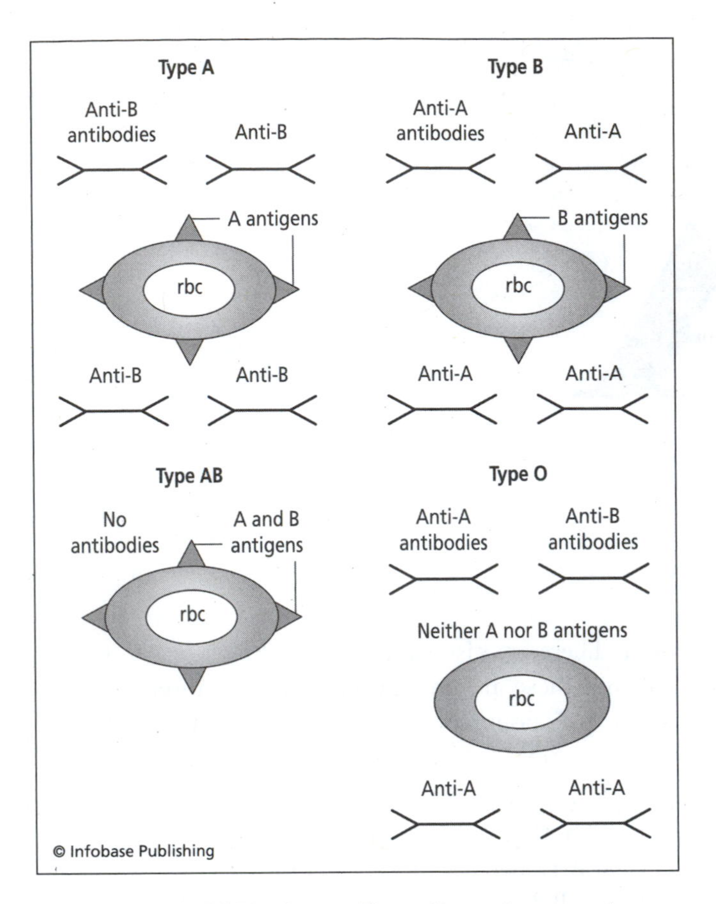

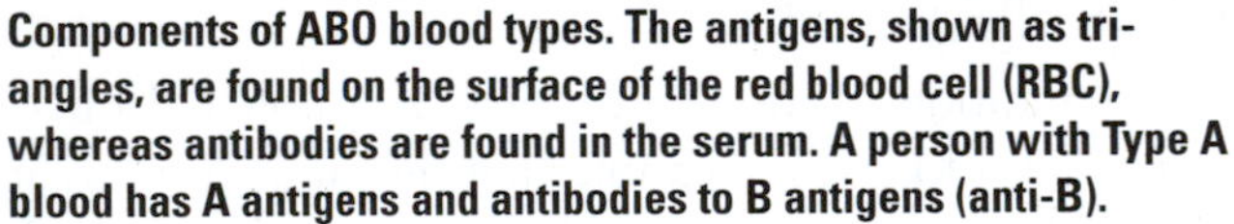

Components of ABO blood types. The antigens, shown as triangles, are found on the surface of the red blood cell (RBC), whereas antibodies are found in the serum. A person with Type A blood has A antigens and antibodies to B antigens (anti-B).

plus modifications and variants, have been applied in forensic serology.

Typing of blood and body fluids in forensic work usually involves stains rather than whole blood, and the stained material may be old and may have been subject to adverse conditions. There are no whole red blood cells left in dried stains since the cell membranes rupture when dried. However, the surface antigens survive and are more stable than the serum antibodies. Absorption-elution typing can be performed on very small samples (threads or fibers) and has been shown to work on stains that are 10 years old and older. It is also more sensitive than absorption-inhibition and is usually the method of choice.

A and B antigenic substances are common in nature, found in plants, animals, and insects. This introduces the danger of FALSE POSITIVE results if other biological material has contaminated a forensic sample. FALSE NEGATIVES are possible if the sample has

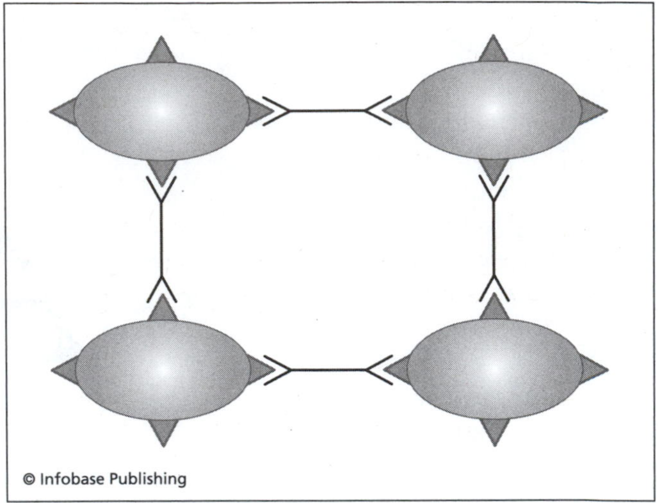

The process of agglutination, in which red blood cells clump in the presence of corresponding antibodies. For example, if Type A serum (containing anti-B antibodies) is mixed with Type B red blood cells, the B antigens will bind with the anti-B antibodies and form linked clumps of cells. The color insert contains a photograph showing agglutination.

been subject to harsh weathering or is very old. This illustrates a common problem in forensic science: often it is not the accuracy and precision of a testing method that determines its success; rather, it is the condition of the sample. Bloodstains are harder to type than whole blood; old, weathered, or damaged bloodstains are more difficult or impossible to type.

See also BLOOD.

Further Reading

De Forest, P. R., R. E. Gaensslen, and H. C. Lee. Chapter 9: "Blood." In *Forensic Science: An Introduction to Criminalistics*. New York: McGraw-Hill, 1983.

Saferstein, R. "Forensic Serology." In *Criminalistics: An Introduction to Forensic Science*. 7th ed. Upper Saddle River, N.J.: Prentice Hall, 2001.

absorption-elution and absorption-inhibition tests

Two tests that are used to type blood and BODY FLUIDS for ABO and other BLOOD GROUP SYSTEMS. Absorption-inhibition was developed in 1923 in Italy by Vittorio Siracusa, and absorption-elution followed in the 1930s. Many modifications and variants have appeared, and the general procedures have been applied to other blood group systems. Although red blood cells rupture when a bloodstain dries, the A and B antigens that are present on the cell surface persist.

As a result, these tests work on whole blood, body fluids (assuming the person is a SECRETOR), and in stains of any of those fluids.

Absorption-inhibition works by reducing the strength of an antiserum based on the type and amount of antigens present in the stain. For example, if the bloodstain comes from a person with Type B blood, the stain will contain B antigens. If an anti-A antiserum of a known strength is added to the stain, nothing will happen. If anti-B antiserum is added, some of the antibodies will bind to the B antigens, reducing the strength of the original antiserum. This reduction of strength is the inhibition for which the test is named. Although effective, absorption-inhibition is less sensitive than absorption-elution and therefore requires larger samples. This presents a problem if the stain is very small or the amount of sample is limited, a situation often encountered in forensic analyses. Consequently, absorption-elution is more common in forensic applications.

The absorption-elution test is illustrated in the accompanying figure, using a stain of Type B blood as the example. Antiserum containing both anti-A and anti-B is added to the stain. Anti-B will bind to the B antigens and will remain behind when the stain is rinsed with cold saline. The stain is then heated, breaking the bond between the B antigens and anti-B antibodies. The solution into which the anti-B antibodies have eluted is then split in half. To one portion, A cells are added and to another, B cells. Agglutination (clumping) will be observed with the B cells, confirming the type of the original stain. Absorption-elution techniques work on samples as small as a single fiber and have been shown to work on stains that are a decade old or older. Absorption-elution is rarely used now given that DNA TYPING has essentially replaced traditional typing techniques.

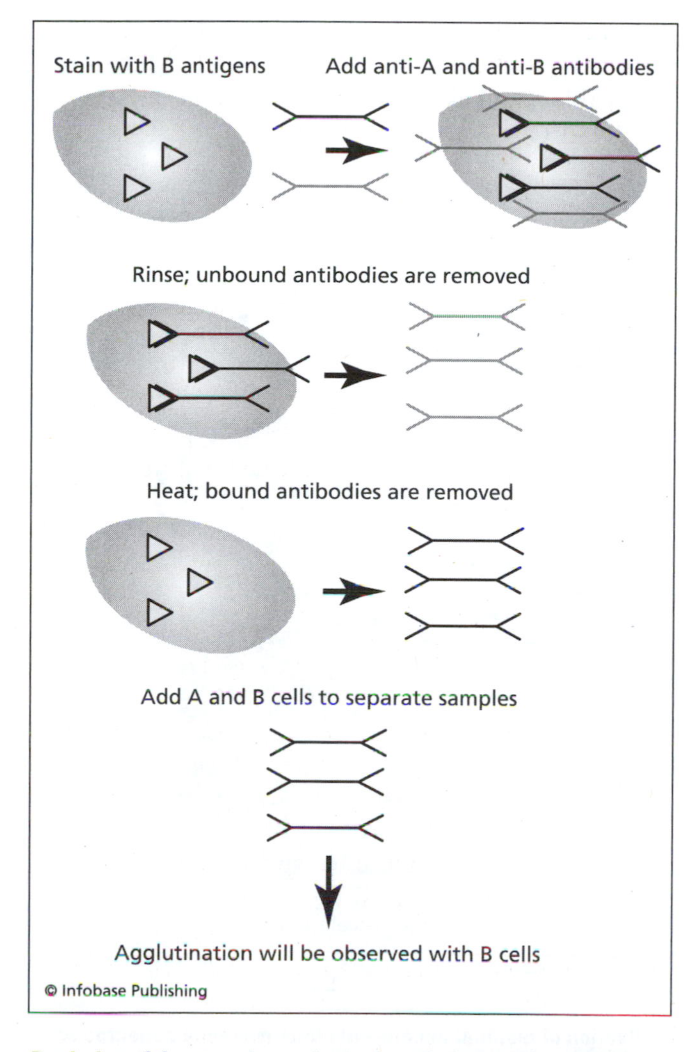

Depiction of the steps in an absorption-elution test for ABO blood type. In this example, the bloodstain is Type B, which would contain B antigens. As a result, the stain will react with anti-B serum. After rinsing and elution, the eluted solution will contain the liberated anti-B antibodies that will agglutinate with B cells.

See also ABO BLOOD GROUP SYSTEM; BLOOD; BODY FLUIDS.

Further Reading

Saferstein, R. "Forensic Serology." In *Criminalistics: An Introduction to Forensic Science.* 7th ed. Upper Saddle River, N.J.: Prentice Hall, 2001.

absorption spectrum A graph that plots the absorbance of ELECTROMAGNETIC RADIATION (EMR) by a selected material as a function of the wavelength of radiation. Although commonly associated with the visible portion of the electromagnetic spectrum where colors correlate with wavelengths, an absorption spectrum can be generated in any spectral range. In forensic analysis, the most common types of absorption spectra used are those in the visible range (VIS), ultraviolet (UV), and infrared (IR). For example, an absorption spectrum of a DYE can help characterize FIBERS and INKS while an infrared absorption spectrum is a standard component of DRUG ANALYSIS.

Instrumentation (generically called SPECTROPHOTOMETER or SPECTROMETER) is required to generate an absorption spectrum. For creating a spectrum of a material in the UV/VIS range, the instrument is called a UV/VIS spectrophotometer, whereas an instrument that

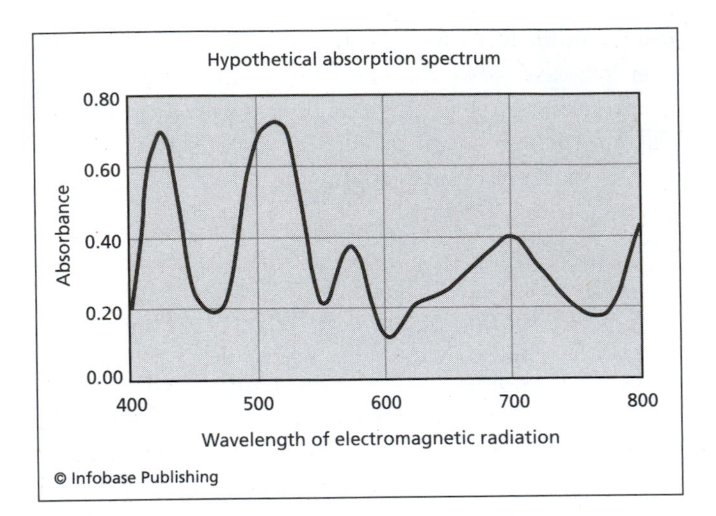

Absorption spectrum is a plot of wavelength of radiation versus the amount of that radiation absorbed by the sample being analyzed. The higher the point (here, the greater the absorbance), the greater the amount of radiation absorbed.

works in the infrared range is called an IR spectrophotometer. In general, the function of the instrument is to break up radiation into individual wavelengths that are then directed at the sample. The amount of energy absorbed by the sample at each wavelength is plotted to produce the spectrum. Different instrument designs exist and many techniques can be used to generate the spectrum, but the result typically is a plot of the absorption versus wavelength, as illustrated in the figure above. The pattern of the spectrum provides valuable chemical information about the sample from which it was generated, and in some cases is sufficient by itself to identify the substance specifically and unequivocally.

accelerant In ARSON cases, an accelerant is the flammable material that is used to start the fire. Accelerants can be solids, liquids, or gases, with gasoline being the most commonly used. Solid accelerants include PAPER, fireworks, highway flares, and black powder. Butane (cigarette lighter fuel), propane, and natural gas are examples of gaseous accelerants, which do not leave any residue at a fire scene. However, gases must be contained and transported, so severed gas lines or spent containers serve as critical PHYSICAL EVIDENCE in such cases.

Liquid accelerants fall into two broad categories: petroleum distillates, which include gasoline and other petroleum products; and nonpetroleum products such as methanol, acetone (used in nail polish remover), and turpentine.

Petroleum distillates are derived from crude oil and are also called hydrocarbons or petroleum hydrocarbons. In crude oil, volatility of the individual components range from extremely volatile substances such as propane (a gas at room temperature) to asphalt, which remains solid even at high temperatures.

Petroleum distillates such as gasoline and kerosene are not single hydrocarbons but mixtures of different components with similar volatilities. The volatility of an accelerant is an important consideration in the COMBUSTION process, determining how much residue will be left and how quickly it will evaporate after the fire is out. Related to volatility is the flash point, defined as the temperature at which a liquid will give off enough vapor to form an ignitable mixture. For gasoline, the flash point is -50°F (-45.56°C). The National Fire Protection Association (NFPA) defines a flammable liquid as one with a flash point of less than 140°F (60°C).

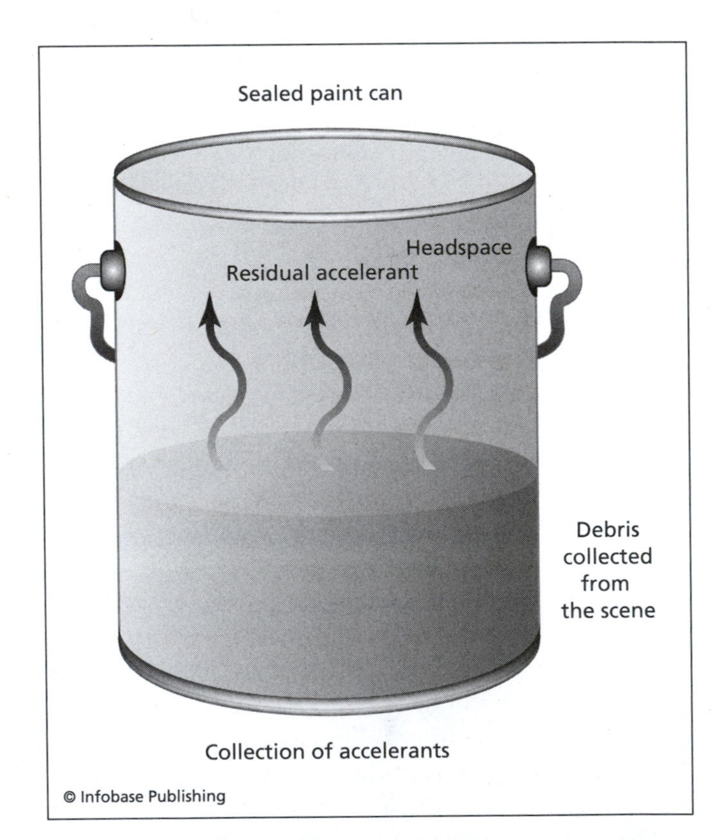

Collection of residual accelerants from evidence collected at an arson scene. The debris is placed in a sealed paint can such that empty headspace remains above it. When the can is sealed, the volatile accelerants continue to evaporate, and the vapors will collect in the headspace, which can then be sampled for further analysis.

Based on volatility and molecular structure, petroleum distillates are often divided into the following categories:

Category	Example
Light petroleum distillates (LPD)	Propane, butane
Medium (MPD)	Paint thinner
Heavy (HPD)	Kerosene, diesel fuel
Aromatics	Benzene, toluene, xylenes
Mixtures	Gasoline

Aromatic hydrocarbons have unique molecular structures and were originally named based on their distinctive smell. Aromatics such as benzene and toluene are found in gasoline.

At a fire scene, the presence of accelerants can be determined using several procedures including trained dogs, chemical color tests, and portable instruments and sensors. Materials such as wood and carpet absorb liquid accelerants, so samples of these materials can harbor valuable evidence. It is important for the investigator to collect CONTROL SAMPLES since carpets and other synthetic materials can interfere with or lead to FALSE POSITIVES during laboratory analysis. Evidence from the scene is usually collected in glass jars or metallic paint cans that are tightly sealed to prevent vapors from escaping. As shown in the figure on the previous page, once the can is sealed, any volatile accelerants present will continue to evaporate into the headspace above the debris.

The primary tool used to detect and identify liquid accelerants is GAS CHROMATOGRAPHY (GC) coupled to either a FLAME IONIZATION DETECTOR (FID) or a MASS SPECTROMETER (MS). Analysis using either instrument produces an output that is distinctive for most common petroleum distillates. Patterns are identified by comparison to standards of known composition. The patterns obtained from evidence can be influenced by weathering and by microbial activity, particularly if the sample is on soil or vegetation. Weathering occurs as lighter (more volatile) components of the accelerant evaporate, and the longer the sample sits before collection, the more severe the weathering effects.

Samples are prepared for introduction into the GC using several methods:

1. *Cold headspace*: The can is punctured and a syringe is used to withdraw a headspace sample that is injected into the GC.
2. *Heated headspace*: Prior to syringe introduction, the can is heated.
3. *Extraction*: The accelerant is extracted from the sample using a solvent such as carbon disulfide or steam. Small portions of the extract are injected into the GC.
4. *Purge-and-trap*: Inlet and outlet holes are put in the can lid. A stream of filtered air is pumped in through the inlet and a charcoal trap is placed on the outlet. The can may be heated, and vapors are trapped on the charcoal. The trapped compounds can be removed using heat (thermal desorption) or solvent extraction.
5. *Charcoal strip/SOLID PHASE MICROEXTRACTION (SPME)*: A charcoal strip or other adsorptive material is lowered into the can or placed on an inlet drilled into the can. A vacuum can be used to draw sample through the trap or a stream of filtered air can be pumped into the can to force headspace to flow out through the trap. The can may be heated, with a thermometer inserted in the can to monitor temperature.

In some cases, the presence of a flammable material in a given area is to be expected and may not be associated with arson. For example, if a fire is started in a garage where a car is parked and gasoline powered equipment such as a snow blower or lawn mower is kept, the gasoline associated with the tools or car is considered to be an incidental accelerant that would normally be present in the area.

See also FIRE INVESTIGATION; INCENDIARY DEVICES.

Further Reading

Midkiff, C. "Laboratory Examination of Arson Evidence." In *More Chemistry in Crime, From Marsh Arsenic Test to DNA Profile*. Edited by S. M. Gerber and R. Saferstein. Washington, D.C.: American Chemical Society, 1997.

accidental characteristics Marks that appear on certain types of evidence such as tires, bullets, or shoes that do not appear on all such evidence. For example, the soles of shoes are mass-produced from a mold, and so all soles made from the same mold should have the same pattern. An accidental mark could appear on one sole if it was accidentally cut at the factory or the mold was somehow damaged. The marks that result are not supposed to be there, but their presence can be invaluable to a forensic analysis for that reason—the marks differentiate that particular shoe (or group of shoes) from the batch. Accidental characteristics can also play a role in questioned document evidence. Rollers in printers and copiers can be scratched or gouged, resulting in a mark on the paper that can be used to link that document to a specific roller and thus a specific printer.

See also WEAR PATTERNS.

accident reconstruction *See* TRAFFIC ACCIDENT RECONSTRUCTION.

accounting, forensic The application of accounting techniques to criminal and civil matters. Forensic accountants study financial records and other financial evidence, prepare analyses and reports, assist in investigation, and like any other forensic professional, can be called on to relate findings to a court of law. Most often, forensic accountants are CPAs (Certified Public Accountants) that specialize in fraud or other investigative accounting. Fraud accounting and forensic accounting are often referred to under the general term "litigation support." In addition to litigation support, forensic accounting can be involved in corporate investigations, insurance claims, and regulatory compliance. Forensic accounting and forensic COMPUTING often overlap. Financial institutions, insurance companies, and governmental agencies, notably the IRS (Internal Revenue Service), FBI, GAO (Government Accountability Office), and SEC (Securities and Exchange Commission), employ forensic accountants, as do some law enforcement agencies. Matters investigated using forensic accounting and related techniques include (among many others) ARSON, bankruptcy, check forgery, check kiting, computer fraud, credit card fraud, contested divorce settlements, embezzlement, financing of terrorism, software piracy, tax fraud, and even divorce settlements if a business is involved. Recent examples of investigations in which forensic accounting was crucial include tracking the money used to finance the attacks of September 11, 2001, and the bankruptcy of Enron, a Houston-based energy trading firm. The Association of Certified Fraud Examiners (ACFE: www.acfe.com) certifies practitioners in the field.

Further Reading

Bologna, G. J., and R. J. Lindquist. *Fraud and Forensic Accounting: New Tools and Techniques*. 2d ed. New York: John Wiley and Sons, 1995.

accreditation *See* AMERICAN SOCIETY OF CRIME LABORATORY DIRECTORS (ASCLD).

acetone-chlor-hemin test (Wagenaar test) One of several chemical tests used to identify blood. Like many PRESUMPTIVE TESTS for blood, it works by forming distinctive crystals with HEMOGLOBIN derivatives (hematin, hemin, and hemochromogen). Procedures for the test were published in 1935, and they are fairly simple. A few drops of acetone (a common ingredient in nail polish removers) are added to a suspected bloodstain followed by a drop of diluted hydrochloric acid (HCl). If hemoglobin is present, characteristic crystals form, which are then observed under a microscope.

ACE-V An acronym for analysis, comparison, evaluation, and verification. This is the process used by LATENT FINGERPRINT examiners to evaluate the patterns of latent fingerprints. The first step (analysis) involves studying the latent fingerprint to determine if it is suitable for comparisons. To do a comparison, the latent fingerprint has to be reasonably clear and a sufficient portion of the finger must have made contact with the surface and left an impression. It is also important to be able to tell what made the impression, be it a finger, thumb, or palm. Once the examiner determines that the print can be compared, the examiner selects what to compare it to, such as a sample taken from a suspect or a print retrieved from a database search. Comparision involves a feature-by-feature examination of each and will ideally result in exclusion (the prints could not have come from the same person or COMMON SOURCE) or indentification (INDIVIDUALIZATION) that conclusively links the two compared impressions to the same person. The results may also be inconclusive, a case in which definititive identification or exclusion cannot be made. The final step is verification of the results by another qualified analyst.

acid phosphatase (AP, ACP, EAP) *See* ISOENZYME SYSTEMS; SEMEN.

acronyms *See* Appendix II.

adenosine deaminase (ADA) *See* ISOENZYME SYSTEMS.

ADME An acronym for the linked processes of absorption, distribution, metabolism, and excretion that occur when a person ingests a substance such as a drug or poison. The most common modes of ingestion encountered in forensic TOXICOLOGY are swallowing, injection, inhalation, and absorption through the skin. Once ingested, the material (also called a

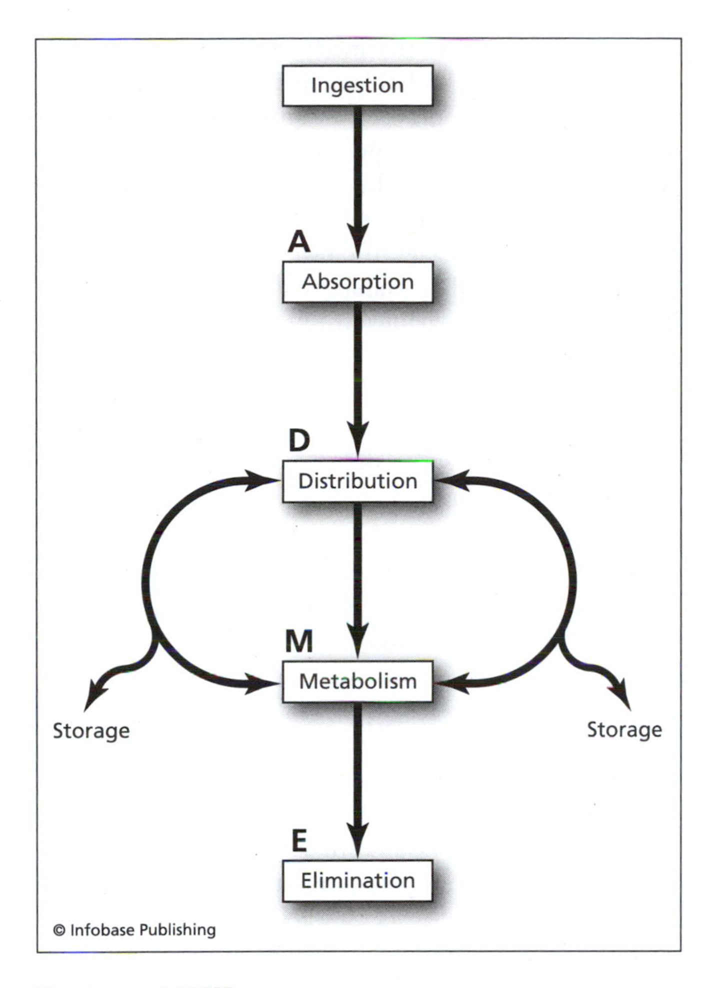

The stages of ADME

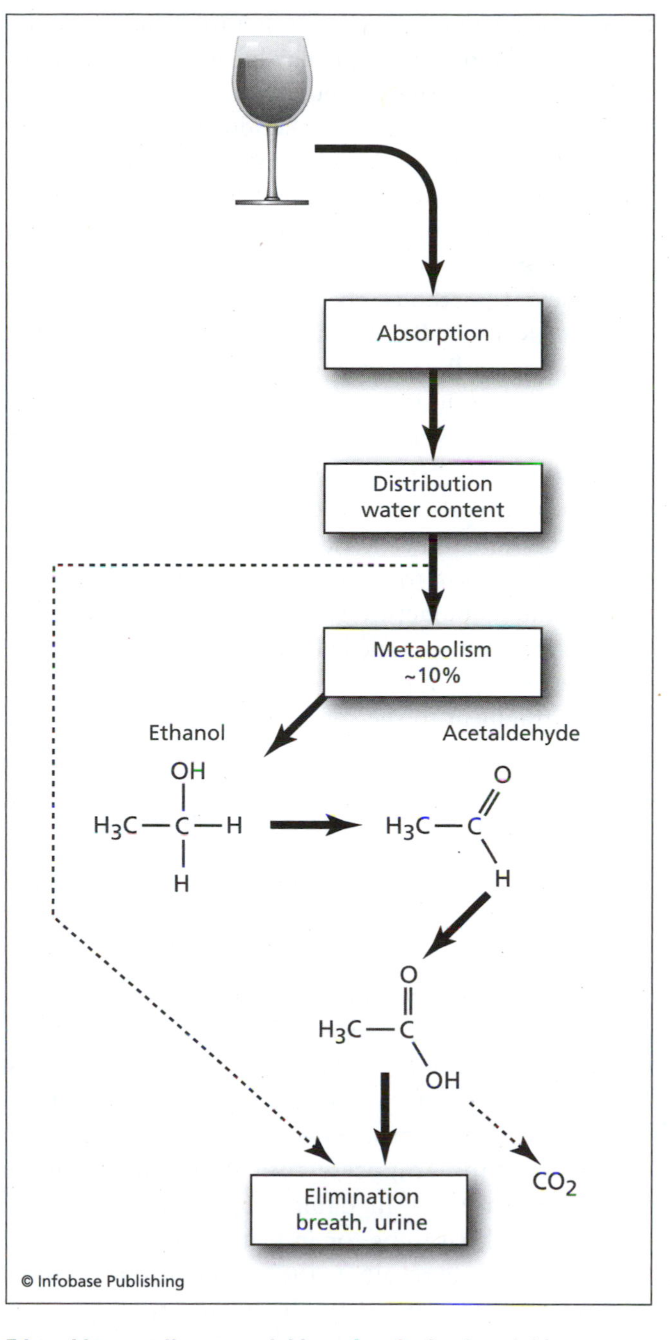

Ethanol is a small water-soluble molecule that is quickly absorbed in the digestive tract and distributed to tissues. In the liver, it is efficiently converted to acetaldehyde, which is excreted in the breath and urine. Some is exhaled in breath as ethanol, and some is metabolized to CO_2 and exhaled. About 10 percent of the original ethanol is excreted in the urine unchanged.

XENOBIOTIC) is absorbed into the tissues. If the xenobiotic is swallowed, this absorption usually occurs in the digestive tract. The absorbed material is distributed throughout the body and tissues depending primarily on how water-soluble it is. Substances are metabolized principally in the liver to form metabolites that may be excreted or otherwised eliminated. Some metabolites may be less water-soluble than the original and thus will tend to linger in fatty tissues. Others may become volatile and can be eliminated through breath, while still other metabolites may be futher metabolized. For example, HEROIN is an illicit drug that can be ingested by swallowing, injection, or by smoking (inhalation). Once absorbed and distributed to the bloodstream, heroin is metabolized to a compound called monoacetylmorphine, which is itself metabolized to morphine.

See also ALCOHOL; METABOLISM AND METABOLITES.

adenylate kinase (AK) *See* ISOENZYME SYSTEMS.

adhesive tape Different types of tape (adhesive, electrical, masking, duct, and so on) may be involved in a crime and become PHYSICAL EVIDENCE. For example, tape might be used to bind a victim or might be wrapped around the handle of a tool used in a burglary. Packing or clear tape may be used to wrap packages containing drugs or bombs. Documents may be found taped to packages or other documents.

Tape consists of a backing material topped with a pressure sensitive adhesive. The backing can be made of a plastic POLYMER, cloth, or paper. The adhesives can be characterized using microscopic, chemical, and instrumental methods such as ATTENUATED TOTAL REFLECTANCE (ATR) INFRARED SPECTROSCOPY (IR), a technique well suited for surface analyses. Other instrumental techniques used include GAS CHROMATOGRAPHY/MASS SPECTROMETRY (GC/MS) and ULTRAVIOLET SPECTROSCOPY (UV/VIS). Physical dimensions of the tape can also be important for comparison and identification; characteristics such as width and thickness can be used to reduce the number of possible sources and manufacturers.

If a fragment of tape is found at a scene, it may be possible through PHYSICAL MATCHING techniques to link the fragment to the source roll, assuming that the tape has been torn rather than cut. Tape is also extensively used to collect evidence such as HAIRS, FIBERS, and FINGERPRINTS using TAPE LIFT techniques.

adipocere A grayish waxlike substance that forms as a result of a slow chemical reaction between body fat and water (hydrolysis) that occurs after death. The word *adipocere* comes from a combination of the words for fat (adipose tissue) and wax, and the consistency of adipocere is like soap. The hydrolysis reaction that produces it is an example of saponification, the chemical process by which fat is rendered into hard soap. Adipocere formation can occur in bodies that are left in damp environments such as mud, wet soil, swamps, or in water. The consistency of adipocere can result in the preservation of lines, shapes, and contours of the body, and the degree of adipocere formation can be useful for estimating the POSTMORTEM INTERVAL (PMI). If formed, it becomes noticeable about eight weeks after death, and the formation process is completed between 18 months and two years.

admissibility and admissibility hearing (*Daubert* hearing) Before any scientific evidence is presented before a court, it must be determined to be admissible. Admissible evidence must be reliable and relevant to the case at hand, and for scientific analysis, the court must be assured that the methods used are scientifically acceptable and reliable. The intent of admissibility proceedings is to prevent the introduction of results obtained by the use of poor science or PSEUDOSCIENCE or to prevent the admission of evidence that has no bearing on the case. Admissibility hearings provide a way for new scientific test methods to be introduced and accepted as viable tools in forensic science. If required, these hearings are held separate from the case presentation.

The standards that courts use to determine admissibility of evidence vary among the jurisdictions. Those following the *Frye* standard (*Frye v. United States*) require that new methods be generally acceptable to a significant proportion of the scientific discipline to which they belong. For example, new chemical tests would have to be generally accepted as reliable among most analytical chemists. Jurisdictions that follow the Federal Rules of Evidence and the *Daubert v. Merrell Dow Pharmaceuticals* decision use more flexible guidelines. Essentially, under *Daubert*, the trial judge is responsible for determining if the scientific evidence is useful and relevant and that the expert presenting it is qualified to discuss the results and offer an opinion. The judge must also determine if the testing method rests on a reliable and reasonable scientific foundation. Such hearings are referred to as *Daubert* hearings.

See also *DAUBERT* DECISION; *FRYE* DECISION.

adulterants *See* CUTTING AGENTS.

affidavit Written testimony taken from an individual who is under oath before an authorized representative of a court. Occasionally, forensic scientists and other expert witnesses offer testimony by way of an affidavit.

affinity *See* PARTITIONING AND AFFINITY.

age-at-death estimation When an unidentified body or collection of remains are found, a critical step in the identification process is to determine the approximate age of the deceased. Three common methods of determining the age are based on skeletal development and measurements, dental development and condition, and AMINO ACID RACEMIZATION (AAR), usually in teeth. Given that skeletal and dental formation follows a known and consistent pattern of development, estimates based on these techniques are reliable to within a year for younger people. However, once growth and development are complete, estimates become more difficult and, in general, the older the person is, the larger the uncertainty in the age estimates. Specialists in forensic ANTHROPOLOGY work with bones and skeletal measurements while forensic dentists (ODONTOLOGISTS) work with dental evidence, with some overlaps occurring.

Skeletal development and ossification (mineralization or hardening) follows a known and consistent pattern from early fetal stages until growth is completed as an early adult. Information can be obtained from X-rays and direct measurements (ANTHROPOMETRY) of specific bones and bone structures such as the knee, wrist, and foot. Once development is complete, skeletal age determinations must rely on measurements taken, such as skull sutures and in the pubis (pelvic) area. Generally, once a person is in their early 40s, degenerative changes in bones and joints will be evident, increasing with age. However, other factors such as health, diet, occupation, and genetics complicate age estimates and drive uncertainties higher. From early fetal development until the mid-teens, dental development is an excellent method of age estimation. Estimates are based on the known rates of emergence of baby teeth and their subsequent loss as adult teeth replace them. Other developmental milestones are indicated in the teeth as well. For example, the stresses of birth disrupt the normal metabolic processes, including those in the cells that form teeth. This disruption creates a line in the dentin called the neonatal line that is easily detectable.

Once the last permanent teeth are in place, age estimation from teeth is complicated by factors including diet, dental care and hygiene, and genetic factors. However, teeth can be aged with reasonable accuracy using amino acid racemization (AAR). This is a particular advantage since teeth are durable and can withstand environmental extremes, including severe trauma and fire.

See also RADIOLOGY, FORENSIC.

Further Reading

Sorg, M. H. "Forensic Anthropology." In *Forensic Science: An Introduction to Scientific and Investigative Techniques.* 2nd edition. Edited by S. H. James and J. J. Nordby. Boca Raton, Fla.: CRC Press, 2005.

airplane crashes *See* TRANSPORTATION DISASTERS.

alchemy An ancient practice that combined science, art, and elements of mysticism. Most ancient cultures that left records practiced alchemy, which grew out of mining, metallurgy, and medicine. The undercurrent, even though the ancients did not recognize it, was chemistry. Alchemy was an odd and interesting blend of science, art, and religion that focused on the concept of purification and of separating material that was considered "pure" such as gold, from the unpure, or whatever it was embedded in. The first mentions of alchemy date to around 400 B.C.E. The Greeks had a word *chyma* that described processes of metalworking, and this might be one origin of the word, but the Chinese and Egyptians recorded similar words also related to metallurgy. All three cultures practiced alchemy and the *al* part appears to have come from Arabic, forming *al-chemy* or "the chemistry." Although analysis and transformation of gold and other materials was part of alchemy, from its inception there were strong religious, spiritual, and mystical branches and aspects to it. It was only in the 16th and 17th centuries that the mystical part superseded the practical, corresponding with the eventual rise of chemistry as a science.

Alchemists were technologists who learned by experience and passed on what they learned to a select few. It was not of particular interest to them why something worked. It did, and that was good enough. As a result, innovation came slowly. From the forensic perspective, the key contribution of the ancient alchemists was in their interest in fire applied to metallurgy and the use of heat as a means of separating materials from one another. Pyrochemistry was to play a role in the first viable tests for ARSENIC. Arab alchemists such as JABIR contributed many advances including the art of distillation and the separation of alcohols. SIR ISAAC NEWTON and ROBERT BOYLE were considered to be among the last of the alchemists, living in the 1600s, when chemistry was emerging from alchemy as a separate and recognized science. After this time, alchemy drifted deeper into mysticism. However, its contribution to CHEMISTRY and TOXICOLOGY was essential to forensic science.

Further Reading

Moran, Bruce T. *Distilling Knowledge: Alchemy, Chemistry, and the Scientific Revolution.* New Histories of Science,

Technology, and Medicine. Cambridge, Mass.: Harvard University Press, 2005.

alcohol Ethyl alcohol (ethanol) is a central nervous system (CNS) depressant that is a factor in approximately 40 percent of fatal traffic accidents. Although alcohols are a class of organic compounds, the use of the term *alcohol,* particularly in forensic contexts, usually refers to ethanol. Other common alcohols such as isopropyl (rubbing alcohol) and methyl alcohol (methanol or wood alcohol) are more toxic than ethanol, although large doses of ethanol can be fatal. Methanol is occasionally encountered as a poison found in homemade or bootleg liquors such as "moonshine."

Ethanol is a colorless volatile liquid that is completely soluble in water. Because it is water soluble, ingested ethanol can move with water in the body and thus quickly diffuses out of the stomach and upper small intestine (duodenum) into the bloodstream and ultimately into the brain where intoxication effects occur. Approximately 20 percent of ingested ethanol is absorbed through the stomach wall and the rest through the walls of the small intestine. Ethanol can be removed from the body by metabolic processes (~90 percent) or by exhalation or in urine, perspiration, or saliva (~10 percent). Ethanol metabolism takes place in the liver, where the enzymes including alcohol dehydrogenase convert ethanol in a stepwise process to acetaldehyde, acetic acid, and to exhalable carbon dioxide and water. Large quantities of acetaldehyde, also known as ethanal, are responsible for many of the symptoms of a hangover.

Ethanol is a product of the yeast-driven fermentation of sugars and is found at percent levels in beer (4–5 percent w/v [weight to volume]), 9–20 percent in wines, and higher concentrations in hard liquors. In hard liquors, the ethanol concentration is given as the proof, which is twice the percentage. Thus, liquor that is 50 proof is 25 percent ethanol (w/v). Although the number and types of drinks ingested over a given time can be used to estimate alcohol concentrations in the blood and brain, many factors determine how fast the alcohol is absorbed and what degree of impairment results. These factors include the presence of food in the stomach, sex and weight of the individual, and rate of elimination.

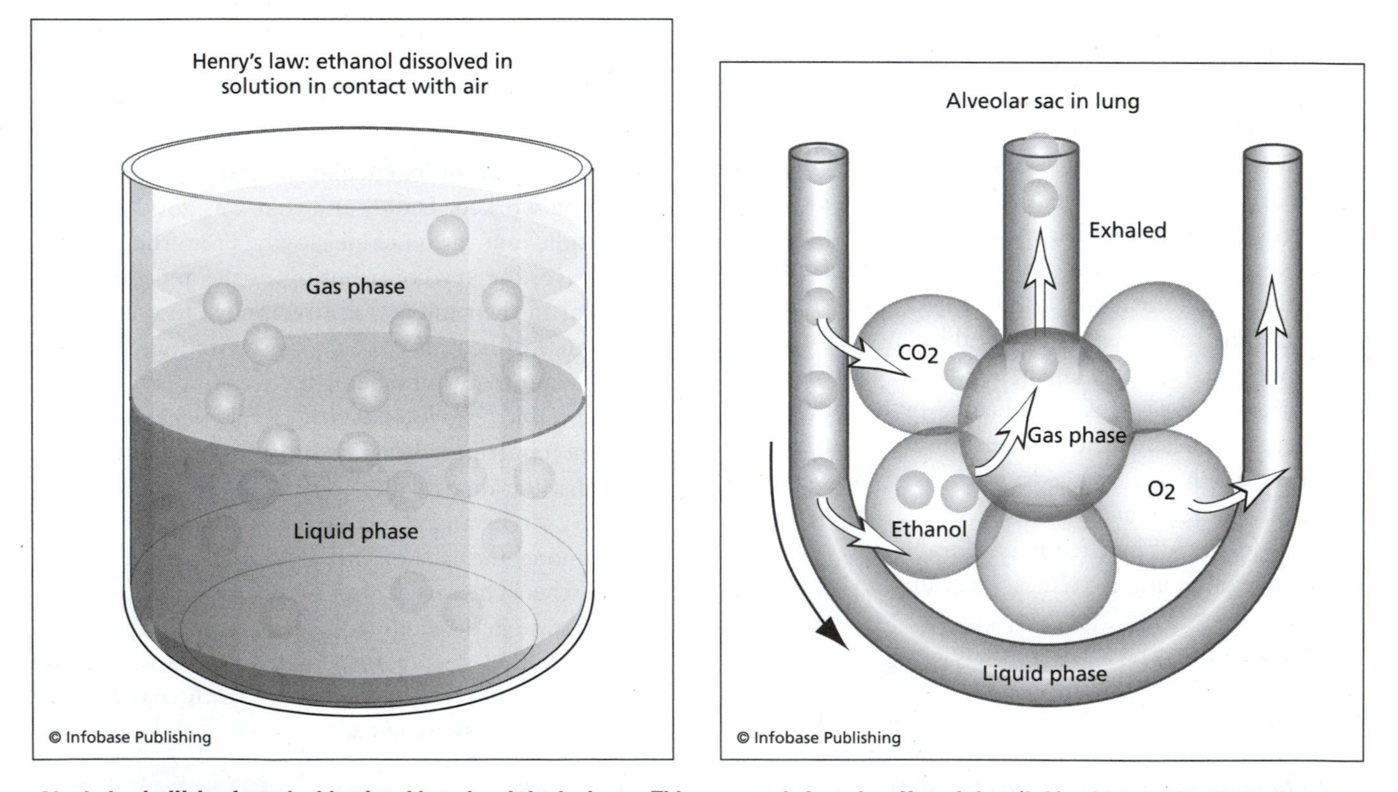

Alcohol volatilizing from the blood and into the air in the lungs. This process is based on Henry's law (left), which applies to systems in which a material (here, ethanol) is distributed across the interface of a liquid (blood) and a gas (air). As ethanol leaves the blood in the lungs, it is exhaled and can be measured.

Given these variables, the degree of impairment must be measured based on the concentration of alcohol in the blood, and by extension the concentration in the brain, where the impairment occurs. The first laws in the United States aimed at intoxicated drivers were passed in 1938 in the states of Indiana and Maine. Since 1939, courts have accepted chemical tests for the determination of BLOOD ALCOHOL CONCENTRATIONS (BAC). Field tests use devices that measure the concentration of alcohol in the exhaled breath, and by extension the concentration in the blood. These quantities can be related to each other based on HENRY'S LAW (illustrated in the figure on the previous page), which states that concentration of a gas dissolved in a liquid is proportional to the concentration of the gas above the liquid, assuming the temperature is constant. This interface of liquid (blood) to gas (air) occurs deep in the lungs. Blood flows adjacent to the walls of the alveolar sacs, allowing carbon dioxide and ethanol dissolved in the blood to escape into the air to be exhaled. Oxygen is also exchanged at this interface. For ethanol in blood in contact with air, the Henry's law ratio has been determined to be approximately 2100:1, meaning that 1 milliliter of blood would contain the same amount of ethanol as 2100 mL of air. Thus, by measuring the concentration of ethanol in the exhaled breath, the BAC concentration can be estimated by calculation. Early standards for legal intoxication were 0.15 percent BAC, then decreased to 0.10 percent, and are currently 0.08 percent in most states. In many European countries, the level is much lower, including 0.02 percent in Sweden. BAC levels above 0.35 percent can produce stupor and coma, and death can occur from respiratory suppression at BACs of 0.45 percent and above.

Depending on the jurisdiction, a blood sample may also be required to determine the exact BAC. Forensic toxicologists use GAS CHROMATOGRAPHY (GC) or a chemical test using alcohol dehydrogenase to experimentally determine the BAC from a blood sample. Collection of the blood sample must be done carefully using ethanol-free disinfectants and the proper procedures including a CHAIN OF CUSTODY, refrigeration, and the addition of anticoagulants and preservatives.

See also ADME; BREATH ALCOHOL; TOXICOLOGY.

Further Reading

Fenton, J. J. Chapter 15, "Alcohols." In *Toxicology: A Case Oriented Approach*. Boca Raton, Fla.: CRC Press, 2002.

Kunsman, G. "Human Performance Toxicology." In *Principles of Forensic Toxicology*. 2nd ed. Edited by B. Levine. Washington, D.C.: American Association of Clinical Chemistry, 2003.

Levine, B., and Y. Kaplan. "Alcohol." In *Principles of Forensic Toxicology*. Edited by B. Levine. Washington, D.C.: American Association of Clinical Chemistry, 2003.

algor mortis *See* BODY TEMPERATURE.

alkaline flame detector *See* NITROGEN PHOSPHORUS DETECTOR (NPD).

alkaloids A class of chemical compounds that are extracted or obtained primarily from seed plants. They were first isolated in the 19th century and were called "vegetable alkaloids." The pure compounds are usually colorless and bitter tasting and are encountered in forensic work as drugs or POISONS. Alkaloids derive their name from the fact that they are basic or alkaline, and this in large measure accounts for their bitter tastes. In addition to carbon and hydrogen, alkaloids contain nitrogen and usually oxygen. CAFFEINE is a typical alkaloid, containing both nitrogen and oxygen in the molecule.

There are three classes of alkaloids commonly encountered in forensic work:

1. Opiate alkaloids: These are extracted from opium poppies and include OPIUM, MORPHINE, and CODEINE.

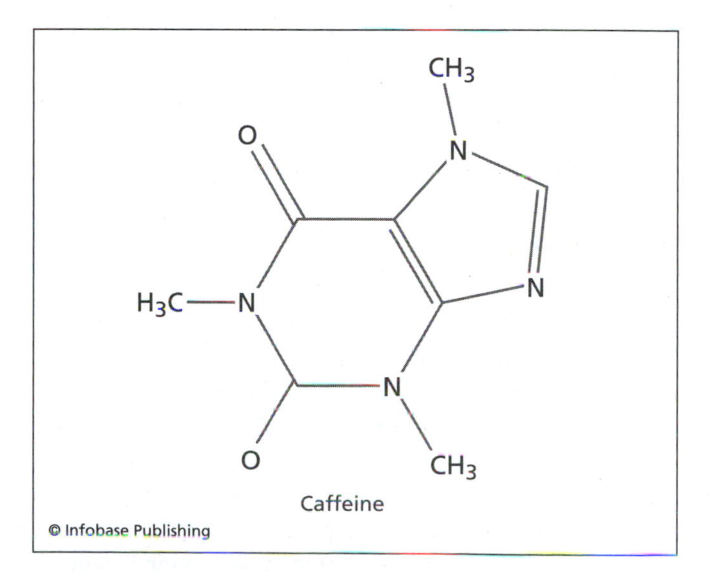

The molecular structure of caffeine, a typical alkaloid

These extracted alkaloids can be used to make synthetic or semi-synthetic narcotics such as HEROIN.

2. Xanthine alkaloids: This class includes familiar compounds such as theophylline, theobromine, and CAFFEINE (previous page), which are found in coffee, tea, and chocolate. Although not illegal or controlled, they are encountered as ingredients in over-the-counter (OTC) pharmaceuticals and as diluents (CUTTING AGENTS) of street drugs.
3. Ergot alkaloids: An ergot is a seedlike pod produced by fungus that can grow on cereal crops such as rye. Many of the ergot alkaloids are poisonous or hallucinogenic, including LYSERGIC ACID DIETHYLAMIDE (LSD).

COCAINE is also an alkaloid and like most can exist in the freebase or salt form. Cocaine freebase is a sticky, gummy substance prepared by extracting the cocaine into a basic solution. On the other hand, cocaine hydrochloride (HCl) salt is a white powder. The freebase form is more potent and is used in freebasing, in which the drug is smoked or injected.

See also DRUG ANALYSIS; TOXICOLOGY.

alternate light sources (ALS) Combinations of lights and filters used in FORENSIC SCIENCE to make evidence such as LATENT FINGERPRINTS, biological fluids, and writing on documents easier to see. The light source itself is usually a strong lamp such as a xenon arc or quartz halogen type that emits an intense beam of light that is channeled through a fiber optic cable. This allows the analyst to focus the bright beam on a small area. With latent fingerprints, the analyst can select one of many chemicals or DYES that will adhere to the print and fluoresce after exposure to the light source.

See also FLUORESCENCE.

Amelogenin gene *See* SEX DETERMINATION.

American Academy of Forensic Sciences (AAFS) The preeminent national and international forensic science professional organization, founded in 1948. Members, currently numbering about 5,000, are assigned to one of 10 sections, each with separate application requirements: ANTHROPOLOGY, CRIMINALISTICS, ENGINEERING, General, Jurisprudence, ODONTOLOGY, PATHOLOGY/BIOLOGY, PSYCHIATRY/Behavior Sciences, QUESTIONED DOCUMENTS, and TOXICOLOGY. The society's journal, *Journal of Forensic Sciences,* is the primary, peer-reviewed publication in the field. The academy is headquartered in Colorado Springs and has an extensive Web site at www.aafs.org.

See also APPENDIX I.

American Board of Criminalistics (ABC) This board was formed in 1989 as a means to develop a national certification program for criminalists. The history of the board traces back to the mid-1970s when the Criminalistics Certification Study Committee (CCSC, funded by a grant from the National Institute of Justice, NIJ) looked at issues and problems associated with a wide ranging examination and certification program. The California Association of Criminalists (CAC) was the first to adopt a formal certification program based on a comprehensive examination administered to any criminalist seeking certification. The ABC program built upon the CAC process and developed the ABC General Knowledge Examination (GKE) as well as specialty examinations in areas such as Fire Debris, Forensic Biology, and Drug Identification.

See also APPENDIX I.

American Board of Forensic Anthropology *See* APPENDIX I.

American Board of Forensic Document Examiners *See* APPENDIX I.

American Board of Forensic Entomologists (ABFE) *See* APPENDIX I.

American Society for Testing Materials (ASTM) An organization devoted to the development of voluntary standards and specifications for numerous materials, systems, services, and procedures. The ASTM was founded in 1898 and now consists of more than 32,000 members organized into committees and subcommittees that develop detailed written documents that are sold to organizations that request them. Selling of these written procedures (called standards) generates most of the funding for this nonprofit group. Currently, there are 129 technical committees including "Amusement Rides," "Concrete and Concrete Aggregates," "Glass," and "Water." The Committee on Forensic Sciences (E30) was formed in 1970 and has subcommittees dealing with CRIMINALISTICS, QUESTIONED DOCUMENTS, PATHOLOGY and BIOLOGY, TOXICOLOGY, ENGINEERING, ODONTOLOGY, jurisprudence, physical ANTHROPOLOGY, PSYCHIATRY and behavioral science, interdisciplinary forensic science standards, long-range planning, terminology, awards, and liaisons. The scope of the committee includes the development of laboratory methods, procedures, standard reference materials, and standard terminology as it relates to physical evidence and forensic science. The organization maintains an extensive Web site at www.astm.org.

American Society of Crime Laboratory Directors (ASCLD) An organization representing crime laboratory directors formed in 1974 to improve crime laboratory operations and procedures. ASCLD coordinates a voluntary accreditation program for forensic laboratories that addresses facilities, management, personnel, procedures, and security, among other things. Membership is open to current and former laboratory managers and forensic science educators.

ASCLD/LAB is the ASCLD Laboratory Accreditation Board that oversees the accreditation of forensic laboratories. The board was formed in 1981 and incorporated in 1988. Accreditation is granted after a laboratory meets strict requirements on analyst education and continuing education and training, laboratory procedures and protocols, evidence handling, QUALITY ASSURANCE/QUALITY CONTROL (QA/QC), and other aspects of laboratory operation. As part of accreditation, analysts participate in proficiency testing, and reaccredidation is required on a five-year cycle. In 2004, the accreditation guidelines were modified and now incorporate elements from the INTERNATIONAL ORGANIZATION FOR STANDARDIZATION (ISO). More than 250 laboratories are now accredited under the older legacy standards and the newer ISO-based standards.

amino acid racemization (AAR) A technique used in ARCHAEOLOGY, GEOLOGY, ANTHROPOLOGY, and forensic science to date materials and to determine AGE-AT-DEATH. All AMINO ACIDS except glycine can exist in two forms (ENANTIOMERS) indicated by the notation d- and l-. Biological processes, including metabolism, favor the l-forms of amino acids and so proteins that are made in the body consist of l-amino acids. However, if a tissue is shielded from the metabolic process, the amino acids will undergo a process of racemization in which some of the l-amino acids will convert to the d-form until a roughly equal mixture of d- and l-forms exist. Each amino acid has a different rate of racemization so the degree of racemization found in a tissue sample can be used to estimate the age. The rate of racemization depends primarily on temperature and moisture; the warmer and/or the wetter the conditions, the faster the process. Thus, if AAR is being used to estimate the age of a deceased person, the environmental conditions in which a sample or body is found are important.

Techniques for AAR were first demonstrated in 1968 using the amino acid isoleucine. Analytical methods vary, but a common procedure is to isolate the amino acid of interest from a sample using ION CHROMATOGRAPHY (IC), and then determine of the ratio of d- and l-forms using GAS CHROMATOGRAPHY (GC) or HIGH PERFORMANCE LIQUID CHROMATOGRAPHY (HPLC). Generally, only a small sample is needed. AAR has been applied to tissues including the discs between vertebrae, the lens of the eye, and parts of the brain, but forensic applications focus on the analysis of the aspartic acid in teeth.

Once a tooth is fully developed, the dentin portion is surrounded by enamel and is effectively isolated from metabolic processes. This is not true of bone, which is continually in contact with blood and other body fluids. Thus, amino acids present in the proteins in the dentin will undergo racemization even while the person is alive. Since body temperature is stable, as are moisture levels, the rate of racemization is fairly constant and the ratio of d-aspartic acid to l-aspartic acid can provide a reasonable estimate of age (within a few years) even when other techniques fail. Although racemization continues after death, the rate slows as body temperature drops.

See also ODONTOLOGY.

Further Reading

Meyer, V. R. "Amino Acid Racemization: A Tool for Fossil Dating." *Chemtech* (July 1992): 412.

Ohtani, S., and K. Yamamoto. "Age Estimation Using the Racemization of Amino Acid in Human Dentin." *Journal of Forensic Sciences* 36, no. 12 (1991): 792.

amino acids The molecular building blocks of proteins. As the name indicates, all of these molecules have at least one acidic site (functional group) as well as an NH_3 ("amino" as in ammonia) group. The figure of a generic amino acid (below) shows the three common methods in which these molecules are drawn. The term *R* indicates different groups that vary depending on which amino acid it is. The carbon that is at the center of the amino acid is called the alpha (α) carbon, and groups that are attached directly to this carbon are called α groups. A beta (β) group would be located two carbons away from the central carbon. For clarity, hydrogen atoms or CH_3 groups are usually not labeled.

Proteins are polymers of amino acids, meaning that they are built by linking many ("poly") amino acids together in a long chain. Twenty amino acids make up the structure of proteins: alanine, arginine, asparagines, aparatic acid, cysteine, glutamine, glutamic acid, glycine, histidine, isoleucine, leucine, lysine, methionine, phenylalanine, proline, serine, threonine, tryptophan, tyrosine, and valine. Each amino acid is distinguished from the others by a different "R" group as shown in the generic amino acid structure.

All amino acids except glycine can exist in two forms called STEREOISOMERS. In stereoisomers, the same functional groups can be attached to the central carbon in the same way to create molecules that are mirror images of each other. As an example, consider the hands—both are the same structure that differ only in the way in which the fingers and thumb are arranged. The right hand is a stereoisomer of the left. Stereoisomers interact with POLARIZED LIGHT in different ways, leading to the term *optical isomers* or ENANTIOMERS. Isomers are named based on the direction in which they rotate plane POLARIZED LIGHT, to the left (levorotatory or -) or to the right (dextrorotatory or +). Thus, the two optical isomers of alanine would be named d-alanine (or +alanine) or l-alanine (or –alanine) as shown. Notice that the only difference is the orientation of the amino group around the α-carbon, which is called a chiral center. A mixture containing equal amounts of the d and l entantiomers is called a racemic mixture, and the process by which one enantiomer converts to another is called racemization. Nature tends to favor the l-forms of amino acids while most chemical syntheses produce a mixture of the d- and l- forms.

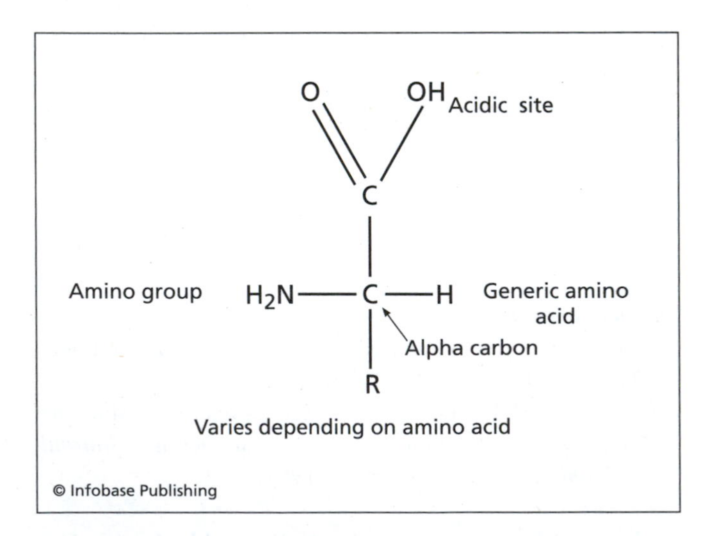

Amino acids, a class of molecules defined by an acidic portion (COOH structure) and an amino (NH_2) site. In this generic amino acid structure, the "R" represents different chemical structures.

amitriptyline *See* ELAVIL.

ammunition For modern firearms, ammunition consists of a projectile (BULLET or pellets) and a CARTRIDGE CASE containing PROPELLANT and the PRIMER that ignites it. A single unit is referred to as a "round" of ammunition and all parts of a round have value as physical evidence. The figures on the next page show the most common types of ammunition encountered in forensic science.

The function of ammunition is to exploit the chemical energy stored in the propellant (gunpowder) by igniting it. The burning releases heat and rapidly expanding gases that are trapped behind the projectile in the breech and barrel of the weapon. When sufficient pressure is built up, the pressure accelerates the projectile forward, giving it kinetic energy (the energy of motion) that is proportional to the weight of the projectile and the speed to which it is accelerated. The equation that describes this relationship is: $KE = ½ mv^2$. Upon impact, the kinetic energy of the projectile is transferred to the target that it strikes. Cartridge ammunition is designed to be self-contained so that all that is needed is a simple mechanical method (linked to the trigger) that can strike the primer and ignite the propellant.

The invention of GUNPOWDER (black powder) has been attributed to many cultures including the Greeks and the Chinese. Black powder contains charcoal (carbon) at about 15 percent by weight, potassium nitrate (KNO_3 or saltpeter) at 75 percent, and sulfur at 10 percent. Although used for centuries on a battlefield, the copious smoke produced by burning gunpowder either quickly obscured the view or gave away the position of those firing. Smokeless powder was developed for use by the French army in 1876 and has replaced

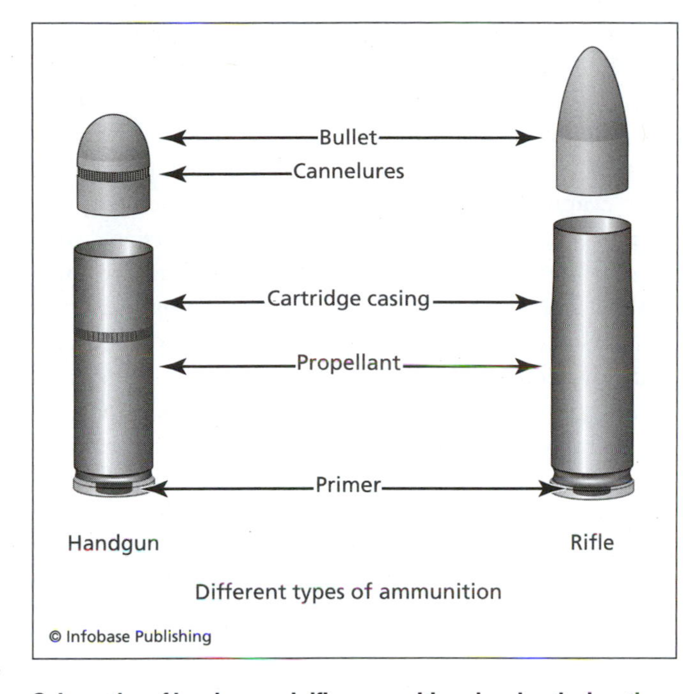

Schematics of handgun and rifle ammunition showing the locations of the primer, propellant, and bullet (not to scale)

black powder in commercial ammunition, although it is still used by collectors and hobbyists. Smokeless powder contains cellulose nitrate and organic stabilizers and is manufactured to carefully control the size of the grains. Powder does not explode when ignited (it is considered a low explosive); rather, it burns very rapidly, and since burning usually occurs at the surface of particles, the size of those particles dictates how much surface area is available and how fast the burning will occur. The term *gunpowder* now commonly refers to smokeless powder even though historically the term has been applied to both smokeless and black powders.

The function of the primer is to ignite the powder. The primer consists of a shock-sensitive material that explodes when struck by the firing pin. Flash holes direct the explosion to the propellant where ignition occurs. Older ammunition and smaller caliber ammunition use a rimfire cartridge in which the primer runs around the circumference of the rim while other rifle and pistol ammunition use centerfire cartridges. The cartridge case itself is usually made of brass and the term *brass* often is used to refer to empty cartridge casings regardless of what they are made of. Brass casings can be reloaded and reused, but casings made out of softer materials such as aluminum are intended for single use.

Bullets vary in composition, coating (jacketing), and shape. Wad cutters, used for target practice, are blunt-nosed lead slugs that are not normally used outside of a shooting range. Bullets may also have a rounded or pointed shape as is common in rifle ammunition. Bullets are made of lead or lead alloys and may be jacketed or semi-jacketed to minimize the transfer of the relatively soft lead to the LANDS AND GROOVES inside of the gun barrel. Jacketing is usually made of copper, copper alloys, or aluminum, all of which are harder than lead. Many other variations exist such as hollow point ammunition and Teflon-coated bullets. The latter are a law enforcement concern given their ability to penetrate body armor. The caliber of a gun (rifle and pistol) is a measurement of the diameter of the gun barrel, so the ammunition must match these dimensions. Caliber is given in both metric units (9-millimeter) and hundredths of an inch (0.38), with some variants. For example, rifle ammunition that is

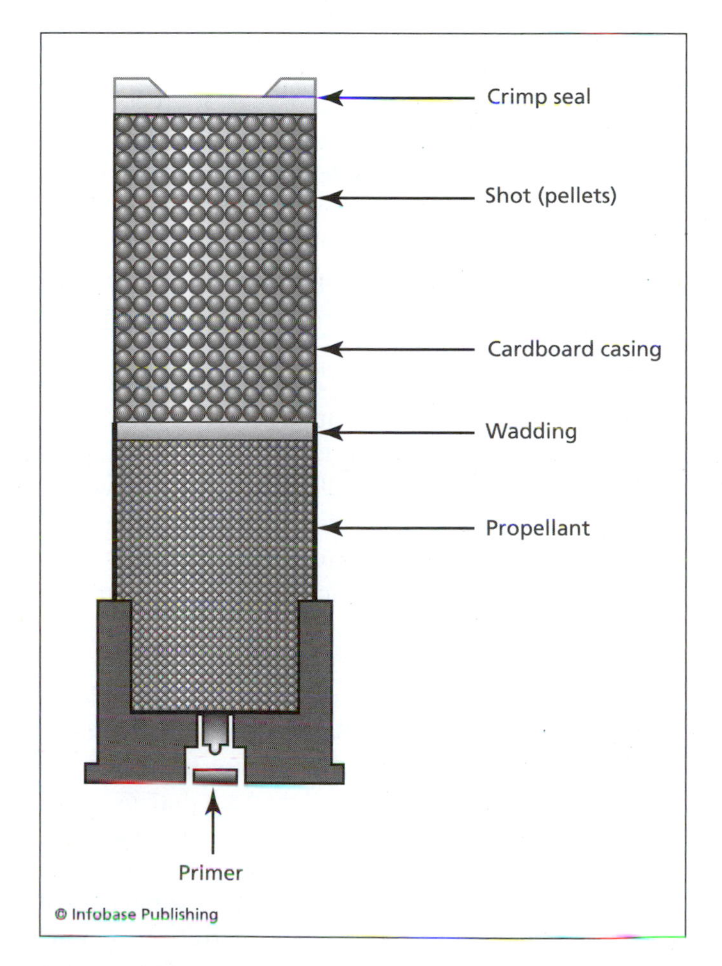

Cutaway view of shotgun ammunition

labeled as ".30-06" ("thirty-ought-six") means that the bullet is for a .30 caliber rifle and that the ammunition type was introduced in 1906.

SHOTGUN ammunition differs in several ways from rifle and pistol cartridges. The cartridge case is made of plastic or cardboard and is crimp-sealed at the top. The projectiles are small lead or steel pellets, the size of which reflects the "gauge" of the weapon. Originally, the gauge of a pellet referred to how many pellets of a given size (the same as the barrel diameter) were needed to reach a weight of one pound. Twelve-gauge pellets were those that each weighed approximately 1/12 of a pound (.45 kg) and would fit in the barrel of a 12-gauge shotgun. Now the term *gauge* is similar to *caliber* and describes the size of the shotgun barrel. Higher gauge numbers mean smaller barrels, so a 12-gauge shotgun has a larger diameter barrel than a 16-gauge, just as 12-gauge shot is larger than 16-gauge shot.

The pellets are separated from the propellant by wadding that can be made of paper or plastic. This wadding material can provide important evidence relating to the manufacturer of the ammunition and its gauge.

See also FIREARMS; GUNSHOT RESIDUE (GSR).

amphetamines Stimulants (amphetamine, dextroamphetamine, and methamphetamine) that were once freely prescribed for weight control, fatigue, and narcolepsy ("sleeping sickness"). Both amphetamine and methamphetamine were used during World War II as a stimulant for troops, and after the war they were used by truckers, dieters, and athletes. As abuse spread, the federal government limited the amount of amphetamines that could be manufactured and removed many types from the market. As a result, illegal demand is now supplied primarily by CLANDESTINE LABORATORIES producing methamphetamine. Street names for the drugs include *speed, ice, crystal,* and *Bennies,* depending on identity and form. Data provided by the 2002–2005 National Survey on Drug Use and Health indicated that approximately 12 million Americans had tried methamphetamine during their lifetime. For high school seniors, the *2005 Monitoring the Future* survey indicated that 2.9 percent had used the drug by the 12th grade.

Amphetamines are psychologically addictive, but debate continues as to the degree of physiological dependence they produce. Amphetamines stimulate the sympathetic nervous system, which controls heart rate, blood pressure, and respiration, and excessive use can lead to severe effects such as hallucinations, convulsions, prickling of the skin, unpredictable emotional swings, extreme aggression, and death. Amphetamines can be taken orally, snorted, injected, or smoked. A dangerous form of methamphetamine, known as "ice," is made by slow evaporation and recrystallization of methamphetamine as a hydrochloride salt, which results in large, clear crystals that can be smoked. Ice is considered to be both toxic and addictive. Other forms of amphetamines include pills ("white crosses" or "Bennies"), liquids, and powders. Occasionally, substances sold illegally as amphetamines are analyzed and found to contain nothing more than sugar and CAFFEINE or EPHEDRINE.

All amphetamines are synthetic and are based on a phenylethylamine (or phenethylamine) skeleton. They are defined as Controlled Substances and are listed on Schedule II of the CONTROLLED SUBSTANCES ACT (CSA). In addition to the three amphetamines already mentioned, this class of drugs also includes MDMA (Ecstasy), phenmetrazine (Preludin), phendimetrazine, MMDA (5-methoxy-3,4-methylenedioxyamphetamine), STP (4-methyl-2,5-dimethoxyamphetamine), and MESCALINE (3,4,5-trimethoxyphenylethylamine). Many of these related drugs such as mescaline are hallucinogens.

As synthetics, amphetamines can be made starting with PRECURSOR chemicals. An older method used PHENYL-2-PROPANONE ("P2P") as the starting point,

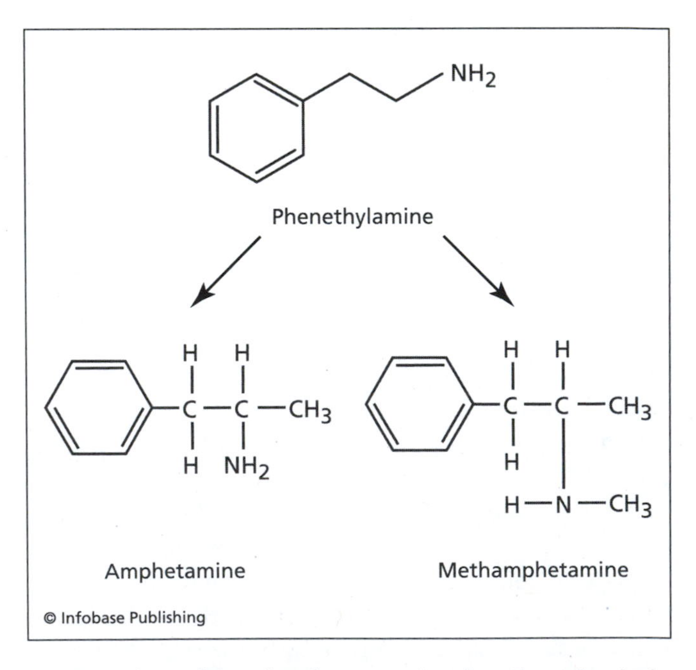

Amphetamine and the related compound methamphetamine, both based on a phenethylamine "skeleton"

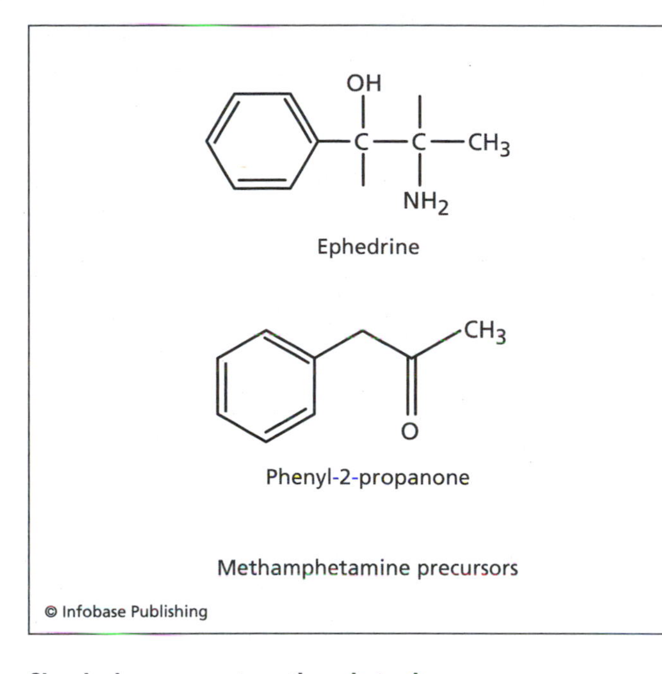

Chemical precursors to methamphetamine

but inclusion of this material on the list of controlled substances reduced its availability and forced clandestine labs to switch to other synthetic routes. The currently favored route starts with EPHEDRINE, an ingredient in over-the-counter (OTC) antihistamine and cold medicines. Other ingredients in this synthetic route can include ammonia, sodium hydroxide (lye), and red phosphorus.

Forensic analysis of suspected amphetamines begins with simple chemical color tests (PRESUMPTIVE TESTS) that indicate the possible presence of the compounds. The analysis progresses through more specific tests including CRYSTAL TESTS, THIN LAYER CHROMATOGRAPHY (TLC), and instrumental techniques such as INFRARED SPECTROSCOPY (IR) and GAS CHROMATOGRAPHY/MASS SPECTROMETRY (GC/MS). Forensic chemists also help identify the synthesis method based on other materials present in the sample. In most cases, it is possible to determine the material with which the material is diluted, materials known as CUTTING AGENTS.

See also PROFILING.

Further Reading

Moore, K. "Amphetamines/Sympathomimetic Amines." In *Forensic Toxicology.* 2nd edition. Edited by B. Levine. Washington, D.C.: American Association of Clinical Chemistry, 2003.

amplified fragment length polymorphism (AFLP) *See* DNA TYPING.

anabolic steroids A class of synthetic steroids related to the male sex hormone testosterone. Widespread abuse of these drugs by athletes, despite the potential for severe long-term side effects, led to anabolic steroids being declared Controlled Substances in 1991. They are listed on Schedule III of the CONTROLLED SUBSTANCES ACT (CSA), along with drugs such as BARBITURATES and some CODEINE preparations. Steroids are considered to be illegal substances by most amateur and professional athletic organizations including the International Olympic Committee (IOC). Although once viewed as primarily a problem in professional athletic and Olympic-level competition, recent surveys indicate that a growing percentage of high school athletes, both male and female, have tried or use steroids. According to the 2005 "Monitoring the Future" Survey, 2.6 percent of male high school seniors had used anabolic steriods, as had 0.4 percent of females.

As a biochemical class, steroids are lipids (fats) that are not hydrolyzable (do not react with water). All steroids are built on a "steroid nucleus" of four rings of carbon atoms, three of which contain six carbons and one that contains five carbons. Common steroids include cholesterol and sex hormones including those used in birth control pills. Anabolic steroids are synthetic steroids related to testosterone, a male sex hormone that promotes the development of secondary male characteristics (called androgen effects) such as deepening of the voice. Testosterone also accelerates muscle growth and promotes rapid healing of injured muscles; these and related effects are called anabolic effects. Synthetic anabolic steroids are designed to minimize the androgen effects, but these cannot be completely eliminated. In athletics, anabolic steroids are used as performance-enhancing drugs that promote development of muscular strength and bulk, reduce recovery time, and allow training at a higher level than otherwise possible. Dangers of anabolic steroid misuse include kidney and liver damage, liver cancer, masculinization and infertility in women, impotence in men, and unpredictable emotional effects including mood swings and extreme aggression. Some of these effects are irreversible. Steroid use is particularly dangerous for adolescents, for whom they can interfere with bone growth.

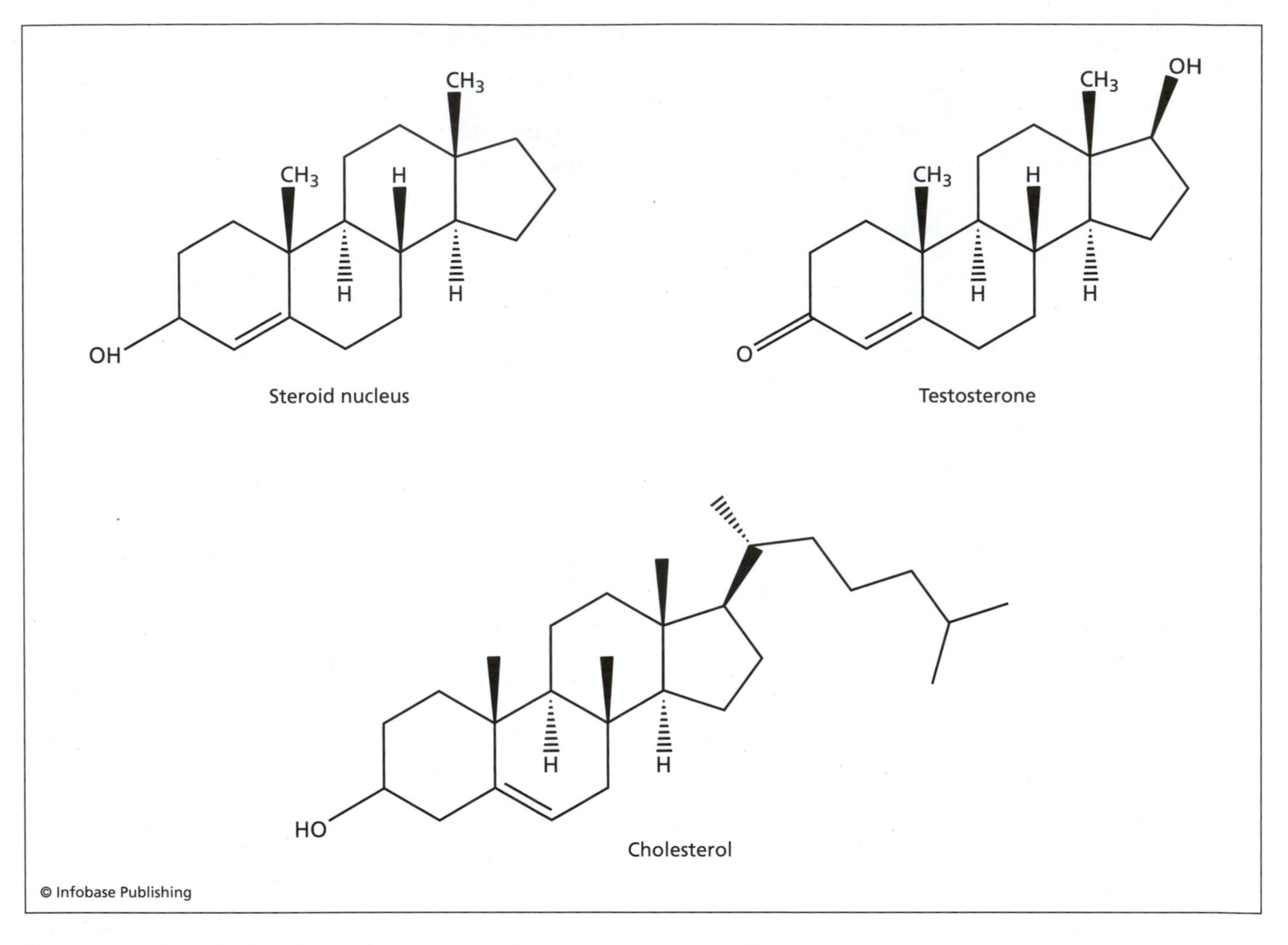

Example structures of different steroids, all based on the same steroid "nucleus"

Anabolic steroids and their metabolic products are detected in blood and urine using GAS CHROMATOGRAPHY/MASS SPECTROMETRY (GC/MS) and other TOXICOLOGICAL techniques. Masking agents are sometimes used to disguise steroid use, but even if the steroids and their metabolites are not detectable, the masking agents usually are. A famous recent case occurred at the Olympic Games in 1988, where Canadian Ben Johnson won the 100-meter dash in record-setting time, only to have his gold medal taken away two days later when his urine tested positive for stanozolol, an anabolic steroid. He was suspended and later returned to competitive track, only to be permanently banned in 1993 when he was again found to have been using anabolic steroids. More recently, professional baseball has adopted stricter rules and testing to address steroid use.

See also "Drug Testing in Sports" essay, p. 116.

analgesics A class of drugs that relieves or reduces pain by depressing the central nervous system (CNS). Aspirin and acetaminophen (Tylenol) are common over-the-counter (OTC) analgesics. Many narcotic drugs, including opiate alkaloids, such as MORPHINE and CODEINE, are powerful analgesics, and abuse can lead to physiological and psychological dependence.

See also DRUG ANALYSIS.

analogs *See* DESIGNER DRUGS.

analysis, comparison, evaluation, and verification *See* ACE-V.

analytical chemistry *See* CHEMISTRY, FORENSIC.

analytical power A term used to express the ability of a chemical analysis to identify a target compound or other substance definitively. Analytical power can refer to a single test or a series of tests designed to narrow down the possible identification of an unknown substance. For example, in forensic DRUG ANALYSIS, a common form of evidence is an unknown white powder. The analyst can apply a series of simple PRESUMPTIVE TESTS to help narrow down the possible identifications of the powder. Using a series of tests provides more analytical power than a single test because with each test, the number of possibilities decreases.

Anastasia and the Romanovs At around midnight on July 16, 1918, Czar Nicholas II of Russia and his family were brutally executed by Bolshevik revolutionaries led by Vladimir Lenin. The killings took place in the basement of a mansion in Yekaterinburg in Siberia and were carried out by several men using guns, bayonets, and clubs. Eleven purportedly died that night, including the czar; his wife, Czarina Alexandra; their son and heir, Alexei; their four daughters, Olga, Tatiana, Marie, and Anastasia; the family physician; a maid; a cook; and a footman. Interest in the case was fueled over the years by rumors that the youngest daughter, Anastasia, had somehow survived and escaped to the West. One of the most insistent Anastasia claimants was a woman named Anna Anderson, who died in the United States in 1984. In the early 1990s, popular attention again focused on the mystery when the Russian government (no longer controlled by the communists) excavated a pit near the town that had long been rumored to be the grave of the Romanovs and their attendants. Portions of nine skeletons, badly damaged by blows, explosives, burning, and acid, were recovered. Using skeletal measurements, dental records, and computer facial projection techniques, the Russians tentatively identified them as the imperial family, but with two missing.

In 1992, a team from the United States, led by the eminent forensic anthropologist William Maples, arrived in Russia to assist with the identification. The team was able to identify the nine and concluded that the two missing skeletons were those of Alexei and Anastasia. This finding fanned survival rumors, at least for Anastasia. Alexei was a hemophiliac and few believed he could have survived the savage attacks of the executioners. Based on interviews with participants in the execution and the aftermath, the consensus of the team was that both had died but that the bodies were burned at a location near the pit where the nine others were buried. The site of the burning was not located.

Further confirmation came in 1993 when Peter Gill and Kevin Sullivan of the British Forensic Sciences Services laboratory conducted DNA analysis. Preliminary work proved that five of the victims were from the same family but could not positively identify them as part of the Romanov family. This was achieved using mitochondrial DNA (mtDNA), which unlike the DNA in the nucleus of the cell is passed directly and unchanged from mother to child. Analysis of the mitochondrial DNA was performed using samples from the pit as well as those from the czar's brother (who died in 1899) and Britain's Prince Phillip. The czarina, like Prince Phillip, was a descendant of Queen Victoria, which provided the investigators with samples of known maternal lineage for comparison. The results showed that the five family members recovered from the burial pit were almost certainly the remains of the Romanovs. The official debate ended in 1998, when the royal family and their servants were laid to rest after a funeral in St. Petersburg.

Anna Anderson, the woman who claimed to be Anastasia, died a decade before the tests were performed and her body was cremated. However, she did have an operation before her death, and the hospital had retained a portion of her tissue. DNA analysis proved that she was not related to the Romanovs. While Anastasia's fate cannot be known for certain, the likelihood is that she died with her family.

Further Reading

Glausiusz, J. "Royal D-Loops." *Discover,* January 1994, 90.

Maples, W., and M. Browning. "The Tsar of All the Russias." In *Dead Men Do Tell Tales.* New York: Doubleday, 1994.

Quinn-Judge, P. "Final Rites for the Czar." *Time,* July 27, 1998, 34.

animal hairs and fibers *See* FIBERS; HAIRS.

anisotropy An optical property of some crystals and fibers useful in the forensic analysis of evidence such as dust, soil, and fibers. A material that is isotropic for a given optical characteristic will have the same value of that characteristic regardless of the direction from which the light is coming. In contrast, anisotropic materials have a nonuniform distribution of such characteristics. Solid materials that are made up of molecules that are randomly placed or molecules that are not symmetric will be isotropic. An example of an isotropic material is table salt, which is made up of simple cubic (symmetrical) crystals of sodium (Na^+) and chloride (Cl^-). Many types of glass are also isotropic. Other kinds of crystals and many POLYMERS (which consist of ordered subunits bonded together) are anisotropic. The term BIREFRINGENCE is also used to describe a form of anisotropy. Birefringence of polymers is important forensically since many FIBERS such as nylon and Lycra consist of synthetic polymers. Birefringence is studied using a POLARIZING LIGHT MICROSCOPE (PLM).

See also DUST ANALYSIS; MICROSCOPY.

Further Reading

De Forest, P. R. "Foundations of Forensic Microscopy." In *Forensic Science Handbook*. Vol. 1. 2d ed. Edited by R. Saferstein. Upper Saddle River, N.J.: Prentice Hall, 2002.

Nomenclature Staff, Royal Microscopy Society. *RMS Dictionary of Light Microscopy.* Royal Microscopy Society—Microscopy Handbook #15. Oxford: Oxford University Press, 1989.

anthrax A deadly bacteria that was sent through the U.S. mail during October and November of 2001. Five people were killed, 13 sickened, and hundreds more exposed and treated. Next to the investigation of the September 11, 2001, attacks, the investigation into this crime was the largest undertaken by the FBI and crossed many traditional jurisdictional lines. Forensic scientists as well as many other scientists from many agencies including the U.S. Army, the Environmental Protection Agency (EPA), and the Centers for Disease Control (CDC) were involved. The bacteria *Bacillus anthracis,* which is found mostly in domesticated animals such as sheep and cattle, cause the anthrax disease. The spores of the bacteria can lay dormant in soil for years and infect humans; the spores can also be manipulated to form a potent biological weapon. There are three types of infection that people can get. First is cutaneous anthrax, which arises when bacteria get into the skin through a cut or similar wound. The skin blackens and dies, although infected people rarely die if they obtain prompt antibiotic treatments. Inhalation anthrax is much more deadly. The early symptoms are similar to a cold or the flu and by the time medical personnel recognize the danger, it is usually too late to prevent death. However, if antibiotic therapy is started soon after exposure, chances of survival are much greater. The third form of anthrax is intestinal, also deadly if not treated early. Tests for anthrax include staining and microscopic study of the bacteria and spores through sophisticated genetic analysis to determine the strain.

See also "Learning from Tragedy: Forensic Science and Terrorist Attacks" essay, p. 128.

Further Reading

Centers for Disease Control (CDC). "Anthrax." Available online. URL: www.cdc.gov/ncidod/dbmd/diseaseinfo/anthrax_g.htm. Downloaded February 29, 2008.

Anthropological Research Facility (ARF) Located at the University of Tennessee at Knoxville, this facility uses donated human remains and animal carcasses to study the process of decomposition. A main goal of the research is to improve estimates of the POSTMORTEM INTERVAL (PMI, time-since-death) and to provide a working laboratory for forensic anthropologists. Dr. William M. Bass founded the facility in 1972. The facility is sometimes referred to as the "Body Farm."

See also ANTHROPOLOGY, FORENSIC; "Forensic Science in Literature" essay.

Further Reading

Bass, Bill, and Jon Jefferson. *Death's Acre.* New York: Penguin, 2004.

anthropology, forensic The analysis and study of skeletal remains that are involved in legal procedures. Anthropology is a wide-ranging field that studies many aspects of human culture and biology from its earliest roots. The discipline can be divided into cultural anthropology and physical anthropology, the branch that examines, among other things, osteology. Osteology is the study of the variability, development, growth, and evolution of the human skeleton, and it is from osteology, as well as ARCHAEOLOGY and medicine, that forensic anthropology has emerged. Physical anthropologists have been assisting law enforcement since the early part of the last century, but the emergence of forensic anthropology as a distinct forensic discipline did not occur until later. One of the pivotal steps was the founding of the Central Identification Laboratory (CIL) by the U.S. Army in 1947. The laboratory, located at Hickman Air Force Base in Hawaii, grew out of work done by physical anthropologists during World War II, where they assisted the army in identification and repatriation of remains. In addition, the Smithsonian has a long history of assisting the FBI with forensic anthropology questions. It was not until the 1970s that the term *forensic anthropology* was widely used and accepted. In 1972, the AMERICAN ACADEMY OF FORENSIC SCIENCES (AAFS) added a forensic anthropology section and in 1977, the American Board of Forensic Anthropology (ABFA), which regulates practices and provides certification for practitioners, was formed.

When studying evidence, the forensic anthropologist must first determine if a material really is BONE or TEETH and if so, whether it is human. It is also important to determine the relative age of the bone as ancient (generally older than 500 years), historic (500–50 years old), or contemporary; however, these time frames vary with context. Once these factors have been established, the remains can be used to determine identification, age at death, sex, race, time since death (POSTMORTEM INTERVAL [PMI]), stature, and manner and circumstances of death. Successful identification of skeletal remains depends on many factors such as what portion of the skeleton is found, its condition, state of decomposition, presence of personal items (such as jewelry), and the extent of records antemortem kept when the person was living. Identification tasks can be greatly complicated in the case of mass disasters (commingled remains), fire and cremation (CREMAINS), and scavenger activity and associated scattering of bones. The common techniques used in identification include the use of dental records (ODONTOLOGY), FACIAL RECONSTRUCTION, and analysis of mitochondrial DNA. Determining the manner and circumstances of death involves looking for evidence of diseases and damage such as bone injury caused by stabbing or firearms. Chemical analysis of the bones can sometimes reveal the presence of poisons such as ARSENIC.

See also AGE DETERMINATION; ARCHAEOLOGY, FORENSIC; DNA TYPING; SEX DETERMINATION; TAPHONOMY.

Further Reading

Iscan, M., and S. Loth. "The Scope of Forensic Anthropology." In *Introduction to Forensic Sciences.* 2d ed. Edited by William Eckert. Boca Raton, Fla.: CRC Press, 1997.

Maples, W. *Dead Men Do Tell Tales.* New York: Doubleday, 1994.

Nafte, M. *Flesh and Bone.* Durham, N.C.: Carolina Academic Press, 2000.

Sorg, M. H. "Forensic Anthropology." In *Forensic Science: An Introduction to Scientific and Investigative Techniques.* 2nd edition. Edited by S. H. James and J. J. Nordby. Boca Raton, Fla.: CRC Press, 2005.

anthropometry The use of body measurements to identify individuals. Most often the term is associated with a system of body measurements developed by ALPHONSE BERTILLON and used for identification purposes until overshadowed by fingerprinting in the early 1900s. The system, also called Bertillonage, was based on the assumption that after the age of 20, skeletal measurements change very little and thus could be used to identify a person unambiguously. Bertillon estimated that the combination of measurements he used would be unique down to about 1 person in every 300 million. However, a major limitation to the system was the care taken in the measurements and errors and inconsistencies introduced by the people doing the measurements. Bertillonage withered as FINGERPRINT technologies advanced, particularly after Bertillon passed away in 1914.

Bertillonage utilized a card-based system that included photographs of the individual, fingerprints, description, and 11 measurements:

- left arm length, elbow to tip of the middle finger
- width of the outstretched arms
- sitting and standing heights
- length of right ear
- length of the left foot
- length of the left middle and little finger
- length, width, and diameter of the skull

Other physical descriptors such as hair color, eye color, and distinguishing scars were included. Bertillon began to develop the system in 1879 in Paris, and, after a trial period, the system was officially implemented in 1883. Another drawback, in addition to measurement errors and inconsistencies, was the time required to complete the measurements. Regardless of the cumbersome nature, the system quickly spread around the world, including the United States. The system was widely accepted until the advent of fingerprint technology about the turn of the 20th century. In the early 1900s there was a period where both techniques were used, with Bertillon strongly advocating his approach. However, he reluctantly added a place for fingerprints from the right hand to his data cards only in 1902. One event contributing to the demise of Bertillonage occurred in 1903 at Leavenworth Penitentiary in Kansas.

An African American, Will West, arrived at the prison, where staff thought they recognized him. A review of Bertillon cards found another inmate with identical Bertillon measurements named William West. Although the men claimed to have been unrelated, subsequent research indicates that they might well have been estranged twin brothers. Regardless, although their measurements were identical, their fingerprints were not, eventually convincing most agencies to switch to fingerprints. It should be noted, however, that there is significant dispute about the veracity and particulars of the now famous "Will West" case.

Body measurements are still used occasionally in forensic ANTHROPOLOGY, where measurements of bones or bone fragments can be used to determine height and stature of the deceased. Measurements are also taken during AUTOPSY, but the goal is no longer identification as it was with Bertillonage.

Further Reading

Cole, S. "From Anthropometry to Dactyloscopy." In *Suspect Identities: A History of Fingerprinting and Criminal Identity.* Cambridge, Mass.: Harvard University Press, 2001.

Nickell, J. "The Two 'Will Wests': A New Verdict." *Journal of Police Science and Administration* 8, no. 4 (1980): 406.

antibodies, antigens, and antisera An antigen is a large molecule (macromolecule), usually a protein or a protein with carbohydrate constituents that, when introduced into an animal, induces the production of a specific antibody. Antibodies are also macromolecules and are proteins. The function of an antibody in the immune system is to destroy or inactivate the antigen it attacks. The study of antigen-antibody reactions is called IMMUNOLOGY.

An antiserum is the purified serum portion of blood that contains antibodies specific to a given antigen. Injecting an antigen into an animal such as a rabbit to simulate an immune response produces antisera.

Immunological techniques are common in forensic SEROLOGY and forensic TOXICOLOGY. For example, typing BLOODSTAINS for ABO type is an immunological procedure in which A and B antigens are detected using antisera containing anti-A and anti-B antibodies. In the area of toxicology and drug analysis, many antibodies have been produced that attack specific drugs. By labeling these antibodies with radioactive isotopes or fluorescent components, drugs can be detected using immunological techniques.

See also BLOOD; IMMUNOASSAY.

anticonvulsants *See* BARBITURATES.

antidepressants *See* BENZODIAZEPINES.

antigens *See* ANTIBODIES, ANTIGENS, AND ANTISERA.

antiserum and antisera *See* ANTIGENS, ANTIBODIES, AND ANTISERA.

Antistius and Caesar In 44 B.C.E., Julius Caesar (~100–44 B.C.E.), Roman dictator, died at the hands of a group of senators wielding daggers and knives. Reportedly, a physician named Antistius examined the body to determine which wound was fatal. He found more than 20 wounds and determined that the second one killed Caesar. The conspirators had apparently hoped that so many wounds precluded a definitive assignment of the fatal one. The practical Roman officials took the only sure route to punishing the true killer by putting all to death. Antistius's determination probably contributed to the rise of the term *forensic,* since he would have presented his findings at the Roman Forum.

archaeology, forensic Although often considered a part of forensic ANTHROPOLOGY, forensic archaeology is emerging as a separate but related discipline. In general, forensic anthropologists concentrate on the analysis of skeletal remains while forensic archaeologists focus on the location and excavation of these remains. As a science, archaeologists have developed detailed and systematic methods of locating, surveying, excavating, and documenting burials and using the information obtained to reconstruct probable activities at the site. Such procedures are ideally suited for processing CLANDESTINE GRAVES and for CRIME SCENE analysis and reconstruction, particularly when the scene goes undiscovered for long periods of time.

See also TAPHONOMY.

arsenic Also known as "inheritance powder" and the "widow maker," this substance was a poison of choice for centuries until reliable chemical tests were discovered to detect it in body tissues. In the absence of these tests, arsenic poisoning was rampant, reaching its zenith, at least in the popular imagination, in Victorian England of the mid- to late 1800s.

History of Arsenic

Arsenic has been used for centuries in medicines, foods, and cosmetics. It is also a trace nutrient, so a tiny amount of it is required for good health. Arsenic is found in high concentrations in seafood and shellfish. Its primary use has been as a pesticide and herbicide (rat poison is probably the most well known), supplanted only recently by synthetic organic chemicals. Other historical uses were in flypaper, as a skin treatment, and as embalming fluid. There were even those who believed that arsenic is an aphrodisiac. Copper arsenate is a wood preservative, and arsenic compounds have been used as pigments in paint, much as lead once was.

Over the centuries, the wide availability of arsenic and arsenic compounds made it a logical choice for poisoning, either for murder or suicide. Arsenic can kill by ingestion of a large single dose, but it also is a cumulative poison, meaning that small doses over time will build up to toxic and eventually fatal levels. Particularly nefarious is the way in which gradual arsenic poisoning mimics many diseases, especially those prevalent in Victorian times and earlier. The symptoms, which include nausea, vomiting, weakness, and diarrhea, are nonspecific and common to food poisoning, cholera, dysentery, and generalized gastric diseases. In 1851, a law called the Arsenic Act was passed in England in an attempt to stop the poisonings. It specified that only a pharmacist could dispense arsenic compounds and that they had to be colored by the addition of soot or indigo. However, the law failed to specify the definition of a pharmacist, so its impact was limited. It was not until definitive and relatively simple forensic tests were developed that arsenic poisonings subsided. Recent cases of arsenic poisoning are primarily accidental, and there is increasing concern that low concentrations of arsenic in drinking water can serve as a means of long-term chronic exposure. The current regulatory limit of 50 ppb (μg/L) is under review, with limits of 10 ppb or lower being considered. Long-term exposure to low levels of arsenic is thought to promote cancers in humans.

Chemistry

Arsenic is a metal and is found in the same chemical family as antimony, another poison. It exists in many forms, all of which are toxic. These include the metal (As), oxides such as As_2O_3 and As_2O_5, acid forms, and even in gaseous forms such as AsH_3 (arsine), which is one of the most toxic forms of the metal. As_2O_3, arsenic trioxide, is a white powder and a favorite choice of poisoners since it looks somewhat like sugar. When heated, the powder smells strongly of garlic, a clue that was used to investigate several cases prior to the availability of definitive chemical tests. Even with chemical testing, the determination of arsenic poisoning was problematical given that arsenic was routinely included in foods, medicines, and cosmetics. Further complicating this was the misperception that immunity to arsenic poisoning could be gained by the ingestion of small amounts of the metal over time. Thus, many people purposely consumed arsenic in hopes of thwarting would-be murderers.

One of the more interesting forms of arsenic is an organic form (bound with carbon), $As(CH_3)_3$, also highly toxic. The latter gas can form under humid conditions when molds in wallpaper paste react with arsenic contained in paint pigments. There has been speculation that NAPOLEON BONAPARTE was a victim of this "death by wallpaper" phenomenon, although that theory is one among many.

Detection and Analysis

One of the first uses of chemical tests for arsenic was during the Mary Blandy trial in England in 1752. Apparently, Mary tried to poison her father using white arsenic by first placing the powder in tea. White arsenic is not very water soluble and, in tea, it probably floated on the surface much as flour would. When this attempt failed, she added the poison to a gruel she prepared for him. The medical examiner in the case, Anthony Addington, noted the odor of garlic in the pan when it was heated, and he offered other testimony supporting the charge. Testing evolved during the 18th and 19th centuries to precipitation tests in which some kind of distinctive solid formed with any residual arsenic. Examples included Green's test and Hume's test, both described in the early 1800s. The most famous chemical test was the MARSH TEST developed by James Marsh, which worked by detecting a thin coating of arsenic metal after a series of reactions and heating. Although sensitive, the method required practice and skill to obtain reliable results. Modern testing methods for arsenic utilize instrumentation in lieu of wet chemical tests and include ATOMIC ABSORPTION (AA) and inductively coupled plasma atomic emission (ICP-AES).

Arsenic persists in hair, nails, and to a small extent in bone, so it is possible to detect cases of arsenic poisoning even in skeletonized remains. Hair is a valuable sample medium since growth rates are known. Thus, depending on what portion of the hair is taken and how long it is, the analysis can provide an indication of arsenic intake over a period of a few days or weeks. This is useful when attempting to determine if arsenic was taken in one large dose or several small doses. If the dose is large, the amount of arsenic deposited in tissues will depend on how long the victim survived after the poisoning. If death is quick (within hours) then there will be insufficient time for arsenic to reach tissues such as hair or bone. On the other hand, if the victim lingers for more than a day, there will be time for arsenic to make its way to tissues. This was a crucial factor in the recent exhumation of Zachary Taylor, the 12th president of the United States. Taylor, known as "Old Rough and Ready," died five days after ingesting a meal and developing what appeared to be food poisoning. A writer, Clare Rising, was convinced by her research that the president had been poisoned with arsenic and that tests on his remains would prove the theory. However, independent analysis at two laboratories showed unremarkable arsenic levels. Since the president had lived for five days, if his meal was poisoned, some of the arsenic would have been detectable.

Further Reading

Emsley, J. *The Elements of Murder.* Oxford: Oxford University Press, 2005.

Maples, W. "Arsenic and Old Rough and Ready." In *Dead Men Do Tell Tales.* New York: Doubleday, 1994.

Saferstein, R., and S. M. Gerber, eds. *More Chemistry and Crime: From Marsh Arsenic Test to DNA Profile.* Washington, D.C.: American Chemical Society, 1997.

arson The act of purposely setting a fire with criminal intent. According to the National Fire Protection Agency (www.NFPA.org), there were approximately 130,000 intentionally set building and vehicle fires in the United States in 2003 resulting in more than 700 deaths and more than $2 billion in property damage. Of these fires, only about 17 percent resulted in an arrest and only 2 percent in convictions. Despite the grim statistics, arson has been declining for several years, although it still results in significant property loss each year. It is the job of the fire investigator to determine whether a fire can be assigned to natural causes, accidents, arson (incendiary), or indeterminate causes. In the case of incendiary fires, the usual motive is profit in the form of insurance fraud. Such fires can be set in structures or in vehicles such as cars, trailers, or boats. Other motives include revenge, vandalism, crime concealment, or psychological disorders such as pyromania.

The role of the forensic scientist in arson investigation focuses on detection of ACCELERANTS such as gasoline, EXPLOSIVES, or INCENDIARY DEVICES that might have been used to start and sustain a suspicious fire. If the fire is set to hide evidence of another crime such as a robbery or homicide, then the role expands and becomes much more challenging. Fire scenes that are also crime scenes are by definition disturbed not only by the fire but also by the action of firefighters and rescue personnel. Bodies have been moved or removed from scenes before it was known that a homicide had occurred, and smoke and water damage of the scene are inevitable. Destruction of evidence, either related to the fire or to the person who set it, is often difficult to find or exploit.

One of the key pieces of evidence for differentiating arson from other causes is the point of origin. Fires burn longest and hottest near the point of origin, reflected in the degree of damage observed. Multiple points of origin frequently point to arson but are not alone conclusive. Other suspicious signs at a scene include signs of forced entry and sabotaged detection and fire suppression systems. On the other hand, assigning the cause of a natural fire can be difficult since the possible causes are so varied. Electrical problems, cooking, smoking, candles, lightning, and creosote build-up in chimneys can all cause fires.

Fatal fires, especially those set to conceal a murder, require more forensic services and involvement. Tasks include determining at AUTOPSY whether the person was alive or dead when the fire started. If a person is alive during a fire, there is usually a deposit of soot in the nose, mouth, or throat, indicating that the person was still breathing during the fire. Elevated levels of CARBON MONOXIDE in the blood also indicate the person was breathing after the fire was burning. Additionally, most victims of the fire itself (as opposed to foul play beforehand) are found face down, except in bed or on other pieces of furniture. In fire deaths, the body is often found in a "pugilistic attitude" in which the arms are drawn in and the hands curled up in what looks like a defensive posture of a boxer. This is a result of the heat of the fire, which can cause the large muscles to contract. Absence of the pugilistic attitude is not alone conclusive but suggests the possibility that the victim had been dead long enough for rigor mortis to set in, thus preventing or lessening the development of the pugilistic attitude.

If the death is a homicide, then evidence such as weapons and BLOODSTAIN PATTERNS may be left at the scene. Although fire alters the appearance of blood and bloodstain evidence, it can survive and be used as evidence. However, the extreme temperatures of the fire can make forensic examinations of the blood and other body fluids difficult or impossible. Despite widespread publicity of a few incidents, suicide by fire (self-immolation) is rare in the United States.

See also COMBUSTION.

Further Reading

Redsicker, D. R. "Basic Fire and Explosion Investigation." In *Forensic Science: An Introduction to Scientific and Investigative Techniques*. 2nd edition. Edited by S. H. James and J. J. Nordby. Boca Raton, Fla.: CRC Press, 2005.

art, forensic Application of drawing, sculpture, and other visual tools to forensic case work. Some of the earliest examples of forensic art were the "Wanted" posters common in the United States and England in the 1800s, prior to availability of photography. Forensic art has expanded since into several different areas and has evolved into a multimedia, computer-assisted forensic discipline. The subareas of forensic art now include

1. composite imagery, in which interviews and witness statements are used to generate a sketch of a missing person or suspect;
2. image modification, in which photographs such as those obtained from surveillance cameras or other visual images are modified, often using computer enhancement techniques;
3. development and aging progressions, in which images are generated to indicate what a child will look like as he or she grows or how an adult's appearance will change as he or she ages;
4. postmortem drawings, in which the artist uses postmortem photos and appearance to create a sketch of what the dead person looked like in life to assist in identification;
5. superimposition, in which computers are used to superimpose photographs of a person's face over the computerized representation of a skull, also for identification purposes;
6. two-and three-dimensional FACIAL RECONSTRUCTION based on a recovered skull; and
7. preparation of graphical or visual information for courtroom presentation.

The INTERNATIONAL ASSOCIATION FOR IDENTIFICATION (IAI) offers professional certification in forensic art.

Further Reading

Taylor, K. T. *Forensic Art and Illustration*. Boca Raton, Fla.: CRC Press, 2001.

art forgeries *See* FORGERY.

asphyxia Death caused by lack of oxygen to the brain. Asphyxia results from suffocation, strangulation, drowning, crushing of the airway, or swelling of the airway in response to injury. Suffocation can occur when the airway is blocked by an object (choking or smothering with a pillow) or in confined spaces where oxygen is depleted or displaced by another gas such as nitrogen or carbon dioxide. Suffocation can also occur when CARBON MONOXIDE gas (CO) is present in high concentrations. In the blood, carbon monoxide displaces oxygen from HEMOGLOBIN, leading to a characteristic cherry-red appearance to skin, blood, and organs. Death by carbon monoxide poisoning occurs in suicides where automobile exhaust is diverted into a car interior or occurs accidentally when an open-flame heater or burner such as a propane heater or barbeque grill is used in a confined space. Because CO is odorless and colorless, victims may have little or no warning before concentrations of the gas reach deadly proportions.

STRANGULATION, a compression of the throat that interrupts blood flow to the brain, can be the result of manual pressure exerted by hands or by use of a ligature such as a belt, rope, scarf, or wire. HANGING is a form of ligature strangulation in which the weight of the body causes the neck compression. Although a common mode of SUICIDE, hanging is rarely seen in accidental deaths or homicides. Ligature strangulation can be suicidal, homicidal, or accidental as in the case of sexual asphyxia.

See also HOMICIDE; SEXUAL ASPHYXIA.

Association of Firearms and Toolmarks Examiners *See* APPENDIX I.

Ativan (Lorazepam) An antianxiety drug in the same chemical family (the BENZODIAZEPINES) as Valium. This family of drugs is widely prescribed and thus often seen in forensic drug analysis.

See also DRUG ANALYSIS; PRESCRIPTION DRUGS.

Atlanta child murders *See* WAYNE WILLIAMS CASE.

atomic absorption (AA) An instrumental technique used to identify and quantify metals including lead (Pb), barium (Ba), antimony (Sb), and copper (Cu) in suspected gunshot residue. Heavy metal poisons such as ARSENIC (As), bismuth (Bi), antimony (Sb), mercury (Hg), and cadmium (Cd) can also be detected by AA. Other terms used to describe this technique include *flame absorption spectrophotometry (FAS)* and *atomic absorption spectrophotometry (AAS)*. In place of a flame, a graphite furnace can be used for heating and atomization. Atomic absorption is one of a family of instrumental techniques designed for the analysis of individual chemical elements in samples (ELEMENTAL ANALYSIS).

In a typical AA analysis, the sample is prepared by extraction with, or dissolving in, an acidic solution. The dissolved sample is drawn into a flame fed by air and a fuel, either acetylene (as in a welder's torch) or a mix of acetylene, and nitrous oxide. In the flame, the solvent evaporates and most chemical bonds are broken, leaving the free metal atoms. In this condition, the atoms are capable of absorbing ELECTROMAGNETIC RADIATION (EMR) in the ultraviolet-visible region, with each element absorbing a different and characteristic wavelength. A hollow cathode lamp, which also contains the metal of interest, shines a beam of light into the flame at the specific wavelength for the metal and the amount of light absorbed by the sample is proportional to the amount of the metal present. The light emerges from the flame and passes through a monochromator (literally "one color"), which filters out extraneous wavelengths of light, allowing the analytical wavelength to pass to the detector. The detector measures how much light is absorbed. A sample containing high concentrations of the metal will absorb more light than a dilute sample. If none of the light is absorbed, then the metal is absent or present in amounts too small to be detected.

A disadvantage of AA is the need for a separate lamp for each element and the ability to test for only one element at a time. Although multielement lamps are available, they are not commonly used for the analysis of trace quantities. As a result, the forensic scientist must know what metals are or might be present ahead of time to select the correct lamp. Accordingly, AA is considered to be primarily a QUANTITATIVE technique rather than a QUALITATIVE one.

See also GUNSHOT RESIDUE; INORGANIC ANALYSIS; INSTRUMENTAL ANALYSIS.

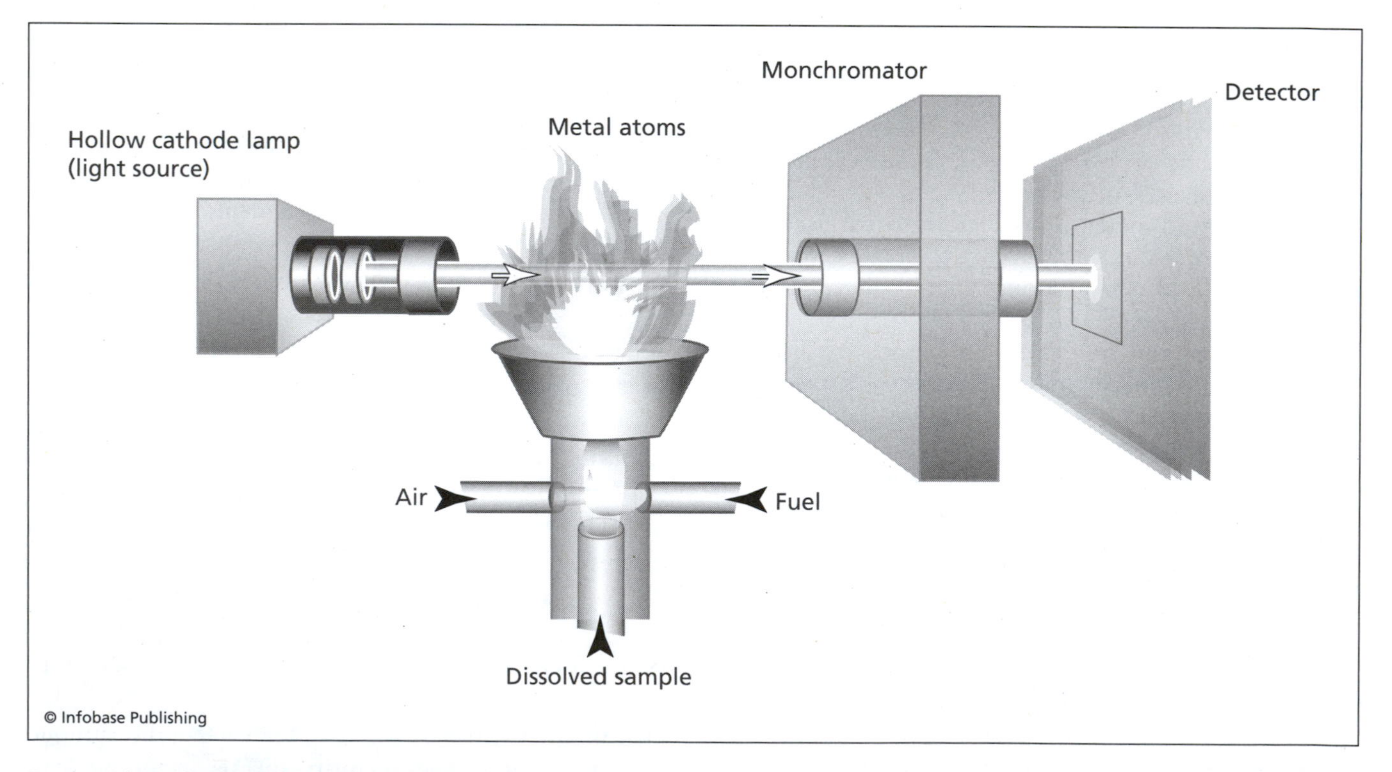

Process of atomic absorption (AA) spectroscopy. Dissolved sample is drawn into the flame, where it is exposed to a wavelength of light emitted by the hollow cathode lamp that contains the metal of interest. If the sample contains that same metal, light will be absorbed in direct proportion to the metal's concentration.

atrocine *See* SCOPOLAMINE.

attenuated total reflectance spectroscopy (ATR) A variation of INFRARED SPECTROPHOTOMETRY (IR and FTIR) used in DRUG ANALYSIS and the analysis of TRACE EVIDENCE such as PAINT and FIBERS. ATR differs from traditional IR in that the ELECTROMAGNETIC RADIATION (EMR) penetrates only a small distance into the sample, making it an ideal technique for surface analysis. Thus, it can be used for otherwise challenging applications such as the analysis of adhesive materials on tape and thin coatings like varnishes, to name a few. ATR requires that the sample be in direct contact with a material with a high REFRACTIVE INDEX (often diamond), which allows for the analysis of solids, liquids, gases, and pastes. The spectra obtained are similar to those from traditional IR techniques but are not identical, so libraries of reference compounds unique to ATR are required for comparison and identification purposes.

See also INSTRUMENTAL ANALYSIS; PAINT.

autoeroticism (autoerotic death, autoerotic strangulation) *See* SEXUAL ASPHYXIA.

Automated Fingerprint Identification System (AFIS) Computerized systems for searching FINGERPRINT databases and identifying suspects. Fingerprints are input as either cards or are taken directly using electronic ("inkless") printing systems. The program locates and identifies major characteristics of the print and searches the database for the closest matches. A fingerprint examiner makes the final decision and identification. Several companies manufacture AFIS systems, but standardized file formats allow prints to be searched across other networked databases. In 1999, the IAFIS (Integrated Automatic Fingerprint Identification System) became operational at the FBI laboratory, allowing automatic searching of the world's largest collection of fingerprints covering in excess of 35 million people.

Further Reading

Jain, R., and S. Pankanti. "Automated Fingerprint Identification and Imaging Systems." In *Advances in Fingerprint Technology.* 2d ed. Edited by H. Lee and R. E. Gaensslen. Boca Raton, Fla.: CRC Press, 2001.

automatic firearms *See* FIREARMS.

automotive finishes *See* PAINT.

autopsy A medical examination, including dissection, that is performed to determine the cause of death and, to the extent possible, the circumstances surrounding the death. The word *autopsy* is derived from Greek and is roughly translated as "to see for one's self" or "to see with one's own eyes," but the term has evolved to the current usage referring to a postmortem dissection. Autopsies are normally performed when a death is suspicious, unattended, or otherwise unexplained. The manner of any death can be broadly classified as natural, accidental, SUICIDE, HOMICIDE ("NASH"), or indeterminate. A medical autopsy is performed in a hospital by a pathologist and is usually done to identify a disease or other condition that might have caused a patient's death. A medicolegal autopsy is usually considered to be a separate procedure and is performed by a physician with specialized training such as a forensic pathologist or MEDICAL EXAMINER. This type of autopsy is more detailed and involves specialized procedures not otherwise performed. An autopsy is an attempt to classify the death and to determine the cause, circumstances, and approximate time of death (and thus the POSTMORTEM INTERVAL (PMI).

Although each autopsy is unique, the procedure usually begins with detailed photography of the body and external examination. Any trace evidence such as fibers or other transfer evidence is collected. Items used to transport the body such as stretchers or litters are also checked for material that might have been dislodged during transport. After the initial exam and documentation, the clothes are removed, preferably by normal methods rather than cutting. If there are stains on the clothes, they are allowed to dry before packaging and transport to the forensic laboratory. At this stage, the body is fingerprinted and checked for any latent prints that might have been deposited on the skin. Laser lights and other illumination techniques may be used. FINGERNAIL SCRAPINGS are obtained and samples such as hair, blood, and vaginal and rectal swaps are collected. The body is then washed and photographed again. At some point, X-rays may be taken, and urine may be collected for drug screening. Urine and vitreous humor (fluid from the eye) is useful for toxicological studies and for determination of blood alcohol concentrations since ethanol can be produced in the blood of the deceased by normal decomposition processes. Measurements of height and weight are also taken.

The medical examiner will carefully examine the body for wounds including bruises, scrapes, needle marks, stabbing wounds, gunshot wounds, and ligature marks (such as from strangulation or being bound). Wound locations are carefully documented and measured, both size and location on the body relative to set anatomical features. Once the external examination is complete, the dissection takes place. The incision in the body cavity is normally a "Y" shape, originating in the armpit areas and descending to the pubic area. Internal organs are examined and sampled, and fluids are collected from the body cavity. Stomach contents are examined and collected and can sometimes provide valuable evidence of when the last food was taken and what it consisted of.

A circular incision in the skull is used to extract the brain for similar studies. Tissue and fluid samples are then examined and subjected to toxicology analysis to add to the information gathered from the examination and dissection. Once the dissection phase is complete, incisions are stitched closed, leaving minimal visible evidence of the procedure. The final results and conclusions of the autopsy are included in the autopsy report and file, which will include all documentation, photos, X-rays, and test results.

Further Reading

Dix, J., and R. Calaluce. *Guide to Forensic Pathology.* Boca Raton, Fla.: CRC Press, 1999.

Knight, B. *Simpson's Forensic Medicine.* 10th ed. London: Edward Arnold, 1991.

Spitz, W. U., ed. *Spitz and Fisher's Medicolegal Investigation of Death.* 3d ed. Springfield, Ill.: Charles C. Thomas, 1993.

Wright, R. K. "The Role of the Forensic Pathologist." In *Forensic Science: An Introduction to Scientific and Investigative Techniques.* Edited by S. H. James and J. J. Nordby. Boca Raton, Fla.: CRC Press, 2003.

B

bacteria *See* MICROBIAL DEGRADATION.

ballistic fingerprinting The process of linking firearms evidence such as BULLETS and CARTRIDGE CASINGS to a specific weapon using an automated search of a large database. While databases of such patterns and marks are not new, existing databases and systems such as IBIS contain data collected from guns used in crimes. In ballistic fingerprinting, the databases would contain images and records from all new guns sold, greatly expanding the size and scope of the databases. The topic came to the public's attention after the D.C. SNIPERS case in late 2002. Eighteen people were shot (13 fatally) by the two snipers before they were caught. Prior to the arrests, police recovered bullets but were unable to match them to a specific weapon because the rifle used had not been traced to a crime and thus was not in existing databases. Although ballistic fingerprinting would increase the utility of the databases in firearms examinations, they are opposed by groups such as the National Rifle Association (NRA), which argue that markings change over the life of a gun and that locating a gun does not necessarily locate a criminal. However, in the wake of the sniper killings, there is increasing interest and study into possible future implementation of some form of ballistic fingerprinting.

See also FIREARMS.

ballistics *See* FIREARMS.

Balthazard, Victor (1852–1950) French *Forensic Scientist* Balthazard served as the medical examiner for the city of Paris and helped advance fingerprint analysis, firearms, and hair analysis at a time when forensic science was emerging as a distinct scientific discipline. He is credited with developing probability models that showed fingerprints were unique, with approximately one chance in 10^{60} (1 followed by 60 zeros) that any two people would have the same patterns. By means of comparison, a one in a million chance would be 10^6, one in a billion 10^9, and one in a trillion 10^{12}. In 1910, he along with Marcelle Lambert wrote the first comprehensive book on hair analysis entitled (translated) *The Hair of Man and Animals*. In it, they advocated MICROSCOPY and the careful evaluation of microscopic structures, which is still the standard today. Balthazard also developed an advanced photographic method of comparing markings on bullets and in 1912 testified in a case using photos and point comparison techniques to identify bullets involved in a fatal shooting. He was also among the first to note other distinctive markings in firearms, including firing-pin impressions and FABRIC IMPRESSIONS that result when a bullet passes through woven fabrics. As further proof of the breadth of his knowledge, in 1939 he made a presentation in Paris discussing the value of BLOODSTAIN PATTERNS as physical evidence. Balthazard played an important role in the history of many of the disciplines found in modern forensic laboratories.

barbiturates A class of drugs based on barbituric acid that act to depress the central nervous system (CNS) and are therefore classified as CNS depressants. Administered primarily by ingestion of pills, barbiturates produce a general feeling of well-being and promote sleep. An overdose of barbiturates can lead to coma and death and are a common choice for suicide, most famously in the case of Marilyn Monroe. Barbiturates are also prescribed to control convulsions in epileptics. On the street, barbiturates are called downers and are usually named for the color of the capsule they come in, such as "reds." Although barbiturates are the oldest class of CNS depressants, their legitimate uses have declined as other safer alternatives such as the BENZODIAZEPINES have been introduced.

Adolf von Baeyer, a German chemist, first synthesized barbituric acid in 1863 and reportedly named the compound after a woman. It was not until 1903 that the first derivative (Veronal) was marketed as a sedative, and several others followed. Barbiturates are classified by how long lasting they are, with pentobarbital and secobarbital being slow acting, amobarbital intermediate, and barbiturates such as phenobarbital being long acting. Abuse of barbiturates can lead to dependence, and an overdose can kill by altering the pH of the blood and disturbing the system that regulates breathing. Since barbiturates are acidic, overdoses can cause inflammation of the stomach lining and small intestine, where absorption takes place. Barbiturates are controlled substances and are listed on Schedule II, III, and IV of the CONTROLLED SUBSTANCES ACT (CSA).

Victor Balthazard, an early pioneer of forensic science who was knowledgeable in areas ranging from forensic medicine to microscopy and blood-spatter patterns ***(National Library of Medicine, National Institutes of Health)***

Forensic testing and identification is similar to other drug classes and starts with PRESUMPTIVE TESTS, moving to definitive identification using GAS CHROMATOGRAPHY/MASS SPECTROMETRY (GC/MS) and INFRARED SPECTROSCOPY (IR). THIN LAYER CHROMATOGRAPHY (TLC) and HIGH PERFORMANCE LIQUID CHROMATOGRAPHY (HPLC) can also be used. Since different barbiturates are classified on different schedules of the CSA, identification of the specific barbiturate is critical.

Barr body A small structure found in the nucleus of female cells that has been used in SEX DETERMINATION. The female sex chromosomes differ from those of males (XX v. XY), and in many female cells, the inactive X chromosome shrivels. This structure is called a Barr body or sex chromatin that male cells will lack. Barr bodies absorb fluorescent dye strongly and under a microscope look something like baseball bats or drumsticks. If an abundance of these structures is seen in a sample, it is possible to determine the sex of the donor. However, given the difficulty of the test and problematic results, forensic use of Barr bodies was limited and has been replaced with DNA TYPING techniques.

Bayesian statistics A method of comparing HYPOTHESES (theories) that takes into account prior knowledge and modifies it using information gathered from evidence. One of the advantages of a Bayesian approach is that it requires the comparison of two scenarios. For example, consider a hypothetical case in which blood is found at a scene and is typed as AB blood in the ABO BLOOD GROUP SYSTEM. A suspect is identified who also has blood type AB. The traditional statistical approach to interpreting these results would involve citing POPULATION FREQUENCIES, which show that about 3 percent of the population is type AB. Alone, this information supports the hypothesis that the suspect is guilty and is based on the prior knowledge of population frequencies. However, in a Bayesian approach, this information could be modified to take into account new information gathered from investigation or analysis. Perhaps the suspect has no wounds or other scars that would support the idea that he lost blood at the scene. This information would decrease the importance of this suspect having the relatively rare AB blood.

Baye's Theorem can be stated informally as follows:

Posterior odds = prior odds × LIKELIHOOD RATIO

In the above example, the prior odds would be reflected in the fact that it is known that 3 percent of the population has AB blood. It is existing knowledge, something that can be expressed using a number. The likelihood ratio compares two possible interpretations, such as the suspect did contribute the blood versus the suspect did not. This information is often more subjective than the prior odds, but it provides a way to at least compare two possibilities as opposed simply to accepting or rejecting one. The posterior odds represent the new estimate.

See also STATISTICS.

bear claw *See* CYSTOLITHIC HAIRS.

Becke line (Becke line method) A method used in microscopic analysis to determine the relative differences in refractive index of two adjacent media, such as a particle and the surrounding mounting media. The Becke line appears as a bright halo of light surrounding a specimen that is immersed in a liquid. When the REFRACTIVE INDEX (RI) of a specimen is the same as the refractive index of the liquid, the Becke line vanishes. This phenomenon can be further exploited by moving the specimen relative to the objective of the microscope. The apparent direction of motion of the line, either into the liquid or into the specimen, is useful in determining how the refractive indices relate to each other. In forensic science, the Becke line method is used in the analysis of particulates such as GLASS, minerals, and FIBERS.

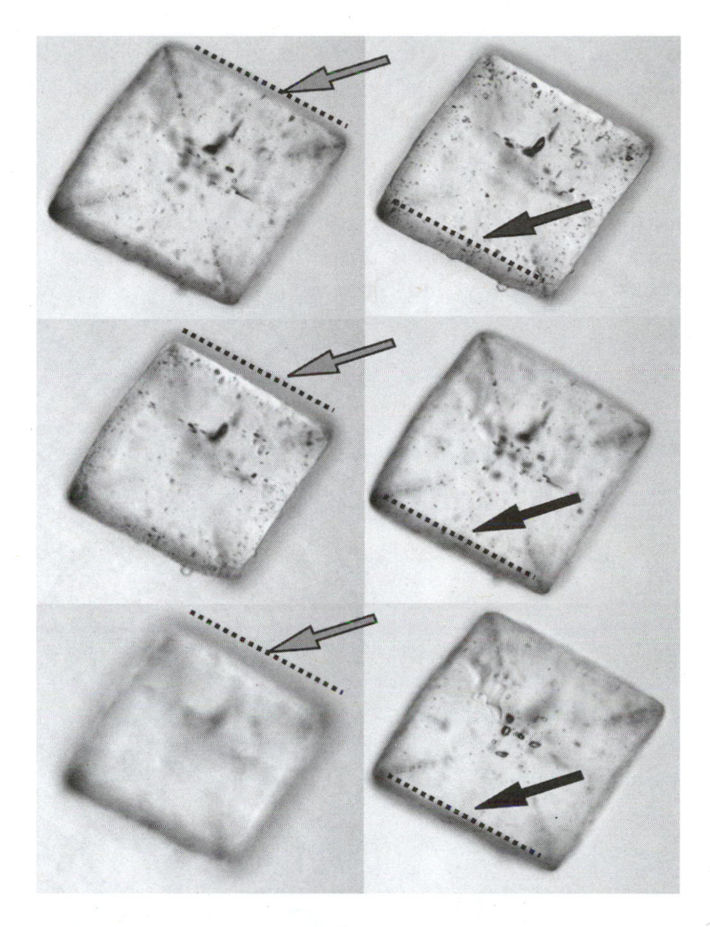

A crystal of silver chloride in two types of immersion oil. The line moves into the media with the higher refractive index. ***(Courtesy of Rebecca Hanes, Bennett Department of Chemistry, West Virginia University)***

Further Reading

Nomenclature Staff, Royal Microscopy Society. *RMS Dictionary of Light Microscopy.* Royal Microscopy Society—Microscopy Handbook #15. Oxford: Oxford University Press, 1989.

behavioral evidence A broad category of forensic disciplines and investigative tools that includes forensic PSYCHIATRY, PSYCHOLOGY, neurology, neuropsychiatry, neuropsychology, DECEPTION ANALYSIS, and POLYGRAPH testing. Some areas are routinely accepted and used by law enforcement and the courts, whereas others are considered less reliable or even questionable in validity and reliability.

benzidine A PRESUMPTIVE TEST for blood that works by detecting the presence of HEMOGLOBIN. The heme group in hemoglobin has the ability to catalyze certain oxidation reactions, and this is called peroxidase activity. The peroxidaselike activity of heme is the basis of the benzidine test, as well as several other presumptive blood tests. In this case, benzidine, which is colorless, is oxidized in the presence of hemoglobin and changes to a bluish color. However, the test is not specific, and many other substances can give a positive result (or FALSE POSITIVE). Since benzidine has been shown to be a potent carcinogen, this test is no longer used.

benzodiazepines One of the most widely prescribed drugs in the world, benzodiazepines are used as mild tranquilizers and as anticonvulsants. Benzodiazepines produce a generalized sense of well-being and reduce anxiety, but unlike BARBITURATES, they do not generally cause as much sleepiness. The most famous member of this family is probably Valium (diazepam); other examples include lorazepam (Ativan), Xanax, Halcion, and Klonopin, which are used to control seizures. Abuse of the benzodiazepines can induce physical and psychological addiction, and the drugs are listed on Schedule IV of the CONTROLLED SUBSTANCES ACT (CSA). Forensic chemists identify them using THIN LAYER CHROMATOGRAPHY (TLC), GAS CHROMATOGRAPHY (GC), and HIGH PERFORMANCE LIQUID CHROMATOGRAPHY (HPLC). To detect these

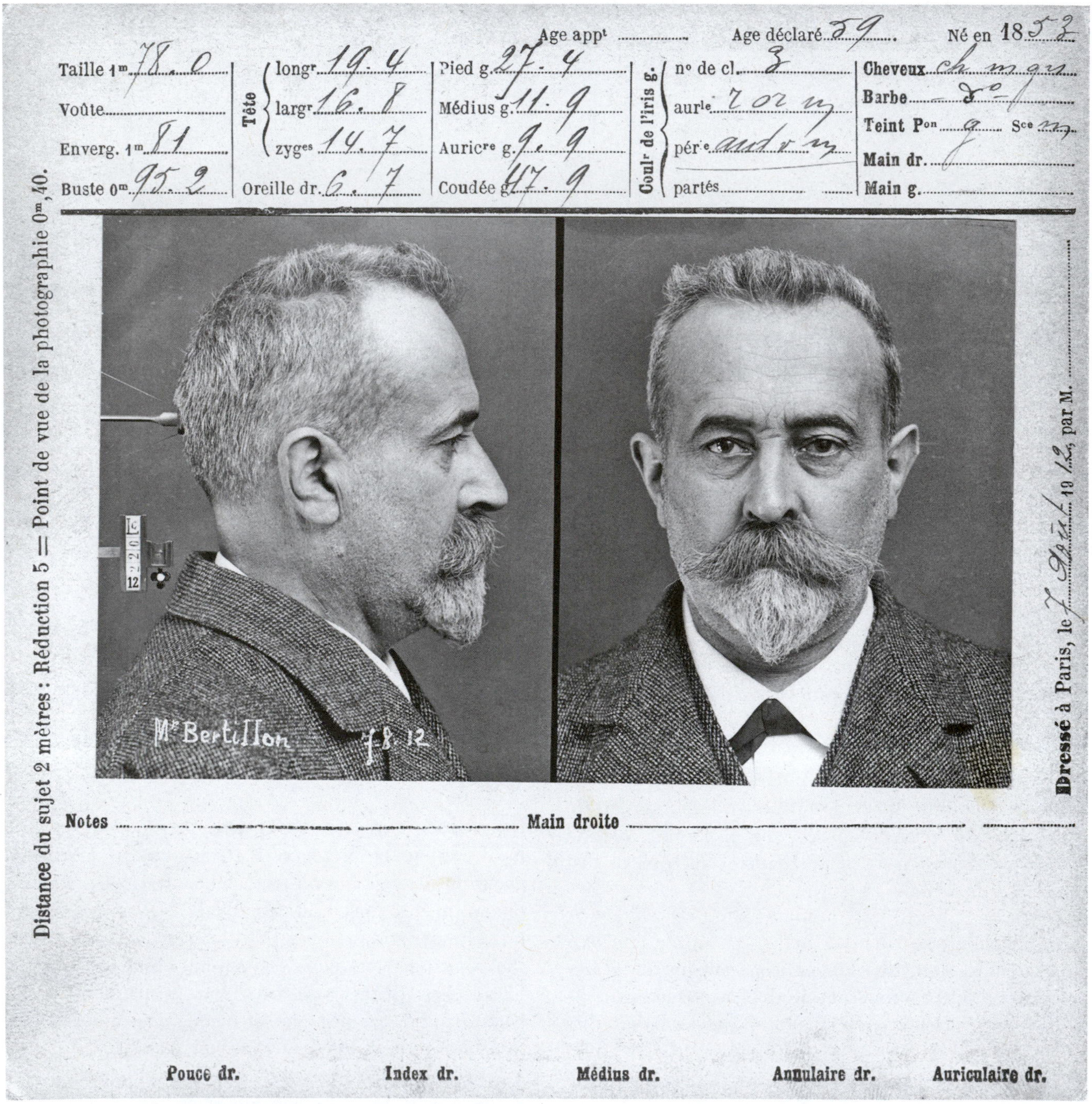

Alphonse Bertillon in a self-portrait using one of his identification cards *(Adoc-photos/Art Resource, NY)*

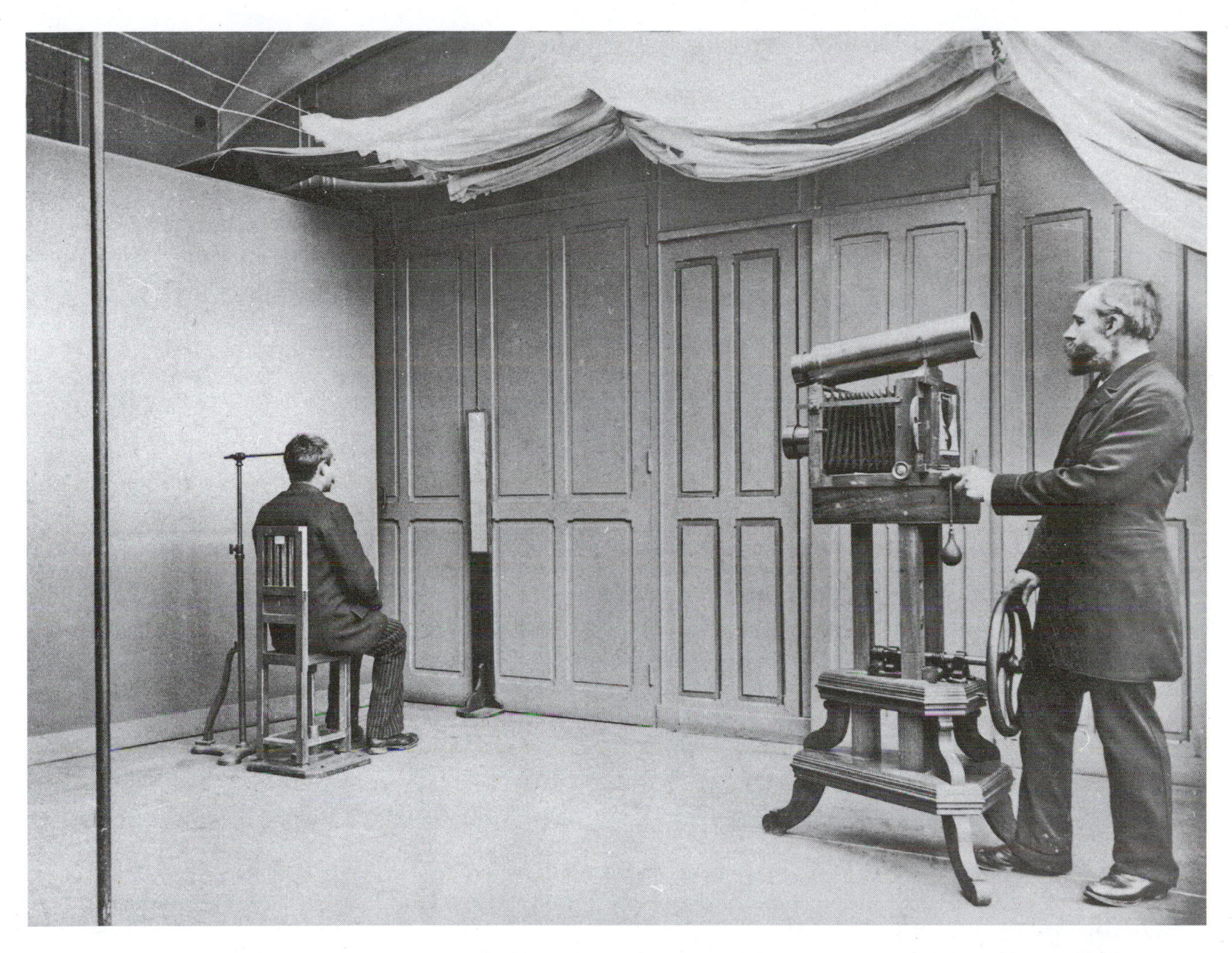

Alphonse Bertillon taking photo and measurements *(© Museé de la Préfecture de Police, Paris, France/Archives/Charmet/The Bridgeman Art Library)*

compounds in blood and urine, additional tests based on IMMUNOASSAYS are employed.

benzoylecgonine (BZ) A primary metabolite (*see* METABOLISM AND METABOLITE) of COCAINE that was first isolated from cocoa leaves (the natural source of cocaine) in 1923. Once cocaine is metabolized to BZ, the metabolite has a half-life (time required to excrete half the remaining compound) of about seven hours, meaning that it can be detected in the urine approximately 48 hours after cocaine is administered. Workplace drug testing focuses on this metabolite and is conducted using IMMUNOASSAY techniques and GAS CHROMATOGRAPHY/MASS SPECTROMETRY (GC/MS).

Bertillon, Alphonse (1853–1914) French *Forensic Scientist* Bertillon developed the first systematic method for the identification of suspects and criminals, setting the stage for fingerprinting, which ultimately replaced it. The system, called ANTHROPOMETRY or Bertillonage, used 11 body measurements along with descriptive information and photographs stored on a card, similar to modern fingerprint cards. An example is shown on page 32. After development and implementation in 1883, the system spread throughout the world while elevating Bertillon to the forefront of pioneering forensic scientists.

Bertillonage was the accepted standard of individual identification until the early 1900s, when problems with the system, coupled with growing inferences that FINGERPRINTS were individually unique, led to a gradual abandonment of the body measurement system. Bertillon resisted the use of fingerprints, although he did add space to his data cards for the inclusion of fingerprint data from the right hand of the individual being cataloged. Ironically, despite his reluctance to accept fingerprinting, Bertillon was the first forensic scientist in Europe to use them to solve a case. In October 1902, he used fingerprints left at a crime scene to identify the murderer, a convicted swindler whose prints had been taken and cataloged on his Bertillon card. The man later turned himself in and confessed.

Despite this success, Bertillon's continued lack of enthusiasm for fingerprints worked against him in a later, more famous case. When the *Mona Lisa* painting was stolen in 1911 from the Louvre Museum in Paris, Bertillon was unable to identify the suspect even though he left prints on the glass covering the painting. Unfortunately for Bertillon, these prints were from the left hand and his cards stored prints only from the right, so even an exhaustive search of his data cards was in vain. It was not until after Bertillon's death that the transition was complete.

In addition to anthropometry, Bertillon is also remembered for his contributions to the development of crime laboratories as distinct entities within law enforcement. He was instrumental in directing the evolution of what amounted to specialized photographic studios into identification bureaus focusing on his system of body measurements. This system was widely adopted as a standard identification tool by many police departments. Although not a crime laboratory in the modern sense, his identification bureau helped pave the way for further improvements and expansion into other areas of crime science and criminalistics. He also advanced work in the fields of questioned documents and crime scene photography.

Bertillonage *See* ANTHROPOMETRY.

bias In scientific analysis, a tendency to obtain results that are offset from the correct or true result. Bias can be psychological in the sense that an analyst may, consciously or unconsciously, expect or desire a certain result for a test. As a result, the probability of obtaining the desired outcome may increase. Bias can also be introduced, purposely or not, when test results are interpreted in a report or a courtroom testimony. Finally, problems with the test itself can unintentionally introduce bias. For example, if a forensic chemist is testing a sample for the presence of a drug but uses laboratory glassware that, unknown to the analyst, is contaminated with that drug, the results of the test will be biased and a FALSE POSITIVE can result. To combat bias, laboratory methods and protocols include the use of CONTROLS and other QUALITY ASSURANCE/QUALITY CONTROL procedures (QA/QC) designed to detect, minimize, and correct biases.

bioassay An analytical technique that takes place in a living organism or using materials derived from a living organism. The goal of a bioassay is to determine how the organism, tissue, or organ responds to a compound or group of compounds. In forensic TOXICOLOGY, a bioassay might be used to determine how an organ, such as the liver, responds to a poison or a drug.

See also BIOMARKERS.

biodegradation and biodeterioration *See* MICROBIAL DEGRADATION.

biological microscope *See* COMPOUND MICROSCOPE.

biological substances Any type of evidence of biological origin. Examples include blood, body fluids, insects, and plant material such as seeds or leaves, wood, algae, feathers, starches, diatoms, and vegetable fibers. Biological substances originate from a living organism.

biology, forensic The analysis of evidence for biological composition or characteristics using biological and biochemical techniques. Forensic SEROLOGY (analysis of blood and body fluids) and forensic ENTOMOLOGY are examples of subdisciplines in forensic biology. The term has become more commonplace since about 1990, with the advent of DNA TYPING. Now, the term *forensic biology* is frequently used interchangeably with *DNA typing*.

biomarkers Compounds found in biological materials that indicate a condition, process, or event. Some diseases, such as cancer, produce unique chemical compounds that can be used to diagnose the presence of the disease. In forensic science, TOXICOLOGISTS can use biomarkers as indicators of ingestion of specific drugs or POISONS. Biomarkers can also be helpful in determining cause of death. Other applications of biomarkers are found in ENVIRONMENTAL FORENSICS, where analysts attempt to identify the source of oil spills based on biomarkers found in the oil.

See also BIOASSAY.

biomechanics, forensic The application of biomechanical principles and practices to legal matters and proceedings. Biomechanics is the study of the mechanics of motion in biological organisms, primarily muscle-driven motion. These considerations can become important in many cases in both CIVIL and CRIMINAL LAW. In assaults, suicides, mass disasters, and homicides, biomechanical investigations can provide detailed information on how injuries might have been inflicted, how an injured person moved or was able to move, or whether a proposed motion was feasible. For example, if a suspect in a murder case claims to have stabbed someone while falling backward, that might indicate self-defense. However, if the injury pattern suggests instead that the suspect was moving forward, that would contradict that statement. In civil cases, biomechanics can be used to study such things as unattended (no-witness) falls. For example, if a person claims to have been injured in a fall down a stairs, forensic biomechanics could be used to determine if the victim's claims are consistent with pattern and severity of injury as well as other physical evidence. As a result, forensic biomechanics is playing an increasing role in liability lawsuits in which an injury or death results. Biomechanical investigations rely on principles of physics and engineering and involve numerous calculations. As such, computer models are used in many cases to model scenarios and to determine which ones were possible, which ones were most likely, and which ones were impossible.

Further Reading

Sacher, A. "The Application of Forensic Biomechanics to the Resolution of Unwitnessed Falling Accidents." *Journal of Forensic Sciences* 41, no. 5 (1996): 776.

Sloan, G. D., and J. A. Talbot. "Forensic Application of Computer Simulation of Falls." *Journal of Forensic Sciences* 41, no. 5 (1996): 782.

biometrics Originally, the use of statistics to measure living organisms. Now, the term is also used to describe the use of unique physical characteristics for individual identification by an automated system. An example of a biometric device is a door lock that requires a person to place his or her thumb on a reader pad for verification of identity. Other physical features that have been studied for use in biometric devices include the pattern of blood vessels in the eye (a retinal scan), the pattern of the iris in the eye, facial features, speech recognition, and the geometry of the hand.

See also FINGERPRINTS.

Birch method *See* NAZI METHOD.

birefringence Many crystalline materials and synthetic POLYMER fibers have two REFRACTIVE INDICES (RI). The refractive index of a material is its ability to bend or alter the path of light. The common example of this is a pencil placed in a glass of water that appears to bend. This is a result of the differences in the refractive index of water and air. The birefringence of a material is defined as the difference between its two refractive indexes and is determined using a POLARIZING LIGHT MICROSCOPE (PLM). The refractive indices of the fiber can be different when observed parallel to the long axis of the fiber from when observed perpendicular to it in polarized light. The calculated difference ($n_{\parallel} - n_{\perp} = B$) between these two values is the birefringence of the fiber. Forensically, birefringence is useful in the analysis of minerals such as those encountered in soil, dust, glass, and fiber evidence.

See also ANISOTROPY; FIBERS; MICROSCOPY.

Further Reading

McCrone, W., et al. *Polarized Light Microscopy.* Ann Arbor, Mich: Ann Arbor Science Publishers, 1978.

Nomenclature Staff, Royal Microscopy Society. *RMS Dictionary of Light Microscopy.* Royal Microscopy Society—Microscopy Handbook #15. Oxford: Oxford University Press, 1989.

bite marks (bitemarks, bite-marks) A type of IMPRESSION EVIDENCE that can be left in the skin of a victim, but also in food, chewing gum, and even in pencils and pens. Given the variability in the dental structure and such things as distance and angles between teeth, missing teeth, fillings and other dental work, and unique wear patterns, bite marks are often considered to be individually unique, although there is controversy surrounding bitemark evidence. Bite marks in victims are common in sexual assaults, homicides, domestic assaults, and child abuse cases, and courts have accepted bite mark evidence since the 1950s.

A bite mark in flesh will have a circular or oval appearance and will consist of lacerations, bruises, scrapes, or depressions. Often a bruised area is seen in the center of the mark, and scrape marks left by the teeth may be evident. However, there may not be visible tooth impressions. Factors such as strength of the bite, clothing, and movement or struggle while the biting occurs can alter the appearance and limit the depth of penetration. Skin itself is a poor media for taking impressions and does not tend to record or retain details well.

Time is critical when bite mark evidence is collected. On a dead person, the marks will fade, and on the living, the wounds will begin to heal, changing the appearance. In addition, saliva deposited by the biter can be valuable for DNA ANALYSIS. Bite marks left in food also require quick action as the food can dehydrate or begin to spoil, which can also alter the appearance and dimensions of the impression. Bite marks are documented photographically and occasionally using UV light (which sharpens details) and IR light, which can illuminate underlying features. Casts or molds may be taken if appropriate. At AUTOPSY, the area of the skin where the mark is located may be excised and preserved for further analysis. Tracings can also be made of the mark as further documentation. Forensic ODONTOLOGISTS (forensic dentists) perform the matching of a bite mark to a suspect.

black powder Also called gunpowder, black powder was used as a PROPELLANT for early FIREARMS. It consists of a mixture of 75 percent potassium nitrate (KNO_3 or saltpeter), 15 percent carbon (charcoal), and 10 percent sulfur. Black powder is a low explosive and is generally prepared by a wet mixing stage, followed by pressing into a cake, drying, and breaking up the residue into granules. When ignited, black powder produces copious smoke and accordingly was replaced in the late 1800s with SMOKELESS POWDER, which is the propellant used in modern AMMUNITION. Collectors and hobbyists, such as Civil War reenactors, still use black powder weapons so firearms examiners occasionally encounter such weapons and evidence. In addition, black powder can be used as an accelerant or as an ingredient in an EXPLOSIVE OR INCENDIARY DEVICE. Often residual grains of black powder can be seen after it is burned, and these grains can be subjected to PRESUMPTIVE TESTS. Examples include a flame test in which the element potassium (a component of KNO_3) emits a lavender color when placed in a flame and a modified GRIESS TEST, which reacts with the nitrate ion (NO_3^-).

blood Classified as an extracellular fluid, meaning that it is found outside of cells in the body. Other examples of extracellular fluids include digestive fluids and lymphatic fluid. Blood has many functions and is a complex mixture of organic and inorganic materials, including electrolytes such as sodium (Na^+), proteins, and several different kinds of cells. The characteristic red of blood comes from the complex formed between HEMOGLOBIN in red blood cells (RBCs) and oxygen.

By spinning a blood sample in a centrifuge, it can be divided into a cellular component (approximately 45 percent of the total volume) and a noncellular component called plasma, which makes up the remaining 55 percent. Plasma can be further subdivided into serum and fibrinogen, the material that forms clots. Serum, a clear straw yellow in color, carries electrolytes, with the sodium ion (Na^+) and the chloride ion (Cl^-) being the most concentrated (sodium chloride, NaCl, is table salt). The bicarbonate ion (HCO_3^-) is also found in high concentrations, and this is the form in which waste carbon dioxide (CO_2) is transported in the blood. Proteins (albumins and globulins) are also carried in the serum. The word *SEROLOGY* is derived from *serum*.

The cellular portion of blood can be divided into three types of cells: red blood cells (RBCs, also called erythrocytes); white blood cells (WBCs, leucocytes); and platelets (thrombocytes). RBCs, which transport oxygen and bicarbonate, are the most numerous and are unique in that they lose their nucleus before entering into the circulatory system. WBCs (several types exist) are the next most numerous and are active in fighting diseases. Platelets are needed for clot formation. Prior to the advent of DNA TYPING, serological techniques were used by forensic scientists to type GENETIC MARKER SYSTEMS in blood. Genetic marker systems are those in which several variants (POLYMORPHISM) exist in the population with known frequencies and that are inherited. The ABO BLOOD GROUP SYSTEM is one example of a genetic marker system in blood. Prior to DNA TYPING, the goal of forensic analysis was to type many genetic marker systems in an attempt to INDIVIDUALIZE a blood sample to the extent possible.

See also BLOOD GROUP SYSTEMS; BLOODSTAIN PATTERNS; BLOODSTAINS; ISOENZYME SYSTEMS.

blood alcohol concentration (BAC) In many jurisdictions, a blood sample is required to prove that a driver is legally intoxicated. This is in addition to any field tests performed based on ALCOHOL in the breath. The blood alcohol test is considered an evidentiary test, meaning the results can be used as evidence in a prosecution; in contrast, field tests produce approximate results and are used only to determine if a BAC test should be performed. Trained medical personnel draw the blood for a BAC measurement, which must be performed carefully using ethanol-free disinfectants and the proper procedures including a CHAIN OF CUSTODY, refrigeration, and the addition of anticoagulants and preservatives. Forensic toxicologists perform the BAC analysis. Statistics for 2004 from the National Highway Transportation Safety Administration (NHTSA, www.nhtsa.gov) reported that of 42,634 automobile fatalities nationwide, an average of 39 percent were related to alcohol.

In the body, the fate of alcohol can be divided into an absorption stage, a distribution phase, and an elimination phase in which the ethanol is oxidized via metabolic processes to carbon dioxide (CO_2). During the absorption phase, concentration of alcohol is higher in arterial blood than it is in the veins; this is important since blood samples are drawn from veins. In contrast, BREATH ALCOHOL measurements reflect the concentration of ethanol in arterial blood, which is more representative of the concentration of alcohol reaching the brain. At the end of the absorption process the two measurements of BAC converge to similar values.

Early laboratory test methods for BAC employed WET CHEMISTRY methods, but these methods have been largely replaced by instrumental methods, primarily GAS CHROMATOGRAPHY (GC). Ethanol is a volatile liquid, and most analytical methods take advantage of that property. Three common procedures are static HEADSPACE, heated headspace, and SOLID PHASE MICROEXTRACTION (SPME). In all three methods, ethanol that has evaporated out of the blood is collected and injected into the gas chromatograph where the identification as ethanol is confirmed and the quantity determined. SPME employs a special material (typically charcoal combined with other materials) to collect ethanol vapors and preconcentrate them on the absorbent fiber, which is then introduced into the gas chromatograph. Blood alcohol concentration is reported as grams per deciliter (g/dl) or as a percentage. Most states have adopted a legal limit of 0.08 percent, meaning that 0.08 percent of a person's blood by volume is ethanol. Anyone with a greater BAC greater is considered to be legally intoxicated.

See also ADME.

Further Reading

Kunsman, G. "Human Performance Toxicology." In *Principles of Forensic Toxicology.* 2nd ed. Edited by B. Levine. Washington, D.C.: American Association of Clinical Chemistry, 2003.

Levine, B., and Y. Kaplan. "Alcohol." In *Principles of Forensic Toxicology.* 2d ed. Edited by B. Levine. Washington, D.C.: American Association of Clinical Chemistry, 2003.

blood chemistry *See* BLOOD; TOXICOLOGY.

blood group systems The discovery of the ABO BLOOD GROUP SYSTEM in 1900 marked the beginning of an era in which SEROLOGY and blood, BLOODSTAIN, and BODY FLUID evidence became invaluable in forensic science. With discovery of each new group, the potential for INDIVIDUALIZING a bloodstain grew greater, although that goal was not reached using blood group systems. Blood group systems are based on ANTIGENS that are POLYMORPHIC, meaning that more than one variant exists, and at known frequencies in the population. For example, in the ABO system, the antigen is located on the surface of the red blood cell (RBC) and can be of Type A or B. Thus, the ABO system is polymorphic with known frequencies: 42 percent have Type A blood, 43 percent Type O, 12 percent Type B, and 3 percent Type AB. Like blood group systems discovered later, the ABO system is a GENETIC MARKER since the genes control which antigens are inherited and thus so is the blood type. For most of the blood group systems, the typing techniques used are similar to those used for ABO typing and include ABSORPTION-ELUTION and ABSORPTION-INHIBITION.

Blood group systems are found in all components of the BLOOD and in the body fluids of SECRETORS as well. Antigen substances are found on the surface of the RBC, white blood cells (leukocytes primarily), and are associated with platelets. None of these antigens are found in such abundance as the AB antigens, so detection and typing of other blood group systems are more difficult and require larger samples. For whole, fresh blood this is not a great concern but in forensic cases where sample quality and quantity is limited, it presents severe limitations. The first blood group system identified after ABO was the MN system, discovered in 1927 by KARL LANDSTEINER and Levine. Soon it was realized that another marker—Ss—was closely related, and so the group is now known as the MNSs system. Frequency in the population is well dispersed, the most common type being MNSs at about 24 percent, ranging down to 1 percent for the NS type. While this seemed promising for forensic use, the system proved difficult to type and interpret.

The Rh system was identified in the 1940s (Landsteiner and Weiner) and has become known for its role in pregnancy and birth. Briefly, if a mother is type Rh-negative and gives birth to an Rh-positive child, she may develop antibodies to the Rh-negative factors. If she becomes pregnant with another Rh-positive child, the mother's antibodies can harm that baby. The name *Rh* was derived from the Rhesus monkeys that were used in the early research. The system proved to be extremely complex with more than 40 antigens. Adding to the complexity was an inconsistent and nonstandardized naming system. In the Fisher-Race system of naming, the six most common antigens are called C, D, E, c, and e. In forensic work, most tests focused on identifying the presence of D. Although the system has many variants and the antigens are fairly stable in bloodstains, the complexity of the system has limited forensic use.

A number of other so-called primary blood group systems have been identified, with the Kidd, Duffy, P, Kell, and Lewis systems having been used in forensic serology. However, none are easy to type in stains, and none are as persistent as the A and B antigens of the ABO system. More than 40 secondary blood group systems have also been discovered, but none have been used forensically. Research into typing techniques for forensic work faltered once the ISOENZYMES were discovered and simple typing techniques using ELECTROPHORESIS were developed for the isoenzyme genetic markers. However, even isoenzyme systems have given way to DNA TYPING, which is much more successful in individualizing blood than isoenzymes or blood group systems ever were.

Further Reading

Lee, H. "Identification and Grouping of Bloodstains." In *Forensic Science Handbook*. Vol. 1. 1st ed. Edited by R. Saferstein. Englewood Cliffs, N.J.: Regents/Prentice Hall, 1982.

Spalding, R. P. "Identification and Characterization of Blood and Bloodstains." In *Forensic Science: An Introduction to Scientific and Investigative Techniques* 2nd edition. Edited by S. H. James and J. J. Nordby. Boca Raton, Fla.: CRC Press, 2005.

bloodstain patterns Often it is the way in which BLOODSTAINS are deposited at a scene that gives investigators the best chance to reconstruct the crime that created them. Stain-pattern analysis is often of greater value in resolving the circumstances of a case than are the DNA results from these same stains. Since the general characteristics of BLOOD are the same within different blood sources (humans, cows, dogs, and so on), it is possible to create general rules about what forces, impacts, and events likely created a given bloodstain pattern. Some components vary between species; therefore, it is recommended to do experimentation with blood of the same species as the crime scene being reconstructed. However, the complexity of patterns requires specialized training and extensive experience to interpret patterns with confidence. Although proven laws and principles of physics, mathematics, and biology underlie the interpretation of spatter patterns, it still involves an element of subjectivity, and it is not uncommon to have different experts on bloodstain patterns examine the same evidence and come to different conclusions as to the specific or most probable actions required to produce specific stain patterns.

History

The analysis of bloodstain patterns was first documented in the 1800s in Europe, particularly in Germany. In 1939, Victor BALTHAZARD made a formal presentation on the analysis of bloodstain patterns at a conference in Paris, bringing bloodstain patterns to the forefront of the emerging discipline of forensic science. Investigators in the United States also began to study and use bloodstain patterns, leading to their use in the famous case in which Dr. SAM SHEPPARD was tried and convicted of the murder of his wife in 1954. Dr. PAUL KIRK of the University of California at Berkeley examined the evidence in 1955 and used it in his analysis of the case. The case was the subject of widespread publicity and was the basis of a television series in the 1960s and a 1993 movie, both entitled *The Fugitive*. The evidence was revisited in the 1990s with some differing opinions being offered. In 1971, Herbert MacDonnell and Lorraine Bralousz published a landmark report entitled *Flight Characteristic and Stain Patterns of Human Blood* (U.S. Government Printing Office), considered to be one of the seminal works of bloodstain-pattern analysis. In 1983, the International Association of Bloodstain Pattern Analysis was formed, and during the 1990s, training courses were developed in pattern interpretation and in the related area of CRIME SCENE RECONSTRUCTION. The INTERNATIONAL ASSOCIATION FOR IDENTIFICATION (IAI) first issued certification in bloodstain-pattern examinations in 1999. The certification is good for five years and is then renewable upon completion and passing of both a written and a practical examination. Since the early 1990s computer programs have been developed to assist crime scene investigators and spatter-pattern specialists in reconstruction events based on evidence found at the scene.

Blood: Movements and Volume

The movement and characteristics of blood can be predicted using laws that apply to fluids such as water. One common misconception is that liquids falling through space retain a teardrop shape, which is not the case. Liquids assume the shape that offers the least resistance, which is essentially spherical, although the shape is not fixed. The teardrop form with its characteristic tail results at the time of impact and is related to resistance encountered at the point of contact compared with those portions of the droplet that have not yet hit the surface. Because of this behavior, the position of the tail can be used to determine direction of travel of the drop prior to impact.

On average, an adult male body holds approximately 1.5 gallons (5.5 l) of blood and the average female body contains 1.2 gallons (4.5 l). The volume of blood spilled at a scene can be useful in determining time of death, how badly victims were injured, how long they survived after the blood was spilled, their potential mobility after an injury was sustained, and the determination of death even when a body has not been located. Among other things, overall spatter patterns can be useful in determining the direction that the blood source was traveling and the direction that the blood was traveling when it was deposited on a surface, relative positions and distances of attacker and victim, sequence of events, the type of weapon used, and the force involved in causing the injury. Spatter patterns can be invaluable in supporting or disproving the stories of people involved in the incident and in reconstructing events that witnesses have recounted.

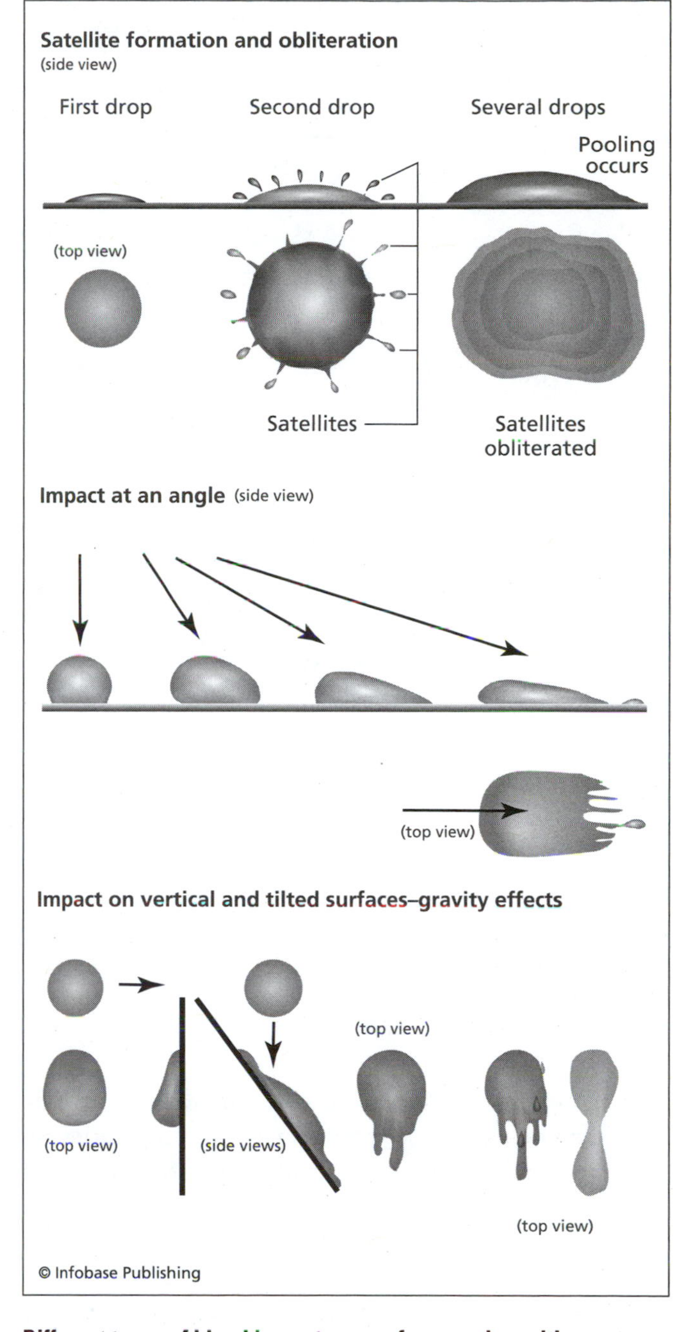

Different types of blood impacts on surfaces and resulting patterns. Variables that determine the shape of the blood-spatter (bloodstain) patterns include volume of blood, speed at which it is falling, direction (angle) of motion, and angle of impact.

Patterns

Spatter patterns can be roughly classified by the force, measured in feet per second, required to produce the drop or droplets that strike the wall, ceiling, floor, or other stained objects. An example of a low velocity pattern would be produced by something like a nosebleed, in which blood drips under the influence of gravity but not otherwise significantly accelerated. Low velocity patterns show a generally circular shape. The stains show a variety of shapes and sizes, depending upon the angle at which it impacts a surface. The individual spots, however, tend to be relatively large.

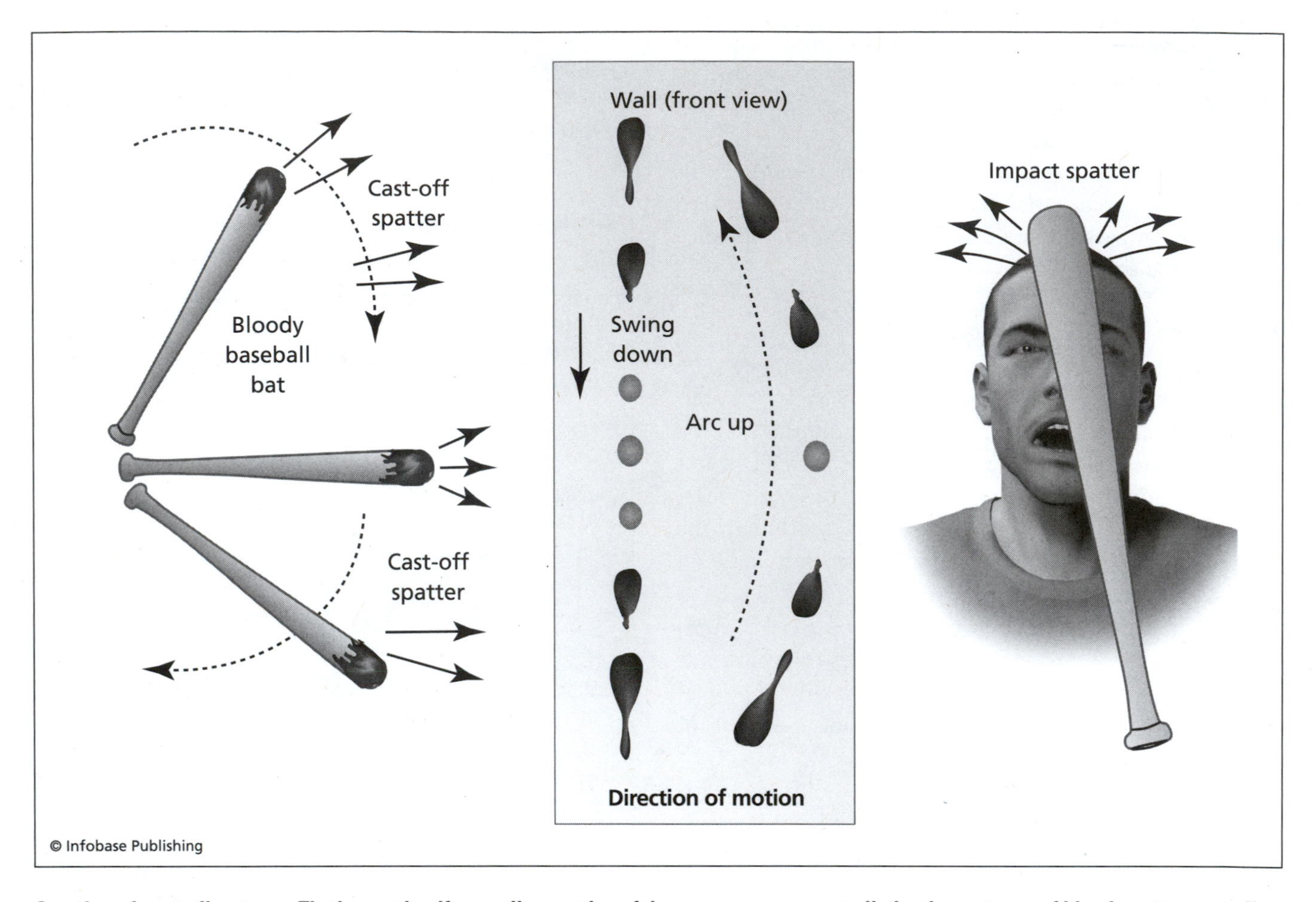

Creation of cast-off patterns. The impact itself as well as motion of the weapon can create distinctive patterns of blood spatter on walls, floors, and ceilings.

If the low velocity blood source is moving, the circular patterns will exhibit serrated, fingerlike projections on one side and become more oval than circular. These multiple tails point in the direction of movement of the blood source. If blood travels faster, it typically creates medium velocity spatter patterns in which the drops and resulting stains are smaller in overall size than the low velocity stains.

One type of medium velocity pattern results from the impact of one surface on which blood is present by a second surface. This pattern is reminiscent of a sunburst and is called impact spatter patterning. A second category, referred to as cast-off patterning, results from blood that has accumulated on the impacting instrument (hammer, bat, knife) and flies off the instrument as a result of the centrifugal force from the swinging action. As shown in the figure above, this pattern exhibits a trail or linear pattern in which both the up and the down swings can be identified. High velocity patterns are typified by a misting or atomizing effect with the preponderance of the stains within the pattern being very small, .04 inches (1 mm) or less in size. It is about the size of a period written with a number two lead pencil. Bullet wounds typically produce high velocity spatter patterns. They can also be seen in cases involving industrial accidents with high-speed machinery or chain saws and are also present in explosive incidents. The overall pattern resulting from these three specific actions are very distinct and thus are not easily confused with one another, making the conclusion of the type of incident relatively easy to determine.

Depending on distance from the point of origin, back spatter (that blood exiting back out of the entrance gun shot wound, traveling back toward the gun) may be found on the weapon and/or person firing the weapon. Likewise, a forward spatter (that blood exiting the exit gunshot wound, traveling with the direction of the bullet) can be deposited on near verti-

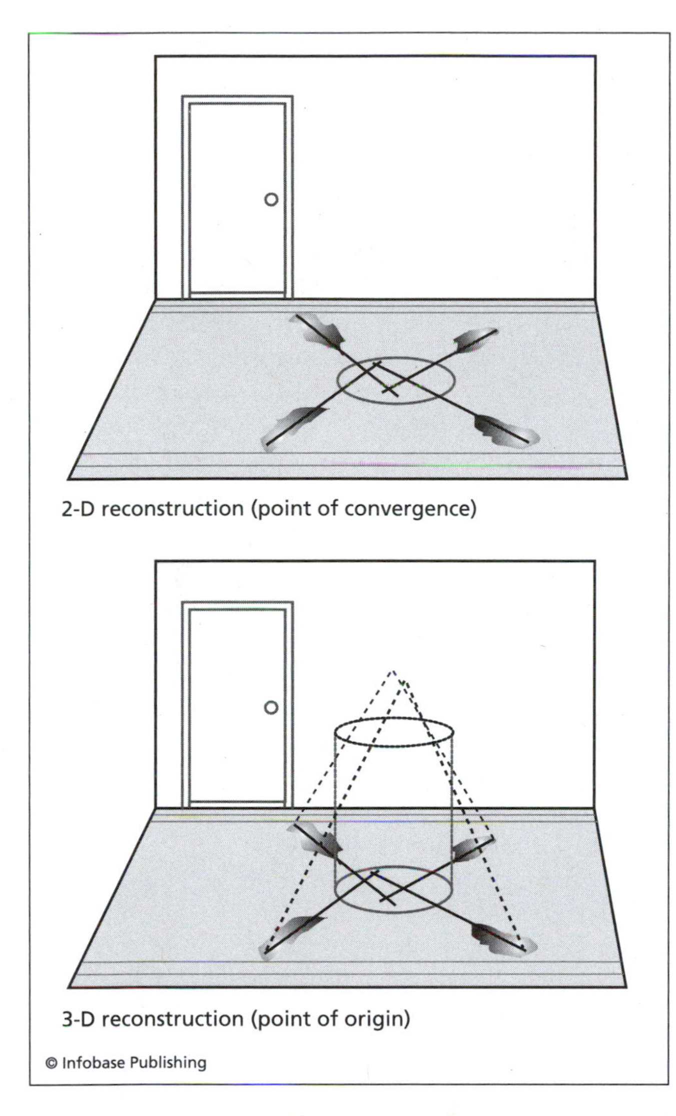

Use of geometric relationships to estimate point of origin of spatter patterns in two and three dimensions

cal or horizontal surfaces. The amount of blood in the forward spatter will be greater than that of the back spatter and will travel a greater distance. Typically forward spatter will travel up to approximately 46 inches (1,168 mm), unless mixed with other tissue, allowing greater distance distribution. Typically back spatter has a maximum range of 2–3 feet (0.6–0.9 m). Both forward and back spatter exit the wound in a conical pattern. Spatter on clothing can help determine who was in close proximity to the blood source at the time of the bloodletting event. By examining the pattern alone, it cannot be determined whether the resulting stain is the result of forward or back spatter, and thus the exact identity of the shooter cannot be determined by bloodstain patterns alone. The high velocity spatter pattern, both forward and back, can be used to estimate the relative position of the victim when the wound occurred.

When a drop of blood strikes a surface, the appearance of the stain will depend on many factors such as direction of travel, the roughness or smoothness of the surface, and its absorption characteristics. The tail(s) on a stain, unless equally distributed around a circular stain, indicate that the drop did not hit the surface at 90°. The tail points in the direction that the drop was traveling prior to impact, not the direction from which it originated. As shown in the figure on page 39, an impact will produce a stain and on some occasions, a smaller drop (splash) may result. This smaller (satellite) stain is not attached to the original (parent) stain. The tail on a satellite always points back to the parent stain. Satellite formation occurs most often in drops of low velocity origin and occurs on all sides of the parent stain in which the blood source is stationary. An example of this would be if a person with a bloody nose stands still, steadily dropping blood that falls in large, low-velocity drops onto the same location, such as a table or the floor. In these cases, blood pooling

Bloodstain patterns typical of arterial spurting ***(Courtesy of Michael Bell, West Virginia University, Forensic and Investigative Sciences program)***

Use of strings to estimate the area of origin of bloodstain patterns *(Courtesy of Michael Bell, West Virginia University, Forensic and Investigative Sciences Program)*

eventually occurs, and this pooling effect overruns and obliterates the satellite stains. For this reason, satellites are not detected as often as they occur. A single drop hitting at an angle may also form a satellite. These satellites are not obliterated and are easily identified. Satellites are always smaller than the parent drop, and the satellite tail always points back to the parent drop. Since the satellites do not travel far, identification of the parent drop is simplified. Gravity will also act on blood, but not the same as it would with water, given the greater thickness of blood. Impacts on a tilted or vertical surface show signs of gravity flow, with collection of blood at the lower end of the stain. The degree of tilt will also affect the stain appearance. An example of the effect of angle is shown in the color insert.

The degree of spatter depends on many things, especially the texture of the material that the drop hits. Impact on a rough surface produces a more irregular stain than impact on a smooth surface such as tile or glass. If the drop strikes a surface at an angle, the drop will become oval shaped, and the more oblique the angle, the more oval to elongated the stain will be and the more likely a satellite stain will be produced. By measuring the width and length of a stain and the application of laws of trigonometry, the angle of impact of a stain can be determined. Based on the shape of stain, the analyst can examine multiple stains and determine, by placing a straightedge through the long axis on or beside the stained surface, the area where the paths cross. A circular area encompassing where the multiple lines cross is called the point of convergence. This is a two-dimensional (2-D) reconstruction and is about the size of a volleyball. The actual location of the blood source is at some undetermined distance above this point of convergence. By knowing and re-creating the angle of impact on each of the selected stains, a circular area (but not a specific point in space) can be identified in the area above the point of convergence, at which the many paths meet. This location is called the point of origin (three-dimensional or 3-D reconstruction) and will place the blood source at an area, the approximate size of a volleyball, in space (i.e., 47 inches [1,194 mm] up from the floor, 17 inches [432 mm] out from the wall). The difference between the point of convergence and point of origin is illustrated in the figure on page 41.

Numerous other types of patterns can be observed. Arterial gushing or spurting occurs when blood is pumped out of the body under pressure, and the resulting spatter is distinctive with large stains deposited in an arc, wave, or zigzag pattern (see page 41). Cast-off patterns (illustrated in the figure) can occur when a bloody object such as a club, knife, or fist is swung or otherwise rapidly moved, flinging blood off in a linear fashion. Transfer stains occur when something such as a hand, a shoe, or hair has come in contact with blood and then contacts another surface. An example of a transfer pattern is shown in the color insert. Void patterns occur when an object (or person) stands between a source of blood and the impact surface. For example, if a shoe is on a carpet when blood is spilled around or across it, the carpet will have an unstained area in the shape of the shoe when the shoe is removed. This is a void pattern. Blood on a surface can show a swipe patterning, such as when bloody hair swings across a surface. A swipe pattern is characterized by a feathery edge at one margin. A wipe pattern occurs when an object moves through an already deposited stain that has not dried completely. If blood is deposited, partially dries, and is then disturbed, a skeleton pattern can result in which the dry blood on the edges of the stain is preserved but the wet blood, more central in the stain, is wiped away or otherwise altered.

Finally, the activity of insects (particularly flies) at a scene can produce stains of a medium or high velocity size. In addition, if the insect has been consuming blood, a PRESUMPTIVE TEST for blood on those stains will produce positive results. This is believed to result from a regurgitating or defecating action of the insect. Moths, just prior to death, will also leave suspicious stains, often larger and exhibiting tail formations. These stains, generally, do not test positive for the presumptive presence of blood. While the size of the stain and/or chemical testing both meet the criteria for bloodstain pattern examination, the overall pattern is missing. If a distinct pattern is absent, extreme caution is advised when interpreting stains based on size only. The complexity of patterns is shown in a photo found in the color insert.

See also ENTOMOLOGY, FORENSIC.

Further Reading

Bevel. T., and R. M. Gardner. *Bloodstain Pattern Analysis with an Introduction to Crime Scene Reconstruction.* 2d ed. Boca Raton, Fla.: CRC Press, 1999.

James, S. ed. *Scientific & Legal Applications of Bloodstain Pattern Interpretation.* Boca Raton, Fla.: CRC Press, 1999.

James, S. H., P. E. Kish, and T. P. Sutton. "Recognition of Bloodstain Patterns." In *Forensic Science: An Introduction to Scientific and Investigative Techniques.* 2nd edition. Edited by S. H. James and J. J. Nordby. Boca Raton, Fla.: CRC Press, 2005.

bloodstains Blood and bloodstains are among the most common and important types of evidence in many crimes, and they are often found at and recovered from CRIME SCENES. Collection of BLOOD, as well as other BODY FLUID evidence, must be done carefully to avoid PUTREFACTION, which is degradation caused by microorganisms. Occasionally, liquid blood may be found at a scene, and it can be collected, mixed with saline, anticoagulants, and preservatives and refrigerated for immediate delivery to the laboratory. Alternatively and more commonly, the blood can be collected on an absorbent cloth, allowed to dry, and then transported to the laboratory. Drying is critical to prevent putrefaction.

More often, the forensic analyst will be faced with the task of identifying and working with bloodstains. At the scene, these stains are not always readily identifiable or visible. Fresh stains show the familiar rust red color with a glossy sheen. As the stain ages, colors change and fade eventually to a grayish tone. Blood can also take up colors from the material on which it is deposited, including painted surfaces and dyes. The substrates will also influence the appearance of the stain based on how much of the blood is absorbed. For example, blood deposited on glass or metal is not absorbed and will look quite different from blood absorbed into a carpet. The perpetrator(s) may have cleaned a scene to remove blood, and so crime scene personnel will carefully check places such as joints in floors and drain traps to see if blood can be found. Even when no traces are visible, PRESUMPTIVE TESTS such as LUMINOL can be used to see where blood might have been. In all cases when blood is collected from a scene, the crime scene personnel must be careful not to alter any evidence contained in the BLOODSTAIN PATTERNS. These patterns are often crucial in re-creating events at the scene, and records of it must be protected as much as possible for as long as possible. Outdoor scenes can be even more difficult as blood may have been absorbed into soil, diluted by rain, covered with snow, or otherwise affected by environmental and weather conditions.

Blood that has dried on a surface such as glass can be collected by careful scraping or by moistening a cotton swab and rubbing it across the stain. Blood on absorbent materials is usually cut out, dried, and delivered in place. CONTROLS must also be collected to ensure that the substrate is not interfering with or altering test results. Thus, if a stained area of carpet is collected, an unstained area is needed as well. When blood is deposited on large objects such as lamps, pillows, or the like, the object itself can be collected and delivered to the laboratory. In all cases, stains must be allowed to dry and must be kept refrigerated or frozen prior to analysis.

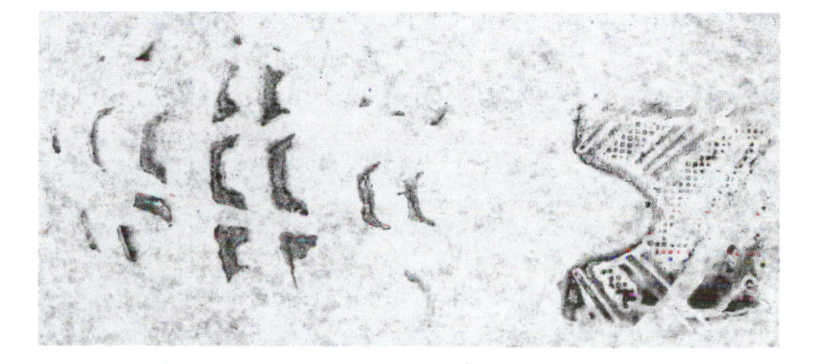

Transfer pattern, bloody shoe on paper ***(Courtesy of Michael Bell, West Virginia University, Forensic and Investigative Sciences program)***

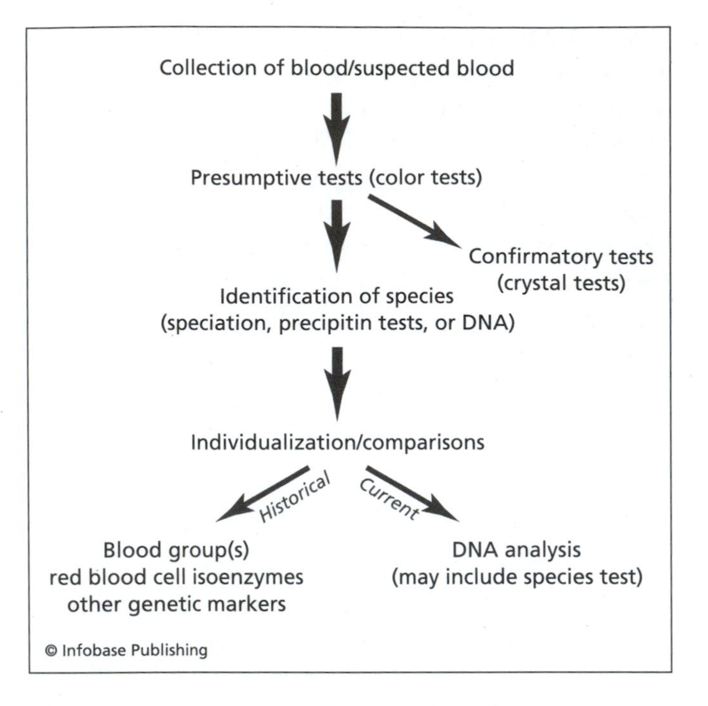

Steps in the analysis of a bloodstain

The illustration shows the general approach used to analyze bloodstain evidence. The first step is to determine if the stain is really blood, and this is accomplished using presumptive tests. These tests involve the use of chemical reagents that change color in the presence of HEMOGLOBIN; however, they are not specific and can yield FALSE POSITIVE results. Common presumptive tests include phenolphthalein (KASTLE-MEYER COLOR TEST), luminol, and ORTHOTOLIDINE. Once a presumptive test indicates that a stain may be blood, a confirmatory test may be performed. These include crystal tests (TEICHMANN, TAKAYAMA, or WAGENAAR tests) or other immunological or spectrophotometric techniques. Confirmatory tests can be accomplished as part of the next step in the analysis, a determination of the species of the blood (human, dog, cow, and so on) using IMMUNODIFFUSION tests. Human blood will react with human antisera but not with antisera from other species, so a positive result at this stage both confirms the presence of blood and that the blood is human.

The final step in the process is to INDIVIDUALIZE to the extent possible the blood and to compare stains to known blood if comparative samples are available (such as blood drawn from a suspect). Until recently, this task was accomplished by typing blood for ABO BLOOD GROUP, ISOENZYME SYSTEMS, and other GENETIC MARKER SYSTEMS as appropriate and feasible. However, DNA TYPING has supplanted ABO and isoenzymes in this capacity. Current DNA methods also can confirm the identification of a substance as blood and that it is human.

Further Reading

Spalding, R. P. "Identification and Characterization of Blood and Bloodstains." In *Forensic Science: An Introduction to Scientific and Investigative Techniques.* 2nd edition. Edited by S. H. James and J. J. Nordby. Boca Raton, Fla.: CRC Press, 2005.

blood types *See* BLOOD GROUP SYSTEMS.

body fluids Body fluids and body fluid stains encountered as evidence include saliva, semen, sweat, urine, feces, vomit, vaginal fluid, and human milk. In 1932, an inherited characteristic was discovered that determined if a person secretes substances such as the A and B ANTIGENS of the ABO BLOOD GROUP SYSTEM into body fluids. Approximately 80 percent of the Caucasian population are secretors, which means that their body fluids (saliva, semen, and vaginal fluids) can be typed using the same techniques used to type blood. With the advent of DNA TYPING, secretor status has become less critical. A person who is a nonsecretor will still have DNA present in their body fluids as long as those fluids contain cells or cellular debris. However, for several decades, secretor status was extensively exploited in forensic SEROLOGY.

The cases most likely to involve body fluids are sexual assaults. Procedures for the analysis of such evidence follow the same general sequence as for bloodstain evidence. The first task is identification of the stain, which is achieved by chemical tests, heating, visual examination, or by presence of salt (perspiration). For example, urine can be identified by odor upon heating or by chemical reagents that react with urea. Saliva is usually identified by the presence of the amylase enzyme, which is present in high concentrations in the mouth and serves to break down simple sugars. Since amylase is present in other fluids, the saliva test must not only detect the enzyme but must also demonstrate high concentrations. Saliva also contains epithelial cells suitable for DNA typing. Since vomit and fecal matter contain cells and cellular debris, it is also possible to perform DNA typing on these samples, although interpretation can be difficult given the wide mix of food material that may be present.

Further Reading

Greenfield, A., and M. M. Sloan. "Identification and Biological Fluids and Stains." In *Forensic Science: An Introduction to Scientific and Investigative Techniques.* 2nd edition. Edited by S. H. James and J. J. Nordby. Boca Raton, Fla.: CRC Press, 2005.

body temperature (algor mortis) The rate of cooling of a body can be used to estimate the time of death or POSTMORTEM INTERVAL, (PMI) and is the best estimator if a body is discovered soon after death. In general, a body will reach ambient temperature in 18 to 20 hours, but the rate of cooling is not necessarily fixed or constant. It depends on many factors including temperature of the environment, humidity, submergence (and if so, water temperature), temperature at time of death (which may not have been 98.6°F [37°C]), clothing or cover on the body, body fat, and the ratio of surface area to weight. Body temperature can be measured rectally or by inserting the thermometer in the liver. Computer models have been developed for estimating time of death, but they involve measurements at multiple locations on the body and at several time intervals. Accordingly, such methods are not commonly employed.

Bonaparte, Napoleon (1769–1821) French/Corsican *General and Emperor* The death of Napoleon remains the subject of heated debate nearly two centuries after it happened on May 5, 1821. After his defeat by the British at Waterloo in 1815, Napoleon was exiled to the island of St. Helena, nearly 1,200 miles (1931.2 km) off the west coast of Africa. Over the years, his health slowly deteriorated and he reported symptoms such as disturbed sleep, alternating diarrhea and constipation, light sensitivity, hair loss, neurological symptoms, mood swings, aches, and pains. After his death, an AUTOPSY was performed. Five autopsy reports by eight doctors reached the same conclusion that Napoleon had died of stomach cancer. This finding came as no surprise since Napoleon's family had a history of stomach cancer. The disease had already claimed his father and sister and was suspected to have claimed several other close relatives. In fact, Napoleon himself suspected he had the same disease when he first noted symptoms in 1819. Regardless, rumors surfaced that he had been poisoned by ARSENIC, with the list of suspects including the British doctors on the island, the island's governor, and Count Montholon, whose wife was purportedly having an affair with the "Little General."

The rumors remained just that for more than a century. Then in the early 1960s scientists in England obtained a hair sample and analyzed it using NEUTRON ACTIVATION ANALYSIS (NAA). The findings suggested elevated levels of arsenic, but the sample was small and there was no definitive proof that the hair had actually come from Napoleon. Regardless, the poisoning theory persisted, advocated strongly by Ben Weider, a Canadian businessman who heads a large fitness company. Additional analysis done by the FBI in 1995 showed some hairs with high arsenic levels, 20–50 ppm (parts-per-million) compared to a range of 1 ppm that is normal in similar modern hair samples.

Since the initial findings, several theories have been advanced to explain the elevated levels, many of which do not rely on a criminal act. Assuming the analyzed hair did come from Napoleon and was not somehow contaminated since collection nearly two centuries ago, one suggestion is that green pigments used in wallpaper in his room (copper arsenites) were converted to a gaseous form by molds in the wallpaper paste, which Napoleon then inhaled. Another possibility is arsenic from medicines that were commonly used in the period. For example, a liquid called arsenicalis was used to treat appetite loss, a symptom consistent with stomach cancer. Finally, the arsenic in the hair might have been added as a preservative and insect killer after the locks were cut from whoever's head they came from. Thus, there remains little definitive or compelling evidence that Napoleon died of anything other than the stomach cancer his doctors described nearly two hundred years ago.

bone Bone is the rigid tissue that supports the body structure. High in mineral content, it is also the site of red blood cell production. It is both durable and light weight, and along with teeth it is the component of the body that endures the longest after death. Forensic ANTHROPOLOGISTS study bone (osteology) and from their observations attempt to determine the race, sex, stature, and age of the deceased. The analysis starts by determining whether the bone is human and a rough estimate of how long a person has been dead and how old the person was when he or she died. For very old bones, carbon dating techniques can be used, but for more recent deaths,

Forensic Science and History

Some of the most interesting forensic cases are not criminal cases but historical mysteries that might have involved a crime. When does a case become "historical"? There is no set standard, but one guideline used draws the line at the year 1900—cases before that are considered history; cases after that are considered criminal. Regardless, the intersection of history and forensic science is an interesting and busy one. Some examples are discussed in the text, such as the possible ARSENIC poisoning of NAPOLEON BONAPARTE and Zachary Taylor and the identification of bones belonging to the czar of Russia (*see* ANASTASIA AND THE ROMANOVS). A few other notable examples are discussed below.

Was LSD Involved in the Salem Witch Trials?

The Salem Witch Trials have interested people for years and have been the subject of numerous books, plays (*The Crucible* by Arthur Miller), and films. In December 1691, between eight and 12 young girls in Salem, Massachusetts, began to suffer from symptoms such as hallucinations, convulsions, and sensations of crawling skin. As the afflictions persisted, suspicions of witchcraft began to surface in the strictly religious Puritan community. The girls eventually implicated several individuals as instigators, leading to arrests, hearings, and trials. Within a year, 19 men and women had been executed, all but one by hanging. By early 1693, the witch craze ended, and ever since scholars have struggled to explain the episode and the subsequent killing of people who were, for the most part, highly respected members of their community.

Linda Caporael put forth a new and intriguing theory in 1976 that suggests that the girls' afflictions could have been the result of poisoning by ergot ALKALOIDS such as LYSERGIC ACID DIETHYLAMIDE, more commonly known as LSD. The symptoms reported were strikingly similar to those that result from ingestion of ergot alkaloids. These compounds are produced by fungus that can grow on grains such as rye, a staple of the diet in Salem. Furthermore, when such outbreaks occur, women and children appear to be the most susceptible to the effects. Caporael determined that weather conditions during the growing season of 1691 were conducive to fungus formation, which correlates with the timing of the outbreak. The following season was drier, discouraging fungus, and also correlating with the end of the incidents. While correlation is not the same as cause and effect, Caporael makes a convincing case based on strong circumstantial evidence and certainly one that is more satisfying than mass hysteria or some of the other explanations offered thus far.

Jamestown: First Colonial Crime Scene?

Arsenic again is at the heart of a controversial theory as to the "starving time" in the first English colony in America. During the winter of 1609–10, 540 of 600 colonists died. Until recently, most accepted that the incredibly hard life of the colonists, coupled with drought and Indian attacks, were to blame. However, recent excavations at Jamestown have renewed the debate and led to alternative possibilities. One is the old standby arsenic, which the colonists used to kill rats. One author has even gone so far as to suggest that Spain was responsible for the poisoning. However, the presence of rats suggested another, more mundane, possibility—plague. Given the drought, it is also possible that the settlers were forced to drink the swampy water surrounding the colony, which could have led to water-borne illnesses

age estimation can be attempted by a microscopic examination of structures in the bone called osteons. Osteons form continuously during the lifecycle, and in human bone, osteons are unevenly distributed. In contrast, the osteons in many animal bones are more orderly. The older an animal or person was at the time of death, the greater the percentage of damaged osteons. Once a bone is determined to be of human origin, the next step is to identify the specific bone or structure. The final stage is to use that information to estimate characteristics of the person such as sex, stature, race, and age. The structural details of a single bone or collection of bone fragments is rarely sufficient to identify the victim specifically. However, INDIVIDUALIZATION of bone and bone fragments can sometimes be accomplished using mitochondrial DNA, which can be recovered from bone marrow and dental pulp. These types of techniques have been used to identify the remains of pilots and soldiers discovered years after the end of the Vietnam War.

Bone analysis is also important in mass disasters such as airplane crashes in which the problem of COMMINGLED REMAINS is encountered. In such cases, bone is often burned, which can provide clues about the accident and fire itself. The texture and color of burned bone can be related to the heat and duration of the fire at the point where the victim was located. For example, prolonged contact can turn bone into a and even potentially to salt poisoning. Unfortunately, the skeletal remains recovered and studied so far have not revealed any definitive answers.

Death and Identification of Pizarro

Francisco Pizarro was a Spanish conquistador who founded the city of Lima, Peru, as the capital of the land he conquered in the early 16th century. During his time in South America, Pizarro had many rivals and made many enemies, and they caught up with him on June 26, 1541. Several men stormed Pizarro's home, and in the sword battle that followed, Pizarro was killed. Other weapons such as daggers and pikes (short spiked lances) were used to attack Pizarro, who was not wearing armor. A knife to the throat inflicted the fatal wound. Pizarro's body was spirited away by loyal servants and buried in a courtyard. After the political winds changed, he was exhumed in 1544 and reburied in a white robe within the cathedral of Lima. No one disputed that the exhumed body was that of the dead conquistador.

In the next century, his remains or portions of the relics stored with them were apparently moved and the trail of the body becomes cloudy, although it appeared to have remained in the church. In 1977, a lead box was discovered in the cathedral with an inscription that indicated it contained Pizarro's remains. A skull was removed and the box set aside. However, workmen apparently mixed up the contents of this box with that of another smaller box stored in the same area, which held remains of more than one individual. Complicating the situation was the fact that mummified remains identified as those of Pizarro had been placed on public display in a box held in a glass sarcophagus back in 1891. The find raised many interesting questions and forensic anthropologists were asked to help determine which set of remains belonged to Pizarro. The key would be an examination of the wounds found on the skull and chemical analysis.

Portions of the skull associated with the lead box displayed wound patterns that would be consistent with blows from a heavy sword. Conversely, the mummified remains showed no sign of trauma. Other wounds on the skeleton were consistent with the wounds that Pizarro reportedly received. In addition, chemical analysis of the skull indicated traces of lead, linking the skull to the lead box where it was assumed to have come from. After intense study, the remains in the leaden box were accepted as those of Pizarro, and they replaced the mummified remains that had remained incorrectly identified for over a century.

Further Reading

Benfer, R. A., and H. L. Restaure. "The Identification of the Remains of Don Francisco Pizarro." In *Forensic Osteological Analysis.* Edited by S. I. Fairgreive, ed. Springfield, Ill.: Charles C. Thomas, 1999.

Caporael, L. R. "Ergotism: The Satan Loosed in Salem?" *Science* 192, no. 2 (1976).

Jamestown Rediscovery Web site. "Jamestown Rediscovery." Available online. URL: http://www.apva.org/jr.html. Downloaded January 31, 2003. Provides general information on the current Jamestown excavations and findings.

Public Television Web site (PBS). "Secrets of the Dead—Death at Jamestown, Background." Available online. URL: http://www.pbs.org/wnet/secrets2/case3.html. Downloaded January 31, 2003.

Public Television Web site (PBS). "Secrets of the Dead—Death at Jamestown, Clues." Available online. URL: http://www.pbs.org/wnet/secrets2/case3_clues.html. Downloaded January 31, 2003.

white ash while shorter exposures may result only in a yellowing or other discoloration.

See also AGE DETERMINATION; ODONTOLOGY; SEX DETERMINATION.

Further Reading

Fairgrieve, S. I., ed. *Forensic Osteological Analysis: A Book of Case Studies.* Springfield, Ill.: Charles C. Thomas Publisher, 1999.

Reichs, K., and W. Bass, eds. *Forensic Osteology: Advances in the Identification of Human Remains.* 2d ed. Springfield, Ill.: Charles C. Thomas Publisher, 1997.

botanicals and plant material *See* BOTANY, FORENSIC.

botany, forensic The analysis of evidence obtained from or related to plants as applied to legal matters. Plant matter is a kind of TRANSFER EVIDENCE, but the scope of forensic botany has expanded dramatically from this simple definition in the past few years. Still, application of botanical techniques is not yet as routine in forensic work as those of ENTOMOLOGY, which relates to insects. Plant matter has been used in civil and criminal cases and has been applied to tasks such as estimation of a time of death and the POSTMORTEM INTERVAL (PMI), identification of plant matter in stomach contents to characterize a last meal, identification of plant poisons, linking of a suspect to an outdoor scene, determining if a body has been moved, determination of whether a person was alive or dead when placed in water, and detection of CLANDESTINE GRAVES. The first court case in which evidence from plant material was accepted (and played a pivotal role) was in the 1935 trial of Bruno Hauptmann, the man accused of kidnapping and murdering the son of Charles LINDBERGH. Testimony offered by a botanist concerning the wood used to construct the ladder that was employed to reach the baby's room was crucial in convicting Hauptmann and eventually sending him to his execution.

Forensic botany is sometimes further divided into related specialty areas. Palynology involves the study of pollen and spores and has been widely used in archaeology to date remains and determine seasons and environmental conditions. Mycology is the study of fungi such as mushrooms, whereas taxonomy is the study of plant structure and classification. Knowledge of plant ecology is used to determine in what areas a given plant is best suited to live and where it is most likely to be found.

Knowledge of plant structure, identification, and ecological niche has been exploited in cases where plant matter is transferred from a scene to a victim or suspect. Stickers, seeds, pollen, and bits and pieces of plant matter from an outdoor scene are often easily picked up by clothing, shoes, and tires. Identification of the plant matter as being characteristic of a certain area or ecosystem can link a suspect to a scene or indicate if a body found in one place has been moved there from somewhere else. Plant identification is also important in cases of poisoning (human or livestock), identification of noxious weeds (nonnative species), and identification of drugs such as MARIJUANA, OPIUM, and peyote.

Like techniques used in forensic entomology, botanical evidence can be exploited to estimate how long a body has been in a specific outdoor location long after decomposition has set in. If a body is dumped on top of living plants, the loss of light will cause the covered plant, leaves, or grass to lose their green color as the chlorophyll fades away. Based on the stage of development of the plants and the degree of fading, estimates of when a body was dumped are sometimes possible. The same is true when plants have started to sprout up and grow through skeletal remains. Related to this is the use of knowledge of plant succession to locate clandestine graves. When the ground is dug up and replaced over a fresh grave, existing plants are destroyed. New plants will spread into the bare area in a reasonably predictable pattern that will contrast to the already established plant life immediately around it. This difference may persist for years, marking the grave to the trained eye, and is another example of the application of plant ecology.

Unlike animal cells, the cell walls of plants are rigid, made up of cellulose and a tough material called lignin. Because humans and most animals cannot digest cellulose, plant material is often identifiable in stomach contents, vomitus, and fecal material. Such evidence has been used in rape cases, murder investigations, and poaching cases.

Finally, botanical evidence can be useful when a suspected drowning occurs. If a person is placed in water, a type of algae called DIATOMS, which are found in freshwater and salt water, can enter the lungs. If the heart is still beating, the circulatory system can deliver the diatoms to remote parts of the body such as the liver. Thus, finding diatoms in tissues far removed from the lungs is evidence that the person was alive when he or she entered the water. Classification of the specific diatom can also help determine where the person entered the water, which may be some distance from where the body is recovered. Although the diatom method is not full proof, it has proven useful in many cases in which decomposed bodies have been recovered from water.

Pollen and spores are a unique and rich source of transfer and TRACE EVIDENCE, and the techniques used are similar to those that have been employed in archaeology. Pollen are small grains of male reproductive bodies found in seed plants. The name comes from Latin and is translated as "fine dust or powder," which describes the consistency of the material. The grains are very small, on the order of 15–50 5m (roughly one one-millionth of a meter), but under a microscope the pollens of different plants can be recognized by differences in size, shape, and other features. However, the fact that pollens are so widespread can complicate using it as evidence. For example, if a long period of time has elapsed since a suspect has been in a location of interest, other pollens not associated with that location will have accumulated and can prove misleading. CONTAMINATION of evidence is always a concern, as is the stability and durability of the pollen itself. Perhaps the most well-known use of pollen evidence was in the study of the Shroud of Turin, in which a burial cloth that was purported to be that of Jesus Christ was ultimately shown to be a clever forgery.

Further Reading

Brock, J. H., and D. O. Norris. "Forensic Botany: An Under-Utilized Resource." *Journal of Forensic Sciences* 42, no. 3 (1997): 364.

Lane, M. A., et al. "Forensic Botany Plants, Perpetrators, Pests, Poisons, and Pot." *BioScience* 40, no. 1 (990): 34.

bovine serum albumin (BSA) A serum derived from the BLOOD of cattle that is sometimes used in serological techniques and DNA TYPING.

Boyle, Robert (1627–1691) An Englishman whose work was critical for later advances in forensic CHEMISTRY and forensic TOXICOLOGY. Robert Boyle began his career as an alchemist, but more than any other individual, he was responsible for extracting chemistry from ALCHEMY. Born to privilege, he had an excellent education and made use of it. He was also a devotedly religious Christian, and one of his many theological titles was *Free Discourse Against Customary Swearing*. Boyle found his way to London and became more interested in chemistry. Eventually, he found an assistant named ROBERT HOOKE (1635–1703), now famous for many discoveries including significant work with microscopes, the first forensic instrument. Hooke was more of a generalist than Boyle, having dabbled in mechanics, physics, and biology as well as chemistry. The microscope served him well in all.

Apart from their individual contributions, their partnership led to fundamental discoveries, such as Boyle's law of gases, which described the inversely proportional relationship of pressure and volume of gases. Boyle realized the fundamental implication—air is not empty space or a vacuum; rather, it has to be made up of something tangible. He christened these little somethings "corpuscles," which could crowd together as pressure increased, causing the volume to decrease. From a forensic perspective, the discovery of the nature of gases was critical. Many chemists of the age, including Boyle, knew of ARSENIC poisoning and were hard at work trying to develop a reliable test for it in human tissues. The first to be successful relied on converting arsenic to gaseous form before plating it out in the metallic form.

Other conceptual breakthroughs attributed to Boyle were the definition of what an element is: a substance that is not combined with anything else, or that cannot be further broken down into any other material. Consider arsenolite, As_2O_3. The Greeks thought that this white powdery mineral was elemental arsenic,

Robert Boyle, ca. 1660 ***(Hulton Archive/Getty Images)***

but Boyle contradicted that in two ways. First, the arsenic is combined with something (oxygen), so the powder cannot be an element. Second, it is possible to generate gaseous arsenic from this material, so it cannot be a pure substance or element. Finally, once the metal arsenic is isolated, it will not break down further using chemical techniques such as heating. Boyle did not have the means at his disposal to test his ideas, but he did state the theory and defined the path of later experiments.

Boyle also studied acids and bases, (then called alkaline materials), both fundamental to later forensic work. He devoted extensive efforts toward developing a test for arsenic based on sublimation, or the direct conversion of a solid to a gas. Some of his tests relied on hydrogen sulfide gas (H_2S, the smell of rotten eggs), but he was unable to generate reproducible and reliable results. He was tantalizingly close to what was eventually used as the first reliable test for poisonous arsenic. Boyle was also the first to advocate the use of solutions chemistry over PYROLYSIS as a means of analysis. Because mineral acids and other methods of getting metals into solution were available during his time, significant progress was made in moving toward dissolution as a means of separation.

brass *See* AMMUNITION; CARTRIDGE CASES.

breath alcohol and breath analysis ALCOHOL (ethanol) in blood can evaporate from the blood into exhaled air deep in the lungs. As such, exhaled air contains a concentration of alcohol that is proportional to the concentration of alcohol in the blood. The concentration is governed by HENRY'S LAW, which states roughly that when a fluid such as blood is in equilibrium with a gas such as air, the concentration of a volatile substance (ethanol) in the gas will be proportional to the concentration in the fluid, as long as the temperature remains constant (as it does in the body). For ethanol in blood in contact with air at body temperature (98.6°F or 37°C), that ratio approximately is 2100:1, meaning that there will be 2,100 times as much alcohol in the blood as in the air. Since this ratio is known, it is possible to relate breath alcohol concentration mathematically to blood alcohol concentration.

Several companies manufacture breath analysis instruments, among them the well-known Breathalyzer. Many of these instruments are based on spectrochemical measurements (SPECTROSCOPY) of reagents that react with ethanol. For example, potassium dichromate ($K_2Cr_2O_7$), which is a deep yellow color, will react with ethanol in the presence of a catalyst ($AgNO_3$, silver nitrate) and an acid (H_2SO_4, sulfuric acid) to produce acetic acid and $Cr_2(SO_4)_3$, which is a green color. The amount of $K_2Cr_2O_7$ that reacts is proportional to the amount of ethanol present in the breath sample. A COLORIMETER (a spectrometer designed to measure visible light, i.e., colors) monitors the concentration of the $K_2Cr_2O_7$ by measuring the intensity of the yellow color. The more ethanol that is present, the more the yellow will disappear. Similar instruments exploit INFRARED SPECTROSCOPY (IR), which is based on the ability of ethanol molecules to absorb infrared radiation. Other instruments employ electrochemistry and fuel cells. In a fuel cell device, ethanol is oxidized to acetic acid (the acid found in vinegar) and a small electrical current is generated as a result. The greater the current, the more ethanol present. A disadvantage of the fuel cell method is that it destroys the sample, while IR methods do not.

Accurate breath alcohol measurements require air expelled from deep in the lungs (alveolar air). Also critical is extensive QUALITY ASSURANCE/QUALITY CONTROL (QA/QC), including accurate and reliable control of temperatures and calibration using solutions with known concentrations of ethanol. There are other volatile compounds that might be present in breath. Notably, acetone may be seen in diabetics. Accordingly, modern instruments are designed to minimize FALSE POSITIVES from potential interfering compounds such as acetone. Special calibration devices are also used to ensure that the readings obtained in the field are reliable and reproducible.

The National Highway Transportation Safety Association (www.nhtsa.gov) maintains an extensive website of information related to alcohol effects and measurement as it relates to highway safety.

See also ELECTROMAGNETIC RADIATION (EMR).

Further Reading

Kunsman, G. "Human Performance Toxicology." In *Principles of Forensic Toxicology.* 2nd edition. Edited by B. Levine. Washington, D.C.: American Association of Clinical Chemistry, 2003.

Levine, B., and Y. Kaplan. "Alcohol." In *Principles of Forensic Toxicology.* 2nd edition. Edited by B. Levine. Washington, D.C.: American Association of Clinical Chemistry, 2003.

breech face markings (breechblock markings) The breechblock is the part of a gun (pistol, rifle, shotgun) that cradles and supports a CARTRIDGE when it is inserted into the chamber prior to firing. When the trigger is pulled, the firing pin strikes the PRIMER, igniting it and the PROPELLANT. The rapid expansion of gas accelerates the bullet down the barrel, but it also drives the cartridge case backward into the breechblock. Since the breechblock is a machined or filed surface, it possesses a pattern of markings that can be transferred to the cartridge case (IMPRESSION

EVIDENCE) if it collides with sufficient velocity. These markings can be examined using a COMPARISON MICROSCOPE in much the same procedure as BULLETS to determine if a cartridge was fired from a specific gun. Complications can arise if the cartridge has been reloaded and fired more than once since each firing will produce a separate set of breechblock impressions.

See also AMMUNITION; FIREARMS.

brucine A highly toxic ALKALOID that has been used as a POISON. Also known as dimethoxystrychnine, it is similar to strychnine in action and has a bitter taste.

building materials Materials used in construction can serve as TRANSFER EVIDENCE (TRACE EVIDENCE) and are often found in burglary cases. A partial list of construction materials that may be encountered includes glass; minerals such as gypsum, found in sheetrock and plaster; mineral fibers such as asbestos; wood; cement and mortars; stucco; brick; insulation materials; and metals. Much of the forensic work on such evidence is done microscopically and may involve PHYSICAL MATCHING. A special subcategory of building materials is the insulation used in safes, which can consist of vermiculites, cements, or diatomaceous earth, all designed to reduce the risk of fire damage to the safe contents. Anyone attempting to cut or saw into such safes will often get this insulation material on their clothes. A famous case that involved building materials was the LINDBERGH KIDNAPPING, in which wood evidence from a ladder was critical.

bullets The projectiles fired from rifles and pistols. The primary component of bullets is lead, but there are many types and configurations of bullets available, varying by shape and degree of jacketing, among other things. The lead that is used also varies depending on the metals alloyed with it. Bullets made of softer leads tend to break up on impact while harder lead alloys resist fragmentation. Common shapes of bullets include round nose, flat nose, variations of pointed noses, variations of tail shapes, and wad cutters, which are flat-nosed slugs of lead or lead alloy used for target practice. Some bullets have cannelures, which are small indentations used to hold lubricant or for crimping a bullet into the proper position on the cartridge.

Fully jacketed bullets ("full metal jacket") consist of a harder metal shell (copper alloys or steel) that encases the lead core. Semiautomatic pistols and rifles use jacketed ammunition to prevent lead fouling of the chambering mechanisms and to increase the ease of bullet feed. Semijacketed bullets have the front portion of lead exposed and are much more prone to fragmentation. Hollow point bullets have the center portion of the nose removed, promoting a mushrooming effect upon impact. Some other bullet variations include soft point, bronze point, and Teflon bullets.

When fired, bullets travel through the barrel and acquire a spin by the rifling of the barrel. This rifling consists of LANDS AND GROOVES machined into the barrel. As the bullet travels down the barrel, it is marked by the lands and grooves, creating IMPRESSION EVIDENCE. These impressions take the form of parallel STRIATIONS that will be characteristic of the gun from which it was fired. A bullet can also pick up impressions from materials that it travels through, including FABRIC IMPRESSIONS from clothing. Upon impact, bullets may fragment or mushroom, depending on the type of bullet, jacketing, and velocity. Fully jacketed bullets and those made of harder leads resist fragmentation; softer and semijacketed bullets often break into pieces.

Collection of bullets as evidence is done with care to prevent imparting any additional marks and to prevent loss of trace evidence. If a weapon is recovered, the firearms examiner will fire it into a bullet trap of some type (usually a large box stuffed with cotton or a water tank) that will stop the bullet without damaging it or marking it with anything other than marks made by the gun itself. Bullets or fragments recovered from a scene or body can be compared to these test bullets. For the examination of bullets recovered as evidence, the sequence begins with an examination for the presence of TRACE EVIDENCE such as fibers, blood, tissue, paint, or other impressions. Next, if no weapon has been recovered, the examiner attempts to identify CLASS CHARACTERISTICS such as caliber (using weight and measurements), rifling characteristics of the barrel such as the number and dimensions of lands and grooves, and their degree and direction of twist. These class characteristics usually are not sufficient to identify a specific make and model of weapon, but this information can dramatically narrow down the number of potential guns that might have been used to fire the bullet. In certain cases, however, the land and groove pattern is gun specific. If a suspect weapon is recovered, the examiner will compare the striations from the test bullets to the questioned bullets. If the striations can be matched, the bullets in question were fired from the same gun as the test bullets. This examination is conducted using a COMPARISON MICROSCOPE, with a test bullet mounted on one side and the questioned bullet or fragment on the other. This matching process is a form of INDIVIDUALIZATION, which links the bullet to a single weapon. Thus, in these cases, the firearms examiner can return one of three results—the bullet was fired from the suspect gun, the bullet was not, or the results are inconclusive.

Several factors can complicate examination of bullets. Each gun barrel will impart unique markings to bullets, but these markings can change over time, either as a result of normal wear and tear or purposeful alteration. Barrels can be filed down, cut, or in the case of many weapons, replaced altogether. Guns that are recovered from water may be rusted, and rust can create striation marks as well. Crimp marks on bullets can result from the ejector mechanism or loading/reloading process, and silencers can mark bullets as well. The examiner, in drawing final conclusions, must consider all these factors. Computerized databases are playing an increasing role in firearms cases. DRUGFIRE, a program developed by the FBI, and the IBIS system (Integrated Ballistic Identification System), developed by the BUREAU OF ALCOHOL, TOBACCO, FIREARMS, AND EXPLOSIVES (ATF) contains data on CARTRIDGES and bullets. ATF provides the IBIS system and equipment through the National Integrated Bullet Identification Network (NIBIN) program. These databases have proven useful for linking seemingly unconnected shootings by identifying the use of the same gun in more than one crime.

A bullet that has been fired through a gun. The outer metal is copper covering a lead core. The marks left by lands and grooves in the barrel are visible. ***(Courtesy of Michael Bell, West Virginia University, Forensic and Investigative Sciences program)***

When bullets fragment, it becomes more difficult to identify class characteristics and to link a fragment to a specific weapon. Badly fragmented bullets often cannot be analyzed using the normal procedures. Even determining caliber can be difficult, although sometimes it is possible to gather all fragments and at least estimate a minimum weight of the original bullet. INSTRUMENTAL ANALYSIS techniques have also been used to create elemental profiles of bullets in attempts to identify a common source. The elements that are characterized include copper (Cu), arsenic (As), silver (Ag), antimony (Sb), bismuth (Bi), cadmium (Cd), and tin (Sn). The instrumental techniques that have been used include NEUTRON ACTIVATION ANALYSIS (NAA) and INDUCTIVELY COUPLED PLASMA methods (ICP-AES and ICP-MS). NAA was used to characterize the bullets recovered during the KENNEDY ASSASSINATION.

See also BALLISTIC FINGERPRINTING; FIREARMS.

bullet wounds Contrary to popular belief, it is not always easy to identify bullet wounds, nor is it always easy to distinguish entrance wounds from exit wounds. Once identified (usually by the MEDICAL EXAMINER or pathologist at AUTOPSY), information can be gathered from the wounds, including estimates of distance of the shooter from the victim and relative positions of each. High-powered weapons such as rifles can produce massive damage, but injuries from smaller weapons such as pistols and revolvers can leave identifiable and characteristic wound patterns.

When a bullet strikes flesh, the skin is first stretched and then broken as the projectile penetrates. As it enters, material on the surface of the bullet such as dirt and dust, lubricants, powder and primer residue, and lead will be wiped onto the skin in a pattern called bullet wipe or smudge ring. The bullet will also scrape off skin cells, creating an injury called a contusion ring. These features may be obscured or altered by the presence of clothing, and in some cases the bullet-wipe pattern may obscure the contusion ring. The shape of the bullet wipe and contusion ring can provide clues about angles and relative positions; in the case of straight-on shooting, these features will be roughly circular but can be more oval shaped if the shot comes from an angle or is offset from center. Beyond the bullet wipe and contusion ring there will be a dispersed deposit of material (GUNSHOT RESIDUE, or GSR) that contains flakes of unburned powder and other residues. The concentration of these residues and how much they are spread out will depend primarily on the distance between the shooter and the victim. In general, the closer the two are, the smaller the dispersal will be. The pattern can be altered by conditions at the scene; for example, if it is an outdoor scene and there is a wind blowing, residues may not reach the skin as they would in an indoor scene or on a calm day.

Once the bullet passes through the skin, its path is not predictable. The bullet may break up and create several paths, or the bullet (or fragments) may strike bone and be further diverted or damaged. Fragmentation can result in several exit wounds. In the case of jacketed bullets or those that stay intact, the exit wound may be characterized by a ragged, exploded appearance, but this is not always seen. Exact trajectories are determined at autopsy, using dissection and X-ray techniques (RADIOLOGY). At the scene, BLOOD-STAIN PATTERNS can also be useful. If a weapon is fired at close range, a back spatter of blood may be deposited on the shooter and/or objects nearby. Similarly, a forward spatter can be produced when the bullet or fragments exit the body. There is usually more forward spatter than back spatter, and the different directions of travel can impart distinctive characteristics to the spatter patterns.

The appearance of the entrance wound depends on the distance the gun is from the victim. Contact wounds from pistols and revolvers, particularly to the head, are often distinctive. A muzzle print may be imparted to the skin, and since the barrel is pressed against the skin, gases are forced under the skin and expand and can tear the skin away from bone structures beneath. This tearing results in a characteristic

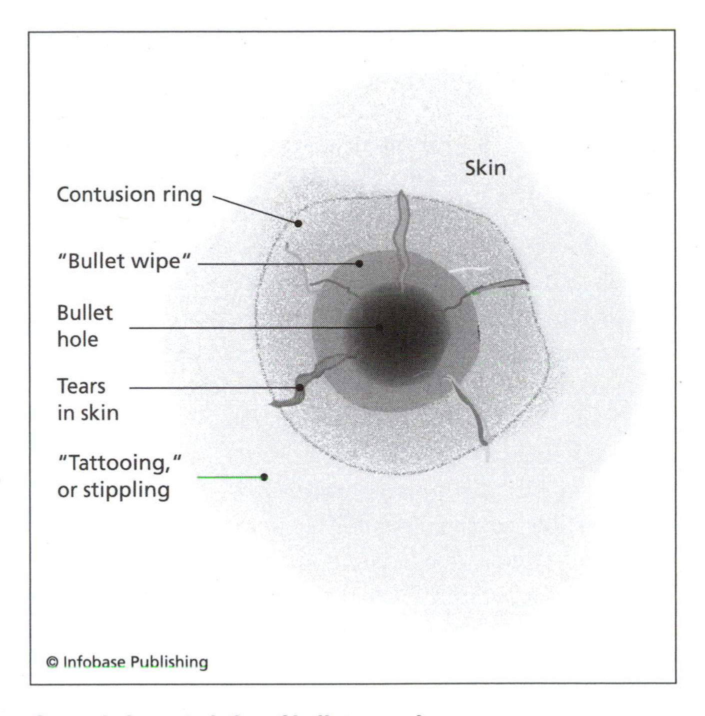

General characteristics of bullet wounds

star-shaped pattern in the wound. Close-range shots, considered to be those taken from approximately 12–18 inches typically show the pattern illustrated in the figure. The farther the two are apart, the more dispersed the tattooing pattern will be. Distance shots taken from more than 18 inches (457.2 mm) often leave no visible GSR around the wound, although sensitive chemical and instrumental tests may still reveal the presence of residues.

Wounds from shotguns are different in that the projectiles fired are small circular shot. Distance between the shooter and victim can be estimated based on the dispersal of shot and how much has struck the victim. A close or contact wound produces tremendous damage, but as the distance increases, so does the dispersal of the shot.

See also AMMUNITION; FIREARMS.

Bureau of Alcohol, Tobacco, Firearms, and Explosives (ATF) A federal agency in the U.S. DEPARTMENT OF JUSTICE that is responsible for the collection of taxes and regulation of the three industries. As a result, ATF supports a wide variety of investigative and forensic activities. Historically, enforcement of liquor, tobacco, and firearms revenue collection in the United States fell to divisions within the Internal Revenue Service (IRS) and its predecessors. Not surprisingly, the passage of the Eighteenth Amendment to the Constitution, ushering in the era of Prohibition, was a busy time for the passage of liquor laws and for the organization of governmental entities to enforce them. During that time and the years after repeal of Prohibition in 1933, responsibility for regulation of the liquor industry was repeatedly moved and altered. In 1935, the Federal Alcohol Act (FAA) was signed and the Treasury Department assumed the duties of regulating alcohol. Six years later, alcohol enforcement in the Justice Department was merged with those functions in the Treasury Department, consolidating these activities in one place.

The firearms industry boomed (literally and figuratively) during the Prohibition years and the rise of organized crime, and in the mid-1930s, federal firearm regulations and taxes were enacted. In 1942, enforcement of these taxes was allocated to the Treasury Department unit responsible for alcohol. Ten years later, this entity, the Alcohol Tax Unit assumed duties for alcohol, tobacco, and firearms. In 1968, this responsibility expanded once again to include explosives. The Bureau of Alcohol, Tobacco, and Firearms (ATF) was officially formed in 1972 as an agency independent of the Internal Revenue Service but still under the Treasury Department. In 1982, ATF's responsibilities expanded again to encompass arson. Today, the bureau is responsible for regulations, issuance of permits, tax collection, and enforcement in alcohol, tobacco, firearms, explosives, and arson.

Analytical and forensic laboratory services in ATF and its predecessors date back to 1886, when chemists performed simple analyses of butter to determine if it had been adulterated with margarine. The ATF laboratory system currently consists of several branch laboratories and includes mobile laboratory vehicles that can be dispatched to crime scenes or mass disaster sites such as TRANSPORTATION DISASTERS or locations of known or suspected terrorists acts. Similar to the FBI, the ATF is active in training state and local law enforcement personnel including forensic examiners in the areas of arson, fire investigation, explosives, and firearms identification. The National Firearms Examiner Academy provides rigorous and intensive training for new firearms and TOOLMARK analysts while the Fire Research Laboratory concentrates on fire and arson investigation. In 1985, the ATF laboratory system was the first federal forensic laboratory to be accredited by the AMERICAN SOCIETY OF CRIME LABORATORY DIRECTORS (ASCLD). As part of the Homeland Security Bill passed in 2002, ATF was split into two bureaus. The Bureau of Alcohol, Tobacco, Firearms, and Explosives resides in the Department of Justice (DOJ), while the Tax and Trade Bureau (TTB) remained under the Treasury Department. The ATF maintains a Web site at www.atf.treas.gov.

burden of proof The responsibility for presenting evidence and testimony to support a position. In the American legal system, the burden of proof is on the prosecution, meaning that the prosecution must prove the charges are true. If the burden of proof were on the defense, it would mean that the defendant would be responsible for disproving the charges.

C

cadaver dogs Specially trained dogs used to locate buried, concealed, or scattered human remains hidden in CLANDESTINE GRAVES. This is a challenging task since each stage of the decomposition process is characterized by different smells given off principally by microbial activity. As a result, dogs must be trained to recognize a large number of odors. Environmental conditions are also a factor, as remains may be buried or partially buried or obscured under a layer of snow, leaves, or other debris. Also, the remains may be scattered due to scavengers or other processes, further complicating the task. Cadaver dogs can also be used in a related task, locating bodies or portions of bodies in mass disasters such as airplane crashes.

Further Reading

Komar, D. "The Use of Cadaver Dogs in Locating Scattered, Scavenged, Human Remains—Preliminary Field Results." *Journal of Forensic Sciences* 44, no. 2 (1999): 405.

caffeine An ALKALOID and stimulant found in coffee, tea, and cola beverages. It is a white powder. Caffeine is sometimes encountered in DRUG ANALYSIS when it is used as a CUTTING AGENT (diluent) for drugs such as COCAINE, AMPHETAMINE, and methamphetamine. Occasionally, a powder sold on the street as amphetamine or other illegal stimulant is just caffeine or a mixture containing caffeine, sugar, EPHEDRINE, or other adulterants.

caliber Originally, this term meant the diameter of the barrel of a rifled pistol or rifle; however, the term can also refer to the size of CARTRIDGES used in FIREARMS. In a rifled firearm, LANDS AND GROOVES are machined into the barrel, which gives a bullet spin as it leaves the barrel. As shown in the illustration, caliber is measured from the top of the lands and is given in hundredths or thousandths of an inch or in millimeters. Common calibers include .22, .38, .40, .45, and 9 mm for pistols and .22 and .30-06 for rifles. The caliber of a gun is considered to be a nominal measurement, meaning that the actual barrel

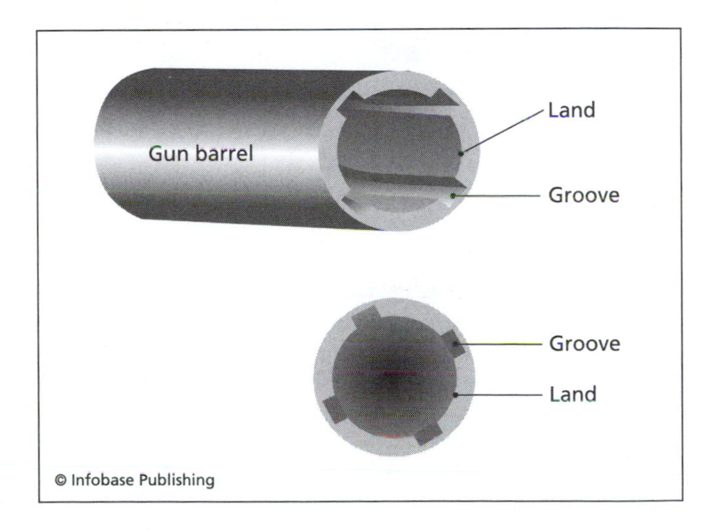

Measurement of caliber. The nominal caliber of a gun is the measured distance between the tops of two opposing lands.

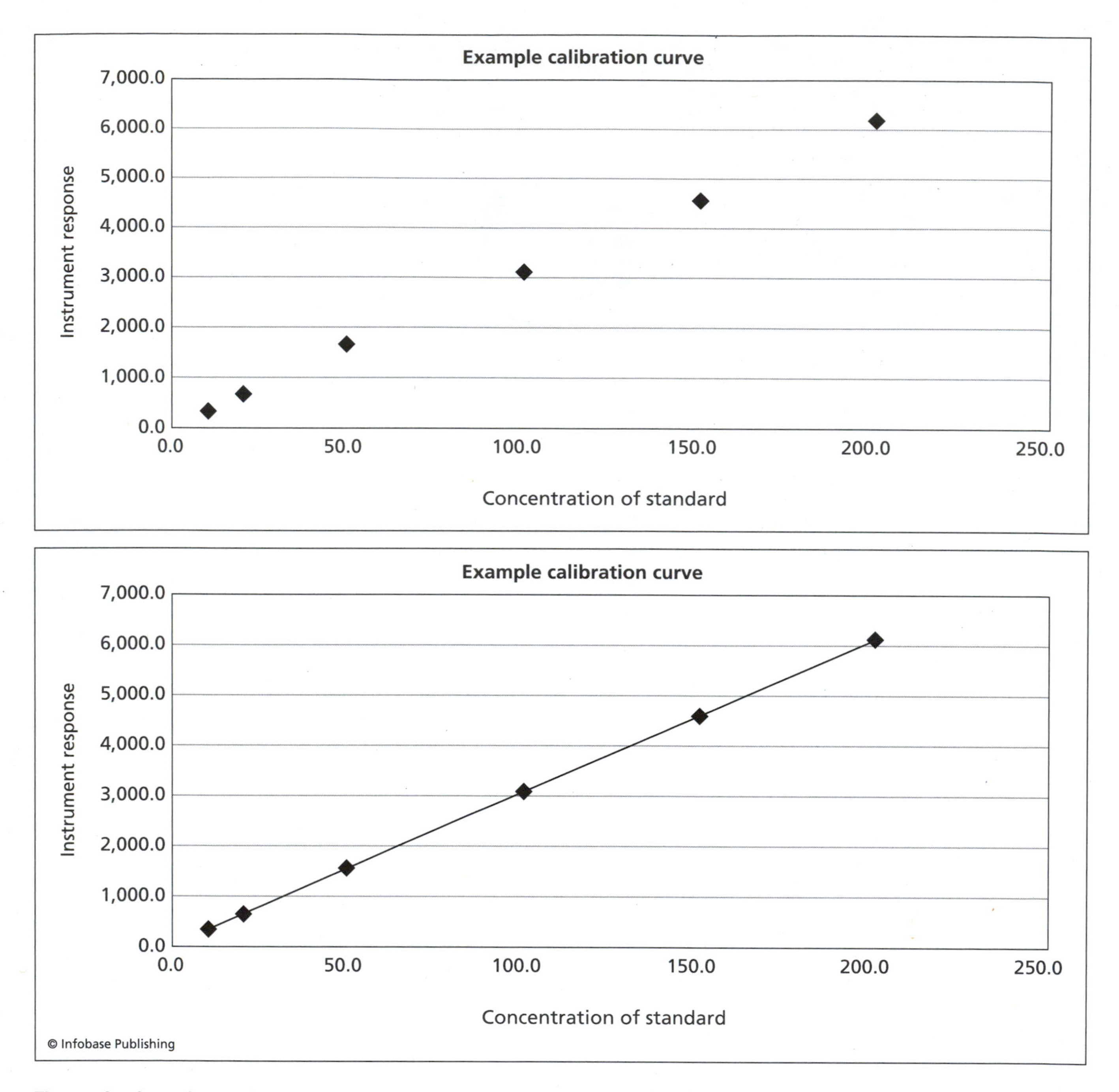

The top plot shows the raw data obtained for a hypothetical instrument calibration. The x-axis shows the concentration in units such as parts-per-million and the y-axis, the instrument response in units of peak area, voltage, or other quantity based on the instrument. The lower plot shows the equation of the line that best describes the calibration response. When an unknown is analyzed, the instrument response can be plugged into the equation to obtain the concentration.

diameter may vary slightly from the caliber measure used to describe it.

calibration The process of linking the response of an instrument or other measuring device to a concentration of a chemical or material that the instrument is designed to measure. BREATH ALCOHOL measurement devices, for example, must be calibrated to ensure the readings obtained in the field are accurate based on the instrument's capabilities. As is typical for calibration, the instrument's response to a series of different ethanol concentrations at standard operating temperatures is essential to know that the instrument's readings are reliable when used with a real person. This series of known samples of known concentrations are called calibration standards. The key to calibration is the quality of the calibration standards used. Most forensic laboratories purchase standards prepared by companies that guarantee their reliability.

Many calibrations are simple but still vital. For example, analytical balances must be checked and calibrated regularly to ensure that the weight displayed for a sample is accurate and reliable. For this purpose, special weight sets that are traceable to standards issued by the National Institute of Standards and Technology (NIST) are used. Glassware and other laboratory equipment such as pipets must be calibrated to be certain of the volumes that are delivered at each setting. A reading of .03 fluid ounces (1.00 mL) delivered for one pipet may correspond to an actual delivered volume of .04 fluid ounces (1.05 mL) for that pipet and .03 fluid ounces (0.98 mL) for another. The process of linking the reading to the actual volume delivered is a form of calibration. Calibration of instruments such as gas chromatographs involves the analysis of a series of standards of known concentrations. The chemist plots the data as shown in the figure to generate a calibration curve that is defined by a straight line. Once the equation of the line is known, the analyst can calculate the concentration of an unknown sample by simple mathematical calculations.

California Association of Criminalists (CAC) This organization, founded in 1954, is the oldest regional forensic science association in the United States. It was instrumental in developing and implementing certification examinations for CRIMINALISTS.

See also AMERICAN BOARD OF CRIMINALISTICS (ABC); CERTIFICATION.

cannabis *See* MARIJUANA.

cannelures Small grooves imprinted by rolling on the base of a BULLET or near the top of a CARTRIDGE CASE. For bullets, cannelures can hold lubricant or they can be used as a seat around which the throat of the cartridge case can be crimped closed. In cartridge casings, the cannelures prevent the bullet from being forced backward into the cartridge case. The absence or presence of cannelures, as well as their number and dimensions, can be used to help identify the manufacturer of a bullet and in some cases an approximate date of manufacture.

capillary electrophoresis (CE) Instrumental techniques used to separate and identify a variety of substances of interest in forensic science. These techniques evolved from ELECTROPHORESIS, carried out on horizontal slabs of gel or other media, and they share the same basic principles. Many types of capillary electrophoresis exist; the ones most used in forensic science are capillary zone electrophoresis (CZE), capillary gel electrophoresis (CGE), and micellular electrokinetic capillary chromatography (MEKC or MECC). These techniques are used to separate and analyze everything from STRS (DNA TYPING) to DRUGS to GUNSHOT RESIDUE (GSR).

The figure on page 58 illustrates the similarities between traditional slab electrophoresis and CGE in which a gel-filled capillary tube is used to carry out the separations of large charged molecules such as DNA fragments. In the CGE system, a sample is introduced via an injection system (usually a syringe) and separation occurs by "sieving"—the larger the molecule, the slower it moves through the gel media. The detector responds whenever one of the separated components passes through it. Many types of detectors exist, including mass spectrometers, those that exploit absorption of ultraviolet light (UV detectors), and those that respond to fluorescent materials found in the sample or chemically attached to its components. Electrochemical detectors are also available. The output of the detector is called an electropherogram, which is a plot of detector response as time progresses. In the example shown, component D is the fastest so it will pass the detector relatively early in the run while component A, which is the slowest, will arrive at the detector last. The relative sizes of the peaks on the electropherogram can be used for QUANTITATIVE ANALYSIS since their size is proportional to relative amounts present. In the example shown, the amount of D is the greatest and so it shows the largest peak while A, the least concentrated, has the smallest peak.

Capillary zone electrophoresis (CZE) is used in applications such as ink, drug, and gunshot residue analysis, the chemistry of which is considered to be "small molecule" analysis. This is in contrast to the analysis of DNA fragments discussed above, which are considered to be large or macromolecules. One of the distinguishing characteristics of CZE is electroosmotic flow, which is illustrated in the figure. In CZE, the capillary tube is filled with an aqueous (water-based) buffer solution containing both positive ions (cations) and

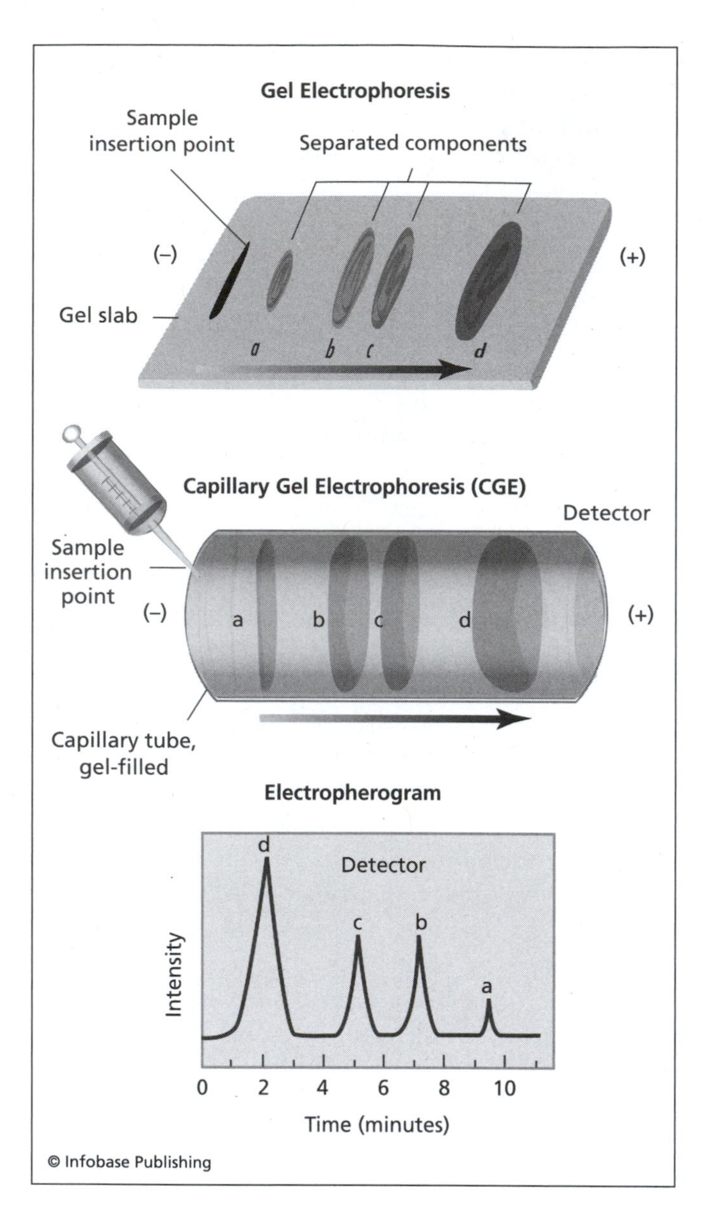

Comparison of gel electrophoresis and capillary electrophoresis. In both techniques, separation is accomplished in a gel and is based on the size and charge of the different molecules. Smaller molecules will move faster and, in the case of CGE, will emerge from the column and encounter the detector before larger molecules. The output of a CGE detector, an electropherogram, is a plot of the emergence time (called the elution time) versus the intensity of the signal. Thus, the height or area of the peaks is proportional to the amount of each species present.

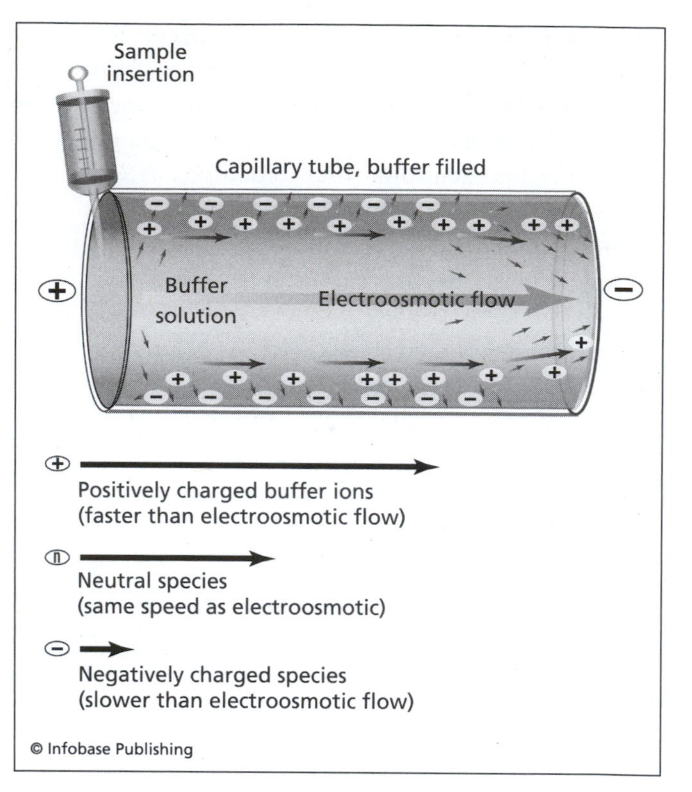

Capillary zone electrophoresis. The electroosmotic flow carries molecules toward the detector, with the relative speed of each different compound dependent on its size and charge.

negative ions (anions). The surface of the capillary has a net negative charge because of Si-O– chemical groups present in the glasslike material. Cations in the buffer are attracted to the negative charge and congregate near the surface. In turn, these positively charged ions are attracted to the negative end of the capillary tube and will migrate in that direction and, in so doing, create a flow of buffer to the cathode (so named because cations flow in that direction). When a sample is introduced into the instrument, any species that are positively charged will move very quickly toward the detector and will separate based on factors such as relative sizes and charges. The electroosmotic flow already moving in that direction further speeds them along. Neutral species will move at the same speed as the electroosmotic flow, and all neutrals will arrive at the detector at the same time. Although the negatively charged species are actually attracted to the opposite end of the capillary (the anode), the electroosmotic flow is sufficient to overcome this attraction, eventually delivering the negatively charged species to the detector. Thus, negatively charged species move slower than the electroosmotic flow, but still in the same direction. Detectors for CZE are similar to those used in HIGH PERFORMANCE LIQUID CHROMATOGRAPHY (HPLC) and include the ones listed previously for CE.

Finally, micellular electrokinetic chromatography (MEKC or MECC) combines elements of chromatography with CE to allow for separation and analysis of neutral (uncharged species). Several variations and modifications of CE and CZE have also been developed and it is the versatility of CE that has led to its increasing use as in forensic applications.

Further Reading

Northrup, D. "Forensic Applications of High-Performance Liquid Chromatography and Capillary Electrophoresis." In *Forensic Science Handbook*. Vol. 1. 2d ed. Edited by R. Saferstein. Upper Saddle River, N.J.: Prentice Hall, 2002.

carbon monoxide (CO) A colorless, odorless, and tasteless gas produced as a by-product of COMBUSTION reactions. Carbon monoxide is also highly flammable and toxic and is a leading cause of poisoning (accidental and intentional) in the United States. The mechanism of poisoning arises from carbon monoxide's ability to bind with HEMOGLOBIN 200 to 300 times as strongly as oxygen. Hemoglobin that is carrying a full load of oxygen (four O_2 molecules) is called oxyhemoglobin; when carbon monoxide displaces the oxygen molecules, that form of hemoglobin is carboxyhemoglobin. High levels of carboxyhemoglobin in the blood result in oxygen starvation of the brain, a form of suffocation. The antidote for carbon monoxide poisoning is breathing pure oxygen for at least two hours, which will eventually convert carboxyhemoglobin back to oxyhemoglobin. Carbon monoxide is found in cigarette smoke as well as car exhaust, a result of incomplete combustion.

Exposure to high levels of carbon monoxide (0.4 percent by volume) can cause death in an hour, with higher concentrations working faster. A common method of suicide is by car exhaust, and in an enclosed space like a car interior or garage, carbon monoxide from a running engine can reach fatal levels in 10 minutes or less. Very high concentrations can cause almost immediate unconsciousness and rapid death. People who die in fires often succumb to carbon monoxide poisoning, or at the very least have elevated levels of carboxyhemoglobin in their blood. CYANIDE gas can also be present, and fire deaths are often attributed to combinations of the two substances. This can be critical information for investigators as criminals may set fires to cover up a murder. If a victim is found after a fire, a low level of carboxyhemoglobin in the blood indicates that the person was already dead when the fire started.

Carbon monoxide poisoning is also seen when open flame heaters (or malfunctioning heaters) are used in an enclosed or improperly ventilated space. Such deaths are often seen in winter, when propane heaters or other devices are used indoors. The danger is especially great since windows and doors are typically closed up tight to keep out the cold. The symptoms of carbon monoxide poisoning are insidious and can masquerade as less perilous problems or illnesses. Since carbon monoxide cannot be detected by smell, sight, or taste, victims often do not realize the danger until they are too impaired to save themselves. Symptoms include nausea, headaches, lightheadedness progressing to sleepiness, muscle weakness and numbness, and finally unconsciousness. Toxicologists typically report carbon monoxide concentrations found in the blood as %COHb, an abbreviation for carboxyhemoglobin. Normal healthy individuals have a %COHb level of 0.4–0.7 percent, smokers 3–8 percent. Above 10 percent, symptoms begin to appear, leading to coma and death at levels of 50 percent or greater, depending on factors such as victim's age, health, BLOOD ALCOHOL CONCENTRATION, and drugs in their system.

The techniques used to detect carbon monoxide in blood are based on determining the concentration of carboxyhemoglobin and %COHb. Preliminary or PRESUMPTIVE TESTS that result in color changes can be used when the %COHb is greater than 10 percent. There are also SPECTROPHOTOMETRIC techniques that allow for a measurement of %COHb relative to oxyhemoglobin levels. GAS CHROMATOGRAPHY (GC) is also used, along with a releasing agent. When added to the blood, the agent frees the CO, which is gaseous. The vapor above the blood is then sampled and injected into the instrument, where any carbon monoxide present is separated, detected, and quantitated.

Further Reading

Kunsman, G., and B. Levine. "Carbon Monoxide/Cyanide." In *Principles of Forensic Toxicology.* 2nd edition. Edited by B. Levine. Washington, D.C.: American Association of Clinical Chemistry, 2003.

carbonic anhydrase (CAII) *See* ISOENZYME SYSTEMS.

career opportunities Careers in FORENSIC SCIENCE are as varied as the many disciplines within it. In a typical full-service forensic laboratory, there will be sections in CHEMISTRY and DRUGS, FINGERPRINTS, FIREARMS and TOOLMARKS, TRACE evidence, SEROLOGY/BIOLOGY, QUESTIONED DOCUMENTS, and CRIME SCENES. MEDICAL EXAMINERS are physicians who specialize in forensic PATHOLOGY, and they constitute examples of the many forensic professionals who do not work in a forensic lab per se. Many others active in forensic science do so as consultants or as part of another job such as artists, archaeologists, entomologists, anthropologists, computer scientists, and so on. Many of these professionals are associated with universities. Thus, there is no single career path or mode of entry into a forensic science career.

For entry-level positions in a forensic lab, a bachelor's degree in a natural science such as chemistry or biology is normally required, with different jurisdictions having different specific educational requirements. With the increasing emphasis on DNA TYPING, many labs require courses in biology such as genetics, microbiology, biochemistry, and molecular biology. The normal course of a career begins with a period of training and apprenticeship under experienced laboratory analysts for several months. Once the trainee satisfies all requirements, he or she is qualified to work cases independently in the area in which he or she trained. Continuing education is essential, primarily through specialized training courses offered by professional associations, instrument and software manufacturers, and the FBI. Depending on the jurisdiction and discipline, analysts may seek CERTIFICATION in their area of expertise. It is not unusual for scientists to move to different areas of the lab during their careers, starting, for example, in drug analysis and then moving to trace and later into firearms and toolmarks. Each change requires another training and testing period in the new area.

See also EDUCATION.

carpet *See* FIBERS.

cartridge cases Also called shells or casings, the cartridge is the part of a round of AMMUNITION that encloses the PROPELLANT. The BULLET is seated at the forward end and the PRIMER at the base of the cartridge case. Casings are usually made of brass (which can be reloaded), nickel-coated brass, or aluminum, which is not designed to be reloaded. Given the composition, cartridge casings are often referred to generically as "brass." In SHOTGUN ammunition, the cartridge case is made of plastic or cardboard and crimp-sealed at the top.

Cartridge casings used in revolvers are manually removed while cartridges fired in semiautomatic and automatic guns are ejected by the gases created when fired. Rifle cartridges are generally distinguished by their longer size and tapering in the neck. Cartridges may have CANNELURES around their circumference to act as a seat for the bullet. Cartridges are also differentiated based on what type of primer is used; rimmed primers are used in smaller caliber weapons such as

Examples of cartridge cases. The case to the right is for a revolver; the others for semi-automatic weapons. ***(Courtesy of Michael Bell, West Virginia University, Forensic and Investigative Sciences program)***

.22 while center-fire cartridges are used in larger caliber weapons. The base of a cartridge is often labeled in some way, such as the manufacturer and/or caliber of weapon that the ammunition is intended for.

When cartridge casings are recovered at a crime scene, where they are found can provide critical information. In the case of automatic and semiautomatic weapons, the direction of ejection of the cartridges, elevation (relatively flat compared to arcing), and side of the weapon discharged from can provide useful information. Depending on the circumstances and other evidence recovered, position of the recovered cartridges can help determine this information and thus point to possible types and makes of weapons. When cartridge cases are examined in the FIREARMS section of the forensic lab, usually the first types of evidence sought are LATENT FINGERPRINTS and TRACE EVIDENCE such as DUST, dirt, HAIRS and FIBERS, BLOOD, and so on, that might be adhering to them. Once this examination is complete, the firearms expert will conduct a microscopic analysis using a COMPARISON MICROSCOPE. If a suspect weapon is available, comparisons can be made using cartridges recovered from test firing.

A typical examination begins with the determination of CLASS CHARACTERISTICS such as stamps on the end of the cartridge and other markings that can indicate a manufacturer, a potential source of the ammunition, and a type of weapon in which the cartridge was used. There are cases where weapons are modified to fire cartridges not designed for them, which can complicate this initial investigation. Another factor that the examiner must consider is the possibility that the cartridge has been reloaded and used more than once in one or more different weapons. Each firing will add additional and distinctive marks atop those already there.

If a suspect weapon is available, the next step is to obtain comparison cartridges from that same weapon using ammunition as nearly identical to that originally submitted as evidence as possible. The goal of this comparison is to INDIVIDUALIZE the cartridge and to determine conclusively if the cartridge was or was not fired from the recovered weapon. Every weapon will leave marks of some type on a cartridge casing, and the list of possibilities includes BREECH FACE MARKINGS, FIRING PIN IMPRESSIONS, EXTRACTOR/EJECTOR MARKS, chambering marks, marks made by or in the magazine, and marks due to the expansion of the casing against the walls of the chamber. Ejector, extractor, and magazine marks are created only in automatic and semiautomatic firearms, and not all cartridges will bear all these types of marks. Under the comparison microscope, the firearms examiner will attempt to determine if the markings from the test cartridges can be matched up to those on the cartridge casings submitted as evidence.

Further Reading

Rowe, W. F. "Firearm and Tool Mark Examinations." In *Forensic Science: An Introduction to Scientific and Investigative Techniques*. 2nd edition. Edited by S. H. James and J. J. Nordby. Boca Raton, Fla.: CRC Press, 2005.

casting A technique to preserve and replicate IMPRESSION EVIDENCE made in soft material such as soil or snow. Casts can be made of TIRE PRINTS, SHOE PRINTS, and occasionally TOOLMARKS and BITE MARKS. When done properly, casting techniques produce excellent positive replicates of the impression, but they are not exact duplicates. For example, plaster of paris casts tend to shrink and will be smaller than the original impression. Thus, it is always preferable to transfer the entire impression to the laboratory. However, in many cases, the impression evidence is in materials such as snow or mud that cannot be preserved or removed for transport. Similarly, bite marks in human flesh or perishable food will not last, and casts are the best way to record the impressions for detailed study.

A key component of casting is a detailed series of photographs taken of the impression from many angles, using different lighting and clear indication of scale. This documentation is essential for studying the original impression in place and for comparing it to the cast made. For impressions in soil such as tire prints and shoeprints, dental stone is the preferred material for casting, although plaster of paris and sulfur have also been used. After photography is complete, the next step is to insert or construct a retaining frame around the impression that will contain the casting liquid. Before the liquid is poured, the impression is usually sprayed with a lacquer material to prevent the weight of the casting material from pressing down and altering features. The cast is poured and allowed to dry for about half an hour before it is lifted out. Twenty-four to 48 hours is needed for the cast to dry completely, and it must be carefully protected during transport to prevent damage or alteration. Foot

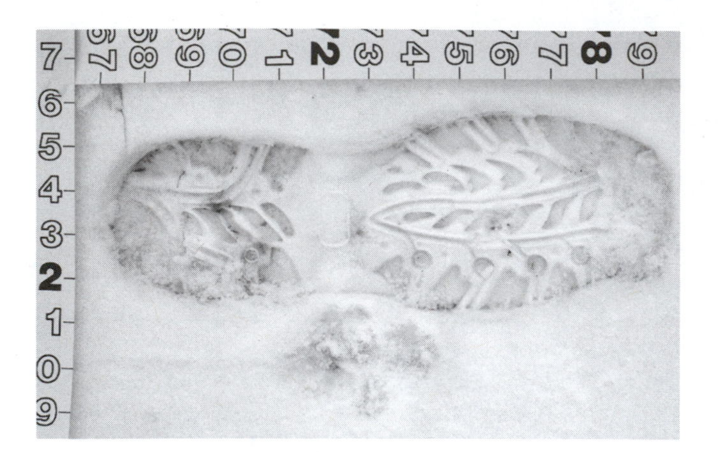

A shoeprint in snow prior to casting with melted sulfur ***(Courtesy of Michael Bell, West Virginia University, Forensic and Investigative Sciences program)***

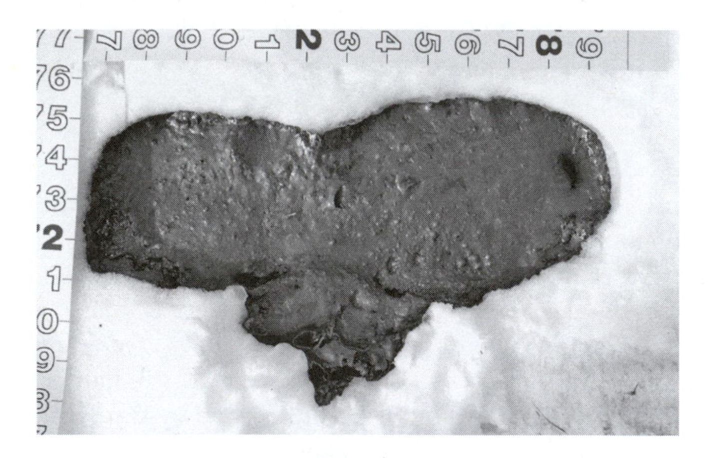

The poured cast ***(Courtesy of Michael Bell, West Virginia University, Forensic and Investigative Sciences program)***

The final cast showing the patterns from the snow ***(Courtesy of Michael Bell, West Virginia University, Forensic and Investigative Sciences program)***

and tire impressions in snow are a special challenge requiring a different approach since casting media gives off heat as it sets (an exothermic reaction), which would cause the snow to melt and alter the impression. Specialty products such as Snow Print Wax are used in such cases.

Bite marks are another type of impression amenable to casting. For bite marks in skin, an additional consideration is the possible presence of saliva, so the area is swabbed before the casting. The cast is made of silicon or other rubber casting material, usually after the area is photographed and traced to provide additional information and documentation. For bite marks in food, the type of food must be considered; the approach to casting a bite mark in a chocolate bar (water soluble and easily melted) is different from casting a bite mark in an apple, for example.

Further Reading

Bodziak, W. J. "Casting Three-Dimensional Footwear Impressions." In *Footwear Impression Evidence.* New York: Elsevier, 1990.

cause of death The immediate cause of a death. In criminal matters, a medical cause of death is not necessarily the legal cause. For example, if a victim is stabbed and dies as a result of complications such as infection, the legal cause of death remains the stabbing, and the case remains a homicide even though the immediate cause of death was infection. Cause of death is also distinct from the circumstances of death, which comprise the situation and conditions that led up to the fatal encounter. It is the responsibility of the attending physician, CORONER, or MEDICAL EXAMINER to determine the cause of death. Depending on circumstance, an AUTOPSY may be performed, but it is not mandatory.

cellulose A biopolymer based on glucose. Biopolymers come from plants, animals, and other natural sources. Aside from proteins, the most important biopolymers in FORENSIC SCIENCE are the polysaccharides such as cellulose. These polymers consist of carbohydrate monomers, for example glucose. Sugars are classified as carbohydrates, and the formula of simple sugars such as glucose, $C_6H_{12}O_6$, can be expressed as "hydrated carbon" or $C_6(H_2O)_6$. The simplest polysaccharides are dimers (disaccharides), many of which are familiar. Table sugar is a disaccharide consisting of fructose linked to glucose. Saccharide monomers such as glucose exist in an open-ring and closed-ring conformation, with the closed ring being preferred.

Cellulose is composed of a linear chain of glucose molecules. The way glucose molecules are connected imparts strength and rigidity to cellulose that is not seen in other similar biopolymers. Cellulose is insoluble in water and functions in plants as structural support in the form of microfibrils. Wood, the raw material of paper, is about 20–40 percent cellulose and cotton is nearly 90 percent. Although it is a glucose polymer as is starch, many animals, including humans, are not able to digest it because they lack the enzyme necessary to break the linkages.

Cellulose is closely related to other biopolymers found in wood and plants. Together these compounds are the raw materials used in a number of products encountered in forensic chemistry. For example, paper is based on wood chips derived from soft or hard woods, plants, or recycled paper stocks. Of interest in paper production are the polysaccharides found in wood: lignin, hemicellulose, and cellulose. These compounds are part of the lignocellulose complex found in biomass such as wood. The lignocellulose compounds impart strength and varying degrees of rigidity to plants. For example, a tree that stands several feet tall needs a different support framework than mosses or grass, the latter two of which are of little use in modern paper production.

Natural Fibers Derived from Cellulose

Cotton fibers are the most important of the natural fibers encountered in forensic laboratories. Other natural fibers include kapok and hemp. Cotton is classified as a seed fiber. The cotton plant is a shrub that grows to height of a few feet. It produces pink flowers that fall off and leave behind a seedpod called a boll. The cellulose fibers grow inside until the boll breaks open when ready to harvest. The raw cotton fibers

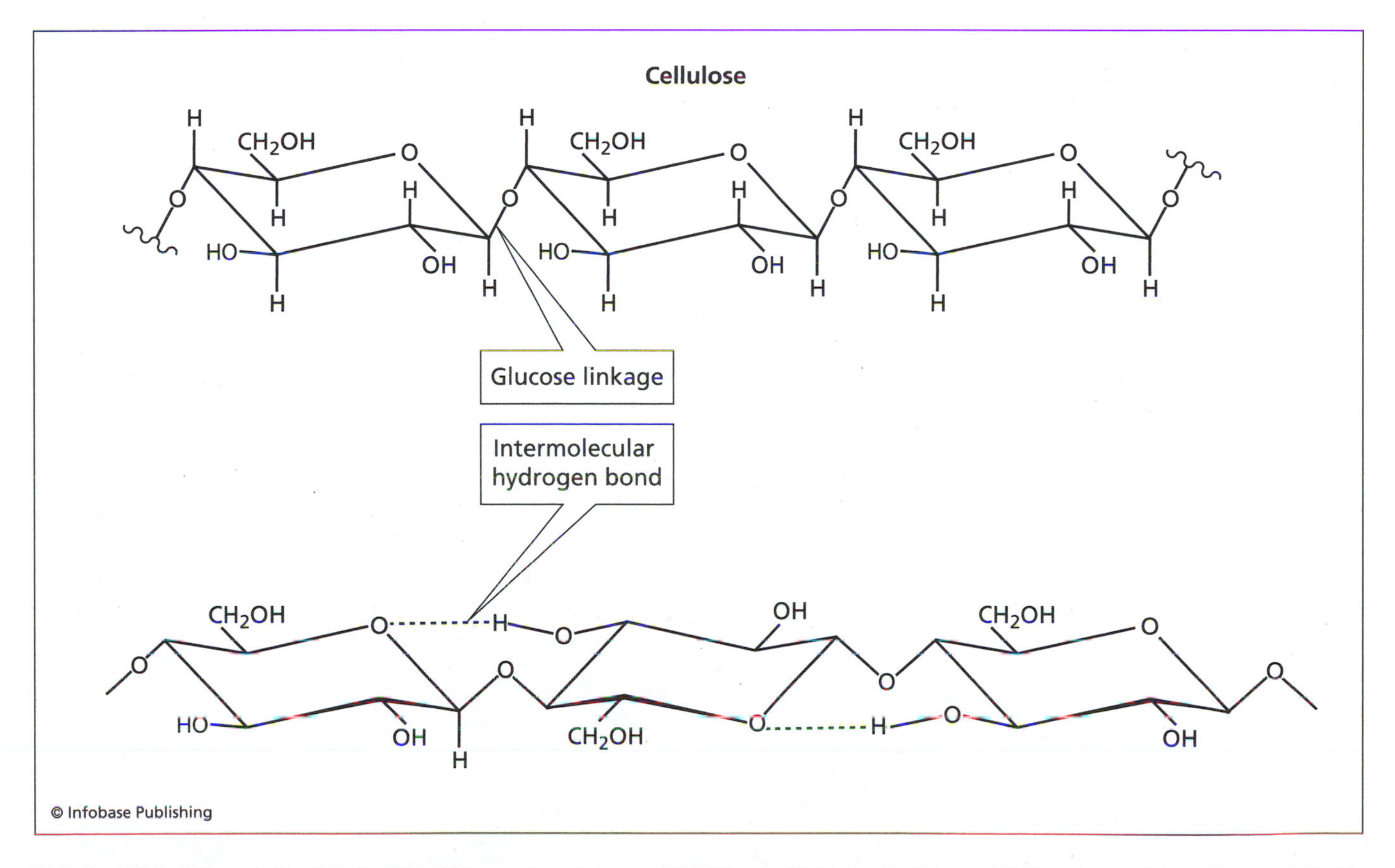

The chemical structure of cellulose. The glucose sugar rings are joined by the linkages as shown, and the strength of the fiber is derived partially from the hydrogen bonds between rings.

are yellowish to white and up to two inches long. A cotton gin removes the seeds, and the baled fibers are shipped for processing such as raking and spinning into yards or fabrics. Because cotton is so common, cotton fibers are typically not very useful as evidence. Another forensically important use of cotton is as guncotton or nitrocellulose (NC). To make this highly flammable material, cotton is treated with nitric and sulfuric acid, which causes nitration of some of the active -OH sites in cellulose.

Cotton fibers are easily identified using polarizing light microscopy and appear as thin twisted ribbons. Although cotton lacks a uniform crystalline structure, cellulose has ordered regions that will interact with polarized light. However, these regions are randomly located, so cotton lacks distinctive extinctions or birefringence. Given the ease of identification using optical techniques, there is little reason for further instrumental analysis unless other classifications are possible based on dyes or other treatments.

Regenerated and Reformulated Cellulose: Semi-synthetics

The first synthetic fibers produced were simply regenerated forms of cellulose referred to as semi-synthetics. Cotton is a versatile fiber, but because the fibers are short, cotton yarns and fabrics consist of shorter fibers spun together rather than of single contiguous fibers; as a result cotton is not a strong fiber. However, the -OH functionality of cellulose provides a chemical handle that is exploited for generation of contiguous cellulose fibers. In the simplest process, raw materials like wood are treated with a strong base to partially disrupt the polymer form. Carbon disulfide is added to create cellulose xanthate, a thick viscous solution that can be extruded through an orifice as a single long fiber. The extruded fiber enters a coagulating solution containing sulfuric acid and sulfate salts. The outer surface shrinks and wrinkles as the inner cellulose reforms. The surface is much smoother than that of cotton fibers. This regenerated material is highly reflective and appears shiny if not treated with delustering agents. The characteristic sheen is the origin of the name "rayon," referring to the shine as giving off rays of light. In addition to fibers, the viscous solution can be cast as a thin film, producing cellophane.

Both cotton and rayon are cellulose, but because of the extrusion process, their properties are significantly different. Even though rayon exists as a long fiber, the degree of polymerization of rayon is about 10 times lower than that of cotton cellulose. Because of these properties and the silky sheen (luster) of rayon, it is used for undergarments and other delicates. One variant of rayon is hollow viscose made by adding Na_2CO_3 to the extruded solution. When placed in the acidic coagulation bath, any Na_2CO_3 trapped inside the fibers is converted to CO_2, producing a void in the structure.

See also FIBERS; PAPER.

Central Identification Laboratory (CIL) *See* ANTHROPOLOGY, FORENSIC.

certification A process employing written and laboratory testing that ensures the proficiency of forensic analysts and the reliability of the data and results they produce. Proficiency assurance can be divided into two areas: the certification of individual analysts through written and practical testing and laboratory accreditation. The goals of both accreditation and certification are ultimately the same—to ensure that the analyses being conducted in forensic laboratories are done properly and that the results produced are accurate and trustworthy.

For practitioners working in many of the subdisciplines, certification is obtained through professional associations. For example, forensic engineers can be certified through the National Academy of Forensic Engineers (NAFE). Analysts in the core forensic disciplines, those found in typical full-service laboratories, are certified under the auspices of the AMERICAN BOARD OF CRIMINALISTICS (ABC) through a combination of written and laboratory examinations. The GENERAL KNOWLEDGE EXAMINATION (GKE) is exactly what the title implies, a written test that covers the general principles of forensic science and CRIMINALISTICS. Passing the GKE gives the analyst the title of diplomat and allows him or her to take specialized exams in specialty areas such as drug analysis and DNA typing. These tests are written and require yearly lab proficiency testing, and successful completion awards the analyst with fellow status in the ABC.

Further Reading

Inman, K., and N. Rudin. "Ethics and Accountability—The Profession of Forensic Science." In *Principles and Practic-*

es of Forensic Science, The Profession of Forensic Science. Boca Raton, Fla.: CRC Press, 2001.

Netzel, L. R. "The Forensic Laboratory." In *Forensic Science: An Introduction to Scientific and Investigative Techniques.* 2nd edition. Edited by S. H. James and J. J. Nordby. Boca Raton, Fla.: CRC Press, 2005.

chain of custody Procedures and documentation used to ensure the integrity of evidence from collection to courtroom presentation and through to final disposition or destruction. It is also referred to as the chain of evidence, chain of possession, or continuity of evidence. The chain of custody (recorded on a written form) begins when evidence is collected and includes marking of the evidence (or the bag/container it is placed in), documentation of where and how the evidence was found, and who collected it. The evidence or container is sealed (usually with evidence tape) and marked such that any opening will be obvious. Typically, this involves placing initials over the tape seal and the package such that any breakage of the seal will be easily detected. The officer or investigator who collects the evidence is then solely and completely responsible for that evidence, including security and insuring that no alteration or tampering occurs. If anyone other than the responsible person opens, transports, or examines the evidence, this is documented on the chain of custody form. Any transfers (such as officer to crime lab personnel) are indicated on the form, and the crime lab analyst then assumes responsibility for the evidence. The paramount goal of the process is to avoid any breaks in the chain, which would bring into question the reliability of the evidence and the link between the evidence and the scene where or person from whom it was obtained. One of the first issues addressed when evidence is introduced into court is to examine the chain of custody and to ensure that the evidence has not been altered or tampered with. Problems with the chain of custody played a role in the outcome of the O. J. SIMPSON case.

chemical analysis Techniques used to identify the chemical components present in a material (QUALITATIVE ANALYSIS) and often the amounts or concentrations of some or all of those components (QUANTITATIVE ANALYSIS). Forensic chemists and toxicologists employ chemical analysis in a variety of cases such as those involving the characterization of drugs, fire debris (ARSON), PAINTS, polymers, and FIBERS, among many others. Chemical analysis can be roughly divided into two categories, traditional "WET CHEMICAL" methods and INSTRUMENTAL ANALYSIS. Wet chemical methods include PRESUMPTIVE TESTS for certain drugs, where a reagent is added to a drug and a color change is observed if the drug is present, presumptive tests for blood, and THIN LAYER CHROMATOGRAPHY (TLC).

Since the 1960s, instrumental analysis has assumed the central role in chemical analysis, displacing many of the older methods. As the name implies, these techniques rely on instrumentation to obtain qualitative information, quantitative information, or both for a given sample. GAS CHROMATOGRAPHY/MASS SPECTROMETRY (GC/MS) is now widely used for the analysis of drugs, as is INFRARED SPECTROSCOPY (IR), ULTRAVIOLET SPECTROSCOPY (UV/VIS), and HIGH PERFORMANCE LIQUID CHROMATOGRAPHY (HPLC), to mention a few. TRACE EVIDENCE analysis increasingly exploits SCANNING ELECTRON MICROSCOPY (SEM) coupled to X-RAY FLUORESCENCE (XRF) instruments. CAPILLARY ELECTROPHORESIS (CE, CGE) is used in DNA sequencing. Current trends in chemical analysis include increasing miniaturization and automation such as robotics, which can allow forensic laboratories to process far more samples than was previously possible.

See also X-RAY TECHNIQUES.

chemical properties Properties of matter that cannot be determined without carrying out a chemical reaction. This is in contrast to physical properties, which can be observed or measured without changing the original composition of the material. Examples of physical properties include density, color, size, mass, and boiling point, whereas flammability is a chemical property. Consider octane, a component of gasoline. The density and color of octane can be measured directly without altering it, but to determine the flammability of octane, it is necessary to ignite it. This is a form of COMBUSTION, and octane is transformed into carbon dioxide and water. In any chemical change, the chemical composition of the original substance is changed. By contrast, in a physical change, the chemical composition is unchanged. Freezing liquid water to ice is a physical change (and freezing point is a physical property) because the chemical composition of water (H_2O) is not changed by the freezing process. Determination of both physical and chemical properties constitutes an important part of many forensic analyses.

chemistry, forensic The application of the principles and techniques of chemistry, particularly analytical chemistry, to situations in which the legal system is, or may become, involved. The busiest section in most forensic labs is the section that analyzes drugs, perhaps the most well-known application of forensic chemistry. The roles of forensic chemists and forensic toxicologists often overlap, and, depending on the size and organization of the lab, the chemistry section may be responsible for drug analysis and TOXICOLOGY including blood alcohol (BAC) testing. Other areas of forensic chemistry include analysis of ARSON evidence and fire debris, EXPLOSIVES, PAINT, FIBERS, GUNSHOT RESIDUE (GSR), and other kinds of TRACE EVIDENCE. Forensic chemistry employs a large arsenal of sophisticated chemical instrumentation including GAS CHROMATOGRAPHY/MASS SPECTROMETRY (GC/MS), HIGH PERFORMANCE LIQUID CHROMATOGRAPHY (HPLC), INFRARED (IR), VISIBLE, AND ULTRAVIOLET (UV/VIS) SPECTROPHOTOMETERS, and SCANNING ELECTRON MICROSCOPES coupled with X-ray diffraction (SEM/XRD). However, much of the work still relies on traditional WET CHEMICAL techniques such as PRESUMPTIVE TESTS and color tests.

chiral separations Chemical separations of optically active isomers of drugs and metabolites. The ability to separate these isomers is becoming more important in DRUG ANALYSIS and forensic TOXICOLOGY. The drug COCAINE is an example of a chemical molecule that is optically active. It has two forms, one that rotates POLARIZED LIGHT to the left (the l-form) and the other that rotates polarized light to the right (the d-form). Cocaine extracted from cocoa leaves is the l-form, but if the drug is made synthetically, it will be the d-form or some mixture of the d and l forms depending on the synthetic method used. METHAMPHETAMINE, a drug made in CLANDESTINE LABS, can exist as the d or l form or as a mixture. Knowing which form(s) are present in evidence can reveal significant information about how the methamphetamine was made and using what precursor chemicals.

Separation of chiral molecules, such as separating the d-form of methamphetamine from the l-form, is challenging because chemically, the molecules behave nearly identically. Most chiral separations utilize CHROMATOGRAPHY and rely on chiral stationary phases such as cyclodextrin. Cyclodextrins used in this way assume a cylindrical shape with optically active sites found in the interior of the cylinder. Chiral forms of the same molecule will undergo slightly different interactions with these chiral sites and as a result, can be separated from each other.

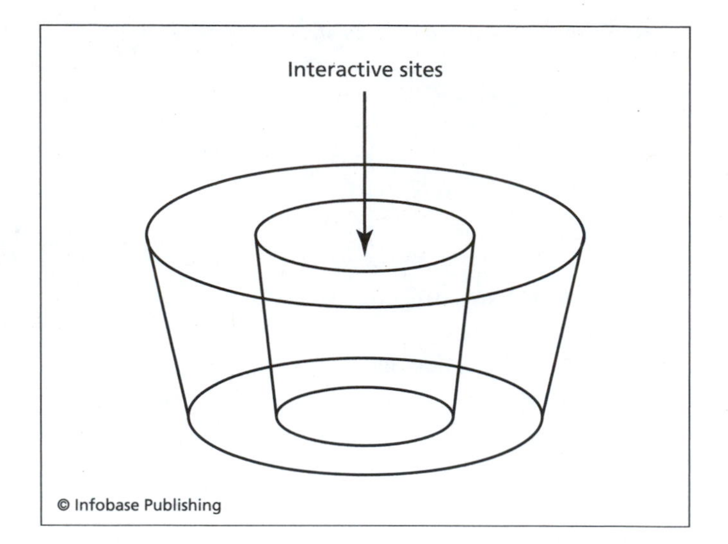

The form of cyclodextrin used in chiral separations

See also DIASTEREOISOMERS; ENANTIOMERS; STEREOISOMERS.

choline *See* SEMEN.

chromatogram The printed or digital output of a chromatographic instrument. A chromatogram is a plot of the response of the detector as a function of elution time as shown in the illustration. In the example shown on the next page, a sample containing three components was introduced into the chromatographic system. Component 1 emerged from the chromatographic column first (eluted first), and the detector recorded a response at 1.3 minutes. This time is assigned based on the top, or apex, of the peak and is called the retention time (t_r). Component 2 had a retention time of 2.8 minutes and component 3, 4.6 minutes. The sizes of the peaks recorded on a chromatogram are proportional to relative concentrations, so here, the sample had the highest concentration of component 2 and the lowest concentration of component 3. Instruments used in forensic science that produce a chromatogram

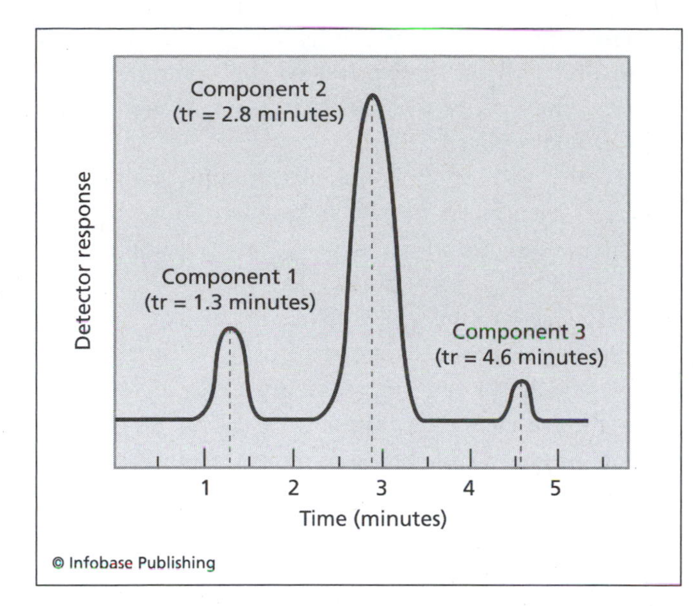

Features of a generic chromatogram. The plot produced by a chromatographic instrument plots a detector response as a function of time; the size of each peak is proportional to the concentration of each component. The notation t_r stands for retention time of a given component in the mixture being analyzed.

include GAS CHROMATOGRAPHY (GC) and HIGH PERFORMANCE LIQUID CHROMATOGRAPHY (HPLC).

chromatography Literally, "color writing," a class of separation techniques used extensively in forensic science. The name color writing originated from early development work in which plant pigments were separated into their components, producing bands of color in the separation column. Similar color bands can be seen in the modern version of THIN LAYER CHROMATOGRAPHY (TLC) when applied to inks. Using TLC, black ink can be broken down into blue, green, red, and other colored components that are easily visualized. Fundamentally, all types of chromatography, from TLC to ion chromatography, share the same principles and the same goal—separation of a mixture into individual components, which can then be identified (QUALITATIVE ANALYSIS) and quantified (QUANTITATIVE ANALYSIS). Chromatography is often used as the front end to sophisticated detectors such as mass spectrometers, creating a class of instrumental analysis techniques known as hyphenated techniques. For example, a gas chromatograph coupled to a mass spectrometer is referred to as a GC-MS (or GC/MS). Samples are introduced into the chromatograph, which separates them into individual components that the detector then sees one at a time.

Chromatography is based on the idea of PARTITIONING between two phases, each of which can be a solid, liquid, or gas. For example, oil and vinegar are both liquid phases, but they remain distinct because neither is soluble in the other. Common table salt is soluble in water but not soluble in oil. Based on these solubilities, if salt is introduced into a jar containing oil and vinegar and the jar is shaken, the salt will dissolve and stay in the water phase. This process is called partitioning and, in this example, partitioning occurs based on different solubility in different liquid phases. Salt is much more soluble in water than in oil, so it will partition preferentially into the water phase and not the oil phase. All chromatographic separations are based on partitioning, although the mechanisms and details vary based on the specific application.

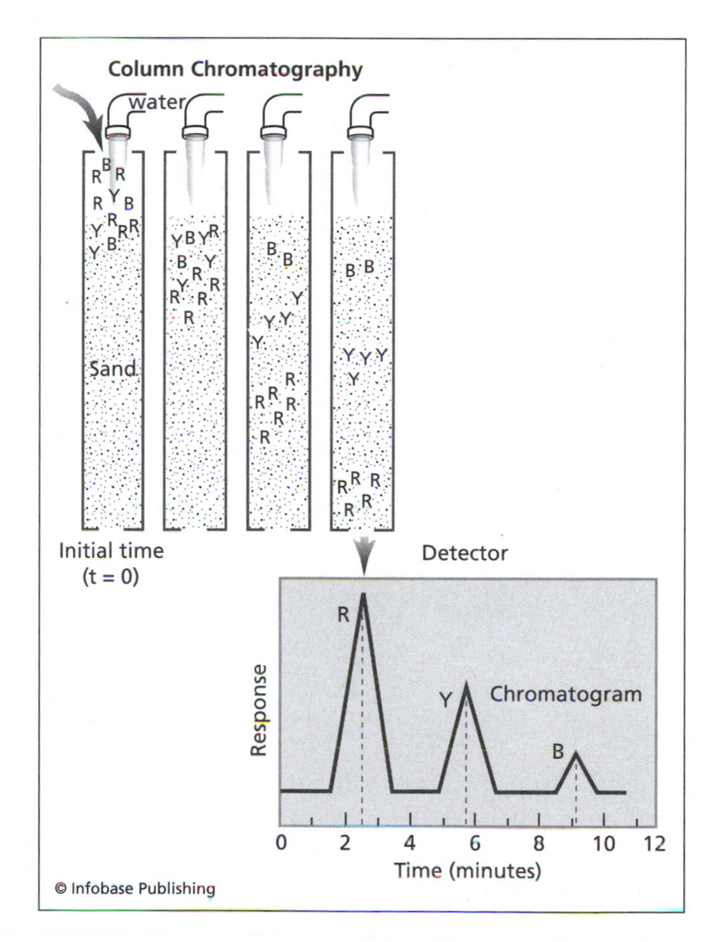

Depiction of simple column chromatography used to separate a colored mixture containing red (R), yellow (Y), and blue (B) components. The red component interacts the least with the column material and will elute from the column first. It is also present in the largest amount and would produce the largest peak on the resulting plot (chromatogram).

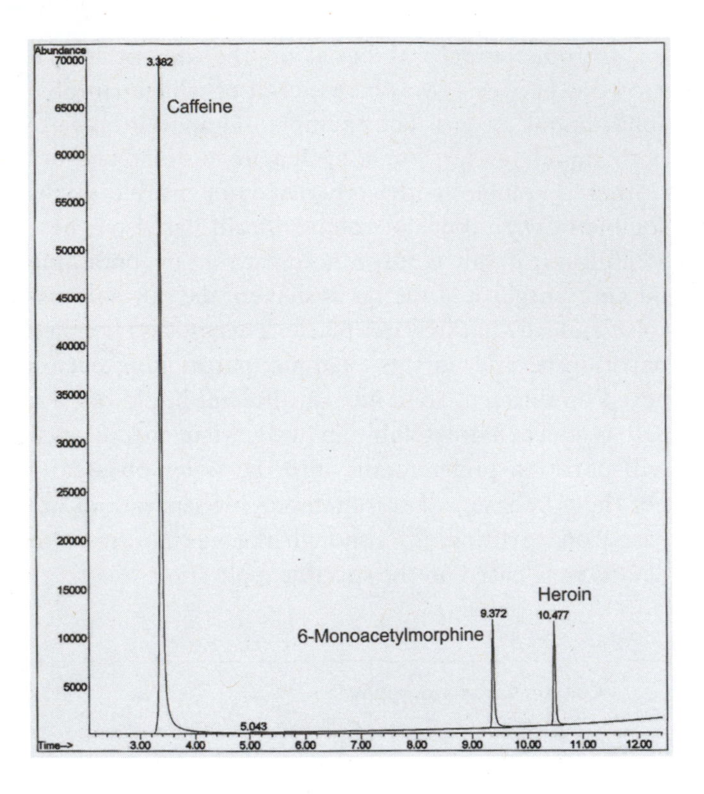

An example of a chromatogram obtained from a gas chromatograph/mass spectrometer ***(photo by Suzanne Bell)***

As previously mentioned, early work with chromatography involved the separation of colored plant extracts. Like ink, these plant extracts are composed of many different colored pigments that can be separated using column chromatography. As shown in the figure on the previous page, a plant extract containing red (R), blue (B), and yellow (Y) components is introduced to the top of a column filled with a solid such as sand or calcium carbonate. Water is continually poured into the column, forcing the extract to migrate downward through the solid phase. At the molecular level, the molecules of pigments are interacting with two phases, the water and the solid. The liquid water being poured down the column is called the mobile phase, and the sand or carbonate is called the stationary phase, since it does not move. The molecular structure of each pigment is different and so will interact with the solid phase differently. The pigments that interact more with the solid phase will travel slowly through the column, while the pigments that interact more strongly with the water will move much faster. Thus, at the end of the column, the pigments will have separated into individual components or colors. In the example shown, the red pigment interacts the least with the stationary phase and will reach the end of the column fastest, whereas blue will be the slowest, since it interacts with the stationary phase the most.

At the end of any chromatographic column, a detector (mentioned above) is required to sense and record the passage of the components. Although anyone looking at the column in the example could see the colors separating and moving toward the bottom, a detector will produce an output called a CHROMATOGRAM, which is a plot of time versus detector response. Here, the red component is not only the fastest but also the one present in the highest concentration. Thus, the detector will record a response early in the run for the red component, and the size of the response (peak height) is the largest of the three since the original mixture contained more red than blue or yellow. The yellow pigment is next to arrive at the detector, with a smaller response, followed by blue, which emerges from the column last and is, in this example, the least concentrated. As a result, the peak for the blue component is the smallest.

The mainstays of forensic chromatography include TLC (DRUG ANALYSIS and INKS), GC/MS (drugs, TOXICOLOGY, ARSON, and TRACE EVIDENCE), GC with other specialized detectors, HIGH PERFORMANCE LIQUID CHROMATOGRAPHY (HPLC, used in drug analysis and trace evidence), and micellular electrokinetic chromatography (MEKC; see CAPILLARY ELECTROPHORESIS). Column chromatography is occasionally used for sample clean up and preparation in drug analysis and toxicology.

See also PARTITIONING AND AFFINITY.

CIE color system Color is a difficult concept to quantify or describe in a common language. One person's version of "red" may be different from another person's version. A descriptor such as "fire engine" red or "stop sign" red makes it easier to imagine a color, but each person perceives color differently. One person may be a bit more sensitive to reds and less sensitive to blues than another, so what the eye sees and what the brain registers is different and inherently impossible to describe with words. Color can be quantitated based on spectral characteristics, removing viewer subjectivity from color description. The CIE system is one method of accomplishing this task, and it relates to how the human eye detects and inter-

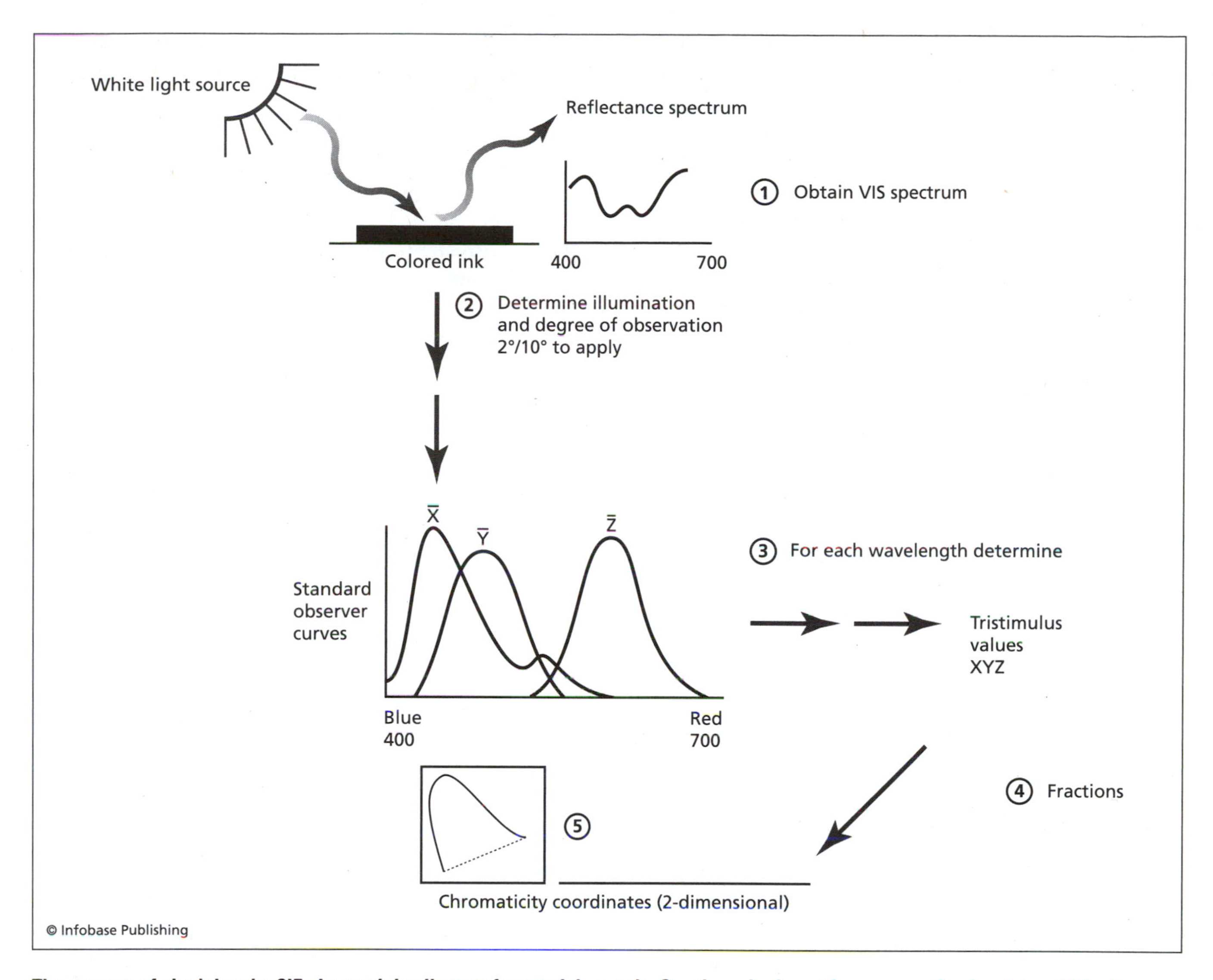

The process of obtaining the CIE chromaticity diagram from an ink sample. See the color insert for an example of a chromaticity diagram.

prets color. To simplify, the eye has structures that respond to three colors: red, green, and blue (RGB). Each person's eyes are slightly different in how they respond to each color, and as a result color perception is subjective. In the forensic context, the ability to describe color without subjectivity is vital. The CIE color system allows for a description of color independent from the variations inherent in human color perception. All that is needed is an instrument for performing reflective visible spectroscopy, or COLORIMETRY of reflected light.

To convert a perceived color into a standardized quantitative equivalent, three elements are needed: a sample such as an ink, a source of illumination, and an observer. Samples such as an ink spot are viewed under white light, and color is perceived based on which spectral components are reflected and which are absorbed. However, not all light sources are equivalent. Daylight has a different spectral spread of wavelengths and intensities than does an indoor fluorescent lamp. Different spectral intensities affect absorbance, reflectance, and color perception. Either a standardized light source must be used, or more practically, results must be normalized to a standard.

One method of standardizing illumination is to consider it relative to the radiation released by an object called a blackbody radiator. Blackbody radiation correlates to the spectral emission profile of a perfect blackbody radiation source when it is heated to a given temperature. The term *white hot* is taken in the same vein; if an object such as an iron rod is heated sufficiently, it glows red, then yellow, and then at the hottest, white to blue. When light has the same spectral spread of a blackbody emitter at a given temperature, the light is said to have that "temperature." Here temperature is a descriptor, but it does not have any physical correlation. A filament in a light bulb is hot, but the actual filament temperature is not the same thing as the temperature of the emitted light.

To standardize the observer variable, methods that are more elaborate are used such as using three lights. In this approach, a viewer (observer) is assumed to be looking at a white screen inside of a box protected from stray light. A lamp illuminates an image area on a screen. On the opposite side of an opaque partition, three lamps emitting the primary colors illuminate another image area on the same background screen. The viewer controls the intensity of the three colored lamps and adjusts the contribution of each until the observed color on both sides of the barrier appear identical. In some cases, this requires alteration of the source light. As long as the alteration is known (for example, adding red to the source), it can be accounted for in the final calculations. Images with different colors can be illuminated and color appearance matched. An examination of the source light spectral characteristics compared to the three combined light spectra allows the observed color to be broken down into three components (called standard observer curves) that roughly correlate to the contribution of red, green, and blue to the perceived color.

The system was first formalized in 1931 by the International Commission on Illumination (CIE). Additional calculations using tables generated by the CIE allow for any spectrum recorded to be converted into a two-dimensional plot called a chromaticity diagram.

As useful as the chromaticity diagrams are, there are limitations. Because the chromaticity parabola is asymmetric, calculation and comparison of color differences are not uniform. Correction factors address this distortion. This technique is applied in the CIE L*a*b* system that was introduced in 1976. This color space also has some inherent distortion but far less than simple chromaticity diagrams. The CIE system is widely

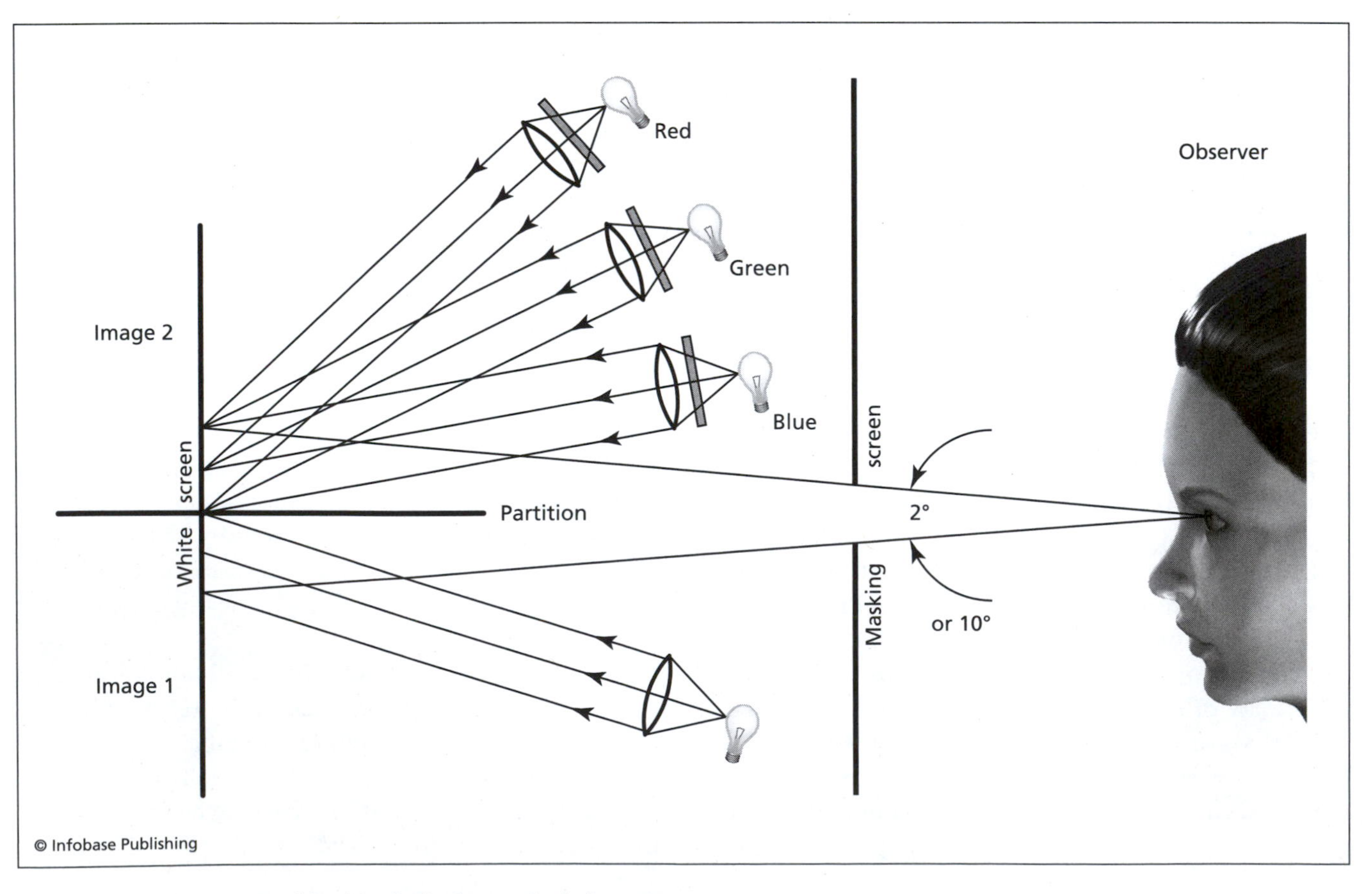

How the RGB colors are combined to obtain the standard observer curves

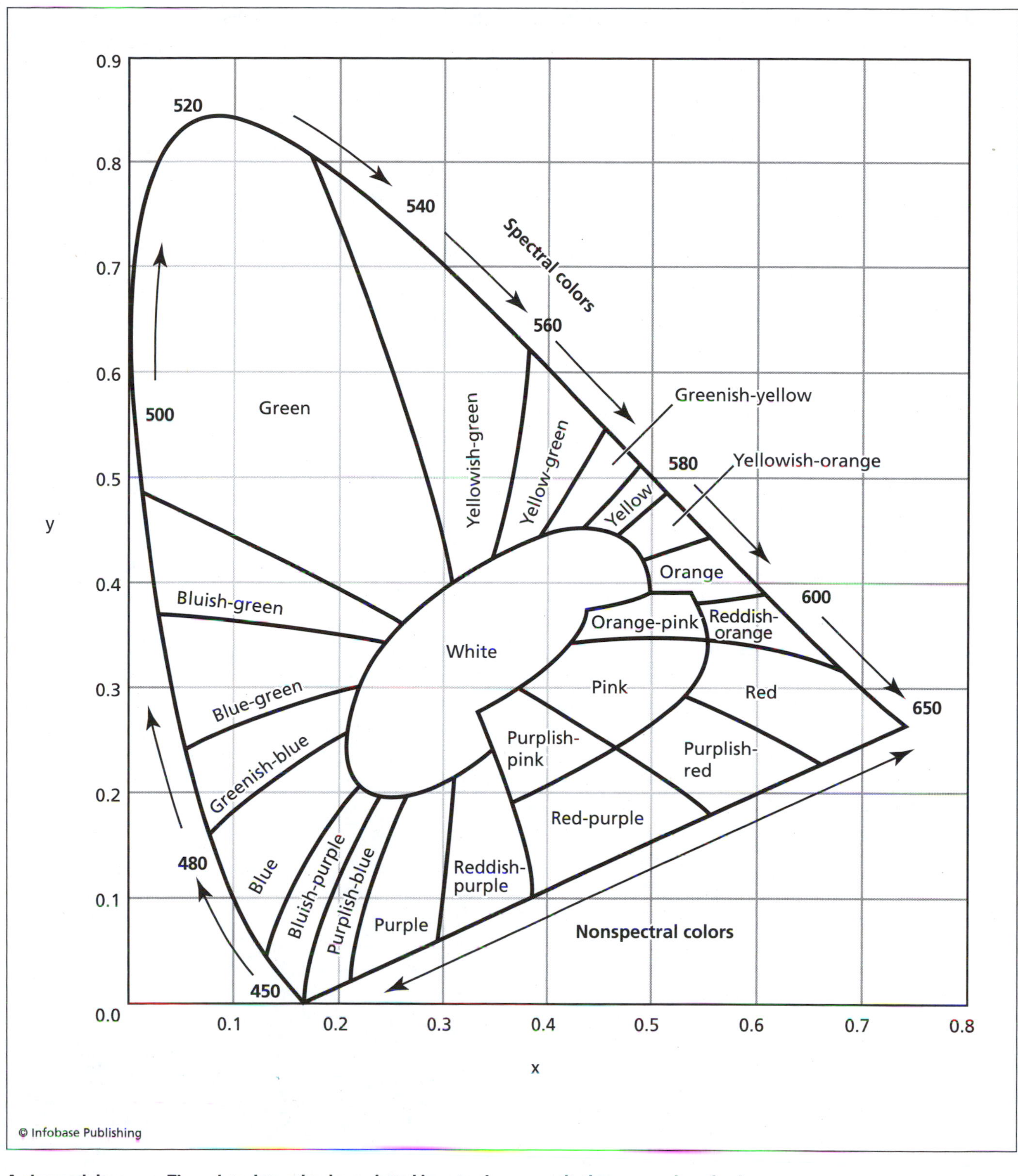

A chromaticity curve. The points determined are plotted here to give a quantitative expression of color.

used in forensic analyses to describe color of INKS, PAINTS, and FIBERS.

See also COLOR AND COLORANTS; MUNSELL COLOR SYSTEM.

cigarettes Cigarette butts and packs are frequently found at crime scenes and can be important evidence linking individuals to the scene. Discarded packs may contain FINGERPRINTS, and in addition to brand information, some packs may contain numerical codes that occasionally can be used to determine an area where the pack may have been sold. Cigarette butts can also be a source of fingerprints and brand information. In addition, saliva residues can often be typed for ABO BLOOD GROUP (if the smoker is a SECRETOR) or for DNA TYPING. Finally, other kinds of tobacco (pipe, cigar, or chewing tobacco) may be found at scenes and analyzed in similar ways as appropriate.

circumstantial evidence Evidence that does not reflect directly on the question at hand, but rather evidence that reflects on the circumstances in question. Much of the evidence produced during forensic analyses is circumstantial evidence. For example, if a bullet is recovered at a crime scene and after analysis is found to have been fired from a gun found at a defendant's house, that is circumstantial evidence. It is not definitive proof that the suspect committed the crime. In contrast, if a forensic chemist conclusively identifies a white powder confiscated from a suspect as cocaine, that is direct evidence and not circumstantial. There is no doubt that the suspect had an illegal substance. There is a common misconception that circumstantial evidence is unreliable evidence, but this is not necessarily the case. In the bullet example, the evidence is compelling and at the very least would demand further investigation. If the suspect's fingerprints were found on the gun, the gun was found to have been recently fired, and investigators determine that the suspect and the victim had recently had an argument, the case has become much stronger, even though, taken individually, the findings are all circumstantial evidence.

Civil Aeronautics Board (CAB) This agency was created by Congress in 1940 and was charged with investigating aviation accidents, determining probable causes, and issuing findings and recommendations to prevent recurrences. The CAB was merged into the newly formed NATIONAL TRANSPORTATION SAFETY BOARD (NTSB) in 1967, which was associated with the Department of Transportation (DOT). In addition to aviation accidents, the NTSB assumed responsibility for all types of transportation accidents on highways and railways, along pipelines, and in marine operations. In 1975, the NTSB became completely independent of the DOT.

civil law *See* CRIMINAL LAW V. CIVIL LAW.

clandestine graves Clandestine graves are among the most challenging crime scenes to locate and then to process, since the sites are often remote and may not be located until months or years after burial. Numerous procedures have been used and are being researched for use in clandestine grave location. Each technique has advantages and disadvantages, and, to be successful, searches for clandestine graves must include personnel from many disciplines, including botanists, entomologists, and geologists. The climate of a region, amount of rainfall, and time of year that the burial occurred make each site unique and further complicate location and identification. Many of the techniques used to locate clandestine graves rely not on detecting the body directly but rather on finding soil disturbances above it.

As shown in the figure, whenever a grave is dug and filled in, ground is disturbed, and existing vegetation is destroyed. Plants will recolonize the site in an identifiable succession pattern distinct from the surrounding undisturbed vegetation. Forensic botanists can study this succession to help identify areas where ground disturbance has occurred, and this is an example of forensic ecology. However, any disturbance in soil can initiate a new succession, so not all successions indicate a clandestine grave lies beneath. Arial photography using a variety of imaging techniques has been shown useful in detecting disturbed areas and has the added advantage of being able to search large areas. If a grave is relatively new, forensic entomologists can

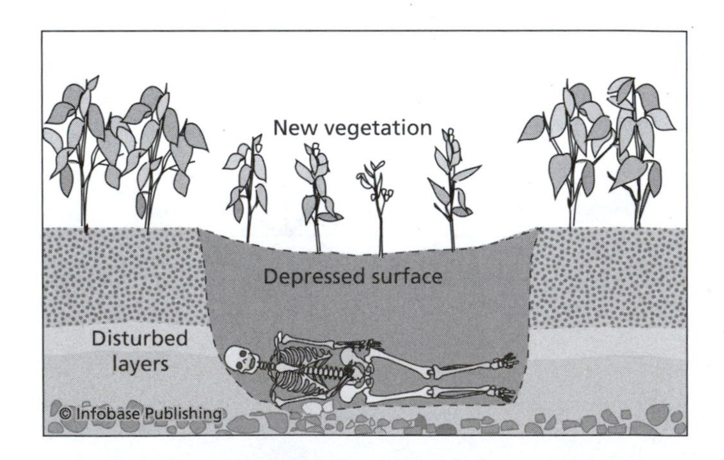

A clandestine grave. Digging a grave produces a localized disturbance that extends as deep as the grave. Material originally dug from the hole is used to backfill it, but often this fill will compress, leaving a depression over the gravesite. Vegetation is also disturbed and must re-establish itself in phases.

assist in a way analogous to botanists, by looking at the succession of insects that are colonizing the area and the body. Underground imagining techniques have also been used including ground penetrating radar (GPR), magnetic field sensors, and metal detectors. Detection of odors associated with decomposition has been evaluated using portable GAS CHROMATOGRAPHY (GC). CADAVER DOGS are also used, but their performance depends somewhat on temperature and other weather conditions.

Older gravesites are often well suited to the techniques of forensic ARCHAEOLOGY, in which most of the other clues such as vegetative change and odor have vanished or been greatly reduced by time. Archaeologists and anthropologists can also assist in the excavation once a clandestine grave is located.

Further Reading

France, D. L., et al. "A Multidisciplinary Approach to the Detection of Clandestine Graves." *Journal of Forensic Sciences* 37, no. 6 (1992): 1445.

clandestine labs Illicit laboratories established to manufacture or process illegal drugs. Although the types of drugs made vary by region and by current availabilities, the most common type of clandestine labs are those making AMPHETAMINE and METHAMPHETAMINE. Other labs make PHENCYCLIDINE (PCP), DESIGNER DRUGS, and LSD. Some drugs, such as methamphetamine, are relatively easy to make using widely available "recipes," while the preparation of LSD is much more complex and requires highly skilled chemists. Clandestine labs can be located nearly anywhere, in rural or remote areas or in city neighborhoods. Size varies as well, from large operations down to bathrooms, kitchens, or even the trunks of cars. Clandestine labs are among the most dangerous types of CRIME SCENES since many of the solvents and chemicals present are flammable, toxic, or corrosive. Explosions are a constant danger, and many clandestine labs are first discovered by fire departments. More than any other type of scene, the presence of forensic chemists or other specially trained personnel is essential to ensure safety, collection of the proper kinds and quantity of evidence, and minimization of environmental consequences. Often, a hazardous material crew is required to ensure proper disposal of the hazardous materials found, and this must be closely coordinated with the process of evidence collection.

Occasionally, a clandestine lab will be found, but none of the final product will be present. This presents a challenge to forensic chemists, who will have to use their knowledge of drug synthesis to prove that all the materials found at the scene—chemicals, PRECURSORS, equipment, by-products, and intermediates—are consistent with the production of the drug in question. Samples must be taken of many containers, which are often unlabeled and potentially booby-trapped. Once back at the lab, analysis proceeds using the same tools and techniques common in drug analysis, including GAS CHROMATOGRAPHY/MASS SPECTROMETRY (GC/MS), spectroscopy such as infrared (IR) and attenuated total reflectance (ATR), and HIGH PERFORMANCE LIQUID CHROMATOGRAPHY (HPLC).

The federal Chemical Diversion and Trafficking Act of 1988 limits the sale of the precursors for many drug syntheses, which has affected the ability of clandestine labs to operate. However, illicit chemists can devise new synthesis methods or resort to stealing to find the raw ingredients needed. Unfortunately, many of these ingredients have legitimate uses, so controlling all potential raw materials is impossible.

class characteristics and evidence Much forensic work involves assigning physical evidence to classes. This is in contrast to INDIVIDUALIZATION, in which evidence is assigned to one COMMON SOURCE. A class can be considered a discrete group or subgroup of items (or individuals) that are similar due to reproducible processes used to make them. The process of classification of evidence is a process of assigning it to these groups or categories. Members of a given class will share the same class characteristics, whereas evidence that can be individualized will possess characteristics that make it unique.

Consider a bloodstain that is typed and found to be human blood with an ABO BLOOD GROUP of Type B. The fact that the stain is blood, that it is human, and that it is Type B are class characteristics. Class characteristics are not enough to individualize the sample or link it to one individual. Fibers are another example; they can be classified by what they are made of (cotton, rayon, nylon, and so on), how they are used (carpet, clothing, and so on), and by color, but it is generally not possible to individualize fibers since they are mass produced and since different manufacturers and processes produce them in so many different places. Handwriting also has distinctive class characteristics.

To write a capital *A*, two slanted lines meet at a point and are linked by a horizontal line about half way up; conversely, a *Q* involves a circular stroke. However, everyone writes the letters slightly differently, imparting individual characteristics. Almost every type of physical evidence can be classified in some way, but not all evidence can be individualized.

classification *See* CLASS CHARACTERISTICS; EVIDENCE.

clearance rate In forensic TOXICOLOGY, this refers to how fast a drug, toxin, or metabolite is eliminated from the body. If a drug has a rapid clearance rate, such as the predator drug GHB, it is quickly eliminated from the system and difficult to detect only hours after the initial dose. Specifically, the clearance rate refers to the rate at which the compound is removed from the plasma per minute, hour, or other unit of time. The clearance for GHB is about .16 fluid ounces per minute per pound (10 mL/min/kg), meaning that the equivalent of 10 mL of plasma is cleared of GHB per minute per kilogram of body weight. For comparison, the clearance of acetaminophen (Tylenol), a drug that when taken as directed is effective for six to eight hours, is about .08 fluid ounces per minute per pound (5 mL/min/kg). The means by which a substance is removed from the plasma can be through excretion or by conversion to another compound through METABOLISM. For example, ethanol can be removed from plasma by excretion in the urine, exhalation through the lungs, or by metabolism of ethanol to another compound such as acetaldehyde.

See also ADME.

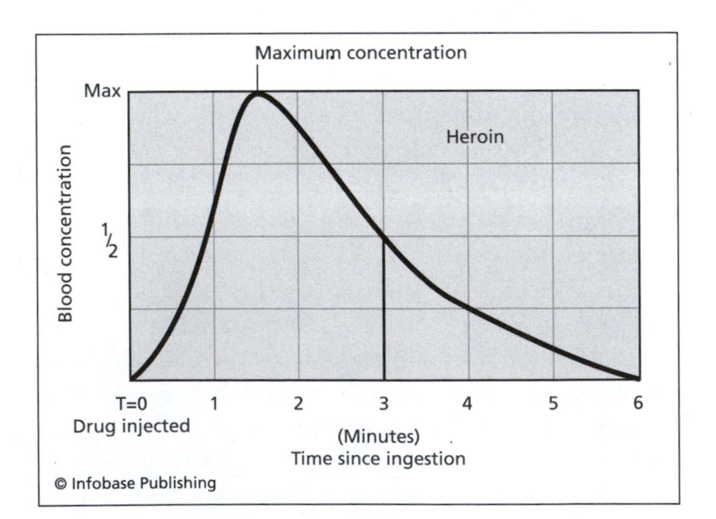

An example of clearance. When heroin is injected, it reaches a peak concentration in the blood quickly and then is eliminated (cleared) by metabolism.

clothing A valuable source of evidence in many cases, and a rich source of TRANSFER EVIDENCE. Clothing tends to collect and hold a number of different kinds of TRACE EVIDENCE including DUST, FEATHERS, and FIBERS. Because clothing is an efficient collector of materials, the sooner it is collected after an incident the better. As time passes, additional transfers unrelated to the incident in question can occur (CONTAMINATION), and relevant evidence can be lost. Trace evidence on clothing can be collected in many ways, one of the most common being the use of ADHESIVE TAPE (TAPE LIFTS). Careful packaging and storage is also essential, keeping each piece of clothing in a separate bag and protected with clean paper such as butcher's paper. If clothing is folded, paper should be inserted between folds to prevent clothing surfaces from contacting each other.

BLOOD and BODY FLUID stains are often found on clothes as well, and they can be rich sources of information beyond the obvious information provided by DNA TYPING and other serological techniques. Position and number of stains can be important in determining what type of weapon was used and the manner in which an attack occurred. BLOODSTAIN PATTERNS on clothing can be used in crime scene reconstructions and to support or refute different stories about what occurred. Complications can arise when garments are cut away or otherwise hastily removed by ambulance crews, paramedics, or emergency room personnel. Whenever clothing is stained, it must be thoroughly dried and properly stored before transport to the laboratory. Cutting and tear patterns associated with wounds can also be useful in determining type of weapon, and damage analysis of clothing can be used to assist in reconstructions. For example, the damage to clothing caused by a stabbing motion will be different from that caused by a slashing motion.

Bloodied or otherwise stained clothing can also be the source of transfer patterns and imprints; if an assailant wearing jeans kneels in blood and then comes in contact with another surface such as a wall or another piece of clothing, the bloody jeans can leave a distinctive imprint of denim on that surface. Fabric imprints (also called textile imprints or patterns) have even been identified on bullets fired through clothing. Finally, GUNSHOT RESIDUE (GSR) caught on clothing can be used to help in distance estimation between the victim and the assailant when the shot was fired. As with stain and damage patterns, GSR on clothing can help sort out conflicting accounts of a shooting.

Recent interesting work has been done on individualizing denim jeans based on the pattern of ridges and folds that develop along the seams and lower cuffs. An analysis of surveillance photos taken of a bank robber was compared to jeans recovered from the suspect's home, and the results were used in court.

Further Reading

Taupin, J. M. "Testing Conflicting Scenarios—A Role for Simulation Experiment in Damage Analysis of Clothing." *Journal of Forensic Sciences* 43, no. 4 (1998): 891.

Vorder Bruegge, R. W. "Photographic Identification of Denim Trousers from Bank Surveillance Film." *Journal of Forensic Sciences* 44, no. 3 (1999): 613.

coatings *See* PAINT.

cobalt thiocyanate A chemical reagent used as a PRESUMPTIVE TEST for COCAINE. There are three common variations of the test. In the first, the reagent is prepared by dissolving cobalt thiocyanate (CoSCN) in distilled water, sometimes with additional ingredients such as ammonium thiocyanate or glycerin. When this is added to cocaine powder, a blue solid (blue precipitate) is formed. Like other ALKALOIDS, cocaine can exist in the basic form ("freebase") or in the form of a hydrochloride (HCl) salt, so for a positive result, cocaine freebase must first be acidified with hydrochloric acid. Obtaining a blue color change does not mean that cocaine is present, only that it might be. Other compounds, such as procaine, can produce color changes as well. To increase the specificity of the test, a variation called the Scott or Ruybal test has been used in which three reagents are employed. The first is a cobalt thiocyanate solution containing glycerin, the second is a hydrochloric acid solution, and the third is chloroform, an organic solvent that does not mix with water (similar to vinegar and oil). If cocaine is present, addition of the first reagent will cause the blue precipitate to form, and the addition of the second will cause the blue to turn to a clear pink color. Adding the final reagent, chloroform, will cause two layers to form in the test tube, the aqueous layer (the water-based solution) and the chloroform layer. If cocaine is present, a blue color will reappear in the chloroform layer.

cocaine A powerful central nervous system stimulant derived from the leaves of the coca plant (*Erythroxylon coca*). Cocaine can also be synthesized, but the process is difficult and expensive and so far has not replaced the coca plant as the primary source in the illegal drug market. The coca plant grows in the Andes Mountains and in some parts of Asia, and the largest source of raw coca is South America, principally Colombia. Early natives of the region chewed on the leaves and brewed teas that they used medicinally. Sigmund Freud, who used it and wrote glowingly about the drug in the 1880s, popularized cocaine in the modern era. Extracts of the coca leaf were also used as ingredients in Coca-Cola and other medicinal preparations in the early part of the 20th century.

Cocaine has legitimate medical uses as a topical anesthetic in the eyes, nose, and throat, producing the same numbing effect as Novocain (procaine). However,

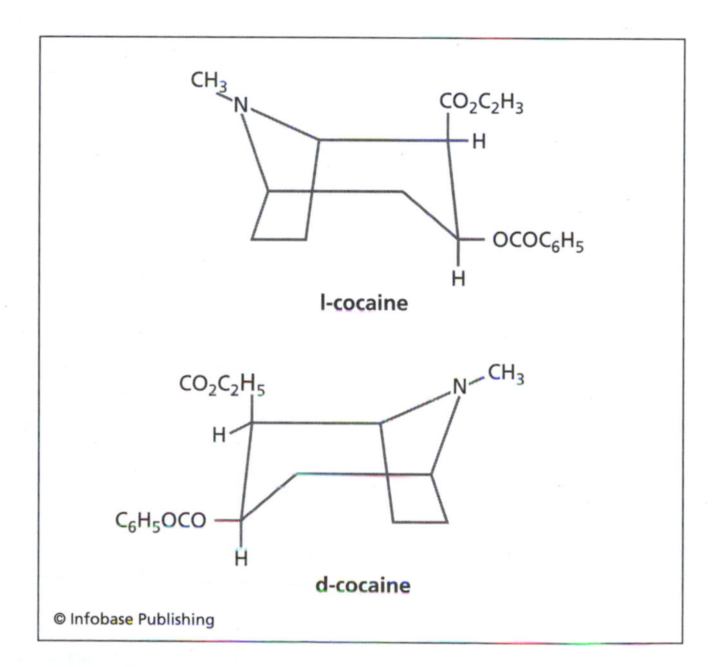

The two optical isomers of cocaine, named for the direction in which they will rotate plane polarized light

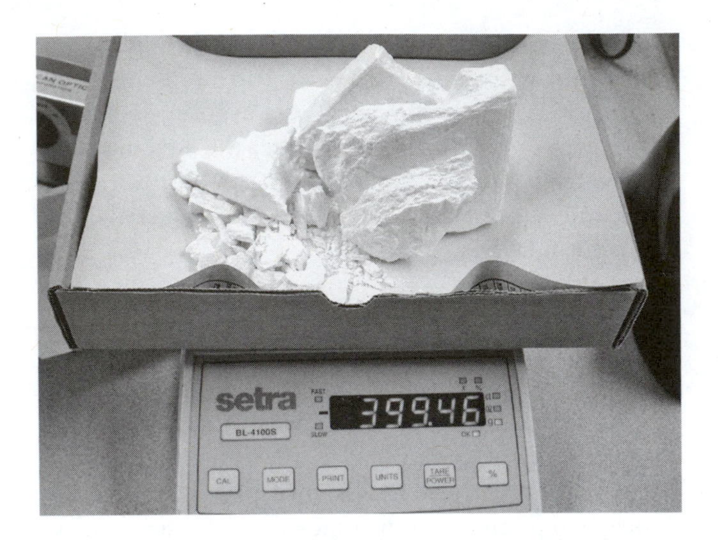

A large cocaine chunk being weighed in the lab ***(Courtesy of Aaron Brudenelle, Idaho State Police)***

other related compounds have largely replaced cocaine in these roles. Illicit cocaine is supplied in the form of the hydrochloride salt, a crystalline white powder, usually diluted with CUTTING AGENTS such as lidocaine, sugars, or caffeine. The hydrochloride salt can be converted to the freebase form ("freebase" or "crack") by dissolving the salt in water and heating gently with baking soda or ammonia. The freebase form is typically smoked, producing an immediate response that is almost as fast as that obtained by injection. The effects of cocaine on the user are similar to those of AMPHETAMINE and METHAMPHETAMINE, elevated heart rate, rapid breathing, and a feeling of alertness and well-being. The primary source of cocaine for the U.S. market is Colombia and smuggling through the southern borders in the states of Arizona, Florida, California, and Texas. Usage of cocaine has declined from its peak in the 1980s, but data from the *National Household Survey on Drug Abuse* still showed that, in 1999, approximately 2 percent of the U.S. population used cocaine. The number of people regularly using the drug is estimated at 1.5 million.

The analysis of cocaine involves the use of PRESUMPTIVE TESTS such as cobalt thiocyanate, CRYSTAL TESTS (gold chloride and platinum chloride), THIN LAYER CHROMATOGRAPHY (TLC), GAS CHROMATOGRAPHY/MASS SPECTROMETRY (GC/MS), HIGH PERFORMANCE LIQUID CHROMATOGRAPHY (HPLC), and INFRARED SPECTROPHOTOMETRY (IR). Because cocaine can be prepared synthetically as well as extracted from a plant, interesting issues have arisen concerning the particular chemical form of the cocaine. There are two STEREOISOMERS of cocaine, meaning that there are two forms of the molecule that differ only in that they are mirror images of each other. Hands exists as such stereoisomers; the left hand and the right hand are mirror images of each other. Cocaine extracted from plants is in the l-form, meaning that it will rotate plane POLARIZED LIGHT to the left. Synthetic cocaine contains d-cocaine, which rotates plane polarized light to the right. Although the formulas and structures of the two cocaine molecules are identical, they are mirror images of each other, as shown in the illustration. Legal issues have been raised about the differences and effects of the two forms, and as a result many labs routinely test cocaine to determine if it is the l-form or the d-form. This can be accomplished using melting point tests or with a polarizer (polarimeter), which determines the direction that the sample rotates plane polarized light.

Further Reading

Drug Enforcement Agency (DEA) Web site. "Cocaine." Available online. URL: http://www.dea.gov/concern/cocaine.htm. Downloaded January 31, 2003.

codeine An opiate alkaloid that is found naturally in OPIUM at concentrations of approximately 0.7 percent to 2.5 percent. Like MORPHINE, codeine is considered a narcotic, although it is only about one-sixth as powerful. Codeine is taken orally as an ANALGESIC (pain reliever) and as a cough suppressant, and it is usually synthesized from morphine rather than extracted from opium. Many codeine mixtures and preparations are listed on Schedule III of the CONTROLLED SUBSTANCES ACT (CSA), although in many countries codeine preparations are available over the counter. The most common forms of codeine are the salts codeine phosphate and codeine sulfate. In the forensic lab, codeine is identified using THIN LAYER CHROMATOGRAPHY (TLC), GAS CHROMATOGRAPHY/MASS SPECTROMETRY (GC/MS), HIGH PERFORMANCE LIQUID CHROMATOGRAPHY (HPLC), and INFRARED SPECTROPHOTOMETRY (IR and FTIR).

CODIS **(Combined DNA Indexing System)** A national program coordinated by the FBI that assists state and local labs in establishing DNA databases from unsolved crimes, missing persons, and convicted offenders. The system electronically stores the results of a specific set of DNA TYPING results and allows networked laboratories to share and compare DNA results. The system has proven to be particularly valuable in sex offense cases in which DNA typing has been able to link suspects to multiple crimes within an area as well as across state lines.

Further Reading

NIST. "FBI Core STR Loci." Available online. URL: www.cstl.nist.gov/biotech/strbase/fbicore.htm. Downloaded January 21, 2003.

Rudin, N., and K. Inman. "The DNA Databank." In *An Introduction to Forensic DNA Analysis.* 2d ed. Boca Raton, Fla.: CRC Press, 2002.

coincidental match A match, association, or linkage, based on PHYSICAL EVIDENCE, that occurred by chance and is irrelevant to the case in question. Often, but not always, coincidental matches can be resolved by additional analysis. For example, if some kind of transfer evidence such as pollen or DUST is found at a crime scene and also on someone's clothing, it could be a coincidence rather than evidence of participation in a crime. Additional study of the pollen or dust might indicate that there are additional INDIVIDUALIZING characteristics that clarify the interpretation of the evidence.

color and colorants Color is how the human eye senses ELECTROMAGNETIC ENERGY in the visible range. Colorants are substances or materials that can absorb or emit electromagnetic energy in the visible range; two types of colorants (DYES and PIGMENTS) are of particular interest in forensic CHEMISTRY. Humans perceive color as a result of two types of interaction with a sample: reflection or transmission. Looking at a painting on a wall results in light interacting with the PAINT and being reflected back to the viewer who perceives color. Looking though a glass of cola is an example of light passing through the liquid, interacting by selective absorption, resulting in the viewer perceiving a reddish brown color. The chemical basis of the colors seen is the same and is based on how the sample (painting or cola) interacts with ambient light to yield color.

If a sample powder appears white, it is reflecting all wavelengths of visible light in the 400 to 700mm range. No visible light is absorbed, and no color is observed. Light from the sun or other type of broad-spectrum light sources emits a mix of the visible wavelengths and is referred to as white light. If a substance or surface reflects the light diffusely or at random angles, it appears white. Conversely, if the material absorbs all wavelengths, it will appear black. Between these extremes is the color gray, which is what is seen if some constant percentage of intensity at all wavelengths is absorbed. White, gray, and black are achromatic—literally, lacking color. If a specific color is perceived, the light is chromatic.

Absorption of energy in the visible range is governed by Beer's Law:

$$A = \in_{\lambda} bc$$

Where $\in$ is the molar absorptivity coefficient at wavelength λ, b is the pathlength, and c is the concentration. Many dyes have high absorbtivity coefficients, in the range of 10^4-10^5 $M^{-1}cm^{-1}$ and as a result are intensely colored. This is important in color testing since the more intense a color, the easier it is to perceive and the smaller the amount of the sample required.

The portion of the molecule capable of absorbing a photon is called the chromophore. This term is generic in the sense that even a UV absorber can be called a chromophore although the human eye cannot detect any color resulting from a UV transition. To have color, the transitions require lower energy photons corresponding to smaller energy gaps. One way to decrease the size of the gaps is through conjugation. Since the gaps are smaller, visible light has enough energy to promote electrons through absorption, imparting color. The more conjugation in the system, the higher the wavelength of light absorbed and the darker perceived color.

See also CIE COLOR SYSTEM; MUNSELL COLOR SYSTEM.

Further Reading

Nassau, K. *The Physics and Chemistry of Color.* New York: John Wiley and Sons, 2001.

colorimetry SPECTROPHOTOMETRY conducted on colored samples, or more generally, descriptions and measurements undertaken to describe or characterize color, such as in the color of a paint chip or FIBER.

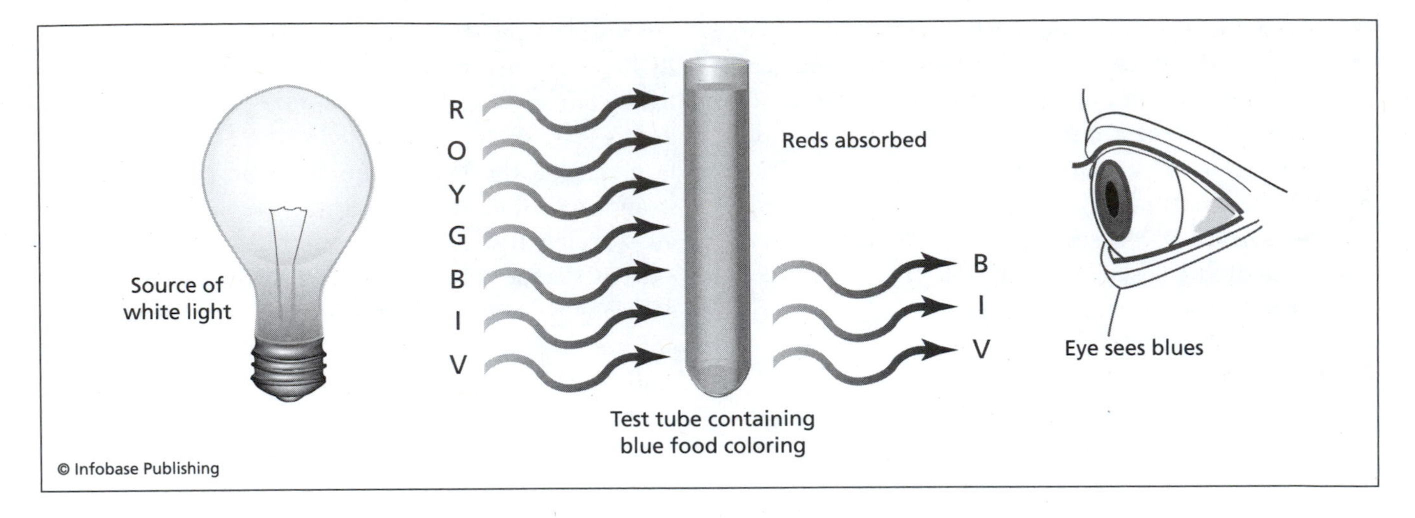

The process of spectrophotometric colorimetry, with the human eye acting as the detector. A test tube containing blue food coloring that is held up to a light will appear blue to the eye because the molecules that constitute the food coloring absorb large amounts of red, orange, and yellow light.

Visible light (colored light) makes up a small portion of the ELECTROMAGNETIC SPECTRUM bordered on one side by ultraviolet radiation (UV) and on the other by infrared radiation (IR). White light can be broken into component colors as seen in rainbows, and these component colors can be separated from each other using a prism or a similar dispersive device. The colors of visible light range from blue to red and are often identified by the acronym ROY G. BIV, which stands for red—orange—yellow—green—blue—indigo— violet, the familiar colors of a rainbow. Colorimetry relies on the principle of complements, which states that the wavelengths of visible light being absorbed by a sample will be the complementary colors of the sample itself. Roughly speaking, a sample of a dark blue food coloring appears as that color to the eye because the red and yellow components of visible light are absorbed by the sample, removing them before they can be seen. The color that is absorbed is the complement of the color seen by the eye. A "color wheel" device used by artists shows what colors are complementary, and colorimetric techniques exploit this relationship to measure the concentration of colored components in a liquid sample by relating the color and intensity of absorbed light to the concentration. In the food coloring example, a more concentrated solution will appear as a very dark blue and absorb more red and yellow light than diluted samples. Accordingly, the amount of light absorbed can be related to concentration. Because the eye can be used as a detector, colorimetry was one of the first forms of spectrophotometry developed. In modern instruments, instrumental detectors replace the eye with devices that convert light to electrical current that can be displayed or further manipulated.

color tests *See* PRESUMPTIVE TESTS.

combings A technique used to collect TRACE EVIDENCE, usually in sexual assault cases. When a rape has occurred, pubic combings are collected in hopes of finding items such as pubic hairs of the perpetrator, fibers, or other transfer evidence that can be used to link the suspect to the victim. Other materials can be combed as well, including beards, with the same intent.

combustion The chemical reaction broadly defined as burning that occurs when a hydrocarbon (a compound containing only carbon and hydrogen) or an oxygenated hydrocarbon is combined with oxygen at elevated temperatures. An understanding of combustion and related processes is critical in the investigation of ARSON and CARBON MONOXIDE poisoning.

At the simplest level, a combustion reaction is one that takes place between a fuel and an oxidant. Heat is important, both as a product of the reaction and to ensure that the reaction has enough energy to be self-sustaining. The ingredients of combustion are often depicted in a "fire triangle" as illustrated in the figure. The oxidant is usually (but not always) oxygen.

A very simple example of combustion is the reaction of hydrogen gas and oxygen, which can be obtained in the atmosphere, which is approximately 21 percent O_2:

$$H_2(g) + O_2(g) \Rightarrow 2H_2O(g)$$

Because all reactants are gas (indicated by the small 'g'), this is an example of gas-phase combustion in which an actual flame is observed. Indeed, the combination of hydrogen and oxygen is extremely dangerous. The only chemical product of the reaction is water vapor, along with an enormous amount of heat energy. Combustion reactions all produce heat, classifying them as exothermic (heat-releasing). Energy is also given off as light seen in the flame.

Hydrocarbons are the usual fuel for combustion reactions of interest in ARSON, where the hydrocarbons are used as ACCELERANTS. Gasoline, the most common accelerant, is a mixture of different hydrocarbons such as hexane, octane, and benzene. The generic reaction for combustion of a hydrocarbon is given by:

$$\text{Hydrocarbon (g)} + O_2(g) \Rightarrow CO_2(g) + H_2O(g)$$

This assumes complete combustion, which is rarely the case. Often, depending on temperature and the amount of oxygen available, significant amounts of by-products are produced, including carbon monoxide (CO). Interestingly, the metabolism of glucose ($C_6H_{12}O_6$), a common sugar, also produces carbon dioxide and water, but not by a burning reaction per se but rather by a controlled, step-wise metabolic process:

$$C_6H_{12}O_6 + O_2(g) \Rightarrow CO_2(g) + H_2O(g)$$

Wood, which does burn in combustion, contains high concentrations of CELLULOSE, which is a compound made up of many glucose subunits (a glucose polymer).

Combustion can be divided into different categories based on where the reaction occurs. In the case of hydrogen gas (H_2) or hydrocarbons, the combustion is a flame combustion that occurs in the gas phase. Gasoline must evaporate (volatilize) in sufficient quantities for flame combustion to occur. For gasoline, the flash point (the temperature at which a liquid will give off enough vapor to form an ignitable mixture) is -50°F (-45.5°C). Other accelerants with higher flash points require hotter temperatures to drive sufficient evaporation for a flame to exist.

A second type of combustion is glowing or smoldering combustion, such as that seen with a backyard charcoal barbeque grill using charcoal brickets. Charcoal is carbon, and, even in a grill, the temperatures are not high enough to evaporate the carbon into a flame. Instead, the reaction takes place on the surface of the brickets, which get hot and have a glowing red appearance. Depending on the fuel, smoldering fires can "burst into flame" when additional oxidant is provided. This occurs when one blows across smoldering embers or grass, which speeds the reaction. This in turn produces enough heat to volatilize more fuel for a flame to form.

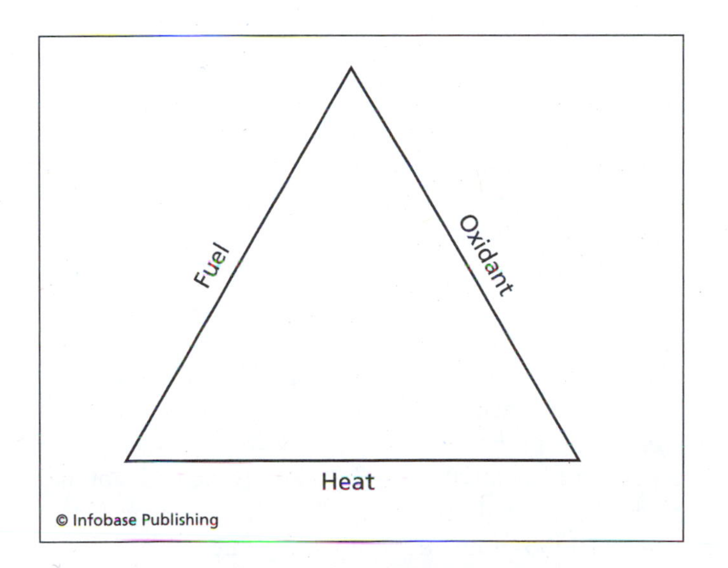

The "fire triangle" showing the need for all three ingredients—a fuel, an oxidant, and heat—for combustion to occur. Removal of any one of these ingredients will result in extinction of the flame (or other combustion process).

commingled remains *See* ANTHROPOLOGY, FORENSIC.

common source Any evidence that can be shown to have come from one unique source or that can be shown to have originally been part of the same structure or unit is said to be evidence with a common source. For example, FINGERPRINTS found at a scene that also match those of a suspect are said to have a unique common source (the suspect), since no two

Classification, Identification, Individualization, and a Common Source

Many forensic analyses, whether of bullets, BLOOD, or FINGERPRINTS, use a similar approach. First, evidence must be identified. Is it a bullet? Is it a BLOODSTAIN? Is it a fingerprint? This is often the easiest part of the analysis, yet the term *identification* can be misunderstood. In a forensic context, to identify a fingerprint is to find an impression, do something to visualize it, such as dust it with powder, and then say, "yes, this is a fingerprint." This does not mean that the fingerprint has been linked to an individual. However, it is common to read accounts in the media where "identification" of a fingerprint means that it has been linked to an individual. In forensic terms, such a linkage is called INDIVIDUALIZATION or finding a COMMON SOURCE. A fingerprint cannot be individualized until a comparison fingerprint is available; identification does not require one.

For most physical evidence, the steps in the analysis follow a similar pattern. After evidence is identified, it is examined for CLASS CHARACTERISTICS, which allow the analyst to assign the evidence to a category. A bullet recovered at a crime scene can be classified by caliber and weight; blood can be classified as human or from another species. Once evidence is assigned to a subgroup (or class), the next task is to attempt to INDIVIDUALIZE the evidence by linking to a COMMON SOURCE through comparison. To individualize the bullet recovered at the crime scene to a specific gun, a suspect gun must be recovered, and the firearms analyst must fire it and recover the test bullets for comparison. A match shows that both bullets were fired from the recovered gun, and the bullet from the crime scene has been individualized. However, if the gun had never been recovered, the firearms examiner could only assign class characteristics to the bullet.(caliber, mass, materials, and so on). While this information is valuable, the same type of bullet could have been fired in hundreds or thousands of different guns, each creating a different pattern of striations on the bullets. The uniqueness of the impressions (the IMPRESSION EVIDENCE) allows for individualization.

The idea of common source and individualization is pivotal in PHYSICAL MATCHING and may involve materials that would seem to be anything but unique. As an example, in a hit-and-run accident a piece of broken GLASS may be transferred to the clothing of the victim and recovered during the investigation. Although glass can be assigned class characteristics such as color and thickness, glass itself is not particularly distinctive. There is no way to examine the glass chemically or microscopically and conclude without question that the source was the car involved. If the fragments of a shattered headlight are reassembled like a puzzle, however, and the glass recovered from the victim fits perfectly into that puzzle, the physical match is proof of a common source. Since the impact and the circumstances that caused the glass to be transferred to the victim were unique, the fitted fragments could not be generated by any other chance event. The physical match individualizes the glass recovered from the victim and assigns it to a common source—the headlight on the suspect vehicle. Physical

individuals have the same fingerprints. When DNA TYPING is performed, blood and body fluid evidence can often be shown to have a common source, assuming that the probability of a chance duplicate is sufficiently small. Common sources are also encountered in TRACE EVIDENCE as a result of PHYSICAL MATCHING. For example, consider a glass window that is broken during a burglary and reassembled using glass shards recovered from the shoes of a suspect. The physical match of the glass fragments into the reassembled whole indicates a common source of the glass, proving it all was once part of a single window. The glass itself is not unique; rather, it is the physical match that indicates a common source.

Identification of a common source is part of the process of INDIVIDUALIZATION in forensic science, which begins with general CLASSIFICATION and ideally leads to the identification of a common source. Consider the analysis of a bloodstain, where classification begins by first confirming that the stain is blood and then determining species (human or not). Further classification can occur if the blood is typed for ABO BLOOD GROUP and ISOENZYMES. Identification of a common source (individualization) requires a blood sample from an individual (the suspect, for example), and DNA TYPING, which usually can show that the blood either came from the suspect or it did not, to a reasonable degree of scientific certainty. In other words, the probability of a duplicate combination of types must be small enough to be considered unique. If the DNA analysis proves that the suspect's blood and the bloodstain are both the same, then a common source has been established and individualization achieved.

Further Reading

Inman, K., and N. Rudin. "Classification, Identification, and Individualization—Inference of Source." In *Principles and* matching is used frequently in firearm, toolmark, and trace evidence analysis.

This example illustrates an interesting phenomenon important in forensic science. Bullets are mass-produced, as are all consumer items. Modern manufacturing techniques are automated, accurate, and reproducible. Consequently, often there is nothing unique about the products, at least at a level that is reasonably detectable. Even so, uniqueness can be acquired through use and unique wear patterns. Two bullets that come off the assembly line at the same time will have the same class characteristics and, for all intents and purposes, they will be identical. However, once those bullets leave the factory, they take divergent paths. Different people in different places will fire the bullets in different guns, and, as a result, the bullets will no longer be indistinguishable.

A more familiar example would be a clothing factory that mass produces jeans. The jeans may share identical class characteristics such as manufacturer, size, and color, but each person who buys a pair will treat them differently. One person might wear them every day to work at a construction site, while another might put them in a drawer and wear them once a month. Another person might wash them twice a week, yet another person, never. One pair might get torn on the knee, another pair along a seam. Thus, the wear pattern for every pair of jeans is different, meaning that at some level, each pair is unique even though they left the factory as nearly identical as modern assembly lines can make them. For the forensic scientist, this uniqueness is invaluable.

The analysis and typing of blood evidence introduces yet another aspect of uniqueness and common source, one that has changed dramatically with the advent of DNA TYPING. Prior to that, the analysis of blood evidence followed the same scheme as most other types of evidence. First, class characteristics were assigned by answering several basic questions: Is this substance blood? Is it human blood? What is the ABO blood type? None of this information was sufficient to individualize a blood sample. It might be enough to exclude someone as the source, but it never was enough to say conclusively who *was* the source to a reasonable degree of scientific certainty. Even with comparison standards, there was no way to individualize blood and body fluid evidence. DNA typing has changed that completely. The catch, as always, is that a comparative sample is needed.

In and of itself, a DNA type means nothing, just as a fingerprint by itself means nothing. What makes the uniqueness valuable is having something to compare it to. A fingerprint found at a crime scene has to match one kept in a file or recovered from a suspect if the common source is to be determined. Similarly, blood at a scene can be linked to an individual only if blood from that individual has been or can be typed and shown to match the blood type already determined. Identifying a common source—individualizing—always requires that a comparison or physical match be made, and made successfully. Thus, in many cases, it is not a forensic scientist's knowledge of any one discipline that is his or her most valuable asset. Rather, it is his or her skill at comparison, individualization, and the ability to recognize and prove a common source.

Practice of Criminalistics: The Profession of Forensic Science. Boca Raton, Fla.: CRC Press, 2001.

comparison microscope A microscope that consists of two COMPOUND MICROSCOPES linked together in such a way that two different objects can be examined side-by-side. The analyst looking through the viewer sees a circular area divided down the middle by a thin line, with the image from the left microscope stage on the left and the image from the right on the right of the viewer. There are two types of comparison microscopes, differentiated by the type of lighting used. For comparing objects that are opaque (BULLETS, CARTRIDGE CASES, and tools, for example), a vertical or reflected light source is needed, whereas for objects that transmit light, such as FIBERS, a transmission light source is employed. Flexible fiber optics can also be used, which allow the examiner to illuminate the objects as needed. In the 1920s, CALVIN GODDARD

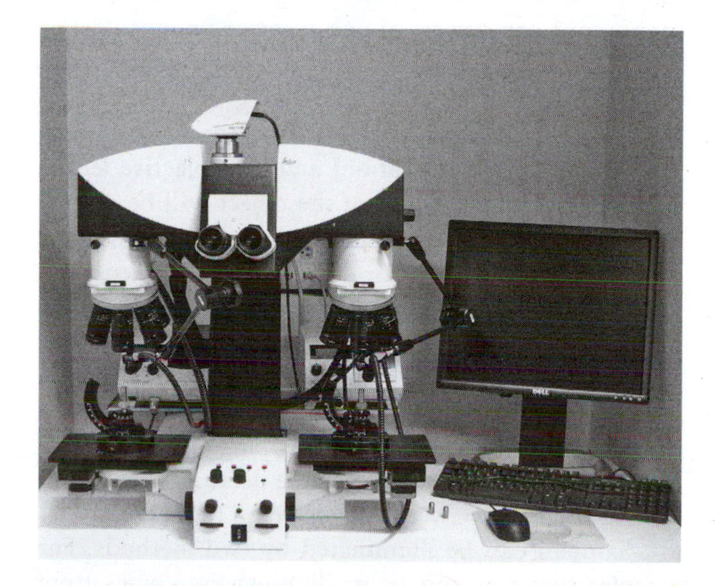

A forensic comparison microscope such as used for firearms, toolmarks, and fibers ***(Courtesy of Michael Bell, West Virginia University, Forensic and Investigative Sciences program)***

pioneered the use of the comparison microscope for the examination and INDIVIDUALIZATION of bullets. Comparison microscopes are also used to examine other FIREARM and TOOLMARK evidence as well as hairs and fibers. Since two different microscopes are used to compare items, it is critical that they be closely matched in every characteristic. This ensures that observations are not influenced by differences in the optics between them.

composite sketch *See* ART, FORENSIC.

compound microscope A compound microscope (also sometimes called a biological microscope) is one of the most versatile tools in forensic MICROSCOPY. It was also one of the first instruments to be routinely used, becoming widespread by the late 1800s. In SEROLOGY, the results of blood group typing are observed using the compound microscope. In trace evidence analysis, it can be used for an initial examination of HAIRS and FIBERS, POLYMERS, and paint chips, to name just a few applications. Forensic chemists also use microscopes to identify certain kinds of CUTTING AGENTS.

A simplified design of a compound microscope is shown in the figure. The magnification is accomplished by two lenses, the objective lens (the one closer to the sample) and by the ocular or eyepiece lens. The objective is often mounted on a platform called the nosepiece with two or more lenses that can be rotated into position above the sample. The view through the microscope can consist of one eyepiece (monocular) or two (binocular). Each lens has a magnification associated with it; total magnification is the product of the two. For example, if the ocular lens is selected as 10× (ten times magnification) and the objective lens is selected as 45×, the image of the object will be magnified by a factor of 450. The strength of magnification is not the sole criterion for determining what lenses to select. As the magnification increases, the size of the lens decreases, leading to a smaller field of view. Thus, if larger areas must be examined, higher magnification is not practical. Likewise, as magnification increases, the depth of focus decreases, meaning that the depth of the object that is clearly visible decreases. Selection of magnification power depends on all these factors.

Samples can be illuminated by two methods. For samples that transmit light, illumination comes from below and is called transmitted light. If the sample is opaque, vertical light (also called incident or reflected light) is used.

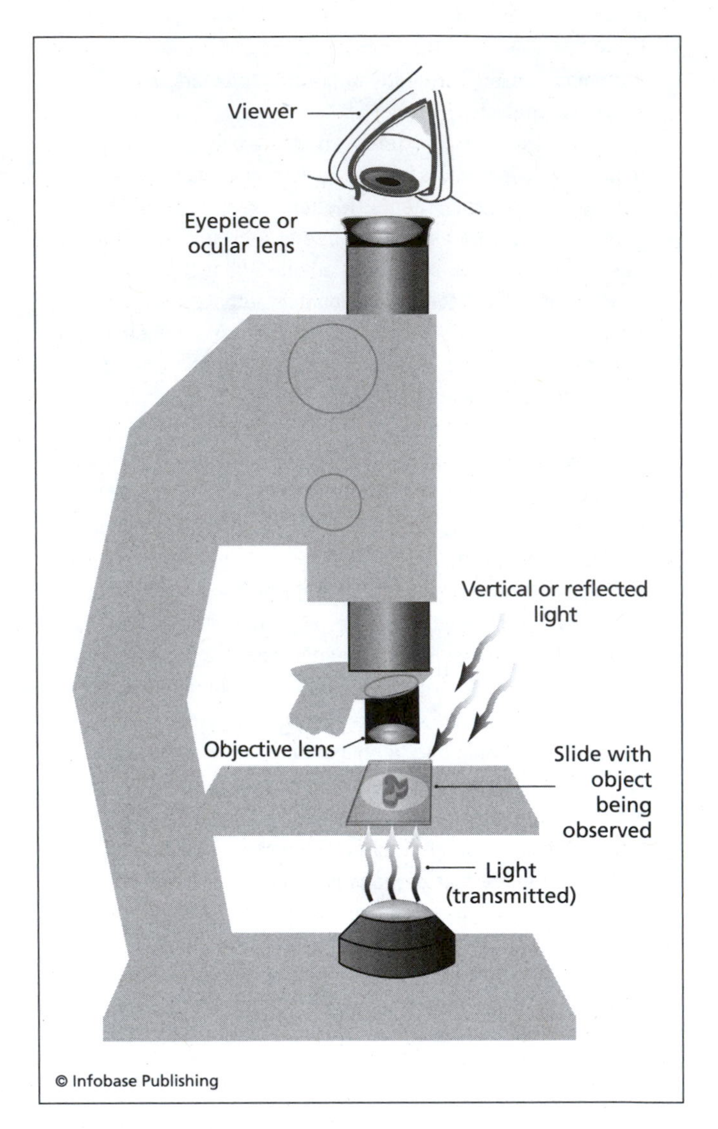

Simplified light path of a basic compound microscope. Here, the light shines from below through a sample via transmitted light that is magnified and focused by the lenses in the body of the microscope.

Further Reading

Nomenclature Staff, Royal Microscopy Society. *RMS Dictionary of Light Microscopy,* Royal Microscopy Society—Microscopy Handbook # 15. Oxford: Oxford University Press, 1989.

computing, forensic (computer forensics, cyberforensics) The application of the techniques of computer science and digital technology to electronic forms of evidence called digital evidence. Fundamentally, all digital evidence can be reduced to exactly that—a stored collection of ones and zeros. Since it is stored and must physically exist in at least one place, digital evidence is still physical evidence, although certainly a unique type. For example, digital evidence can be copied, so there is no one unique source, as there would be for a murder weapon or bloodstain. A printout or copy of a document created on a computer is not an original in the same sense that a typewritten document is. Thus, the foundation of admissibility of electronic evidence is often more complicated than for traditional physical evidence and paper documents. However, digital evidence does have some unique features that are helpful to investigators. Since it can be copied, extensive analysis can be done on the copies without damaging or destroying the original. It is also relatively easy, using specialized tools and techniques, to determine if evidence has been altered or tampered with. Finally, it is very difficult for all traces of digital evidence to be erased or destroyed. For example, e-mail may be erased from a personal computer, but a backup may exist in at least one other place.

Techniques of forensic computing and investigation can be applied to a myriad of evidence including e-mail, video and sound recordings, chat room interactions, and text files. Often computers themselves or portions of them such as disk storage devices are seized and examined. Traffic on networks can be monitored and extracted as evidence. Many types of devices aside from the personal computer can become the subject of investigation, such as larger and mainframe computers, computer networks, and any type of device with computing capability such as electronic organizers, personal digital assistants (PDAs), and sophisticated calculators.

Some of the tasks of forensic computing include recovering data from disks and other storage media that have been erased, formatted, damaged, or tampered with; finding hidden, disguised, or encrypted information; and breaking passwords and other security devices used to hide evidence. When found, data must be safely recovered from disk or other storage device and done so in a manner that does not alter the original evidence. At all stages of recovery and examination, the analyst must ensure and be able to prove that data have not been lost, altered, or tampered with. For traditional physical evidence, this is accomplished by the CHAIN OF CUSTODY, but for digital evidence, additional steps, documentation, and safeguards are often required. The analyst may also follow a "cybertrail" to find sources of evidence such as e-mail or hacker attacks.

Types of investigations that would involve forensic computing include software piracy, e-mail and Internet fraud, Internet and e-mail stalking, computer virus attacks, and any investigation where evidence is stored electronically. Because computers have become integrated into modern society, digital evidence is commonly encountered, even when the crime is not a computer crime per se. For example, a murderer may have sent e-mail to a victim, or a drug dealer may keep records of transactions on a computer spreadsheet. Thus, forensic computing is emerging as one of the most important and rapidly developing forensic disciplines of the 21st century.

Further Reading

Casey, E. *Digital Evidence and Computer Crime: Forensic Science, Computers, and the Internet.* London: Cambridge University Press, 2000.

Stephenson, P. "Investigation of Computer-Related Crimes." In *Forensic Science: An Introduction to Scientific and Investigative Techniques.* 2nd edition. Edited by S. H. James and J. J. Nordby. Boca Raton, Fla.: CRC Press, 2005.

condoms Condoms and the associated lubricants and spermicides are an increasingly important form of evidence in sexual assault cases. Several properties are of interest including the material (POLYMER) of which the condom is made, composition of the lubricant, type of powder (used to keep the rolled condom from sticking), and spermicide used, if any. Vaginal swabs, taken as part of SEXUAL ASSAULT KITS, can be analyzed and if traces of lubricant components are found, can indicate that a condom was used during the assault. A number of techniques have been employed for this purpose including spectrophotometry (infrared [IR] and nuclear magnetic resonance [NMR]) and GAS CHROMATOGRAPHY/MASS SPECTROMETRY (GC/MS). The latex material used to make condoms has also been analyzed with similar instrumental techniques. Semen can often be obtained from used condoms and analyzed by DNA TYPING.

contamination The inadvertent or unintentional transfer of foreign materials or other alteration of evidence that can ruin the evidence or lead to incorrect interpretation of results. Contamination is a serious concern in any forensic analysis and can be introduced at any stage, from collection through the actual analysis. In the case of biological evidence such as blood and body fluids, improper collection procedures, microbial activity, or improper handling can introduce contamination. Materials and chemical reagents used in the analysis can also become contaminated, as can glassware and other equipment. In the analysis of TRACE EVIDENCE, inadvertent transfers of foreign materials such as HAIRS and FIBERS are always a concern. DRUG ANALYSIS and other chemical tests can also be impacted by contamination of samples, glassware, equipment, and instrumentation. At any stage, contaminants can influence the outcome of tests and can lead to FALSE POSITIVE or FALSE NEGATIVE results.

The threat of contamination can be reduced or eliminated by use of proper procedures, careful handling by all who touch the evidence, and the use of QUALITY ASSURANCE/QUALITY CONTROL (QA/QC) throughout the analyses. Examples of QA/QC procedures include the collection and analysis of CONTROL SAMPLES, background samples, and samples of known composition. If these control samples show no contamination, then it is less likely that the actual evidence was contaminated. Each forensic discipline has specialized procedures designed to eliminate contamination or at least detect it if it does occur.

The danger of contamination is particularly high at a CRIME SCENE. Ideally, once the crime has occurred, nothing at the scene should be altered until documentation and evidence collection. This is very difficult and sometimes impossible, but proper steps taken at the scene are critically important to ensuring the integrity of the evidence collected. Securing the scene and limiting access are simple but crucial steps necessary to reduce opportunities for contamination to occur.

control samples **(controls or known samples)** Samples with a known composition, identification, type, or source used for comparison or to ensure that an analysis is being properly performed and that the results are reliable. Controls are part of QUALITY ASSURANCE/QUALITY CONTROL (QA/QC) procedures that forensic labs follow to ensure the reliability of their data and to eliminate incorrect results including FALSE POSITIVES and FALSE NEGATIVES. Controls are also frequently referred to as "knowns" since their composition and identification has already been established. Examples of control samples include known drugs such as cocaine that are analyzed along with suspected cocaine samples, blood of known types or DNA type, and known accelerants for arson cases. There are a number of sources for known materials including commercial vendors, manufacturers (for PAINTS, FIBERS, AMMUNITION, and so on), and other casework, as long as identification is absolute.

Another type of control used in forensic chemistry is a "blank," a sample that does not contain the substance of interest. The idea behind analyzing a blank is to ensure that equipment, instruments, or implements are not already CONTAMINATED with the substance of interest, leading to false positive results. For example, if a syringe is used to prepare and inject a suspected cocaine sample into an instrument for analysis, but the syringe has been contaminated by a previous sample, the results will be positive even if there is no cocaine in the sample being analyzed. If instead a blank is injected, one that is known to be cocaine-free, the positive result will identify the contamination and allow the analyst to correct the problem before any casework is compromised.

Finally, the terms *exemplar* or *comparison sample* are sometimes used to describe a specific kind of known sample. For example, in QUESTIONED DOCUMENTS, the examiner often needs handwriting samples from a person for comparison purposes, and these are called exemplars. Exemplars are also used for other types of comparisons such as in FIBER identification. An exemplar would be a fiber of known origin that was used as the basis of a comparison. In forensic GEOLOGY, a questioned sample would be one recovered from a crime scene or in some other way linked directly to a crime. The controls or comparison samples would be those taken for comparison from sites that could be potential sources for that soil.

controlled substances **(Controlled Substances Act [CSA])** The federal law governing dangerous drugs, analogs, and raw ingredients (PRECURSORS) used in CLANDESTINE LAB synthesis. The CSA was passed as part of the Comprehensive Drug Abuse and Prevention Control Act in 1970. The act classifies drugs into one of five schedules (indicated with a Roman numeral) based on accepted medical use and the danger of physical and psychological dependence. Each schedule also specifies penalties, how the drug can be obtained, and whether a permit is necessary. Precursor chemicals used in clandestine labs are also listed. In 1984, an amendment was added as part of the Comprehensive Crime Control Act that permitted the DRUG ENFORCEMENT ADMINISTRATION (DEA) to temporarily add substances to a schedule without having to wait for a more formal and lengthy procedure to take place. This allows the DEA to address newly discovered precursors and designer drugs synthesized by clandestine laboratories. In the Anti-Drug Abuse Act of 1986, a provision was added to address analog drugs, synthesized by clandestine labs to act similarly to other controlled substances.

Schedule I drugs are those that have a high potential for abuse and no accepted medical applications. HEROIN, LSD, and MARIJUANA are listed on this schedule. Schedule II drugs (MORPHINE, COCAINE, and METHAMPHETAMINE) are similar, but they do have accepted medical uses. Abuse can lead to severe addiction and dependence. Schedule III drugs have less potential for abuse and addiction and include ANABOLIC STEROIDS and some CODEINE and BARBITURATE preparations. Schedules IV and V reflect decreasing risk and increasing legitimate uses, with drugs such as VALIUM on Schedule IV and over-the-counter (OTC) cough medicines with codeine on Schedule V.

Further Reading

Drug Enforcement Agency (DEA) Web site. "Controlled Substances Act." Available online. URL: http://www.dea.gov/pubs/csa.html. Downloaded January 31, 2003.

cordage *See* ROPE AND CORDAGE.

coroner Latin for "crowner," an elected or appointed official who is charged with determining the CAUSE OF DEATH in cases in which the death appears to be the result of foul play, was unattended, or occurred under questionable or suspicious circumstances. The position of coroner was a remnant of Roman law instituted in England during the 12th century. At that time, the coroner had a number of duties that included handling of the assets belonging to felons and suicide victims. The system evolved, and by the 1800s the coroner's only duty was to investigate suspicious deaths using an inquest system. The coroner system was common in the United States until 1877, when Massachusetts became the first state to abolish it in favor of a MEDICAL EXAMINER (ME). New York City followed in 1915, and many jurisdictions have since converted to MEs. However, there is no national standard in the United States, so both systems are found.

The modern coroner is a judicial officer who is elected or appointed but whose tasks are mainly administrative. The job of the coroner is to determine the cause of death using whatever resources are necessary, which can include ordering an autopsy by a pathologist or forensic pathologist. The coroner often does not perform the autopsy, and scientific or medical training is not necessarily required for the position. This represents one of the disadvantages of the coroner system. Given the advances in and complexity of modern forensic science and pathology, even a well-intentioned coroner (elected or appointed) may lack the background or experience to properly administer the office.

Further Reading

Moenssens, A. A., J. E. Starrs, C. E. Henderson, and F. E. Inbau. *Scientific Evidence in Civil and Criminal Cases.* Westbury, N.Y.: The Foundation Press, 1995.

Fisher. R., and M. Platt. "History of Forensic Pathology and Related Laboratory Sciences." In *Spitz and Fisher's Medicolegal Investigation of Death.* 3d ed. Edited by W. Spitz. Springfield, Ill.: Charles C. Thomas Publisher, 1995.

Wright, R. "The Role of the Forensic Pathologist." In *Forensic Science: An Introduction to Scientific and Investigative Techniques.* Edited by S. H. James and J. J. Nordby. Boca Raton, Fla.: CRC Press, 2003.

corpus delicti Literally, "the body of the crime." This term can refer specifically to the thing or person on which a crime was perpetrated, such as a corpse, but more commonly it refers to the "foundation and substance" of a crime. According to *Black's Law Dictionary* (6th ed., 1990, West Publishing Co.), this definition of the corpus delicti is the criminal act along with the criminal agency associated with it. In modern usage, the corpus delicti refers to the entire body of evidence that leads to the conclusion that a crime has been committed. Interestingly, homicide cases have been successfully tried even when no corpse has been found because the "body of evidence" was overwhelming even without it.

corroborative reconstruction *See* CRIME SCENES.

Cotton, Mary Ann (1832–1873) British *Poisoner* One of the most famous post-Marsh test ARSENIC poisonings was that attributed to Mary Ann Cotton. Her victims, at least 20, were her close relatives, including her mother, several children and stepchildren, three spouses, and one lover. Her motive was insurance money and her downfall, excess. She killed her children (whose deaths were initially attributed to gastric distress) when their existence interfered with impending relationships and marriages to obtain the money. The last child she killed was her son Charles. The attending doctor was suspicious and obtained samples from the boy's body. Using the Reinsch test, he found arsenic, leading to Mary's arrest. The trial itself had to await the birth of yet another child, who owed his life to fortunate timing, a timely conviction, and the gallows. The jury rejected Mary's suggestion that arsenic vapors given off from wallpaper killed her son.

See also REINSCH, EGAR HUGO.

counterfeiting *See* QUESTIONED DOCUMENTS.

courtroom procedures Most of the cases forensic scientists work on never go to court. Reasons for this include failure to apprehend a suspect, charges not being filed, or the suspect reaching a plea agreement (plea bargaining) or a STIPULATION agreement, in which the expert's report is accepted as fact and his or her testimony is not required. When cases do go to court, usually the case is tried in criminal rather than civil court. However, forensic scientists in any discipline and particularly those in areas such as forensic engineering may frequently testify in civil matters.

For a typical forensic scientist employed by a local, state, or federal laboratory, the procedures leading to courtroom testimony are summarized in the accompanying figure. This is not universal, and the exact sequence may vary depending on many factors, but it does provide a reasonable overview of the most common sequence. Once the laboratory analysis is completed, a report is written and sent to the officer or agent who submitted the evidence. The report is then shared with prosecuting attorneys and, if charges are filed, with the defense. If a trial is scheduled and the forensic scientist is required to testify, a SUBPOENA is delivered to the scientist that states when and where the trial is to be held. The analyst is required to appear under penalty of law.

After the scientist takes the witness stand and is sworn in, the first task is to establish that he or she is qualified to offer expert testimony to the court. This is accomplished by the VOIR DIRE, where the prosecution introduces the scientist and then has him or her describe his or her qualifications including academic background, training, and experience. The defense can ask questions of the expert or request additional information. Normally, at the end of this procedure, the forensic scientist is accepted as expert by the TRIER OF FACT and allowed to testify about the case being tried.

Next, the prosecution begins the direct examination, which can take a descriptive or question-and-answer format. One of the goals of the direct examination is to lay the foundation for the admissibility of the evidence in question. The defense has the opportunity to attack the admissibility of the evidence, much as they have the opportunity to attack the qualifications of the expert during the voir dire procedure. If the evidence is not ruled admissible, the expert is not allowed to testify about work done on it or conclusions reached.

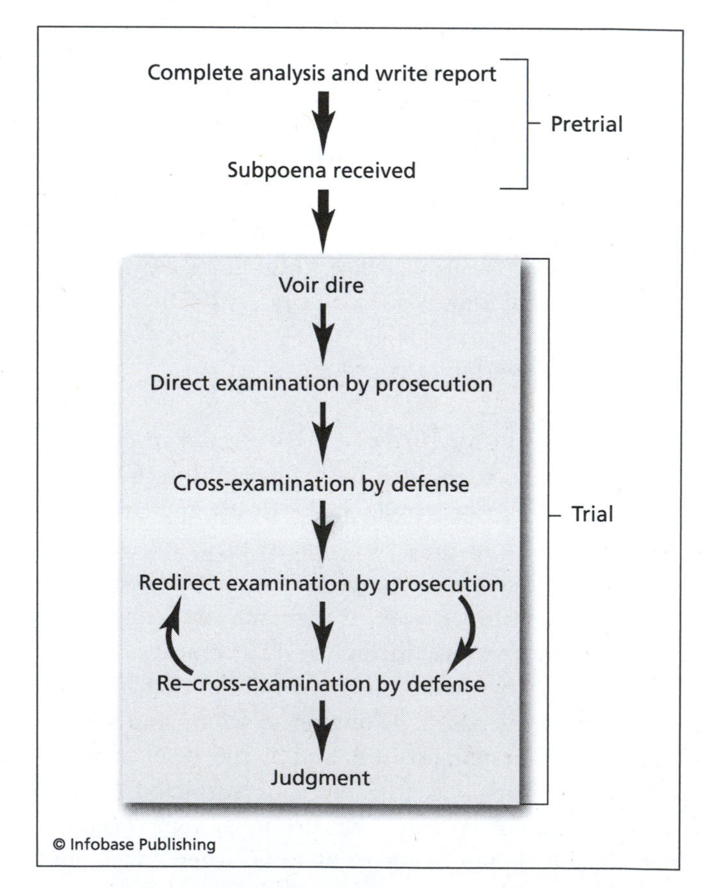

A typical process that a forensic scientist goes through when testifying in court. Several iterations of questioning (re-cross and redirect) are possible.

At the conclusion of the direct examination, the defense has the opportunity for cross-examination of the witness if they so desire. After the cross-examination, the prosecution can again ask questions in the redirect, followed by re-cross. This cycle continues until both parties are satisfied and there are no further questions. The duration of expert testimony varies based on the complexity of the case and the laboratory analysis involved. Many simple drug analysis cases may involve only a few minutes, while more complex cases may require days on the stand. Once the final re-cross is complete, the witness is excused and is either allowed to leave (dismissed) or told to remain available for possible recall.

See also COURT SYSTEMS; EXPERT WITNESS TESTIMONY.

court systems Within the United States, courts are organized based on the jurisdiction, which can be local (city or county), state, or federal. Where a case is tried depends on many factors, including what law enforcement agency or agencies were involved in the investigation. While the federal system is uniform, state and local courts are organized in different ways. In the federal system, the district court acts as the gateway. If a jury is to be the TRIER OF FACT, the panel will consist of 12 people and the verdict must be unanimous. The first level of appeal above the federal district court is found in circuit courts, which consist only of panels of judges that review district court decisions that are appealed based on claims of procedural errors. No new evidence is introduced; the role of this court is only to rule on the appeal. Possible outcomes of this appeal process include upholding of the lower court's ruling, reversal of the earlier decision, or ordering a new trial. The next higher level of appeal in the federal system is the Supreme Court.

State courts are often modeled on the federal system, but there are no uniform standards. County and city courts (often called municipal courts) are similarly diverse.

See also COURT PROCEDURES; EXPERT WITNESS TESTIMONY.

creatine and creatinine Creatinine is a natural component of urine, present as a result of muscle action. When forensic toxicologists test urine for drugs, the creatinine level may be measured to determine if the subject has consumed large quantities of water in an attempt to dilute drug residues and metabolites to undetectable levels. Thus, an abnormally low creatinine level in a urine sample suggests that the subject may be attempting to hide drug use by excessive fluid intake. Recently, this dilution test has been complicated by the use of creatine, a legal supplement used by athletes and body builders. Creatine breaks down to creatinine, resulting in elevated levels in the urine. As a result, there is concern that taking creatine supplements can hide urine dilution. This problem stretches beyond athletic drug testing to workplace and clinical settings, where urine screening is routinely performed.

See also essay, "*Drug Testing in Sports,*" p. 116.

cremains The material left over when a body is burned or incinerated. Cases such as misplaced remains, commingling of remains, cremation to hide evidence of a crime, and illegal or improper disposal of cremains are becoming more common as the practice of cremation gains popularity. Forensic anthropologists are typically the specialists called upon to analyze the cremains, which consist of burned bone, ash, and other noncombustible materials. Cases are particularly challenging given that commercial cremation often involves a pulverization step that further destroys the material remaining after incineration. In a typical commercial cremation, the body is placed in some type of container (ranging from cardboard box to coffin) that is then incinerated at temperatures of around 1500°F (815.5°C). Magnets are used to remove ferrous metals, followed by the pulverization step, if used. Further complicating such cases is the variability in procedures depending on the funeral home or crematorium performing the work. If positive identification is needed, forensic anthropologists examine bone fragments that remain as well as materials not destroyed in combustion. Often, sufficient bone fragments remain to allow for identification of sex as well as estimations of age and stature. In addition, metallic dental work and components of pacemakers survive the process and can be used to assist in identifications.

Further Reading

Murray, K. A., and J. C. Rose. "The Analysis of Cremains: A Case Study Involving the Inappropriate Disposal of Mortuary Remains." *Journal of Forensic Sciences* 38, no. 1 (1993): 98.

Warren, M. W., and W. Maples. "The Anthropometry of Contemporary Commercial Cremation." *Journal of Forensic Sciences* 42, no. 3 (1997): 417.

crime According to *Black's Law Dictionary* (6th ed., 1990, West Publishing Co.), a crime is defined simply as "a positive or negative act in violation of penal law; an offense against the state or the United States." A crime is therefore a violation of a code or written law that was created to protect society. Crimes are first divided based on whether they are classified as felonies or misdemeanors. There are many types of crimes including crimes against property (burglary, for example), crime of omission (failing to stop and aid victims of an accident), crimes of passion, crimes of violence, and white-collar crime such as fraud, bribery, and software piracy.

crime labs (forensic labs) The first crime laboratory was established in 1910 by EDMOND LOCARD in Lyons, France. Although poorly equipped with little more than a microscope, the lab was successful, and Locard went on to become one of the key figures in the development of forensic science. More labs sprang up in Europe following World War I, and since many of the early pioneers had medical backgrounds, forensic science was often considered to be a branch of medicine. The first crime lab in the United States was established by the Los Angeles Police Department in 1923, headed by AUGUST VOLLMER. In 1932, the FBI organized a forensic laboratory that has evolved to become the largest in the world. Labs continue to grow and have become an integral part of law enforcement.

All labs are different and have different sections depending on jurisdiction, size, and many other factors. A typical full-service laboratory will have sections for CHEMICAL ANALYSIS (drugs and ARSON primarily), TOXICOLOGY, BIOLOGY/SEROLOGY, trace evidence, FIREARMS and TOOLMARKS, QUESTIONED DOCUMENTS, and FINGERPRINTS. Other services that may be provided include polygraphs, voiceprints, computers and digital evidence, photography and image analysis, and crime scene processing/evidence collection. Some forensic labs are associated with the MEDICAL EXAMINER'S (ME) office, while others serve state patrols, county sheriff's departments, city police departments, or federal agencies. There are also a number of private labs in operation. In the United States there is no national standard or organization that oversees forensic laboratories, although accreditation and certification programs have been established through the AMERICAN SOCIETY OF CRIME LABORATORY DIRECTORS (ASCLD) and the AMERICAN BOARD OF CRIMINALISTICS (ABC).

Further Reading

Netzel, L. "The Forensic Laboratory." In *Forensic Science: An Introduction to Scientific and Investigative Techniques.* 2nd edition. Edited by S. H. James and J. J. Nordby. Boca Raton, Fla.: CRC Press, 2005.

crime scenes, crime scene investigation (CSI), and crime scene reconstruction The scene of a crime is the starting point of physical evidence, and the care with which a scene is processed is crucial to the success of any investigation based on it. Unfortunately, crime scenes deteriorate rapidly, so a primary goal of the crime scene investigator is to preserve the scene in both time and space, allowing for it to be processed in as nearly pristine a condition as possible. This is extraordinarily difficult, especially in high-profile cases that attract a great deal of attention. Investigators also face the possibility of more than one scene for the same offense. For example, a murder may be committed in one place (the primary scene) and the body dumped somewhere else (the secondary scene). Crime scenes are easily contaminated by unauthorized and unnecessary traffic, and once lost, a scene can never be recovered.

At a scene, the investigator—a police officer, forensic scientist, or evidence collection specialist (also called crime scene technician or crime scene investigator)—has many goals and tasks. First and foremost, he or she must secure, protect, and preserve the scene, which often means limiting access and keeping a careful record of who is at the scene and how they interact with it. If injured people are present, they must be attended to, but all possible precautions have to be taken to ensure the integrity of evidence that may be disturbed during this process. If there is a body, activities must be coordinated with the MEDICAL EXAMINER OR CORONER who will ultimately receive the body and perform the AUTOPSY. Other specialists are called depending on the type of case and evidence and can include identification and FINGERPRINT specialists, CRIMINALISTS, and, in some cases, the DISTRICT ATTORNEY. The processing of major crime scenes is a team effort, with a designated crime scene supervisor or coordinator overseeing all activities.

Crime scenes are processed with several goals in mind. The first is to determine the extent of the crime and to identify the elements and evidence associated with it (the CORPUS DELECTI). The scene will be

Myths of Forensic Science

A number of misconceptions and misunderstandings hover around forensic science, many arising from treatments in novels, movies, and TV shows. Many of these inaccuracies have been repeated so often that the public accepts them as fact. This list uncovers some of the more prevalent forensic science myths and explains why they are fiction, not fact:

1. How many scenes have you seen or read in which an undercover police officer identifies a strange powdery substance simply by putting a tiny pinch on his or her tongue? There are several problems with this scenario. First, taste is not a specific test for any drug. Cocaine is in the same chemical family as Novocain, an anesthetic used by dentists and doctors. True, touching cocaine to one's tongue will produce a numbing effect, but so will any number of related compounds, including Novocain, lidocaine, and benzocaine. In fact, illicit cocaine is often diluted with lidocaine or procaine to foil the taste test. Similarly, heroin and other drugs taste bitter, and a number of substances such as quinine can mimic that sensation on the tongue. Finally, beyond these scientific problems is the obvious danger that a taste test invites. The unknown white powder could be lye or cyanide as easily as it could be an illicit drug.
2. Another fixture of forensic fiction is the ubiquitous chalk line around a body, a practice rarely used in modern crime scene investigations. For one thing, the act of marking around a body disturbs the scene and can destroy critical evidence. Second, thorough documentation of a crime scene by drawings, photography, and video should clearly and unambiguously show where the body was. Third, there are many surfaces on which chalk will not write, such as carpet and dirt. Thus, while the technique is sometimes used, it is not used with the frequency most people would expect.
3. Guns are often found at fictional crime scenes, and the audience sees or the reader reads how the detective carefully picks up the weapon by inserting a pencil into the barrel and lifting gently. He or she does so to prevent damaging or obliterating any fingerprints. The motivation is good but the practice is a terrible mistake. The barrel may contain trace evidence such as hair or fibers that can be dislodged and lost. The condition of the barrel (rust, powder residues, etc.) provides important information to the firearms expert, all of which would be irrevocably altered. A pencil can also mark or damage the barrel itself and affect any subsequent test firings. Thus, the gun should be picked up carefully by the trigger guard.
4. It used to be fingerprints, now it's DNA—the solution to every crime. Find a fingerprint at a scene, and the case is closed. Of course, the reality is quite different. Finding a latent fingerprint at a scene is one step of a process that requires careful documentation, examination, and an element of luck. The fingerprint database maintained by the FBI contains millions of fingerprints, but not every fingerprint of every person in the world. Millions more people have never been fingerprinted, and if one of these millions is the source of the fingerprint found at the scene, no matches will be found. Second, prints on file are complete 10-finger collections taken under controlled conditions. A latent print at a scene is usually a partial of one or a few fingertips, often found in a place or media that makes it very difficult to document and recover for later study. Partial prints may lack enough features for a fingerprint examiner to work with. Thus, a latent print can be valuable evidence, but finding it alone is no guarantee that it will be.

   The same is true of DNA evidence recovered from blood, body fluid, hair, or other evidence. A DNA type alone is not helpful; there must be something to compare that type to, as well as reliable statistics to support the results of the comparison. As in fingerprint evidence, to establish a COMMON SOURCE for a DNA type a comparison sample or evidence is required.
5. When a villain wants to incapacitate an unfortunate victim, chloroform on a rag held over the mouth is a frequent tool of choice. The victim struggles for a second or two and then loses consciousness. Chloroform was one of the earliest anesthetics used, but the effect of inhalation is not instantaneous. In fact, initially the victim would struggle and possibly convulse. The rag would likely slip, further delaying incapacitation. Once the victim finally slips into a limp pile, he or she is also very close to death, which defeats the purpose of using chloroform in the first place.
6. When a murder occurs, a fictional detective needs to know the time of death (the POSTMORTEM INTERVAL [PMI]). It is not an easy task. Short of a stopped watch, an exact determination is impossible. Estimates can be made, and the best ones integrate several pieces of information including body temperature and cooling (ALGOR MORTIS), rigidity (RIGOR MORTIS), LIVIDITY, stomach contents, and presence of insects (ENTOMOLOGY). The stage of the DECOMPOSITION varies based on numerous factors, and once the body is skeletonized or nearly so, PMI estimates are difficult or impossible.
7. "We'll send it to ballistics," or "We just got the ballistics report," or similar lines are common in forensic fiction,

(continues)

Myths of Forensic Science *(continued)*

but the reference to "ballistics" is technically incorrect. Ballistics is the study of projectiles in motion and covers things such as how to aim an artillery piece so the shell travels the path needed to reach the target. The correct lines of dialogue would be "We'll send it to firearms," or "We just got the firearms report." Even within the profession, the term ballistics is still often used synonymously with firearm analysis, but it is not a preferred usage.

8. A classic objection to forensic evidence is that it is "purely circumstantial," implying that as such, it has limited value. Most scientific analyses produce circumstantial evidence, and a successful case is typically based on the weight of much circumstantial evidence rather than on the merit of one or two pieces. Eyewitness testimony or direct evidence is invaluable, but human memory, especially of stressful events, is not infallible. In the worst cases, witnesses lie. Circumstantial evidence may be the only thing that can point to the truth. For example, finding a bloodstain at a crime scene that matches the suspect's type is circumstantial. Such evidence indicates only that the stain likely came from the suspect, not that he or she is guilty. If the crime scene is also the suspect's home, finding their blood there is hardly noteworthy and could be perfectly innocent. Only other evidence and testimony can shed further light on the meaning of the finding.
9. Another common misconception and misnomer in fiction is the designation of all drugs as narcotics. Some of this confusion may arise from the designation of police officers involved in drug investigations as being involved in "narcotics" or being designated as "narcs." Strictly speaking, narcotics are a class of drugs that relieve pain and induce sleep. HEROIN is a narcotic, but COCAINE is not. The confusion is not limited to fiction and has produced some legal problems in terms of how a drug is defined.
10. Many people think that trace and transfer evidence is useless unless it can be INDIVIDUALIZED and linked to a COMMON SOURCE. While it is very rare that a hair (except with DNA), fiber, or speck of dust can unambiguously be linked to one source, it does not mean the evidence is not useful. If a hair or fiber is inconsistent with a comparison sample, that can be invaluable information and may provide EXCLUSION EVIDENCE. If such samples are found to be consistent, that too is information that will be integrated into the investigation. Furthermore, class information such as type of fiber and color identify CLASS CHARACTERISTICS that are often valuable in narrowing an investigation.
11. "It's a precise match!" the detective shouts, "We have identified the perpetrator with a high degree of precision." However, there is confusion about the terms *precision* and *accuracy* in a forensic context. The accuracy of a scientific analysis is defined as how close the result obtained is to the correct or true result. The term *precision* applies only when replicate measures or comparisons are made. High precision means that replicate tests obtained the same or similar results, which can be expressed using simple statistics. To illustrate, consider an example: a white powder containing 50.0 percent cocaine is submitted to three laboratories for testing. The first laboratory obtains a result of 50.1 percent, a reasonably accurate result; the second lab obtains 49.9 percent, also good, while the third obtains 50.0 percent, exactly correct. All three labs have obtained accurate results, and since all three closely agree, the results are precise (reproducible).

    The concepts of accuracy and precision become blurred when the analysis moves away from chemical and biochemical tests to something such as microscopic examination of a hair or fiber. Such observational evidence can be studied, but the concepts of accuracy and precision are not directly applicable. If several analysts come to the same general conclusions about a given hair, the results are reproducible and precise, but defining accuracy when dealing with class evidence is difficult.

documented and evidence collected with the goal of being able to reconstruct the crime for later analysis and study. Evidence will also be sought that can be used to link a person or persons to the scene at the time the crime was committed, primarily through the analysis of TRANSFER (TRACE) EVIDENCE. Although each crime scene is different, the general steps followed for processing are the same.

After the scene is secured, the first step is a detailed documentation using different combinations of notes, photographs (and in many cases, video), and sketches. If the scene is outdoors, notes are made of weather and other environmental conditions. Careful measurements are taken to locate each piece of evidence, documented in photos, notes, and sketches. In cases where sketches are made, the one done in the field is referred to as the rough sketch, which will often be later redrawn to a final or finished sketch. Forensic artists may be called upon to create a finished sketch or model for courtroom use. Careful and complete

12. A plot device frequently used to recover INDENTED WRITING is to have a character rub a pencil across a piece of paper that was underneath the original writing. In real QUESTIONED DOCUMENTS examination, this is never done for the simple reason that rubbing the pencil across the page destroys the evidence. Indented writing can be visualized a number of ways, such as with oblique lighting, that are NONDESTRUCTIVE, and in forensic science, nondestructive tests are always preferred when feasible.
13. Criminal PROFILING is becoming a popular element of detective and mystery fiction, much of it portraying the profiler as a mystic, psychic, or otherwise mysteriously gifted person who "sees" crimes through dreams and visions. While such depictions are colorful and entertaining, most criminal profiling is founded on psychology, statistics, known patterns, and analysis based on databases of previously collected data. Successfully profiling involves no magic, just much hard work, study, experience, and an element of luck.
14. Many complex trials have evolved into highly publicized "battles of the experts," in which opposing forensic analysts take the stand and present dissimilar interpretations of the same evidence. Often the public will assume that one or the other is either lying or is incompetent. Worse, they might assume that one expert has been bought off and will report a result most favorable to their side. While all scenarios are possible, honest and competent examiners can look at the same data and come to different conclusions. The more complex the case and the more subjective the interpretations, the more diverse the opinions may be. However, divergence does not automatically signal deceit; honest disagreements are quite possible.
15. Both the courts and science exist to seek the truth, but the processes they use are not interchangeable. In science, hypothesis and the scientific method are used to determine the truth and reliability of a scientific law, method, or result. In a court of law, adversarial testing is used to determine the validity of a scientific method. These two methods—scientific and adversarial—are not the same. For example, a new test for the analysis of blood may be accepted by the scientific community but rejected by a court; similarly, a court may accept "science" that is dismissed by the majority of the scientific community. Just because a court, through the process of an adversarial hearing, has accepted a technique does not mean that it is scientifically defensible. A current example of this is the increasing debate about the admissibility of fingerprint and other impression evidence (See essay "The History of Fingerprints: Stay Tuned.") Although fingerprints have been accepted as scientific evidence for decades, challenges are being raised to the scientific basis underlying fingerprints. Thus, the overlap between science and the law is not always as seamless as might be expected.

Finally, a few quick myth busters:

- Human beings do not spontaneously combust, but a lit cigarette will burn human fat.
- Worms rarely devour a buried corpse; it is doubtful that they ever "play pinochle on anyone's snout."
- It's not always possible to determine what a last meal was based on stomach contents.
- A dead person's retinas do not retain the last image that they saw.
- Hair and nails do not continue to grow after death. The skin shrinks and makes it appear so.
- Automatic database searches (fingerprints and firearms) do not produce a perfect match. The searches provide a list of possible matches (if any), and the analysts make the final determination.
- A CORONER is not the same thing as a MEDICAL EXAMINER (ME).
- A POLYGRAPH (lie detector) is not infallible (See "The Future of Lie Detection" essay.)

documentation is essential for successful reconstructions and courtroom use. Various computer programs are also available to create such presentations.

After documentation (or in concert with it) comes collection of evidence, which is often a more difficult task than it might at first seem. It is essential that enough evidence be collected to allow for a complete reconstruction, but it is equally imperative that no extraneous or redundant evidence be submitted, to make the best use of laboratory time. It is also critical for the crime scene investigator to recognize evidence when he or she sees it, and not to limit the search and collection to too small an area. Particularly in large outdoor scenes, a systematic search pattern is essential to ensure that all relevant evidence is located, documented, and collected. Several common search patterns are shown in the figure on the next page, and the one selected will depend on the unique characteristics of the scene. For a single person working outdoors, the outward spiral, either clockwise or counterclockwise, is a good choice, whereas a single person working indoors may elect to use a point-to-point method. In this approach, the searcher moves between key locations within a room. Large outdoor scenes are best searched with several people using a strip or modified strip pattern. Searchers stand at arms-length from each other and walk in tandem, thoroughly and efficiently covering a large area.

Evidence collection includes the initiation of a CHAIN OF CUSTODY for all items collected, with the packaging and marking of each piece depending on type of evidence, condition, and so on. Each piece of evidence is packaged individually to protect it and to prevent cross-contamination of other evidence. Control samples are also collected as needed and appropriate. For example, if a body is recovered from a house where there is a pet such as a cat, hair from the cat will be needed. Similarly, fingerprints are needed from all persons at the scene, and these are called ELIMINATION PRINTS. Depending on the case, buccal swabs (from inside the cheek, used to obtain DNA type) and hair samples may be required as well. These control or elimination samples prevent false conclusions and direct the analysts' effort toward the most promising physical evidence. In crimes of violence, BLOODSTAIN PATTERNS are often critical, and they can be the key element in reconstructions. When possible, the entire spattered item will be transported, but when that is not possible, thorough documentation and detailed photography is essential. In addition to standard crime scene photography (distant, medium, and close), blood spatter patterns need to be photographed with a yardstick, scale, and individual stain identifiers at 90° to the stain. Ultraclose-ups of each individual stain are also needed.

Other types of evidence collected can include fingernail scrapings from suspect and victim, as this is

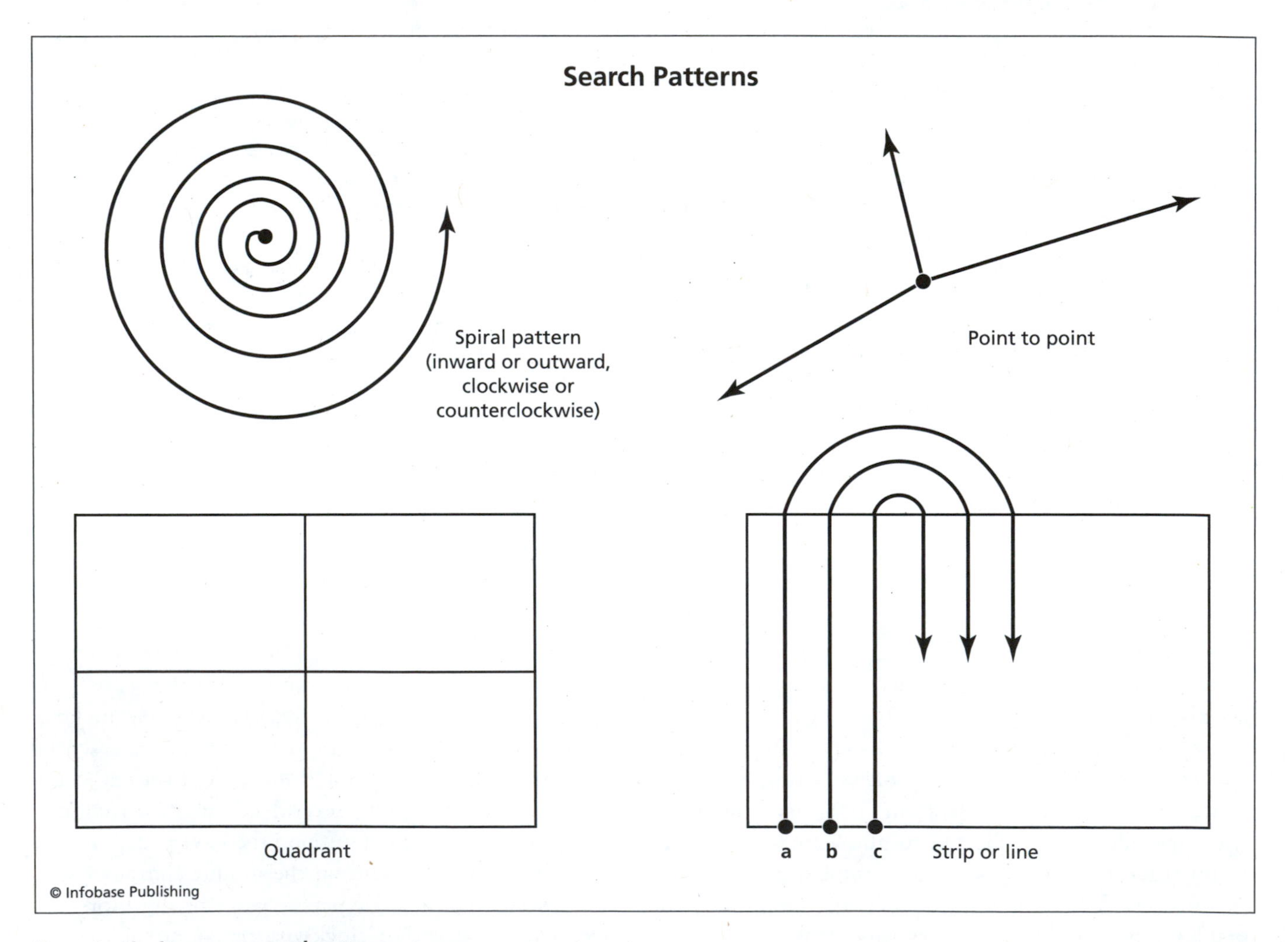

Example of crime scene search patterns

often a source of TRANSFER EVIDENCE. Once documentation and collection are complete, evidence can then be packaged and transported to the laboratory. The final stage of crime scene investigation is the reconstruction based on scene and bloodstain pattern examinations, all reports and interviews, autopsy findings, and completed laboratory analyses. Reconstructions can be used to test scenarios, generate new ideas and hypotheses, and evaluate conflicting stories of events that occurred. Some reconstructions involve physical models, whereas others can be performed using computer modeling techniques, such as in the case of blood-spatter pattern analysis. Fundamentally, the goodness and reliability of any reconstruction depends on how well the scene was processed and documented.

Further Reading

Fischer, B. A. J. *Techniques of Crime Scene Investigation.* 6th ed. Boca Raton, Fla.: CRC Press, 2000.

Miller, M. T. "Crime Scene Investigation." In *Forensic Science: An Introduction to Scientific and Investigative Techniques.* Edited by S. H. James and J. J. Nordby. Boca Raton, Fla.: CRC Press, 2003.

criminalistics Within the profession of FORENSIC SCIENCE, there is debate as to how the terms *criminalistics* and *forensic science* differ, if at all. Fundamentally, a criminalist applies the laws of physical sciences (such as chemistry, physics, and geology) and natural sciences (biology) to the analysis of physical evidence that may be used in criminal or civil proceedings. Some definitions of criminalistics consider the biological sciences (used in forensic SEROLOGY and DNA TYPING) to be a separate part of forensic science, whereas others include all forensic disciplines under the umbrella of criminalistics.

While there is no widespread agreement on the definition, what is certain is that the term *criminalistics* was derived from *kriminalistik,* a word used by Austrian HANS GROSS, an early pioneer of the field. Additionally, the terms *criminalistics* and CRIMINOLOGY are often (and incorrectly) used interchangeably. In contrast to criminalistics, criminology involves the application of social science (psychology and sociology) to the study of crime and criminals. Although the two terms are related, they are not the same. Until recently, criminalists tended to be generalists, performing many different forensic analyses. For example, a criminalist might analyze drugs, type blood, and process crime scenes based on training and experience across several areas within criminalistics/forensic science. Increasingly, however, especially in areas such as DNA TYPING, specialization is becoming more common. Finally, there has been debate as to whether forensic science should be considered a separate profession distinct from chemistry, biology, and so on. In recent years, as forensic science has taken an increasingly important and visible role in criminal justice, the consensus is that forensic science/criminalistics, regardless of the specific definition used, is indeed a distinct profession.

Further Reading

Inman, K., and N. Rudin. "Introduction." In *Principles and Practice of Criminalistics: The Profession of Forensic Science.* Boca Raton, Fla.: CRC Press, 2001.

Criminalistics Certification Study Committee (CCSC) A committee of forensic science professionals from the United States and Canada that was formed as a result of a grant from the National Institute of Justice (NIJ). The CCSC met from 1975 to 1979 and discussed the issue of certification of criminalists. Although their work did not result in a certification program, it formed the basis for certification programs that followed in the late 1980s (California Association of Criminalists, CAC) and the early 1990s (AMERICAN BOARD OF CRIMINALISTS [ABC]).

criminal law v. civil law In criminal cases, the purported incident or action is a violation of a written penal code, the goal of which is the protection of society. On the other hand, civil law relates to disputes between individuals or parties rather than between the state and individuals or parties. In some states, separate courts handle the two types of cases, but usually one court hears both types.

criminology A social science based on the study of criminals, crime, and the penal system. This term is often confused with CRIMINALISTICS, which involves the application of physical sciences (chemistry, geology, physics, and so on) and biological sciences to the analysis of physical evidence. The two terms are related but not interchangeable.

crystal tests (microcrystal, microcrystalline tests) A microchemical test in which a tiny amount of the sample is placed on a microscope slide and reagents are

Microcrystals formed from cocaine and gold chloride ***(Courtesy of Rebecca Hanes, Bennett Department of Chemistry, West Virginia University)***

added. The slide is then viewed under a microscope, with a positive result being formation of distinctive microcrystals. The crystal test is a PRESUMPTIVE TEST (and in some cases a specific test) used as part of DRUG ANALYSIS. There are also crystal tests used to confirm the presence of blood through reactions with hemoglobin as well as tests for EXPLOSIVES. The most common reagents for drugs are gold chloride, platinum chloride, platinum iodide, sodium acetate, and mercuric iodide. In drug analysis, crystal tests are supplemented with other confirmatory techniques such as GAS CHROMATOGRAPHY/MASS SPECTROMETRY (GC/MS) and INFRARED ANALYSIS (IR). The ease of use and specificity of these instrumental techniques have led to a decreasing reliance on crystal tests. For blood identification, the Takayama (HEMOCHROMOGEN), Teichman (HEMATIN), and Wagenaar (ACETONE-CHLOR-HEMIN) tests have been used, but they are also increasingly rare.

Further Reading

Shaler, R. "Modern Forensic Biology," in *Forensic Science Handbook*. Vol. 1. 2d ed. Edited by R. Saferstein. Upper Saddle River, N.J.: Prentice Hall, 2002.

Spalding, R. P. "Identification and Characterization of Blood and Bloodstains" In *Forensic Science: An Introduction to Scientific and Investigative Techniques*. 2nd edition. Edited by S. H. James and J. J. Nordby. Boca Raton, Fla.: CRC Press, 2005.

CSI and CSI effect *Crime Scene Investigation* (*CSI*) is a televison show on CBS that was first broadcast in 2000. It has generated two spin-offs, *CSI: Miami* and *CSI: New York*. The original and *CSI: Miami* are consistently among the top 20 shows on network televison, reaching audiences in the millions. The shows' popularity has led to large enrollments in FORENSIC SCIENCE courses and programs nationwide and has led to increased public awareness of forensic science. One result is what is called the *CSI effect,* a term that refers to the perceived effect of the shows on jurors.

To date, evidence of the CSI effect is indirect and anecdotal. Lawyers and judges are reporting that jurors more often ask about or expect scientific testing even in cases where it is not necessary or appropriate. One study reported that a juror in a case involving complicated DNA evidence stated that it was never that difficult on CSI. The acquittal of Robert Blake, the actor accused of murdering his wife, has been attributed to the CSI effect. The prosecution produced witnesses and testimonial evidence but nothing like scientific tests that were positive for GUNSHOT RESIDUE. Still, there are many in the legal community that feel the CSI effect is not real or not as serious as the stories indicate. Several scientific studies are planned to determine if the effect is real, and if so, how it alters the way juries weigh evidence and decide cases.

See also EDUCATION.

Further Reading

Houck, M. "CSI: Reality." *Scientific American* 295 (1) 2006: 84–89.

Culliford, Bryan (1929–1997) British *Forensic Serologist* Culliford is best known for his book entitled *The Examination and Typing of Bloodstains in the Crime Laboratory,* which was published in 1971 while he was with the Metropolitan Police Laboratory in London. The manual became a primary reference in forensic serology worldwide and brought together information on blood group typing and typing ISOENZYME SYSTEMS. Until the relatively recent ascension of DNA TYPING, typing of blood groups and of isoenzyme systems were the primary tools available for classifying blood and identifying possible sources of stains.

cutting agents Materials used to dilute drugs such as COCAINE and HEROIN. Cutting agents, also known as diluents or adulterants, range from CAFFEINE to

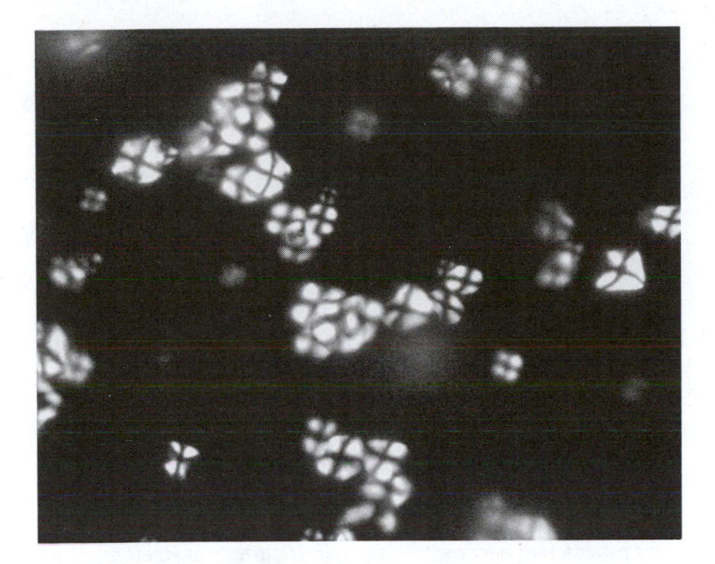

The appearance of starch granules as seen under polarized light conditions ***(photo by Suzanne Bell)***

flour, starch, and sugars such as table sugar (sucrose), dextrose, mannitol, inositol, and fructose. Many can be identified microscopically, while others such as caffeine require chemical analysis for definitive identification. Characterization of cutting agents can be important in linking drug seizures and in recognition of patterns and potential sources. By diluting drugs, the supply can be stretched, as can profits. Cutting agents are often selected because their color, smell, or taste is similar to that of the drug they are mixed with. For example, heroin is sometimes diluted with quinine, which has a similar bitter taste. Cocaine is often cut with lidocaine, which produces the same numbness as cocaine. Caffeine is also a common cutting agent because it is a white powder and because of its stimulant effect.

cyanide A deadly poison that can kill by ingestion, inhalation, or absorption through the skin. The cyanide ion consists of a carbon atom and a nitrogen atom bonded by a strong triple bond and symbolized as CN-. The most common forms of cyanide encountered in forensic science and TOXICOLOGY are the powders sodium cyanide (NaCN) and potassium cyanide (KCN), which can react in the presence of carbon dioxide or acids to form the acid HCN, also known as hydrogen cyanide, hydrocyanic acid, or prussic acid. HCN can be produced by the combustion of materials containing nitrogen, and it is not uncommon to find elevated cyanide (and CARBON MONOXIDE) in the blood of people who have died in home, car, or aircraft fires. HCN has a distinctive odor described as bitter almonds, and as little as two one-thousandths of an ounce (50–100 mg) can be fatal. For the powdered forms, the lethal dose is about twice as high.

Since it has about the same density as air, HCN spreads quickly, permeating any room in which it is produced or released. For this reason, HCN gas was used in gas chamber executions. In Chicago in 1982, powdered cyanide was added to several bottles of Tylenol, resulting in the death of seven people and classification of PRODUCT TAMPERING as a federal offense. Product packaging was redesigned to include seals as a way to make tampering much more difficult. Although cyanide can be and has been used in homicides, as a POISON it has the undesirable characteristic of causing rapid death, something many poisoners wish to avoid. This contrasts with a poison such as ARSENIC, which is cumulative and causes symptoms that can mimic natural disease processes. Spies and criminals have also used cyanide as a means to elude capture or interrogation. Although cyanide is a fast-acting poison, death from ingestion of the powdered forms is not always instantaneous, as is often depicted in the media.

Toxicologists routinely screen samples for cyanide using SPECTROPHOTOMETRY and wet chemical methods. GAS CHROMATOGRAPHY (GC) and electrochemical methods are also used. Minute amounts of cyanide are normally found in the blood as the result of vitamin B_{12} (cyanocobalamin) or from smoking, but these levels are easily distinguished from instances of fatal poisoning, whether homicidal, suicidal, or accidental.

Further Reading

Kunsman, G., and B. Levine. "Carbon Monoxide/Cyanide." In *Principles of Forensic Toxicology.* 2nd edition. Edited by B. Levine. Washington, D.C.: American Association of Clinical Chemistry, 2003.

cyanoacrylate Also known as Super Glue, a chemical (specifically cyanoacrylate ester) used to visualize FINGERPRINTS on a variety of surfaces including plastic, wood, leather, and human skin. First introduced in the early 1980s, cyanoacrylate fuming is easy to perform and has become one of the most common methods of developing latent fingerprints. In a lab setting, the fuming is conducted in an airtight container such as a glass tank and requires the addition of sodium

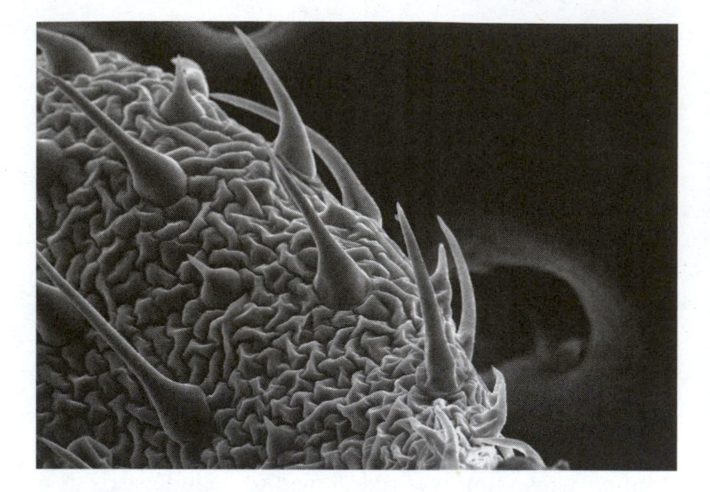

The stem of a marijuana plant as seen using scanning electron microscopy. The crystolithic hairs are the claw-shaped objects. ***(Courtesy of Heather Schafstall, Oklahoma Bureau of Investigation)***

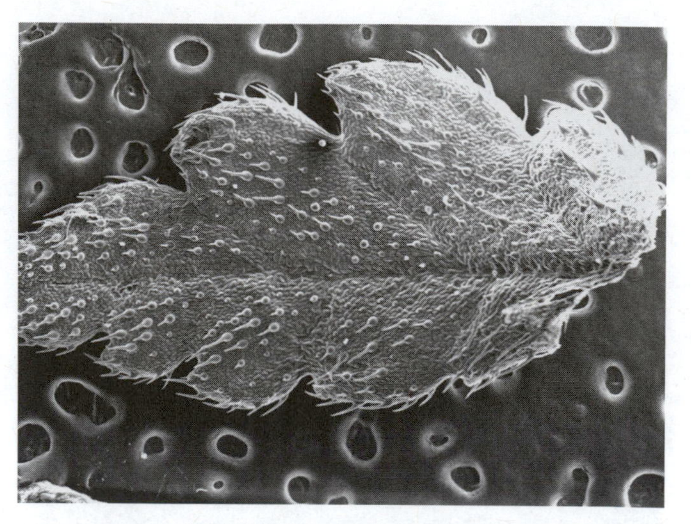

A marijuana leaf with cystolithic hair ***(Courtesy of Heather Schafstall, Oklahoma Bureau of Investigation)***

hydroxide (NaOH or lye) or heating of the cyanoacrylate to create the fumes. The cyanoacrylate causes the polymerization of compounds left behind by contact with skin, resulting in a powdery white print. Additional techniques can be used to enhance the visualization including dusts and fingerprint powders, dyes, and dyes that FLUORESCE. The developed prints can also be lifted in many cases in addition to the normal photography used to document the evidence. Devices have also been developed to allow for fuming to be conducted at crime scenes, as have large "tents" that fit over an entire human body. Using the tents is sometimes successful in developing fingerprints on flesh.

cystolithic hairs A microscopic feature found on the upper surface of the leaves of MARIJUANA (*Cannibis sativa* L). Also known as "bear claws" because of their distinctive shape, cystolithic hairs have a relatively broad oval base supporting the clawlike structure that encases an aggregation of calcium carbonate ($CaCO_3$, the cystolith). Since it is found on the leaf, cystolithic hairs are indicative of marijuana and HASHISH, but since marijuana is not the only plant that has cystolithic hairs, it is not a conclusive test. However, many of the materials submitted to labs as suspected marijuana, such as oregano, do not have cystolithic hairs, eliminating the need for further analysis. Once cystolithic hairs are identified under a microscope, an acidic solution can be added. Because they are composed of calcium carbonate, the cystolithic hairs will emit carbon dioxide bubbles as they react with the acid and dissolve. Thus, appearance of bubbles in acid can provide additional confirmation of the identification.

D

date rape drugs Drugs used to render victims unconscious or otherwise unable to resist sexual assaults. For this reason, these drugs are also called "predator drugs." Many drugs have been used in this way, most notably ALCOHOL, but recently several other drugs have come to the attention of law enforcement and forensic chemists. Of these, two are currently of the most concern, Rohypnol and GHB. Rohypnol is the trade name for the drug flunitrazepam, which is not approved for use in the United States but is smuggled in from Europe and other areas. Rohypnol, also known as "roofies," began to appear in the United States in the early 1990s in the form of a pill or crushed powder. In 1984, it was added to Schedule IV of the CONTROLLED SUBSTANCES ACT (CSA). Rohypnol belongs to the class of drugs called BENZODIAZEPINES and is used (outside the United States) to treat insomnia. It is tasteless, odorless, and dissolves easily in liquids, so it is not difficult to slip the drug into a victim's drink and have it go undetected. The drug can cause dizziness, confusion, nausea, blackouts, and loss of memory. The effects are intensified by alcohol and last for eight to 24 hours. Traces of the drug can be detected in urine up to 72 hours after administration, although it is a difficult analysis. Detection is further complicated by the amnesia-inducing properties of the drug. Victims may not remember or be unclear about the circumstances of the rape, resulting in delayed reporting or no reporting at all. In addition to date rapes, Rohypnol is also abused in conjunction with other drugs such as HEROIN and COCAINE, and long-term use can result in physical and psychological dependence.

A second date rape drug is GHB, gamma hydroxybutyrate, which was at one time used as a bodybuilding supplement. It was pulled off health food store shelves in 1990. Easily synthesized, it has been abused in much the same way as Rohypnol. Low doses relieve tension and promote relaxation, but higher doses produce sleep (sometimes suddenly) and nausea, with alcohol enhancing these effects. In recognition of its dangers, GHB was placed on Schedule I of the controlled substances act (CSA) in 2000.

Other date rape drugs include KETAMINE, barbiturates, MARIJUANA, other benzodiazepines (tranquilizers) such as ATIVAN, cocaine, and opiates such as heroin.

See also DRUG FACILITATED SEXUAL ASSAULT.

***Daubert* decision** (1993) A U.S. Supreme Court decision regarding the admissibility of scientific evidence. *Daubert v. Merrell Dow Pharmaceuticals* (113 S. Ct. 2786 [1993]) involved a case in which parents claimed that the birth defects of their children were the result of a morning sickness medication. The ruling of the Court significantly altered the general approach to the admissibility of scientific evidence that had been set forth in a 1923 case known as the *FRYE* DECISION. In *Frye,* the Court had ruled that in deciding whether to accept new or novel scientific methods and the accompanying testimony into evidence, the primary consideration was general acceptance by the scientific community. In the *Daubert* ruling, the Court stated that under the FEDERAL RULES OF EVIDENCE, Rule 702, general acceptance is not an absolute requirement for determining admissibility. Rather, it is the responsibility of the trial judge to determine if scientific evidence is relevant and reliable. This role assigned to the judge is often referred to as the gatekeeper, and the Court offered suggestions for making that determination, while leaving flexibility to the judges. General acceptance was one criterion, as were peer review of the technique, standards for the method, validation of the method, potential errors, and testable (and thus FALSIFIABLE) theories. As a result of the *Daubert* decision, scientific evidence and expert witness testimony now tend to get greater scrutiny than previously under the *Frye* criteria. Although *Daubert* was a federal ruling, states and local jurisdictions often use such decisions as guidelines for admissibility rulings. As a result, the *Daubert* decision has been an important one for forensic science at all levels. In 1999, the Court extended *Daubert* in the case of *KUMHO TIRE V. CARMICHAEL* to cover all expert testimony, not just that of scientific experts.

***Daubert* hearing** *See* ADMISSIBILITY AND ADMISSIBILITY HEARING

***Daubert* trilogy** Three legal decisions that set forth the current federal standards for the admissibility of evidence. The first was the *DAUBERT* DECISION, described in a separate entry. The second case of the trilogy was decided in 1997 in the case of *General Electric v. Joiner* (522 US 136 (1997)). This was another Supreme Court decision applied to a civil environmental case. The plaintiff, an employee of the company, developed cancer and blamed workplace exposure to PCBs and related compounds. The trial court rejected the testimony of experts supporting this view. Later, an appellate court reversed that decision. The Supreme Court sided with the trial court, supporting the discretion of the trial court in such matters when appropriate. This decision also encouraged judges to take advantage of the advice of appointed neutral experts in evaluation of evidence. The Federal Rules of Evidence (2004), Rule 706 codifies this, stating: "The court may on its own motion or on the motion of any party enter an order to show cause why expert witnesses should not be appointed, and may request the parties to submit nominations. The court may appoint any expert witnesses agreed upon by the parties, and may appoint expert witnesses of its own selection . . . Nothing in this rule limits the parties in calling expert witnesses of their own selection."

The final case of the trilogy was a Supreme Court decision, *Kumho Tire Co., Ltd. v. Carmichael (119 S. Ct. 1167 (1999))*. The issue was the cause of a tire blowout and a resulting fatality. At the trial, the judge excluded the testimony of an engineer based on *Daubert* criteria. The reasoning was that an engineer's testimony did not qualify as scientific testimony under *Daubert* standards. The Appeals court rejected that argument and said that all expert testimony could be considered. The United States Supreme Court upheld the appellate decision, emphasizing that the key criteria for admitting or excluding such testimony were relevance and reliability rather than if the testimony qualified as scientific testimony. The effect of *Kumho* was to apply *Daubert* standards to all expert testimony, not just strictly scientific expert testimony.

D.C. snipers The name given to 41-year-old John Allen Muhammad and 17-year-old John Lee Malvo, arrested on October 26, 2002, who were charged with killing 13 people over a 23-day period starting on October 2, 2002, in Montgomery County, Maryland. Subsequent shootings, all by high-powered rifle, occurred in the Virginia–Washington D.C.–Baltimore area, although the pair was later linked to at least one other killing in September. The arrests came after a truck driver who spotted the suspects' car and followed them to a rest area contacted police. A Bushmaster KM15, a .223 semiautomatic rifle, was recovered from the suspects' car and was quickly linked to bullets and cartridge cases found at the various crime scenes linked to the snipers.

The crime spree brought FIREARMS evidence to the public's attention and led some to call for the creation of a database of images of markings, called BALLISTIC FINGERPRINTING, from all guns sold. In addition, the case pointed out the limitations of eyewitness testimony. At one of the early scenes, witnesses reported seeing a white van, a vehicle that was to become the focal point of the investigation. Only late in the investigation was it learned that the suspects were likely in a dark Chevrolet Caprice, the car they were eventually found in and arrested in. In fact, during the spree, Muhammad was found sleeping in the car in Baltimore by police but was released because the car was legally owned and he had no record. Similarly, profilers in the media and in law enforcement focused on the likelihood that the men they were seeking were white; Malvo and Muhammad are black. Finally, the case was notable for its size and scope and was to become the largest investigation ever undertaken for a localized murder case. By the time the arrests were made, 1,000 FBI agents as well as surveillance aircraft supplied by the armed forces had joined local law enforcement from the different jurisdictions.

See also BALLISTIC FINGERPRINTING; BULLETS; CARTRIDGE CASINGS.

Further Reading

Johnston, D., and D. Van Atta Jr. "Retracing a Trail: The Investigation; Miscues in Sniper Pursuit, Then Calls and a Big Break." *The New York Times,* October 27, 2002.

DEA (Drug Enforcement Administration) Law enforcement agency administered by the DEPARTMENT OF JUSTICE (DOJ) responsible for investigation and enforcement of federal drug laws, primarily the CONTROLLED SUBSTANCES ACT (CSA). The DEA is also responsible for maintaining drug intelligence operations, seizure of assets related to drug trafficking, and coordinating federal, state, and local law enforcement activities related to drug crimes. The history of the DEA began in the early 1900s when predecessor agencies were formed within the Department of the Treasury. From 1927 to 1930, the Bureau of Probation handled federal drug enforcement when it was renamed the Bureau of Narcotics, which it remained until 1968. In 1968, the Bureau of Narcotics combined with the Bureau of Drug Abuse Control (under the Food and Drug Administration) to form the Bureau of Narcotics and Dangerous Drugs within the Department of Justice. Further consolidations resulted in the creation of the Drug Enforcement Agency in 1973. Among other functions, the DEA develops and tests methods for forensic CHEMISTRY and DRUG ANALYSIS and conducts research on newly discovered drugs, DESIGNER DRUGS, analogs, and new methods of drug synthesis. The DEA maintains a comprehensive Web site at www.dea.gov.

death; cause, manner, and mechanism The job of a MEDICAL EXAMINER or CORONER is to determine and certify the facts surrounding a death. This is accomplished by conducting an AUTOPSY as part of the death investigation. The cause of death is that event or process such as a disease or injury that led to the cessation of life. The mechanism of death is set in motion by the cause of death. The manner of death is the category that describes the circumstances that led to infliction of the cause of death. The categories used are homicide (intentional killing by someone besides the victim), suicide (self-homicide), accidental, natural (disease), treatment induced (for example, death during surgery or drug interaction), and indeterminate. They do not all necessarily involve criminal activity, although all can have legal consequence.

See also IDENTIFICATION OF THE DEAD.

Further Reading

Wright, R. K. "Investigation of Traumatic Death." In *Forensic Science: An Introduction to Scientific and Investigative Techniques.* 2nd edition. Edited by S. H. James and J. J. Nordby. Boca Raton, Fla.: CRC Press, 2005.

death investigation *See* CORONER; MEDICAL EXAMINER.

deception analysis, forensic The application of psychological principles and techniques to determine the credibility of a person's words or actions as they apply to legal matters. Fundamentally, deception analysis is an attempt to determine if a person is lying or not, be it in statements, actions, or symptoms. An example of the latter would be if a person were pretending to have a mental or physical illness. Although the POLYGRAPH (lie detector) test is probably the most famous tool of the trade, it is hardly the only one, or even the one that is used most often. The examiner also has available his or her education, training, and experience as well as subjective observations of a subject in an interview, a variety of more objective tests, and newer assessment techniques.

In criminal cases, the examiner may be called upon to determine if a person is competent to stand trial, what degree of criminal responsibility the person has, and how dangerous he or she is to the community. Other examples would be to determine if a suspect who is claiming amnesia is faking the problem, or if a person's claim to no longer be using drugs or alcohol is valid. Applications in civil cases are broader and can range from wrongful deaths to workman's compensation claims. In both criminal and civil cases, examiners are often called upon to determine the credibility of child witnesses, a critical issue in child abuse and custody cases. Finally, the examiner may also play a role in determining what treatment is adequate and appropriate for the subject. Criminal profiling, which can be considered part of deception analysis, is treated in a separate entry (PSYCHOLOGY, FORENSIC).

Detection of lying is subjective, and no method is completely reliable. A person who is lying may display a number of verbal and nonverbal signals, which examiners are trained to look for. Responses to questions, history of deception, consistency of the story, amount of detail, and speed of responses all aid the examiner. The polygraph may be used, but this device has problems and limitations of its own. Some examiners use hypnosis, but it has not found general acceptance within the field. Courts have not accepted testimony purportedly obtained in a hypnotic state.

See also NARCOANALYSIS.

Further Reading

Hall, H. V., and D. A. Pritchard. *Detecting Malingering and Deception: Forensic Deception Analysis.* Delray Beach, Fla.: St. Lucie Press, 1996.

decomposition The sequential process that occurs after death and ends with skeletonization. Although the phases of decomposition are known, environmental factors, principally temperature, dramatically affect its specific path and rate. As a result, the progress of decomposition is not alone sufficient or reliable for estimating the POSTMORTEM INTERVAL (PMI). Rather, it is one of many tools used to estimate how long it has been since death occurred. Studies of decomposition have been undertaken at the University of Tennessee in Knoxville at the ANTHROPOLOGICAL RESEARCH FACILITY (ARF), known informally as the "Body Farm."

Death can be considered as two distinct events, starting with somatic death, in which the heart and breathing stop and brain activity ceases. After this comes a process of cellular death and decay called autolysis (cells that burst are said to have lysed), which then leads into putrefaction. At somatic death, body chemistry begins to change rapidly, and certain enzymes become active that promote degradation of molecules such as proteins and carbohydrates. In these early stages of decomposition, microorganisms already present in the gut multiply and spread. This results in gas production that causes a bloating in the abdomen and then generalized bloating that can swell the body to two or three times its original size before collapsing. The bloating can cause the eyes and tongue to swell and protrude as well. The appearance of the skin begins to change, passing through stages of green before reaching black, a stage appropriately named black putrefaction. Bacteria flourish in blood, which is an excellent medium for supporting it, accelerating the breakdown process. During these changes, the characteristic odors of decomposition are given off. Skin and hair can slough off during these processes, and the body eventually becomes unrecognizable. Stages of advanced decomposition can be reached in 12–18 hours in warm and humid environments, but in cold or freezing areas, the process slows dramatically, stretching into weeks, months, or even years depending on how cold it is.

Putrefaction leads next to a stage of dry decay or mummification, again depending on environmental conditions. As skin dries, it shrinks, and this shrinkage has led to the popular misconception that hair and nails continue to grow after death. ADIPOCERE may form in this stage as well. Eventually, most or all tissue will be dried and will decay away, leaving only bones. In addition to the environmental factors, other considerations for the evaluation of decomposition include insect activity and animal scavenging. Insects such as blowflies are attracted to blood and will lay eggs in wounds and orifices such as the nose and ears. Maggot feeding can distort features and accelerate decay to a remarkable degree, and in areas of considerable insect activity, a fresh corpse can be skeletonized within a week or two. Submergence of a body or inundation will also have an impact on decay and decomposition processes.

See also ENTOMOLOGY; TAPHONOMY.

Further Reading

Nafte, M. "The Process of Decomposition" In *Flesh and Bone: An Introduction to Forensic Anthropology.* Durham, N.C.: Academic Press, 2000.

deductive reasoning A common form of logic applied in forensic science. In deductive reasoning, conclusions are drawn based on evidence and facts that are already established. The following statement is an example of deductive reasoning: all animals that have spines are vertebrates. Humans are vertebrates, so it follows that humans have spines. Deductive logic and the complementary INDUCTIVE LOGIC are both part of the scientific method, and they are both used in forensic work.

See also HYPOTHESIS AND THE SCIENTIFIC METHOD.

defense attorney The attorney that represents the defendant (suspect) in a criminal trial or the defendant in a civil action. In criminal cases, the defendant may be appointed a lawyer from the public defender's office (a governmental agency), or they may hire their own attorney. In high-profile cases, it is not uncommon to see a team of defense lawyers. The defense lawyer has the right to question expert witnesses and to ensure their credibility and reliability through a process called VOIR DIRE.

defense experts Defendants in criminal and civil cases are entitled to have physical evidence analyzed by experts other than those employed by the agency responsible for their prosecution. Forensic laboratories strive to preserve at least half of any samples that they analyze for such a purpose. If that is not possible, they may be required to obtain the permission of the DISTRICT ATTORNEY (DA) before proceeding. Defense experts may come from a variety of places, including other government or private forensic labs, or they may be consultants from a university or the private sector. One of the dangers that defense experts must avoid is the practice or even the appearance of conflict of interest. Since the DEFENSE ATTORNEY is paying for the expert, it is natural for people, particularly jurors, to assume or suspect that the expert's payment is contingent on finding the "right" results, the ones most favorable to the defense. As a result, many defense experts work on a flat fee basis with a clear statement that payment does not depend on findings.

Demerol (meperidine) A synthetic opiate ALKALOID that is a narcotic with effects similar to MORPHINE. Demerol is widely used for pain relief in hospitals and is sometimes abused in ways similar to other narcotics such as HEROIN. Meperidine was first marketed under the name Demerol in 1939.

demography (demographics) The statistical study of human populations. In forensic science, demography is important in determining frequencies of blood types (BLOOD GROUP SYSTEMS and ISOENZYMES) and DNA types. In many systems, frequencies differ among Caucasians, African Americans, Hispanics, and Asians, and this difference must be taken into account when blood typing and DNA TYPING analyses are evaluated. For example, the isoenzyme system PGM has three types (disregarding subtypes) of 1, 2-1, and 2. Type 1 is seen in about 59 percent of the Caucasian population and 66 percent of the African-American population. Thus, when a serologist discusses the DISCRIMINATION INDEX or DISCRIMINATING POWER of a given result, it is important that any relevant demographic considerations be taken into account. Different forensic laboratories handle these issues in different ways, depending on their locations and other factors.

density Defined as the mass of an object or substance divided by its volume ($d = m/v$). The standard units of density are grams/milliliters (g/mL). The density of water at 77°F (25°C) is 1.00 g/mL, which is used as a benchmark for comparison. Small amounts of dense materials (such as lead) are heavy, having a density of greater than 1.00, and they will sink when placed in water. Conversely, comparable amounts of low-density materials, such as foam, are light, have a density of less than 1.00, and will float. In forensic science, density of materials such as SOIL samples and GLASS can be used to help classify them. Density columns—tubes filled with liquids of known densities—are used for this purpose. These are also referred to as density gradients.

dentistry, forensic *See* ODONTOLOGY.

Department of Homeland Security (DHS) An agency formed in the wake of the September 11, 2001, terrorist attacks. The department, which combined old agencies and new branches into one unit, was proposed in June 2002 by President George W. Bush. The department consists of smaller directorates called Border and Transportation Security, Emergency Preparedness and Response, Science and Technology, and Information Analysis and Infrastructure Protection. As examples of the reorganization and consolidation, the U.S. Customs Service and U.S. Secret Service, previously housed under the Department of the Treasury, were moved to DHS, as was the Coast Guard. The Transportation Security Administration (TSA), which oversees airport screening among other tasks, moved from the Department of Transportation to the Transportation Security Directorate. Many functions of DHS involve or incorporate FORENSIC SCIENCE research and practice, notably the forensic laboratory of the U.S. Secret Service and transportation security research and development from TSA. The department also sponsors research and development in forensic areas such as EXPLOSIVES detection.

Department of Justice (DOJ) The agency that coordinates federal law enforcement. The Department of Justice houses the FEDERAL BUREAU OF INVESTIGATION (FBI) the BUREAU OF ALCOHOL, TOBACCO, FIREARMS, AND EXPLOSIVES (ATF) and the DRUG ENFORCEMENT ADMINISTRATION (DEA), all of which play critical roles in the forensic science community. The department also administers grants programs that help fund research in forensic science. The DOJ maintains a comprehensive Web site at www.doj.gov.

deposition A sworn statement that is given under oath but not in a regular courtroom setting. The person documenting the deposition is an officer of the court, and the statement is taken by a court reporter. Occasionally, forensic scientists testify in a case by way of a deposition due to traveling, scheduling, or other unavoidable conflicts that prevent or prohibit them from appearing in standard open court. A deposition may also be used as a preliminary statement before testifying. Attorneys or representatives for both the prosecution and defense are usually present, allowing for questioning much as would take place in open court.

depressants *See* DRUG CLASSIFICATION.

dermal nitrate test A PRESUMPTIVE TEST for the presence of GUNSHOT RESIDUE (GSR), also called the paraffin test, that is no longer used. Whenever a person fires a gun, traces of residue are transferred to the hands. The dermal nitrate test was based on detecting the nitrate ion (NO_3), an ingredient in PROPELLANTS used in AMMUNITION. For the test, the hands of the suspect were painted with hot wax (paraffin) that was allowed to dry. The cast was then removed and tested using the reagent diphenylamine combined with sulfuric acid (H_2SO_4). Locations on the cast that showed a blue color indicated the possible presence of nitrate. In addition to gunshot residue, nitrates are common in many materials that could be found on the hands, including tobacco, cosmetics, fertilizers, and urine. This large number of potential FALSE POSITIVES led to the test being abandoned since it was not sufficiently specific to GSR.

designer drugs Illegal drugs synthesized in clandestine labs that are closely related to controlled substances in structure and effect. The first designer drugs were derivatives of methamphetamine and AMPHETAMINE and appeared in the 1970s. MDMA, also known as ECSTASY, is 3,4-methylenedioxymethamphetamine, while MDA is the amphetamine equivalent. Because of their similarity to existing illegal substances, designer drugs are also known as analogs. MDA and MDMA had a brief period of accepted medical use, but other designer drugs have not. Under the CONTROLLED SUBSTANCES ACT (CSA), designer drugs are classified as Schedule I substances.

The other two common designer drugs are based on Fentanyl and meperidine (DEMEROL). Fentanyl is a narcotic analgesic 100 times as potent as morphine, but short acting. Small changes in the molecule can increase both the potency and length of effect, and so a number of Fentanyl derivatives exist. One of the dangers of analogs, aside from increased potency and duration of effect, are impurities and contaminants that result from clandestine synthesis. Also, the extreme potency means that very small amounts are ingested, making detection in blood and urine difficult.

DFO *See* FINGERPRINTS.

diastereoisomers A term used to describe molecules that have the same molecular formula but different arrangements of the atoms in space. Isomers are molecules, such as the two shown in the figure above, that have the exact same molecular formula (C_4H_8) but different arrangements of their atoms. Stereoisomers have the same formulas and same geometries. Diastereoisomers are a special class of stereoisomers in which the mirror image of one cannot be superimposed on top

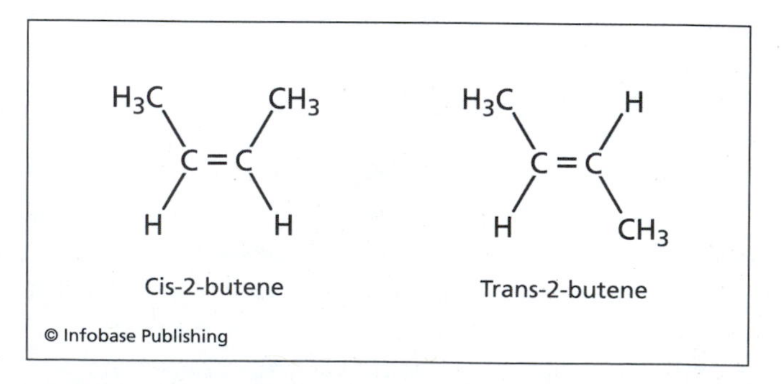

Example of diastereoisomers

of the other. Diastereoisomers have different physical and chemical properties and are sometime an issue in forensic CHEMISTRY (drug analysis), particularly in the case of COCAINE. Cocaine has two enantiomers, d-cocaine and l-cocaine, and they are mirror images of each other. There are three other such pairs of cocaine enantiomers, d- and l-pseudococaine, d- and l-allococaine, and d- and l-pseudoallococaine. These pairs are all diastereoisomers of cocaine.

diatoms A form of algae that can be a source of TRACE EVIDENCE. Diatoms exist in freshwater and salt water and thus can be important evidence in drowning cases, although use of diatom evidence is still somewhat controversial. If the heart is still beating when a person enters or is forced into the water, the circulatory system can deliver the diatoms to remote parts of the body such as the liver, kidney, or bone marrow. Thus, finding diatoms in tissues far removed from the lungs can be interpreted to mean that the person was alive when they entered the water. Since diatoms have a hard outer shell that is resistant to acids, the chemical extraction procedure for isolating diatoms from organs or bones is fairly simple. A portion of the sample is placed in concentrated acid (such as nitric acid, HNO_3) and boiled for several hours. The liquid is then centrifuged and the solid residue rinsed and examined microscopically. Classification of the specific diatom can also help determine where the person entered the water, which may be some distance from where the body is recovered.

Diatoms can also be transfer evidence, either as the algae cell itself or in the form of diatomaceous earth (often called DE), which consists of ancient deposits of the silica-based skeletons of the plant cells. DE has many commercial uses, including as an abrasive, filter media for swimming pools, and insulating material in safes. Thus, DE may become part of DUST or other TRANSFER EVIDENCE and may be useful in linking persons to scenes.

diazepam (Valium) One of the most widely prescribed drugs in the world and one frequently abused. Valium is classified as a depressant and is a tranquilizer that belongs to the BENZODIAZEPINE family that includes ATIVAN and XANAX. Valium comes in small pills and is used to treat anxiety without promoting sleep to the degree that BARBITURATES do. Forensic identification of Valium is accomplished using THIN LAYER CHROMATOGRAPHY (TLC), GAS CHROMATOGRAPHY/MASS SPECTROMETRY (GC/MS), and HIGH PERFORMANCE LIQUID CHROMATOGRAPHY (HPLC).

dichroism A property of materials such as fibers in which the colors that are absorbed (and thus the color that the material appears) are a function of direction. Dichroism is important in forensic MICROSCOPY and in the analysis of TRACE EVIDENCE and TRANSFER EVIDENCE. Specifically, in a dichroic material, the absorption pattern varies depending on the direction or orientation of the material or on the orientation of polarized light that illuminates it. Although often associated with colors and thus visible light, infrared absorption patterns can also be dichroic. To observe dichroism in a material, it is mounted on a microscope slide and examined using polarized light (a polarizing light microscope, PLM). As the stage is rotated, the color of the material will change since the direction of light vibration is changing.

digital evidence *See* COMPUTING, FORENSIC.

digital image enhancement *See* IMAGE ENHANCEMENT.

digital imaging *See* PHOTOGRAPHY.

Dille-Koppanyi test A PRESUMPTIVE TEST for BARBITURATES. It is a modification of an older test called the Zwikker test. It is a two-part test that starts with the addition of a solution of cobalt acetate in methanol to the unknown sample (powdered), followed by a methanolic solution of isopropylamine. A reddish violet color is indicative of barbiturates, but not definitive.

diluents *See* CUTTING AGENTS.

diphenylamine test A PRESUMPTIVE TEST for the presence of GUNSHOT RESIDUE (GSR). A solution of diphenylamine in sulfuric acid (H_2SO_4) is added to suspected residue, and a blue color indicates the possible presence of nitrates (NO_3^- ions) that are ingredients in the PROPELLANTS used in AMMUNITION. The test is not specific, and the reagent will react with nitrites (NO_2^- ions) and other oxidizing reagents. The reagent was used as part of the DERMAL NITRATE TEST, now abandoned due to its lack of specificity. The diphenylamine test can also be used in the analysis of paints, where NITROCELLULOSE can be an ingredient.

direct examination *See* COURTROOM PROCEDURES.

direct transfer *See* PRIMARY TRANSFER.

disarticulation Literally, "disjointed." In forensic applications, the term refers to a body that has been broken up, taken apart, dissected, or separated. As soft tissues such as ligaments and tendons decay or are torn, the joints come apart and the bones can become scattered, either by natural processes or by animals (scavengers). A corpse may also be purposely disarticulated to hide or otherwise dispose of the body. The sequence of disarticulation will be different for a body found in the water, since the water allows the joints to move freely, speeding the process. Wave action and currents add to this effect, and in general the more flexible the joint in water, the sooner it will come apart.

discovery (disclosure) The pretrial process of revealing information to and among the prosecutor and the defendant and his or her defense attorney. Although the rules of discovery vary by jurisdiction, in general, the defendant has a right to see any incriminating evidence against him or her, which can include reports and analyses performed in a forensic laboratory. In addition, the prosecution has an obligation to reveal any evidence to the defendant that is favorable to his or her case. In criminal cases, the defense may have a right to keep some test results confidential, but in civil cases, full disclosure/discovery is generally required of both parties. Many jurisdictions model their requirements on the Federal Rules of Discovery.

discrimination index (discrimination power; probability of discrimination P_d) The ability of a given test to individualize evidence. In forensic science, this term is usually associated with blood and DNA TYPING, but the idea can be applied to any analysis that results in a CLASSIFICATION or categorization of PHYSICAL EVIDENCE. The terms are used in different but closely related ways. First, the discrimination index is the probability that any two people selected at random in a given population will have *different* blood types. The discrimination index is also referred to as the discrimination power. In the U.S. Caucasian population, approximately 42 percent have the ABO BLOOD GROUP type of A, 43 percent O, 12 percent B, and 3 percent AB. The discrimination index of the ABO blood group system is approximately 0.60, meaning that there is a 60 percent chance that any two people selected at random will have a different ABO type. Similarly, the isoenzyme system PGM has 3 types (discounting subtypes), 1-1 59 percent, 2-1 35 percent, and 2-2 6 percent and a discrimination index of 0.52. Thus, PGM has a lower discrimination index and is less likely to distinguish two individuals selected at random. Another way to state this would be to say that PGM has a lower power of discrimination. This is attributable to the fact that nearly two-thirds of the population is PGM type 1-1. Thus, if only one system can be typed, it would be wiser to select the ABO system since it has the higher discrimination index and is more likely to distinguish two individuals.

In contrast, the probability of discrimination (P_d) of a given system looks at the probability that any two people selected at random will have the *identical type*. Thus, a low P_d is desirable since this would mean that the odds of two people having identical types would be low. For the ABO system, the P_d is approximately 0.40, meaning the probability that two people selected at random will have the same ABO type is 40 percent. Note that the discrimination index and the probability of discrimination are related; they must both add up to one. For the ABO system, this calculation would be $0.60 + 0.40 = 1.00$. This makes sense; the chance that any two people selected at random will have either identical types or different types must be 100 percent. This is like flipping a coin—the chances that it comes up either heads or tails is 100 percent since those are the only two possible outcomes.

The more types there are within a given system and the more widely spread the frequencies are, the lower the P_d, and by typing more systems, the P_d increases. For example, the P_d of the PGM system is about 0.48, meaning that there is a 48 percent chance that two people selected at random would have the same type. This is not a very good system by itself, but if both PGM and ABO are typed, the cumulative P_d becomes 0.40×0.48, or 0.19. This means that there is about a 20 percent chance that any two people selected at random will have the same ABO and PGM type. The actual P_d for any system varies by race and other factors, but the calculation method is the same.

Another less confusing way to express the same idea is to look at frequencies in the population and come up with a "one-in" number. For example, assume a bloodstain is typed as ABO type A and PGM 1-1. In the example population used, the combined frequencies are 42 out of 100 for ABO type A and 59 out of 100 for PGM type 1-1. To get the combined frequencies, these values are multiplied together ($^{42}/_{100} \times {}^{59}/_{100}$, or 0.42×0.59) to obtain a value of .25 percent, meaning that 25 people per hundred would be expected to have this combination of types (25 percent). This can be restated as "1-in-4." Because this is a relatively large number, it is not very useful for forensic work. However, it is not unexpected since both types are very common. If a bloodstain were found to have the rarest types, AB and 2-2, the same calculation using those frequencies ($^{4}/_{100} \times {}^{6}/_{100}$) leads to a value of 0.0024, meaning that 2 people per thousand would be expected to have this combination. Thus, discrimination index (discrimination power) and the probability of discrimination depend on the number of different systems typed, the number of variants of each system, and the frequencies of those variants.

DNA typing has largely replaced the typing of blood for groups such as ABO and PGM, but the same principles apply. Using the 13 short tandem repeats (STRs) commonly typed in forensic labs, it is possible to obtain powers of discrimination that exceed the population of the planet by several orders of magnitude. For example, using STR frequencies for African Americans, typing the first three produces a power of discrimination of about one in a million while typing all 13 leads to figures smaller than one in a trillion. This is remarkable considering that the population of earth is less than 10 billion, a hundred times smaller than a trillion. There are many issues related to these assumptions, such as the dependability of frequencies and POPULATION GENETICS, but the reliability of these calculations has been generally accepted.

Further Reading

Sensabaugh, G. "Biochemical Markers of Individuality." In *Forensic Science Handbook*. Vol. 1. Edited by R. Saferstein. Englewood Cliffs, N.J.: Regents/Prentice Hall, 1982.

disguised writing *See* QUESTIONED DOCUMENTS.

distance determination Although this can be a generic term for the process of measurement, in forensic science it most often means determining the distance between a firearm and a target when the weapon was discharged. Such determinations are useful for reconstructions related to CRIME SCENES and for supporting or refuting different stories related to a shooting. Distance determinations are conducted by FIREARMS examiners, who must take into account environmental conditions and other variables when performing these tests. When a pistol (revolver or semiautomatic) is fired, some of the PROPELLANT does not burn completely and is dispersed out the sides and muzzle of the weapon. This material, along with lead from bullets and primers and other components, is called GUNSHOT RESIDUE (GSR), and patterns of GSR around a wound or on clothing can be useful in determining how far away the weapon was when it was fired. Some general guidelines exist for estimating distance based on wound or clothing patterns, but the only reliable method of distance estimation is to obtain the suspect weapon and test fire it using the same AMMUNITION and into the same type of material as was involved in the incident in question. Several test firings are made into targets at set distances, and the patterns of GSR and damage are compared to actual distances. Exact distances can never be determined, but by establishing bracketing ranges, reconstruction of events becomes much easier.

If a suspect weapon has not been recovered, then estimates are made based on evidence (clothing usually) and/or photos of wounds. With pistol shots, generally shots fired from greater than three feet (.91 m) away will deposit little if any visible residue on the target and will make a relatively even, round hole. As the distance decreases, GSR deposits increase and are seen in a roughly circular pattern. In very close shots, fibers may show scorch damage from flames emerging from the barrel. Damage can take on a starlike pattern from blowback resulting from closeness to the target. The GSR is detected by visual examination or visual examination supplemented with chemical developers, the same chemicals used as PRESUMPTIVE TESTS for GSR. The presence of blood or odd coloring of fabrics can complicate pattern identification and interpretation, and occasionally specialized photography such as infrared photography is used to visualize residues.

Shotgun patterns also vary with distance and depend on the gun, ammunition, and choke of the barrel. The choke of a shotgun is the narrowing of the barrel that determines how much the pellets will spread out once fired. As before, test firings with a suspect weapon are essential for accurate distance estimates.

See also BULLET WOUNDS.

district attorney (DA) The attorney appointed by a governmental entity (federal, state, city, or county) responsible for criminal prosecutions in his or her district. Forensic scientists, especially those employed by the same government entity, often work closely with the DA or one of the deputy DAs who handles the cases.

DMORT teams *See* MASS DISASTERS.

DNA (deoxyribonucleic acid) The molecule that makes up genes and carries inherited information. DNA TYPING has revolutionized forensic SEROLOGY and has provided tools that can INDIVIDUALIZE blood and body fluid evidence.

As shown in the figure, DNA is a POLYMER made up of a chain of nucleotides, which consist of bases (cyclic molecules that contain nitrogen) connected to a sugar-phosphate backbone. The sugar in DNA is ribose, which assumes a pentagonal shape in the backbone structure. In DNA, there are four bases (shown) that can be incorporated into a nucleotide: adenine (A), thymine (T), cytosine (C), and guanine (G). The structure of a nucleotide based on adenine is shown in the first figure, and this form is called adenosine monophosphate. This notation means that the adenine is connected to the sugar (ribose), which in turn is linked to a single ("mono") phosphate. Through a process called hydrogen bonding (a weak bond that can form between hydrogen, nitrogen, and oxygen atoms), the bases adenine and thymine are attracted to each other, as are cytosine and guanine. This attraction ultimately leads to the double helix shape of DNA.

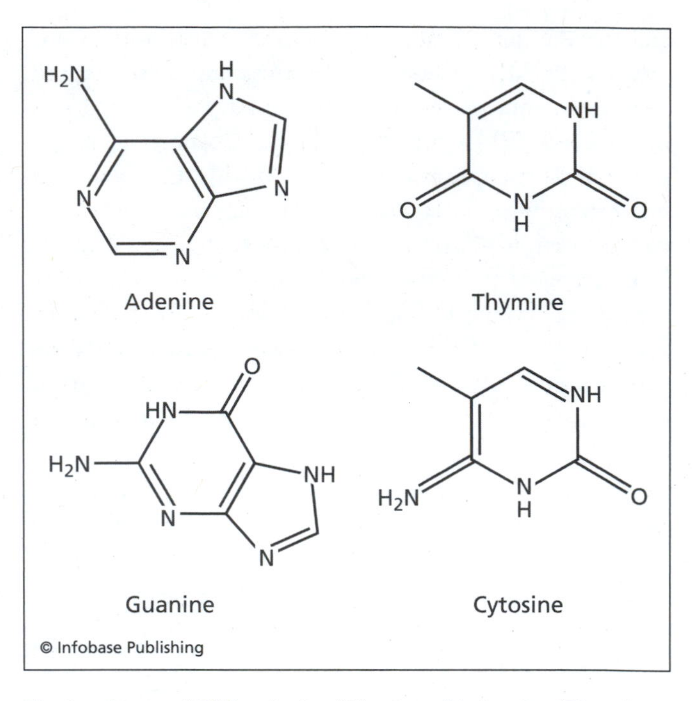

The four bases of DNA: adenine (A) pairs with thymine (T), and guanine (G) pairs with cytosine (C)

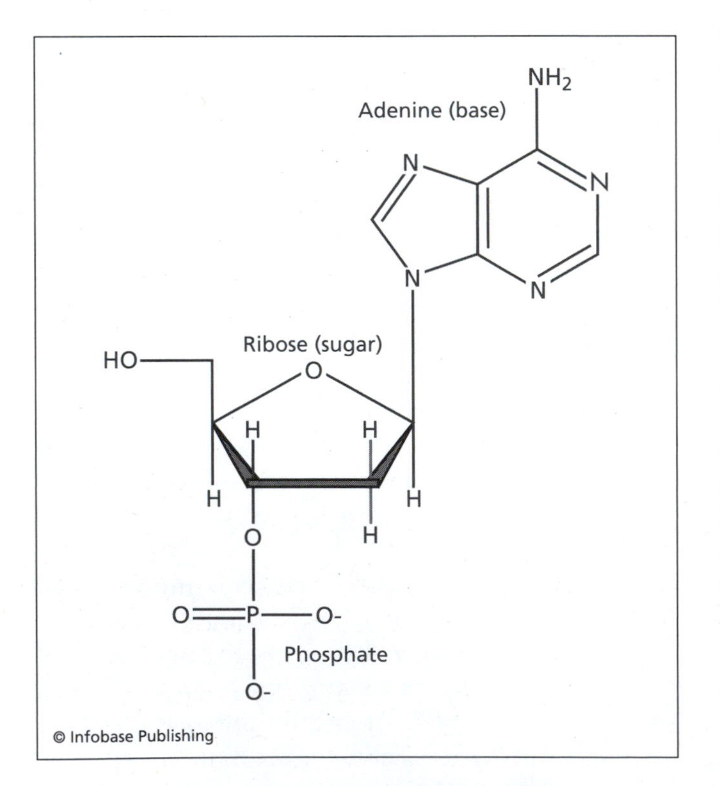

Generic nucleotide structure showing the sugar (ribose), the phosphate group, and a base, here adenine

DNA is composed of complementary strands of nucleotides in which the base pairs organize themselves opposite to their partners. The interactions of the A-T and C-G pairs cause the two strands to twist around each other in a helical coil. The double helix shape of DNA was first proposed in 1953 by Francis Crick (a Briton) and James Watson (an American), although their work was the culmination of much research in the field. As shown in the illustration, DNA can be replicated by first "unzipping" the separate strands, a process that requires enzymes or heat. Once separated, new complementary strands can be synthesized from nucleotides that will arrange themselves in the same opposite pairings as in the original strand, creating two copies of the original, or parent, DNA molecule. In this way, each parent strand acts as a template for a new one. Enzymes are also needed to cause the helixes to reform. One of these enzymes is DNA polymerase, which plays an important role in forensic DNA typing procedures. The polymerase chain reaction (PCR) procedure is used to amplify small amounts of DNA and thus allows forensic scientists to type very small samples.

A gene is the fundamental unit of heredity and consists of a sequence of base pairs along a DNA molecule. Each sequence of three base pairs codes for a specific AMINO ACID, and individual amino acids are joined together to form a protein (polypeptide) molecule. For example, a triplet base pair sequence of C-G-T codes for the amino acid alanine. When the gene is used to direct synthesis of a protein, whenever the C-G-T sequence is encountered, alanine will be incorporated into the protein molecule. Thus, in very simple terms, a base pair sequence in DNA is a gene, and the gene is a code that is used to make a specific protein. In turn, genes are found in chromosomes, which are located in the nuclei of cells. Humans have 23 matched pairs of chromosomes for a total of 46, and it is estimated that each cell contains 50,000 to 100,000 genes. As an example, the gene that codes for one subunit of human HEMOGLOBIN contains nearly a thousand base pairs.

Another aspect of DNA that is relevant in forensic applications is recombinant DNA technology. This technology is referred to as "genetic engineering." In recombinant DNA procedures, enzymes called restriction enzymes are used to selectively remove a specific base pair sequence (gene or genes) from DNA. Restriction enzymes are often called "molecular scissors" because they cut specific segments out of the DNA molecule, with different restriction enzymes recognizing different base pair sequences. In genetic engineering, the excised section of DNA is then inserted in other DNA, where the proteins it encodes for can be produced. Currently, there is a great interest and controversy about genetically engineered crops, in which specific genes are inserted to impart disease or pest resistance to the plant. As another

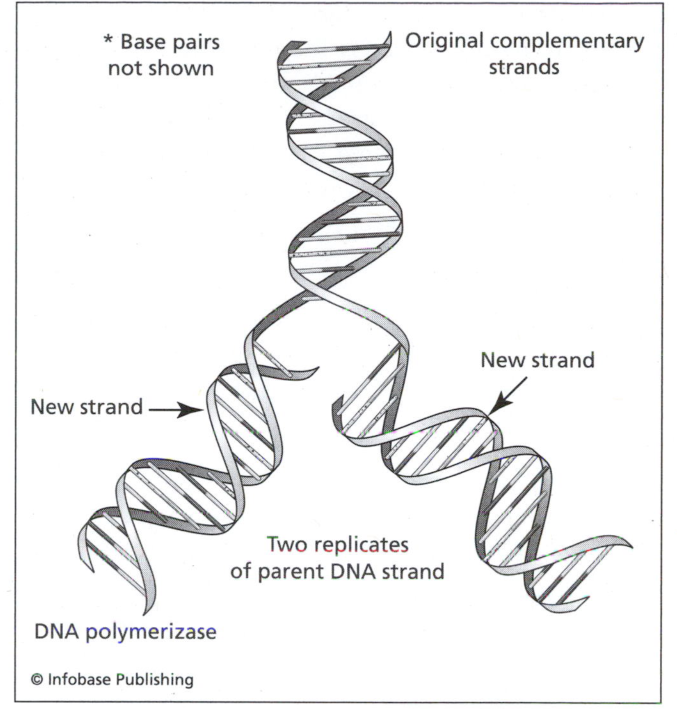

DNA replication. The original paired strand is copied when new bases pair with their complement and paired strands reform.

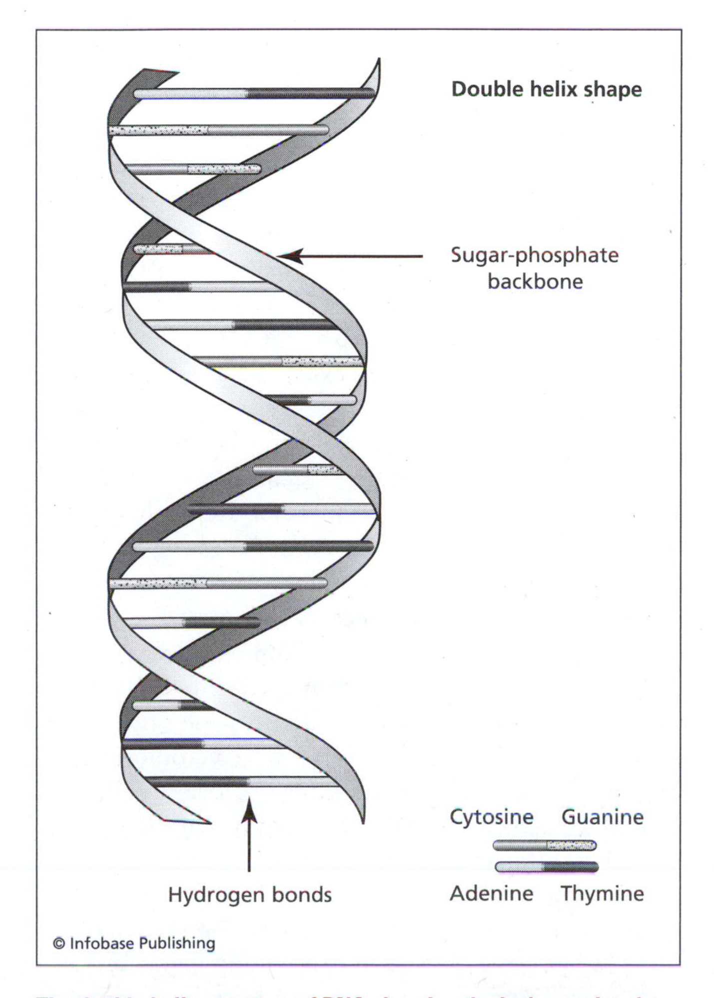

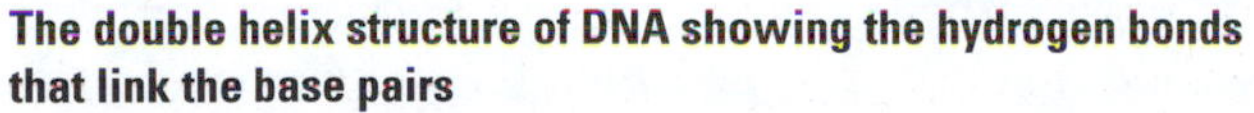

The double helix structure of DNA showing the hydrogen bonds that link the base pairs

example, recombinant DNA/genetic engineering techniques are used to insert a human gene that encodes for the protein insulin into bacteria. The engineered bacteria then produce large quantities of insulin for use by diabetics. Restricted fragment length polymorphisms (RFLPs) were among the first to be exploited in forensic DNA analysis.

Further Reading

Kobilinsky, L. "Deoxyribonucleic Acid Structure and Function—A Review." In *Forensic Science Handbook*. Vol. 2. Edited by R. Saferstein. Englewood Cliffs, N.J.: Regents/Prentice Hall, 1988.

DNA typing **(DNA analysis)** A group of related procedures that has largely replaced traditional blood typing in forensic labs. DNA typing techniques were pioneered by molecular biologists and entered into the forensic arena in the late 1980s and early 1990s. Since then, they have grown quickly into the tool of choice for the analysis of blood and body fluids. Rapid advances in the field continue, so DNA typing applications and techniques continue to change and evolve. Smaller and smaller samples can be analyzed (currently as little as a billionth of a gram of DNA is needed), while the discriminating power (the ability of a typing test or tests to individualize a sample) is increasing. In addition to blood and body fluid stains, DNA typing can be used on a variety of different samples including HAIR, skin scrapings, and even dandruff.

History

The first DNA typing protocol used in forensic science was restricted fragment length polymorphism (RFLP), based on a variation in the number of times a sequence of base pairs in DNA is repeated. Briefly, there are head-to-tail repeat segments (fragments) of bases scattered throughout DNA, the function of which is not clear. However, from a forensic viewpoint, these fragments are extremely valuable since the number of times that a given sequence is repeated usually varies from person to person. For example, consider a hypothetical sequence of bases: G-C-A-A-T, a five base-pair sequence (see DNA for abbreviations and more information about genes, chromosomes, and complementary base pairs). In one person, this sequence may be repeated five times on a given chromosome, while in another person, it might be repeated 15 times. Different types (the polymorphism) arise not from differences in the sequence, but from the number of times it is repeated on a chromosome. In addition, since half the chromosomes are inherited from the mother and half from the father, a person can have two different variants for a given sequence. For example, if a person had five repeats on one chromosome and 15 on the other, their type would be (5,15) in this hypothetical situation. Because the polymorphism comes from the number of repeats, this kind of DNA typing is categorized as VNTR (variable number of tandem repeat) technique. In practice, RFLP examines several different repeating sequences, but the repeat units are bigger, so the DNA fragments are longer.

Prior to RFLP analysis of DNA or biological stain, the material in question must be extracted and isolated from the evidence. Once this is accomplished, the restriction enzymes are added that cut the DNA into fragments in which the repeated segment occurs. The smaller fragments have fewer repeats, while the larger fragment have more repeats. Using gel ELECTROPHORESIS, the fragments are separated based on size, with the smaller fragments traveling farther than the larger fragments. On each gel, controls are used, including an "allelic ladder," which allows the analyst to gauge the number of repeats found in the evidence by comparing the gel position of the evidence sample with the allelic ladder. Once electrophoresis is complete, the fragments have been separated but are still not visible. The next step is to transfer the DNA to a nylon membrane using a technique called Southern blotting. To visualize the bands now on the nylon (which resemble a barcode), radioactively labeled "probes" are placed in contact with it. The probes contain the complementary DNA sequence to the fragment in question. Returning to the previous five base pair example, the repeating sequence of interest was G-C-A-A-T, and so the complementary probe would be radioactively labeled C-G-T-T-A. The probe bonds to the fragments along with the radioactive label in a process called hybridization. The final step is to expose X-ray film over the nylon membrane, which will show bands wherever there is radioactivity. Types are read from the X-ray film, which is called an autoradiograph or autorad. A depiction of an autorad is shown in the first figure.

The discrimination power of RFLP comes from looking at several different fragments. If a given sample shows a pattern expected to be found in one in a thousand people, that alone is not very useful. However, if a second fragment is typed and also is unique to $^1/_{1000}$,

DNA Typing and Privacy

Since 1924, the FEDERAL BUREAU OF INVESTIGATION (FBI) has kept a national database of fingerprints, which has grown to be the largest such collection in the world. Few would argue with the need for such a repository, and none can reasonably argue with its success. When DNA TYPING became available starting in the late 1980s, a new tool of individualization burst onto the forensic scene. However, the approaches used for creating and exploiting databases of DNA types has proven more complex and contentious than did those routinely used for fingerprints.

DNA typing has displaced other methods of analyzing blood and body fluids and, when successful, can reduce the number of possible donors to one in tens of billions, far in excess of the population of the planet. All 50 states collect DNA samples from convicted sexual offenders and those guilty of other selected felonies. The FBI coordinates the CODIS system, a computerized index that allows laboratories to share DNA profiles. This networking creates something similar to the national repository of fingerprint data, in which all connected laboratories have access to data collected by all others. CODIS has proven extremely valuable in identifying crimes committed by the same person and in identifying suspects when other investigative tools have failed to do so. However, critics of the system raise several questions regarding issues of privacy and ethics. Because DNA can reveal things that fingerprints cannot, there are concerns that these databases could be abused. But are these concerns valid or the result of misunderstanding?

Contrary to one common misconception, the CODIS system does not contain a description of the person's genes or their complete DNA sequence. In this sense, the stored DNA profile is just like a fingerprint—a pattern that can be used to distinguish one individual from another, but nothing more. There is no additional information hidden in the profile, just as there is no additional information hidden in a fingerprint. The DNA information accessible through CODIS represents 13 different genetic locations (loci) called short tandem repeats (STR's). These loci do not appear to contain information that might reveal, for example, whether a person has or might develop a genetic disease. What is stored in the system is useful for identification purposes, not for genetic screening in the traditional sense.

The system was intended to assist criminal investigators, to convict the guilty, and to exonerate the innocent, all of which have occurred. However, the power of DNA typing introduces inevitable temptations. One frequently cited example is the potential use of DNA data in cases of disputed paternity, which is a civil rather than a criminal matter. Should data stored in CODIS be available in such cases? To date it has not been, but if the data exists, the potential for misuse does as well. Related to this is the fear that the computerized database could be compromised and made available much as credit card numbers are revealed by malicious hackers. Again, procedural safeguards and tight security have been implemented to prevent this, but there are no guarantees it will never occur.

The analogy between DNA and fingerprints falters when considerations extend to the actual samples. In the case of a convicted criminal, a DNA type is obtained by the analysis of blood or saliva. The laboratory test requires only a small portion of the sample, leaving the bulk of it untouched. So far, there are no standards as to what happens to the remaining sample, which *does* contain a person's entire genetic code. It is this unused sample that could be analyzed to determine if a person is prone to or suffers from a genetic disorder, for example. Furthermore, genetic information not only reveals specifics about the person donating the sample but also about his or her relatives. This is not true of fingerprints. Thus, misuse of genetic data threatens families, not just individuals.

Many labs store samples for years, arguing that more advanced techniques developed at some future date could be applied to stored samples. This foresight has been justified by recent events. In late 2001, Seattle police arrested Gary Ridgeway, a man suspected of being the "Green River Killer" who murdered an estimated 50 women from 1982 to 1984. Early in the investigation, police suspected Ridgeway and questioned him, but lacked sufficient evidence to charge him. Investigators had Ridgeway provide a saliva sample in the hopes that techniques would someday be developed that could help solve the case. DNA analysis of that sample plus of evidence from three of the victims (14 years later) provided the necessary evidence to finally arrest him. If the saliva sample had not been taken and stored, he might never have been caught. Thus, from a law enforcement and public safety perspective, there are compelling arguments for indefinite storage of the actual biological samples rather than just a record of their CODIS profile.

The other side of the argument is that because such a sample contains a person's entire genetic blueprint, the danger of abuse overshadows the potential for solving a relatively small number of open cases. Several tests already exist to identify specific genes, many of them associated with diseases. For example, a genetic test can determine if a woman has a predisposition to breast cancer. If an insurance company were to test an archived sample and find such a gene, the woman might be denied health or life insurance. This could occur even though having the gene does not automatically mean she will develop the disease, only

(continues)

DNA Typing and Privacy *(continued)*

that the likelihood is greater. Similarly, a potential employer might elect not to hire her based on the increased chance that she might become ill. Complicating such a scenario is the fact that there are many cases in which a person is a carrier of a genetic disease but not a victim of it.

As if these concerns were not enough, some people believe that the way in which samples are obtained for routine DNA typing violates privacy rights. Drawing blood is far more invasive than obtaining fingerprints, either by inking or with digital technology. Objections have been raised stating that the act of drawing blood represents an unreasonable search and seizure, which is forbidden by the U.S. Constitution. However, drawing blood is not the only way in which samples can be obtained. Buccal swabs obtained from the inside of the cheek and saliva are amenable to DNA typing and do not require invasive procedures. Court rulings have generally supported states in obtaining blood or other materials for typing from convicted criminals. Thus, this aspect of the debate is not so much about the invasiveness of the collection as it is about its scope. If samples are collected from sexual offenders, then why not from all people convicted of a felony? Why not those convicted of certain misdemeanors? Where to draw that line continues to be debated.

As more samples are collected, storage and analysis of archived materials present practical problems. A recent estimate (2001) of the current backlog of samples awaiting DNA typing numbers between half and three-quarters of a million, requiring the services of 10,000 new forensic scientists within the next few years. DNA typing is a complex analytical technique that requires a scientific degree plus additional training and practice. This long "pipeline" slows potential growth of a qualified workforce even if labs were sufficiently funded to do the work, which they are not. Although laboratory automation continues to make inroads, typing is still labor and time intensive. Further complicating matters is the need to store samples in freezers until they can be analyzed, requiring space, fail-safe equipment (a thawed sample is a ruined sample unless immediately dealt with), detailed and accurate record keeping, tight security, and strict control of access to the samples. A problem in any one area could bring sample integrity into question, tainting results with potentially disastrous consequences. In addition, typing techniques target small samples that are copied (amplified) thousands of times, increasing the potential for contamination by foreign DNA, even in minute quantities. Extensive quality assurance and quality control procedures are essential to prevent contamination and to identify it if it does occur.

Recognizing these concerns, many agencies have adopted policies to avoid worst-case scenarios. Laboratory procedures are carefully monitored and executed. Increasingly, laboratories and analysts obtain certification and other credentials to further ensure that the data produced is accurate and reliable. Access to the CODIS database is strictly controlled, and stored samples are kept in secured areas and protected by rigorous CHAIN-OF-CUSTODY measures. Thus, it is not the procedures that present challenges so much as the number of samples that are daily added to already daunting backlogs.

These healthy disputes about DNA typing and privacy will play out in the courts in the years to come, and it is one that has been repeated many times. New technologies emerge that greatly assist law enforcement in protecting society, yet if carried to an extreme could threaten the foundations on which they stand. There was controversy when fingerprint collections were merged and the list of people routinely fingerprinted increased, yet a reasonable system of access and use has been worked out. No doubt the power of DNA databases to convict the guilty and to exonerate the innocent will be incentive enough for a similar evolution to take place in the 21st century.

Further Reading

Dawkins, R. "Arresting Evidence." *The Sciences,* November/December 1998, p. 20.

Pollard, R. "Crime Genes." *Technology Review* 103, no. 5 (2000): 29.

Sankar, P. "Topics for Our Times: The Proliferation and Risks of Government DNA Databases." *American Journal of Public Health* 3, no. 87 (1997): 336.

Shultz, W. "Chemists Needed for Forensic Analysis." *Chemical and Engineering News* 79, no. 46 (2001): 51.

Stevens, A. P. "Arresting Crime: Expanding the Scope of DNA Databases in America." *Texas Law Review* (March 2001): 921.

the discriminating power is now 1000 × 1000, or one in a million. If four fragments are typed with similar rates, the power increases to one in a trillion, which greatly exceeds the population of earth. Compare these figures to the ABO BLOOD GROUP SYSTEM, where Type A is found in 42 percent of the population, or roughly one-in-two.

RFLP is a powerful discriminating tool, but there are some disadvantages. Since the fragments are relatively long (thousands of base pairs), they are subject to degradation. Older cases or cases in which blood or body fluid evidence has decayed are questionable for RFLP. In addition, the technique is time and labor intensive, taking two weeks to complete for each probe used. Given the backlog of cases in forensic laboratories, this is a serious problem. Finally, RFLP requires a relatively large sample, a stain about the size of a dime if blood is used. Many cases hinge on blood spots much smaller than that. Given these limitations, coupled to advances in other techniques, RFLP has been replaced by the PCR-based STR typing techniques discussed below.

PCR Techniques

Polymerase chain reaction (PCR) techniques are a second category of forensic tests based on amplification techniques (see DNA). In PCR techniques, very small samples of DNA are copied before being typed using different systems and locations (loci) on the chromosomes. PCR cannot be used in RFLP typing because those fragments are too long for reliable amplification, but smaller repeating fragments are suitable; currently 13 loci are used in forensic work. The first locus used was DQA1, which has 28 different types and subtypes. The amplification step of PCR is carried out in a thermal cycler instrument, and each cycle begins by heating the DNA sample, causing the paired strands to unravel. A primer is added to isolate the segment of interest, and then complementary bases pair with those on the original strands, creating another potential sister strand. Adding DNA polymerase zips up the new strands. Repeated heating and cooling under the proper conditions allow many copy cycles. Each cycle doubles the amount of DNA and takes about two minutes. As few as 25 cycles can amplify the amount of DNA available well over a million times. Thus, PCR techniques are very sensitive and can be used on trace amounts of DNA such as those found in saliva on cigarette butts. However, with increasing sensitivity, the potential problems related to CONTAMINATION also increase, and procedures are undertaken with care to ensure that no foreign DNA mixes with that obtained from the evidence or other samples.

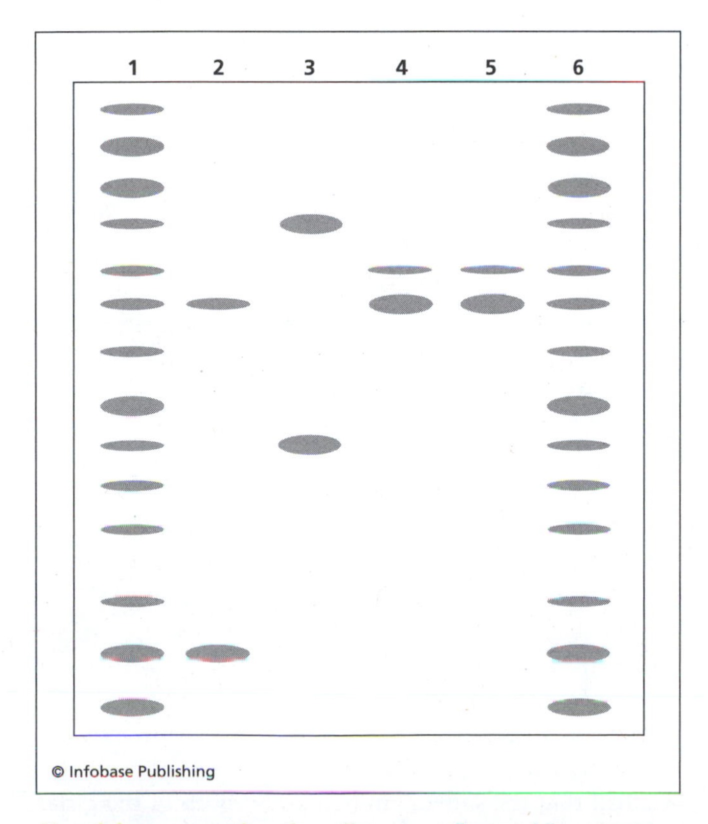

Pictorial representation of a radiogram, or "autorad," produced as part of the RFLP typing process

Once the DNA is amplified, it can be added to spots on commercially prepared strips as shown in the figure. Each dot contains complements of the fragment for that type, and a positive result is visualized as a blue color. Note that each type consists of two values since one is inherited from the mother and one from the father. There are also subtypes, and these are indicated by notations such as 1.2, which would be a subtype of the group 1. In the example shown, the second strip reveals a sample of type 2,3. The control dot is a control sample used to gauge the success of the procedure. Newer commercial systems such as Polymarker have added more loci, increasing the discriminating power of these dot-blot PCR techniques. A slightly different PCR typing tool is found at the D1S80 locus, which is a VNTR segment much like those used in RFLP except they are smaller. Since the fragments are smaller (hundreds of bases rather than thousands), PCR can be used to amplify the DNA. This technique is often referred to as Amplified Fragment Length Polymorphisms (AMP-FLP or AFLP).

The latest techniques in DNA typing, and the ones used almost exclusively, revolve around short tandem repeats (STRs), which consist of only three to seven base pairs and total fragment lengths of about 400. Since they are small, degradation is not as serious a problem as it is for longer fragments, and PCR can be used to amplify the sample. STRs are abundant in DNA, and although

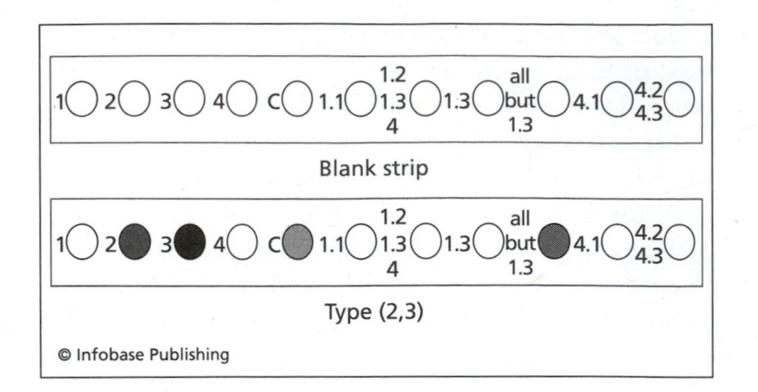

Appearance of DNA DQA1 typing strips

each locus has only a few variants, typing several loci at once dramatically increases the discriminating power. Thirteen loci have been selected as the standard in the United States and include loci such as TH01, D7S820, and D3S1358. Typing three systems leads to a discrimination power of about one in 5,000, while typing all 13 pushes this number into the trillions. An added advantage is that commercial STR kits provide a method to sex samples, a procedure that was cumbersome and difficult using techniques such as staining BARR BODIES. The amelogenin gene, which codes for dental pulp, is found on both the X and Y chromosomes but is six base pairs shorter on the X. Women, who are XX, will show one type, since both variants are the same, but men (XY) will show two, one shorter than the other. An added advantage to sexing a stain or other evidence is that the results eliminate approximately half the population as a potential source with one test.

STRs have become a standard typing test in most forensic labs, and data collected is being integrated into a national database system coordinated by the FBI. The CODIS system (Combined DNA Indexing System) stores DNA typing data from convicted felons and sexual assault cases and has been useful in linking crimes and identifying assailants. Key to the success of the database is consistent procedures and other QUALITY ASSURANCE/QUALITY CONTROL (QA/QC) measures to ensure that results from different laboratories are reliable and comparable. The larger the databases grow, the more useful they will become, much as large FINGERPRINT collections have become.

Mitochondrial DNA

There is one other DNA typing technique that is used in specialized circumstances where DNA from the nucleus is not available or comparison samples cannot be retrieved. All the techniques previously discussed type DNA found in the nucleus of cells. However, DNA does exist in one other place in the cell, in the mitochondria. These structures are located outside of the cell's nucleus. The mitochondria are key in cell energy production and have their own DNA, which is inherited solely from the mother. Thus, even if a person is not available to provide comparison samples (deceased, missing, and so on), comparison standards can be obtained from anyone in the mother's family. Because it is found in even greater abundance than nuclear DNA, it can be used on very small samples, and problems of degradation are reduced. Mitochondrial DNA (mtDNA) was used in the examination of remains thought to be those of ANASTASIA AND THE ROMANOVS, the imperial family of Russia murdered in 1918.

The mtDNA is circular, and there are two regions where variations occur called HV1 and HV2 for "hypervariable regions" 1 and 2, which can be amplified using PCR techniques. The discrimination power of mtDNA is not as great as STRs, the procedures are difficult and expensive, and potential contamination is a significant problem. Thus, few labs routinely perform this kind of DNA typing. However, when there is no nuclear DNA available, mtDNA is often the only choice.

First Case

DNA evidence was first applied to a case in 1987 in England. Four years earlier, a young girl was raped and strangled, and her body was dumped along a path near the town of Leicester. Although serological evidence was recovered, the pool of suspects was large, and little progress was made in the investigation. Three years later, a second rape-murder occurred, and the girl's body was dumped within a mile of where the first victim had been left. The serological evidence recovered indicated that it was likely the same assailant who had killed both girls. A mentally deficient young man confessed to the second murder, but the serological evidence (ISOENZYME type in this case) was not consistent. Police turned to Alec Jeffreys at Leicester University for help. Jeffreys (who would later become Sir Alec Jeffreys) performed an RFLP technique and was able to exclude the young man who had confessed. He was also able to confirm that the same man had killed both of the girls. However, by the end of 1986, police were left with no suspects. Their next step was to draw blood from every man in the area between the ages of 16 and 34 in an attempt to locate the killer. However, the lab procedures were slow, and by the middle of 1987, less than half of the analyses had been completed.

The break in the case came in August, when a local woman reported overhearing a conversation in which a man claimed that another, Colin Pitchfork, had paid him to give blood in his name using fake credentials. In September, police arrested Pitchfork, and the DNA analysis proved that he was the killer. He pled guilty in early 1988 and was sentenced to life in prison.

See also SINGLE NUCLEOTIDE POLYMORPHISM.

Further Reading

Duncan, G. T., M. L. Tracey, and E. Stauffer. "Techniques of DNA Analysis." In *Forensic Science: An Introduction to Scientific and Investigative Techniques*. 2nd edition. Edited by S. H. James and J. J. Nordby. Boca Raton, Fla.: CRC Press, 2005.

Inman, K., and N. Rudin. *An Introduction to Forensic DNA Typing*. 2d ed. Boca Raton, Fla.: CRC Press, 2002.

Shaler, R. "Modern Forensic Biology." In *Forensic Science Handbook*. Vol. 1. 2d ed. Edited by R. Saferstein. Upper Saddle River, N.J.: Prentice Hall, 2002.

DNA wars The first application of DNA evidence in the United States occurred in 1987, but the late 1980s and early 1990s were a time of controversy, contradiction, and courtroom drama collectively referred to as the DNA wars. An important chapter occurred in New York City in 1989, involving another double murder and another confession. In this case, someone had stabbed a woman and her two-year-old daughter in their apartment. One of their neighbors, a man named Castro, was the suspect, and a drop of blood found on his watch was a key piece of biological evidence. A private company called Lifecodes performed the DNA analysis, using the protocols then in use (RFLP) and determined that the type of the blood on the watch matched one of the victims. At trial (*People v. Castro,* 144 Misc. 2d 956 545 N.Y.S. 2d 985 (Sup. Ct. Bronx Co. 1989)), the defense pointed to lack of standardized procedures to question the work and the results.

The court permitted the scientific experts from both sides to meet as a group to discuss the science involved in DNA TYPING. The meeting went well by all accounts. There were eight experts in all, four of whom attended the meeting. They emerged with a two-page consensus report that stated that in effect, an admissibility hearing was the wrong place for scientific peer review, and that the adversarial nature of such a proceeding was incompatible with scientific peer review method. In *Castro*, there was no body of precedent for the judge to fall back on, so the Lifecodes results were not allowed. The prosecution's case evaporated. Castro later confessed to the murders, a vindication of sorts for the science. However, the need for procedural controls was set in concrete by the document written by Richard Roberts and his colleagues.

The larger effect of the *Castro* case was recognition that DNA was a powerful method, but that set procedures and strict quality assurance and quality control procedures would have to be agreed upon by the FORENSIC SCIENCE and the broader molecular biology community before such evidence could expect routine acceptance. The FBI was already moving in that direction, and the National Research Council, a body representing the National Academy of Sciences, was working on a comprehensive report regarding DNA evidence. The first report, "DNA Technology in Forensic Science," appeared in 1992. It generated significant criticism and more debate in the DNA war period. A follow-up report appeared in 1996, "The Evaluation of Forensic DNA Evidence," which addressed some of the contentious issues. Also discussed in the NRC reports were laboratory quality assurance, testing, and how results should be stated. However, as with serology a decade before, the RFLP methods were about to have the rug pulled out from under them, and so this specific debate was self-limiting.

RFLP targeted relatively large sequences. The longer the sequence, the more fragile it is and the more likely it is to degrade. Given the condition of most forensic samples, RFLP was bound to fail when evidence had degraded or aged. The solution lay in targeting smaller repetitive regions called single locus polymorphisms, sometimes abbreviated as SLPs or more recently as STRs (short tandem repeats). STR loci are similar to VNTR loci in that they have replication of the same sequence. The variability is, like VNTRs, in the number of repeats, not the sequence itself. The difference is size; STRs are much shorter than VNTRs and as such, more robust and less prone to damage during analysis. While each STR loci has fewer variants than the VNTRs that had been used, several loci can be typed and combined just as different blood group types could be chained together to increase discrimination.

The first locus studied forensically was the DQα system in the late 1980s, and by 1990 a commercial kit was available through Cetus. Other systems followed, including applications to VNTR markers. The impact of PCR technology spread to forensic applications in the 1990s, both in the United States and Europe. The FBI worked on amplified fragment length polymorphism (AMFLPs) loci such as D1S80, and some of these were used in Europe for a time. While DNA typing became more familiar to the forensic community and more analysts became skilled in the procedures, work continued on selecting what methods and loci to use. Debate and discussion in the forensic community revolved around how robust the loci were (i.e., could they be reliably extracted from old and degraded samples), how much sample was needed (less of a problem since PCR but still significant), and how variable the location is across the population. In recognition of problems with reading VNTR gels, the results had to be numerical and quantitative rather than fuzzy gels read by visual means. This reduced subjectivity and more important, facilitated databasing.

By the mid-1990s, the community zeroed in on STR loci, which were short repeating sequences of three to five bases, the total length of which was around 300 bases, versus thousands for the RFLP regions. Initially, loci were grouped into what were called multiplexes, which contained three or four different loci each, reminiscent of the Group I and Group II approach to isoenzymes and serum proteins. This is not coincidence; it reflects the necessity in the forensic world of getting the most information for each analysis. As the decade wore on, the debate moved on as to which loci to use. In the United States, 13 loci became the standard. With this, the DNA wars had ended.

Further Reading

Enserink, M. "FBI's Top Scientist Takes the Lead in Forensic Biology." *Science* 300 (2003) 41–42.

The Evaluation of Forensic DNA Evidence. Washington, D.C.: National Research Council (NRC), 1996.

Doyle, Sir Arthur Conan (1859–1930) British *Author* Doyle created the immortal SHERLOCK HOLMES character, who has become a widely recognized symbol of scientific sleuthing. Many forensic scientists were inspired to choose their careers by his adventures, published between 1887 and 1927 in the form of short stories and novellas. Doyle, a physician, described many tests and techniques in his stories that would not become common tools of forensic science until years later. Early pioneers in the field, particularly EDMOND LOCARD, praised Doyle's works and cited them as personal inspiration.

Doyle was born in Edinburgh, Scotland, in a poor household. At the age of 18, he enrolled in medical school at the University of Edinburgh, and many of his experiences there appear to have been influential in his writing career. The first story in which Holmes appeared, *A Study in Scarlet,* described a PRESUMPTIVE TEST for human blood, a technique not fully developed until after the turn of the 20th century. Holmes also employed ANTHROPOMETRIC techniques, used a microscope, used dust and mud to link suspects to scenes, and appreciated the concept of TRANSFER EVIDENCE. He also understood the special role and dangers of CIRCUMSTANTIAL EVIDENCE. Perhaps the most famous Holmes quote is found in "The Adventure of the Beryl Coronet," when he stated, "When you have eliminated the impossible, whatever remains, no matter how improbable, must be the truth."

Further Reading

Doyle, Sir Arthur Conan. *The Complete Sherlock Holmes: All 4 Novels and 56 Short Stories.* New York: Bantam Books, 1998.

Gerber, S. "A Study in Scarlet: Blood Identification in 1875." In *Chemistry and Crime: From Sherlock Holmes to Today's Courtroom.* Edited by S. Gerber. Washington, D.C.: American Chemical Society, 1983.

Liebow, E. "Medical School Influences on the Fiction of Arthur Conan Doyle." In *Chemistry and Crime: From Sherlock Holmes to Today's Courtroom.* Edited by S. Gerber. Washington, D.C.: American Chemical Society, 1983.

DRIFTS A specialized form of Fourier transform infrared spectroscopy (FTIR) that examines ELECTROMAGNETIC RADIATION (EMR) that is reflected from a surface rather than radiation that is absorbed. The initials stand for *d*iffuse *r*eflectance *i*nfrared *F*ourier *t*ransform *s*pectroscopy, and it is a surface analysis technique similar to ATTENUATED TOTAL REFLECTANCE (ATR). DRIFTS techniques are more sensitive than are FTIR and can operate over a larger concentration range and thus can be very useful in forensic applications in TRACE EVIDENCE analysis. Sample preparation is easy, and both solids and liquids can be analyzed. The apparatus needed for DRIFTS can be obtained as an accessory for many FTIR instruments. Among others applications,

DRIFTS has been used in the analysis of DRUGS, PAINTS, PLASTICS, and QUESTIONED DOCUMENTS.

drug absorption *See* ADME.

drug classification A system of categorizing illegal, abused, or controlled drugs (listed in the CONTROLLED SUBSTANCES ACT) based on their manner of physiological activity. The classes used and examples are listed below.

1. Narcotics: substances that relieve pain (analgesics) and promote sleep. Contrary to popular belief, not all illegal drugs are narcotics, and this confusion has led to misclassification of drugs. Cocaine is often listed as or considered to be a narcotic even though, by function, it is a central nervous system (CNS) stimulant. Narcotics abuse can lead to physical and psychological dependence (addiction). Heroin and the opiate ALKALOIDS are narcotics, as are synthetics such as OXYCODONE and DEMEROL.
2. Depressants: substances that depress the CNS and can produce effects including loss of coordination, impairment of judgment, and sleep. ALCOHOL is a depressant, as are the BARBITURATES, tranquilizers, BENZODIAZEPINES such as Valium, INHALANTS, and QUAALUDE.
3. Stimulants: substances that stimulate the CNS, producing a feeling of wakefulness, decreased fatigue, decreased appetite, and general well-being. Cocaine, amphetamine, methamphetamine, Ritalin, caffeine, and nicotine fall into this category. In higher doses, many stimulants can also act as hallucinogens.
4. Hallucinogens: substances that alter visual and auditory stimuli and produce hallucinations. LSD, MESCALINE, MARIJUANA, HASHISH, and ECSTASY are hallucinogens.

Drug Enforcement Administration (DEA) *See* DEA.

drug-facilitated sexual assault (DFSA) A rape or other sexual assault in which the victim is incapacitated by a drug. Often the drug is given to the victim without his or her knowledge or consent, such as when added surreptitiously to an alcoholic beverage. In typical scenarios, the victims leave their drinks unattended for a short time and after returning and consuming their beverages, they report memory loss, waking in strange places, finding themselves partially clothed, and feeling generally sleepy and confused. Sleepiness may persist for many hours after an incident. This, along with memory losses, often prevents victims from reporting the assault, or if they do report it, often too much time has elapsed since the drug was consumed for testing to detect it. Many drugs, including alcohol, can be used to facilitate a sexual assault, but this term more often refers to predator drugs such as GHB, ketamine, or Rohypnol. Other drugs that have been used as predator drugs include the BARBITURATE scopolamine, BENZODIAZEPENES, COCAINE, MARIJUANA, AMPHETAMINE and METHAMPHETAMINE, OPIATES, chloral hydrate, and antihistamines. The number of DFSE instances is unknown given the difficulty of detecting them and the reluctance of victims to report possible cases of DFSA.

See also DATE RAPE DRUGS.

Further Reading

Fitzgerald, N., and J. R. Riley. "Drug-Facilitated Rape: Looking for the Missing Pieces." *National Institute of Justice Journal* (April 2000): 9–15.

LeBeau, Marc A. "Toxicological Investigations of Drug-Facilitated Sexual Assaults." *Forensic Science Communications* 1 (1999).

Drug Recognition Experts (DREs) Police officers trained to evaluate drug impairment and to determine what drug or drugs a person has taken to become intoxicated. The Los Angeles Police Department organized the first such drug evaluation and classification program (DEC) in the late 1980s. A DRE test is not a roadside evaluation, but rather occurs after an arrest has taken place and the intoxicated person has been moved to a safe and controlled environment. The evaluation has 12 parts, including a BLOOD ALCOHOL CONCENTRATION (BAC) test, physical examination and testing, and an interview with the arresting officer. Based on these tests, the DRE provides an opinion of what type of drug is involved (depressant, stimulant, narcotic, hallucinogen, PCP, inhalant, or marijuana). The opinion is always confirmed using toxicological tests of urine, blood, or both.

Further Reading

Kunsman, G. "Human Performance Toxicology." In *Principles of Forensic Toxicology*. 2nd edition. Edited by B. Levine. Washington, D.C.: American Association of Clinical Chemistry, 2003.

Drug Testing in Sports

Drug testing for athletes is a niche of forensic TOXICOLOGY that presents unique challenges to toxicologists, athletes, and sports organizations. While many of the drugs banned by such organizations as the International Olympic Committee (IOC), National Collegiate Athletic Association (NCAA), and the National Football League (NFL) are illegal substances, others fall into the gray areas of performance enhancers and supplements. Athletes use these substances hoping to increase strength and endurance or to reduce recovery time from injuries or tough workouts. However, many supplements such as steroids have numerous known side effects, many of them dangerous and irreversible. Consequently, there is heightened concern about the use of drugs and performance enhancing supplements by athletes as young as high school age.

There is also a spirited debate within professional and Olympic sports as to what supplements should be banned and how to interpret record-setting performances obtained by athletes taking such supplements. Furthermore, the deaths of several famous athletes such as Florence Griffith Joyner, a track and field champion who died in 1998, and Korrey Stringer, a football player who died in 2001, have raised public awareness of the problem. Although drug use by either athlete was never proven, suspicions linger that drug use played a role in their untimely deaths.

Some argue for purity in sports and believe that all performances should be done without the assistance of any drugs. Others argue that is an unreasonable and unenforceable position. For example, the NCAA bans CAFFEINE if its concentration exceeds 15 parts-per-million (15 μg/mL) in urine even though there is no prohibition against caffeinated beverages or supplements. Caffeine is also an ingredient in some over-the-counter headache remedies. Thus, the question involves an aspect of intent since a caffeine pill is a different matter from that of one too many cups of coffee. This issue has become more pressing with recent findings of banned substances in supplements that do not list them as ingredients on the label. Regardless of the outcome of these debates, enforcing the rules that result depends on the ability of forensic toxicologists to detect the targeted drugs and their metabolites in blood and urine.

Substances Used

The type of drugs or supplements that athletes take depends on what type of events they are training for. Sports that emphasize strength and short, intense bursts of activity generally favor CREATINE, ANABOLIC STEROIDS, and other hormones such as human growth hormone. Sprinting events, football, and baseball are examples. Baseball player Barry Bonds, renowned for his home-run hitting prowess, has admitted using creatine, which is not a controlled substance, nor is it banned in sports. Whether its use constitutes cheating has yet to be decided.

Endurance athletes (e.g., runners, swimmers, and rowers) whose sports require long periods of exertion favor drugs such as EPO (erythropoietin) and techniques collectively referred to as "blood doping." All of these techniques are geared toward increasing the numbers of red blood cells and thus the ability of the blood to carry oxygen. Training at high altitude can create this effect naturally, but the process takes weeks and the effects wear off when the athlete returns to lower elevations.

Blood doping involves withdrawing some of the athlete's own blood and separating out and preserving the red blood cell portion. The cells are reintroduced just prior to competition. No drugs are used per se, but a blood test that determines the red blood cell count can indicate if doping has likely occurred. The compound EPO occurs naturally in the body, complicating any toxicological tests designed to detect misuse. As a prescription, it has been used to treat liver and kidney problems; as a supplement it stimulates the production of red blood cells, producing the same beneficial

drugs and drug analysis The largest workload in most forensic science labs is related to drug analysis. Drug analysis also represents the bulk of the work done by forensic toxicologists. Drug analysis in forensic labs deals with physical evidence such as plant material or powders seized from suspects and evidence seized from CLANDESTINE LABS. Toxicologists focus on biological samples such as blood, urine, and samples collected at AUTOPSY. Drug testing is also common in the workplace, athletics, and the military, and toxicologists perform these analyses.

Forensic drug analysis proceeds in the same manner as other forensic evidence, starting with simple PRESUMPTIVE TESTS (such as the Duquenois-Levine test for marijuana) and moving toward specific identification. Which drugs are tested for depends primarily on the CONTROLLED SUBSTANCES ACT (CSA), which lists controlled drugs and PRECURSORS and the penalties associated with them at a federal level. The classes of drugs tested for include narcotics, stimulants, depressants, hallucinogens (including MARIJUANA), INHALANTS, ANABOLIC STEROIDS, and DATE effects as blood doping. The Tour de France bicycle race was marred by allegations of EPO use in 1998.

Application of Forensic Toxicology

The challenges to sports and to the toxicologists who monitor drug use in athletes are formidable. There is always a lag time between the discovery that a substance is being misused and the development of a sensitive and reliable test to detect it. Once such a test is available, violators have had time to develop strategies to beat the tests or have moved on to other substances and supplements. In addition, testing is usually done at the time the events take place. Thus, athletes can adjust doses and schedules to ensure that all traces of banned substances are gone when testing is likely. Since most drugs are taken over a long period to build strength and endurance, the beneficial effects will linger after substances are no longer detectable in blood or urine. Most tests are based on urine since this approach has been deemed less invasive than drawing blood. However, for many of the toxicological tests, blood is the preferred medium. In addition, it is nearly impossible for a violator to adulterate or substitute a blood sample, while it is comparatively easy to alter or switch a urine sample. Prior to the Sydney Olympics of 2000, the IOC changed its rules to allow blood testing, but most other sports, such as professional football, still use urine. Related to these problems is the complicity of some sports doctors and even clinical testing laboratories in enabling the use of banned substances.

Recent research has highlighted yet another problem facing toxicologists as well as athletes. Nutritional and performance supplements such as creatine are available to the public and are essentially unregulated by the FOOD AND DRUG ADMINISTRATION (FDA). Given this lack of oversight, label claims can be deceptive, particularly to younger aspiring athletes eager for any edge they can get. The relationship between actual composition and labeled ingredients is loosely controlled, and wide variations are seen. Worse, some of these supplements have been found to contain banned substances such as steroids, ephedrine, and caffeine, none of which were listed on the label. Thus, athletes may take legal supplements, and toxicological analysis will still find banned substances in their blood or urine. Chemical analysis of the original substance can help to detect mislabeling, but only if such a sample can be found.

Forensic toxicologists have worked to improve testing methods, but there will always be analytical problems in cases where banned substances (such as EPO and hormones) are naturally present in the body. Chemically, synthetic molecules are identical to their natural counterparts, so testing must rely principally on detecting abnormally high amounts. In turn, this requires extensive research to determine what the range of normal is. All these considerations lead inevitably to the fundamental issue—cost. Millions of dollars would be needed in a sustained research and development effort to make sure that the verification procedures keep pace with the ingenuity of the violators. Since many of these substances are not illegal, questions are inevitably raised about the allocation of resources to police sports when so many other research needs go unmet. The only certainty is that the debate is sure to continue.

Further Reading

Freeman, M. "Sports of the Times: NFL Shows Trust Is Won by Verifying." *New York Times,* July 20, 2002.

Longman, J. "Unbelievable Performances: A Special Report; Widening Drug Use Compromises Faith in Sports." *New York Times,* December 26, 1998.

NCAA Sports Sciences. "2001–2002 NCAA Banned-Drug Classes. Available online. URL: http://www.ncaa.org/sports_sciences/drugtesting/banned_list.html. Downloaded on July 24, 2002.

Sandomir, R. "OLYMPICS; Banned Substances Found in Many Food Supplements." *New York Times,* October 12, 2001.

RAPE (predatory) DRUGS. BLOOD ALCOHOL CONCENTRATION (BAC) analyses are conducted in toxicology laboratories.

Analytical procedures have been developed by the DRUG ENFORCEMENT ADMINISTRATION (DEA) for use by forensic chemists, but the specific combination of tests used for a given case will vary among labs and depending on the form of the evidence. When an unknown powder or plant material is analyzed, the first steps involve presumptive tests using chemical reagents that produce characteristic color changes. These are screening tests only, but they are useful for narrowing down possibilities. If the evidence submitted is commercial pills, tablets, or other preparations, the analyst will consult reference guides such as the *PHYSICIAN'S DESK REFERENCE (PDR)*, which describes commercially available drugs and how they are supplied. Based on this initial evaluation, an analytical scheme will be implemented. If the evidence is drug paraphernalia such as pipes or syringes, additional sample preparation is required, including EXTRACTIONS.

Techniques used beyond the initial screening tests can include CRYSTAL TESTS but are primarily instrumental, with the exception of THIN LAYER CHROMATOGRAPHY (TLC). The procedures used for specific identification and in some cases QUANTITATION include GAS CHROMATOGRAPHY/MASS SPECTROMETRY (GC/MS), FOURIER TRANSFORM INFRARED SPECTROSCOPY, and HIGH PERFORMANCE LIQUID CHROMATOGRAPHY (HPLC). When taken together, screening tests through INSTRUMENTAL ANALYSIS provide definitive identification of the drug.

Additional tasks in a drug analysis can include identification of CUTTING AGENTS and, in the case of COCAINE, tests to determine what optical isomer is present. Quantitation is often performed with powders such as cocaine and HEROIN, where it is necessary or desirable to know what percentage of a powder or sample is actually the drug and how much is cutting agents. Chemists may also perform tests to help investigators determine if a different drug sample came from the same source.

See also DRUG CLASSIFICATION.

Further Reading

Christian, D. R. Jr. "Analysis of Controlled Substances." In *Forensic Science: An Introduction to Scientific and Investigative Techniques*. Edited by S. H. James and J. J. Nordby. Boca Raton, Fla.: CRC Press, 2003.

Duffy blood group system *See* BLOOD GROUP SYSTEMS.

Duquenois test (Duquenois-Levine test) A PRESUMPTIVE TEST for MARIJUANA and HASHISH. Also called the Duquenois-Levine test or the modified Duquenois-Levine test, it consists of three reagents that react with the active substance in marijuana, THC. The reagents are a 2 percent solution of vanillin and 1 percent acetaldehyde in ethanol, concentrated hydrochloric acid (HCl), and chloroform. Dried plant matter, seeds or seed extracts, or other material to be tested is placed in a test tube and the reagents added in order. The sample is shaken and a positive result is indicated by a purple color in the lower chloroform layer. This is not a definitive test for marijuana, as there are other oily plant extracts that will yield FALSE POSITIVE results. However, the Duquenois-Levine test is an integral part of marijuana analysis. A sequence of photos showing the test is found in the color insert.

dust analysis Dust is one of the most common forms of TRANSFER EVIDENCE that can be used to link a person to a location. Dust is a source of TRACE or microtrace evidence and generally consists of particulates of various sizes and compositions that settle out of the air and onto surfaces, clothing, skin, and so on. Attempts to INDIVIDUALIZE dust (to show an origin from one unique source) depend on characterizing the types of materials that it consists of and their relative amounts. The importance of dust was recognized by key figures early in the history of forensic science. Both HANS GROSS and EDMOND LOCARD wrote about the significance of dust as transfer evidence in the 1920s. Even SHERLOCK HOLMES, the fictional father of forensic science, appreciated the ability of dust to link a suspect to a crime scene. Now, as then, the principal tool of dust analysis is MICROSCOPY. Dust can also serve as a medium for impressions. Latent FINGERPRINTS and other IMPRESSION EVIDENCE such as shoeprints can be left in dust, requiring specialized lifting and preservation techniques.

Dust can be made up of virtually anything that is small enough to become airborne. A partial list of possible components includes hair (human or animal), fibers (natural and synthetic), wood products, building and construction materials (plaster, sheetrock, sawdust), pollen and spores, algae, diatoms, leaves, mineral and glass, soil, starches, flour, spices, sugars, powdered foods (dry milk, cake mix), tobacco and ash, paper, flakes of metal or paint, rust, dried blood or skin, insects and insect parts, and feathers. Dust can be divided and categorized a number of ways. One is by size of the particulates that compose it; a second is to consider the fibrous components such as hairs and FIBERS separate from the particulate components such as glass and soil. Yet another scheme is to break down the components based on origin such as botanical (plant), zoological (animal), anthropogenic (a human source), geological, and food (grains, seeds, and so on).

Because dust is everywhere and because it is so easily transferred, special precautions are needed to avoid CONTAMINATION once a sample has been taken. Conversely, dust is by its nature mobile, so precautions are also taken to ensure that once it is collected, evidence is not lost. Although the exact procedures will vary based on the case and the analyst performing the analysis, dust characterization is done primarily using microscopes such as a STEREOMICROSCOPE, COMPOUND MICROSCOPE, and POLARIZING LIGHT

MICROSCOPE (PLM). The analyst may elect to use a preliminary examination to sort and separate different types of particles and then subject the sorted materials to more specific tests. One invaluable tool for dust analysis is a book entitled *The Particle Atlas,* which as the name suggests contains information needed to identify particulates. In some cases, microscopic examination may be supplemented with microchemical tests, microcrystalline tests, microspectrophotometry analysis (FTIR), and elemental analysis using scanning electron microscopy (SEM) coupled with energy dispersive or X-ray diffraction capability (SEM/EDS or SEM/XRD).

Further Reading

Petraco, N. "Forensic Dust Analysis." In *More Chemistry and Crime, From the Marsh Arsenic Test to DNA Profile.* Edited by S. M. Gerber and R. Saferstein. Washington, D.C.: American Chemical Society, 1997.

dyes compounds that are used as colorants. The term *dye* usually refers to an organic compound (synthetic or natural such as henna), while the term *pigment* refers to inorganic colorants such as zinc oxide, which is used as a sunscreen. A dye is soluble in whatever solvent is being used, and a pigment is not. Thus, the definition of a dye versus a pigment often depends on the solvent. In forensic science, dyes can be important in the analysis of FIBERS, INKS, and dyed HAIR. Fabrics and textiles are colored with one or more dyes, which penetrate the fibers and can be characterized by chemical and spectroscopic analysis. The most common method is MICROSPECTROPHOTOMETRY, in which the color can be characterized by creating an ABSORPTION SPECTRUM. If enough sample is present, chemical separations and comparisons can be attempted using THIN LAYER CHROMATOGRAPHY (TLC), where individual dye components are separated from each other and can be compared. Instrumental techniques such as HIGH PERFORMANCE LIQUID CHROMATOGRAPHY (HPLC) have also been used for dye analysis.

In the area of QUESTIONED DOCUMENTS and forensic CHEMISTRY, the dyes that make up inks can be important in evidence analysis, using similar techniques to those described above. Black inks that are so common in pens and other markers are combinations of different dyes that vary between companies, years, and types of writing instrument. Thus, an ink comparison can be useful in determining if two different pens were used to write a document such as a forged check or in detecting counterfeit currency. The U.S. Treasury has been cataloging inks for more than 30 years, and some companies are now adding tagging materials to inks and changing them on a regular basis. As a result, it is sometimes possible to date a writing based on the composition of the inks used in writing it.

See also COLOR AND COLORANTS.

dynamite *See* EXPLOSIVE; NITROGLYCERIN.

E

ecgonine methyl ester One of two primary metabolites of COCAINE that can be found in BLOOD and URINE after the drug has been ingested. The other important metabolite is BENZOYLECGONINE.

ecology, forensic *See* ARCHAEOLOGY, FORENSIC; ENVIRONMENTAL FORENSICS.

ecstasy A DESIGNER DRUG that is an analog of METHAMPHETAMINE. The complete chemical name of the drug is 3,4-methylenedioxymethamphetamine, commonly abbreviated as MDMA. Unlike other designer drugs, MDMA did have a brief history of legitimate uses before its popularity as an illegal drug. MDMA was originally synthesized in 1914 as an appetite suppressant, although it was never marketed as such. Psychotherapists also experimented with it for a brief time in the 1970s and 1980s, but it was quickly abandoned. Ecstasy is increasing in popularity among adolescents and young adults, becoming a favorite "party drug" used at "raves." Ecstasy produces a sense of euphoria and heightened empathy, and it can lead to hallucinations. It is listed on Schedule I of the CONTROLLED SUBSTANCES ACT (CSA). Most of the drug is synthesized in CLANDESTINE LABS in Europe.

EDTA A preservative used for liquid blood that will be subject to DNA TYPING or other serological analyses. EDTA stands for ethylenediaminetetraacetic acid, and it is a common additive in food and consumer products such as shampoo. EDTA works as a preservative by forming strong, water-soluble complexes with ions such as calcium and magnesium (Ca^{2+}, Mg^{2+}) and many other similar ions that are needed by microbes. When the ions are bound up in EDTA complexes, they are unavailable to microbes, which inhibits their growth.

See also SIMPSON, O. J.

education The education of a forensic scientist/criminalist can be divided into two segments, a college degree program and continuing education once on the job. The usual requirement for entry-level positions into forensic labs is a science degree (B.S. or B.A. degree). In addition, many labs specify minimum numbers of college credits in biology, chemistry, physics, and math. Laboratories now often require advanced biology courses in areas such as genetics, immunology, and/or molecular biology, the background needed for DNA typing. A few institutions offer undergraduate degrees in forensic science, forensic chemistry, or criminalistics; even fewer offer graduate education leading to an M.S. in forensic science or related forensic discipline. One of the most valuable components of an education in forensic science is an internship in a working forensic laboratory, and many programs offer or encourage such experiences.

Once hired, extensive training is required before an analyst can begin doing casework independently. Much of this is accomplished while working under the supervision of senior scientists in an apprenticeship. Continuing

Careers in Forensic Science: A Reality Check

As forensic science garners more publicity both from the media and through fiction, interest in forensic science careers and enrollment in forensic science college degree programs has skyrocketed. However, many students pursue their dream job with unrealistic expectations and ideas about careers in forensic science. In some ways the field is overestimated, and in others it is underestimated, particularly in breadth. For prospective students, a little research can go a long way toward eliminating false assumptions.

Forensic science has come to encompass dozens of specialties that continue to increase as the role of science and technology increases. What started out as a tiny and obscure corner of medicine has grown into a worldwide profession including everything from forensic ACCOUNTING to forensic X-rays (RADIOLOGY), with dozens of areas in between. Forensic COMPUTING followed computers into everyday life; forensic ENGINEERING and accident reconstruction are part and parcel of modern air and ground transportation. It was forensic seismologists who detected the explosion aboard the Russian submarine *Kursk* in 2000 and studied the shock waves generated by the collapse of the twin towers of the World Trade Center in September 2001. Even in their wildest dreams, Sir Arthur Conan DOYLE and his fictional Sherlock HOLMES could never have imagined how far forensic science has spread given that the first forensic science lab (formed by EDMOND LOCARD) had little more than a crude spectrophotometer and a microscope. The continually expanding extent of forensic science opens more career doors than many might realize.

As the breadth of the field has increased, so has the depth, and no area better illustrates this than the analysis of blood. Around the turn of the 20th century, the scientific analysis of blood was limited to a few PRESUMPTIVE TESTS. Since little could be done with bloodstain evidence, correspondingly little specialized education and training was needed to analyze it. However, in 1901, KARL LANDSTEINER discovered the ABO BLOOD GROUP SYSTEM, and in 1915, LEON LATTES introduced a test that could be used to type stains. Forensic SEROLOGY was born. Serologists needed a strong background in biology, biochemistry, and immunology as well as exemplary skills in the laboratory. Advances continued through the 1980s, adding techniques for typing ISOENZYMES and SERUM PROTEINS. Blood evidence expanded to include body fluids as understanding of SECRETORS and nonsecretors grew. But these advances paled in comparison to the effects of the introduction of DNA TYPING, a transition that continues. With DNA typing, forensic serology began an evolution to forensic BIOLOGY, a specialization that would have been unrecognizable to a forensic scientist working a century ago. Similar expansion has occurred in nearly every forensic discipline. The analysis of QUESTIONED DOCUMENTS encompasses e-mail and computer printers; FINGERPRINTS are visualized with lasers; CRIME SCENE reconstructions are done using software; TRACE EVIDENCE is examined with SCANNING ELECTRON MICROSCOPES (SEM) that can magnify an image a millionfold. As a result of these advances, it is becoming increasingly difficult for forensic scientists to be generalists, demanding instead highly trained specialists in many forensic areas.

Another consequence of the increasing scientific and technical rigor of forensic science is a corresponding increase in the education required to enter the field. Increasingly, the minimum requirement for an entry-level job in a traditional crime laboratory is a B.S. degree in a natural science such as chemistry or biology with a complement of math, physics, and statistics courses. As college programs

education is also an important part of forensic science. Agencies, professional societies, and institutes such as the McCrone Institute, which specializes in forensic MICROSCOPY, offer many training opportunities. Continuing education helps analysts keep abreast of the latest technical advances and can also provide the training necessary to allow them to switch or expand into other forensic disciplines.

For specialties such as forensic ANTHROPOLOGY and forensic ENTOMOLOGY, the educational track is different. In these areas, students normally obtain a Ph.D. degree in the field and do graduate and postgraduate work with a forensic anthropologist or forensic entomologists. These forensic scientists do not work in a typical local, state, or federal laboratory but rather participate in investigations as consultants. In many instances, forensic investigations are just one aspect of their career, and so the educational training is directed primarily toward the larger field (anthropology or entomology in this example).

See also FEPAC.

EDXRF (electron diffraction X-ray fluorescence) *See* SCANNING ELECTRON MICROSCOPY (SEM); X-RAY TECHNIQUES.

Ehrlich, Paul (1854–1915) German *Chemist* Although not a forensic scientist, Paul Erhlich's work was in forensic science become popular and more applicants are available to labs, this minimum may creep upward to a master's degree. Outside of the traditional areas of local, state, and federal laboratories (drug analysis, fingerprints, firearms and toolmarks, trace evidence, photography, and serology/DNA), the requirements are even more demanding. To practice forensic ANTHROPOLOGY, an area for which there is great interest but relatively few jobs, the minimum requirement is usually a Ph.D. in anthropology with additional training in forensic applications. There are more applicants for graduate school openings than there are spaces available, and the chances of any one student breaking into the field are slim. MEDICAL EXAMINERS (ME) are physicians, so they must get into and complete medical school and then undertake years of additional training in PATHOLOGY and forensic medicine. A similar route is required of forensic ODONTOLOGISTS (dentists) and forensic psychiatrists. For those considering a career in forensic science, discovering the education needed helps to distinguish between a passing interest and passion.

An applicant for a position in a typical full-service laboratory must go through a series of steps before walking into work for his or her first day. If the applicant has the required educational background, the next hurdle in many jurisdictions is a written test. Basic chemistry, biology, math, and other related subjects are fair game. Those who obtain scores deemed adequate are passed to the laboratory director for review and an interview. If the applicant passes this stage, an intensive background check and polygraph test typically follow. Only after successfully navigating this maze can a person be hired. The first several months on the job are devoted to study, observation, and working with senior examiners in an apprenticeship that takes the trainee through the steps necessary to become an independent forensic analyst. Another component of this ongoing training can be preparation for professional certifications, a task that can take years. Continuing education, training, and refresher courses are a fixture of life in a forensic lab.

As a work environment, forensic labs are dynamic, lively places. Office space is often shared and small; lab space may be as well. Most labs, particularly full-service ones, are located in large cities, often downtown and often in close proximity to a police station or other facility. Once an analyst is doing independent casework, he or she can expect to be SUBPOENAED to testify on his or her work. Subpoenas offer no flexibility, and appearances are rarely rescheduled to meet the analyst's needs, wants, or schedule conflicts. If a court issues a subpoena, there is no conflict; the analyst must appear or be subject to contempt of court charges. The amount of time testifying will vary depending on the area of forensic science the analyst practices, their caseload, and other factors. Analysts who also go out on crime scenes can be asked for help at any time and may be required to live close to the lab to shorten response time. Because of these numerous demands, to work part-time in a forensic lab is nearly impossible.

Even with these considerations and caveats, forensic science programs continue to draw a large pool of enthusiastic students who stay the course and join the profession. As of May 2007, 15 academic institutions offered programs accredited by the Forensic Education Programs Accreditation Commission (FEPAC) in forensic science at the bachelor's and master's degree level. Many other institutions offer forensic programs that are not accredited as of this date. Students interested in forensic science should carefully research academic programs to ensure that the one they select offers the best courses based on realistic educational and career goals.

important to forensic CHEMISTRY and also to TOXICOLOGY as it emerged in the late 19th century. He introduced an organic form of ARSENIC in 1901 as a replacement for toxic mercury-based treatments and won the Nobel Prize for this and for many other achievements in 1908. In the 1930s, arsenic became one of the first chemotherapeutic agents used, and physicians still use some forms in modern cancer therapy today. Ehrlich was one of the more productive and colorful of the color chemists who worked on the forefront of the new DYE industry in the late 1800s. He often trundled a box of cigars under one arm, of which he smoked around 25 a day. He was also reported to eat little and to frequent beer halls where he got into spirited discussions and debates. Despite, or because of, this eccentricity, he won a Nobel Prize in medicine in 1908 for work related to both medicine and dyes. He worked with stains used to color tissues and microorganisms. His work in the late 1880s was the foundation of the Gram staining procedure still used today to differentiate bacterial types. In the 1890s, Ehrlich turned his attention to immunological work and later to medicine and PHARMACOLOGY. He screened hundreds of compounds for a substance that would kill spirochete, a bacteria that was known to cause syphilis. After screening more than 600 compounds, Ehrlich successfully identified one (that contained arsenic) and it became known as *Salvarsan*. Prior to his work, mercury was the treatment, leading to many fatalities and poisonings.

Ehrlich's test A PRESUMPTIVE TEST used in DRUG ANALYSIS to detect LSD and other related ergot ALKALOIDS. The reagent contains p-dimethylaminobenzaldehyde (p-DMAB) in a solution of sulfuric and hydrochloric acids and is often referred to as the p-DMAB test or by the older term Van Urk test. In the presence of LSD or other ergot alkaloids, the reagent will turn purple, but the test is complicated by the fact that LSD is often found in levels so low that the test can produce a FALSE NEGATIVE.

Elavil (amitriptyline) A tricyclic antidepressant that is in the same class of prescription drugs as Prozac.

See also DRUG ANALYSIS; PRESCRIPTION DRUGS.

electromagnetic radiation (EMR)/electromagnetic energy and the electromagnetic spectrum "Light" as normally thought of is visible light or white light, which can be broken down into component colors using prisms and other devices. The result is a spectrum of colors familiar as the colors of the rainbow. However, visible light is one small portion of the electromagnetic spectrum that represents all different kinds of electromagnetic radiation (EMR), also referred to as electromagnetic energy. The term "radiation" as used here is *not* the same thing as radiation associated with nuclear chemistry, nuclear power, and nuclear weapons. EMR, including visible light, is composed of both electrical and magnetic energy. Instruments built to study the interaction of EMR and matter (such as drug samples) are called SPECTROPHOTOMETERS, and such instruments are critically important in forensic science.

EMR can be described as both a wave and a particle. When describing it as a wave, two terms are important, the wavelength (symbolized as λ) and the frequency (ν). The idea can be visualized by imagining dropping a rock into a still pond. Energy ripples away from where the rock was dropped in waves that can be described by their wavelength (distance from crest to crest) and by their frequency (how many waves per second pass a fixed point). Electromagnetic radiation that is high energy has short wavelengths and high frequency; low energy EMR has long wavelengths and low frequencies. When EMR is thought of as a particle, the source of the energy can be thought of as a gun that "fires" discrete packets of energy, which are referred to as photons. The energy of a photon is described by the relationship $E = h\nu$, where h is constant (Planck's constant) and ν is the frequency. Thus, the wave model and particle model are related through the frequency.

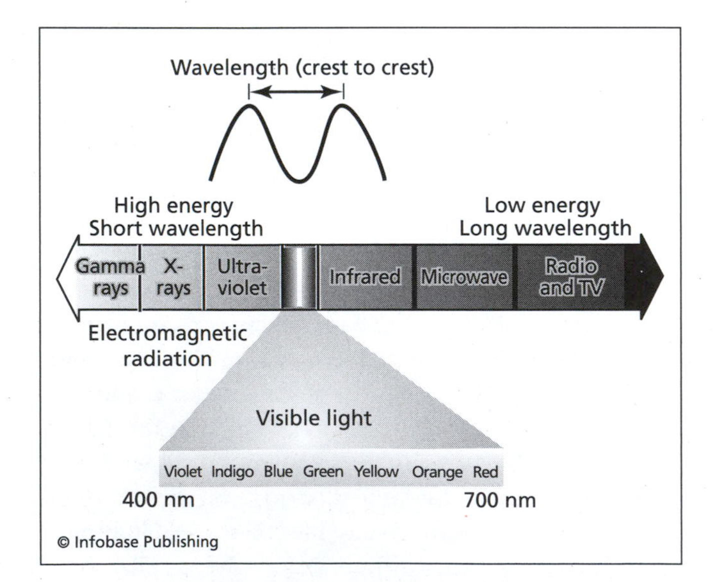

Spectrum of electromagnetic energy. Higher energy corresponds to a shorter wavelength and higher frequency, while lower energy corresponds to longer wavelengths and lower frequency. The visible portion of the spectrum is shown; not to scale.

The figure shows the spectrum of electromagnetic energy. At the high-energy end are cosmic rays and X-rays, leading into ultraviolet (UV) radiation. UV radiation from the sun is known for its ability to cause skin cancer. The next lower energy portion of the spectrum is visible light (VIS), the only type humans can sense directly. As shown, visible radiation consists of colors that can be described by their wavelengths, usually given in nanometers (one nanometer is one one-billionths of a meter long). Visible radiation progresses from violet (nearest the UV range) to red, and then progresses into the infrared region (IR), which is not visible. Lower energy EMR is found in the microwave region (exploited in microwave ovens) and into the TV and radio range. On a conventional radio tuner, radio stations are defined by the frequency of the signal that they broadcast.

The regions of the spectrum most often used by instruments in forensic science labs are the UV, VIS, and IR range. Typically, an instrument will detect how much EMR in the particular range is absorbed and will relate that absorbance to concentration. In the visible range, these techniques are referred to as COLORIMETRY. Infrared absorbance and instrumenta-

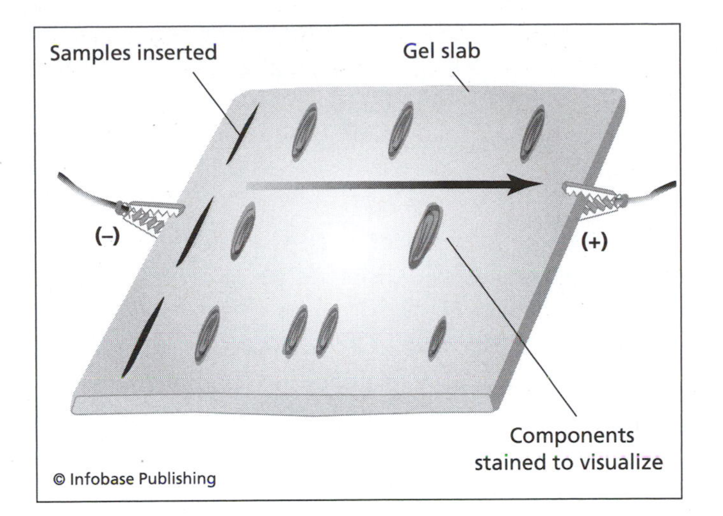

Simplified depiction of gel electrophoresis. The charged protein molecules move through the gel at a rate determined by their size and charge.

tion is used extensively in DRUG ANALYSIS and TRACE EVIDENCE ANALYSIS.

electron microprobe (EMP) *See* SCANNING ELECTRON MICROSCOPY (SEM); X-RAY TECHNIQUES.

electrophoresis A technique of separating large charged molecules such as proteins based on their mobilities in an applied electrical field. In forensic science, electrophoresis was extensively used up until the mid-1990s for the typing of ISOENZYMES in BLOOD and BLOODSTAIN evidence. Although a few labs still type isoenzymes, electrophoresis is now used primarily in the form of CAPILLARY ELECTROPHORESIS in DNA TYPING and other chemical analyses such as DRUG, GUNSHOT RESIDUE (GSR), and INK analyses.

As shown in the figure, electrophoresis is carried out in a gel medium (or similar material such as polyacrylamide), which is attached to an electrical power supply. The ends of the gel are in electrical contact with an aqueous (water-based) buffer solution by means of a wet sponge or similar material. In the example shown, the sample is inserted into a slit at the extreme left-hand side of the gel. In the case of bloodstained fabric, threads from the stain could be inserted directly. The power supply is activated creating an electrical field with a positive and negative end. The proteins, which usually carry a net negative charge, migrate to the positive pole since unlike charges attract. Different proteins in the samples are made up of different subunits and thus will differ from each other in many ways such as size, charge, and shape. Because of these differences and the structure of the gel, each component moves at a different speed. As a result, these components will separate from each other over time. Once the run is complete, the components must be visualized. Depending on what isoenzyme system is being typed, this can be accomplished by use of a developer or by viewing the gel under UV (black) light. Although this form of electrophoresis is rarely used in forensic applications anymore, the fundamental principles apply to capillary electrophoresis (CE).

electrostatic detection apparatus (ESDA) A device used to visualize INDENTED WRITING, which is a type of evidence encountered in QUESTIONED DOCUMENT analysis. Indented writing is writing that is transferred to a sheet of paper underneath the one on which the original writing occurred. For example, if someone writes something on the top sheet of a notepad, the pressure exerted by the pen or pencil is often sufficient to indent the paper beneath. ESDA takes advantage of the same technology used in copying machines and laser printers to help visualize the indented writing. A sheet of thin plastic is placed over the paper that has the suspected indented writing, and both are placed in a vacuum chamber. A charge of static electricity is imparted to the plastic, and then toner powder is applied to the surface of the plastic. An image of the indented writing will be created on the plastic, with the added advantage that the original paper is not altered or damaged. ESDA methods can also be used to lift footprints in DUST.

electrostatic lifting *See* IMPRESSION EVIDENCE.

elemental analysis Chemical testing (using primarily INSTRUMENTAL ANALYSIS) aimed at detecting individual chemical elements. Both QUALITATIVE and QUANTITATIVE information can be obtained. The most common forensic applications of elemental analysis include GUNSHOT RESIDUE (GSR), where elements such as lead, copper, barium, and antimony are of interest. Another example is the analysis of pigments in PAINT, where elements such as iron and titanium are important. The techniques used for elemental analysis include ATOMIC ABSORPTION (AA), ICP TECHNIQUES, and X-RAY TECHNIQUES, including those associated with SCANNING ELECTRON MICROSCOPES (SEM).

elimination prints FINGERPRINTS collected at a CRIME SCENE of all personnel who were in and at the scene and who might have inadvertently left prints on evidence. In addition to fingerprints, palm prints and footprints are often collected.

ELISA (enzyme linked immunosorbent assay) *See* IMMUNOASSAY.

emission spectroscopy A form of SPECTROPHOTOMETRIC analysis. In forensic labs, the most common type of emission spectrometry is inductively coupled plasma atomic emission (ICP-AES), a type of ELEMENTAL ANALYSIS. Emission techniques are based on the analysis of the ELECTROMAGNETIC RADIATION (EMR) emitted by atoms when they have been heated to extremely high temperatures. The energy created in these environments promotes electrons within atoms from lower energy levels to higher energy levels, a state called the "excited state." When these atoms relax out of the excited state and return to their original state (called the "ground state"), electromagnetic energy is released, usually in the visible or ultraviolet regions. By determining the wavelengths of the emitted energy and its intensity, it is possible to obtain QUALITATIVE and QUANTITATIVE information about the sample being analyzed. Early forms of emission spectrometry used flames or sparks to initiate the emission processes, but modern instrumentation relies on inductively coupled plasmas that can reach much hotter temperatures, nearly 14,000°F (8,000°C).

EMIT (enzyme multiplied immunoassay) *See* IMMUNOASSAY.

empirical evidence Evidence obtained from experimentation, analysis, and/or observation. Almost all forensic analyses yield empirical evidence.

See also HYPOTHESIS AND THE SCIENTIFIC METHOD; SCIENCE.

enantiomers A pair of molecules that are related by being nonsuperimposable mirror images of each other. A common example of enantiomers is the right and left hands—they have the same arrangement of thumb and fingers but the mirror image of the right hand is not superimposable on the left hand. In forensic science, enantiomers are an important consideration in DRUG ANALYSIS, where enantiomers and DIASTEREOISOMERS can be very important, particularly in the case of COCAINE.

energy dispersive spectroscopy (EDS) *See* SCANNING ELECTRON MICROSCOPE (SEM).

engineering, forensic The application of engineering knowledge and techniques to legal matters and those that involve courts of law. Forensic engineering includes traffic accident reconstruction, failure analysis (as when buildings collapse), investigation of industrial accidents, product liability issues, and the investigation of transportation disasters such as airline crashes. Unlike other forensic disciplines, forensic engineers often find themselves involved in the investigation of incidents that are not crimes, although criminal activity may be a contributing factor. Forensic engineers try to understand why and how materials or machines failed and what can be done to avoid such failures in the future. The analysis of engineering failures is sometimes referred to as "pathology," an analogy to pathological studies of biological tissue for signs of injury or disease. A professional society devoted to forensic engineering, the National Academy of Forensic Engineers (NAFE), was formed in 1982.

In addition to finding the root cause of failures, the investigation often includes an evaluation of policies and procedures that might have contributed. Thus, forensic engineering analyses are wide-ranging and extend beyond the mechanics of the accident or incident in question. A famous example is the explosion of the space shuttle *Challenger* on January 28, 1986. The hardware that failed was an O-ring that leaked in the unusually cold conditions that morning. However, procedural problems at NASA were identified as contributing factors to the tragedy. Similar findings appear likely in the investigation of the *Columbia* disaster of February 2003. Other famous cases involving forensic engineering include the accidents at the Chernobyl and Three Mile Island nuclear power plants, the collapse of the Tacoma Narrows Bridge in 1940, and the collapse of a walkway in the Hyatt Regency Hotel in Kansas City in 1981 that killed 114 people.

In cases of bombings or suspected bombings, forensic engineers often work alongside other forensic scientists to reconstruct events and to extract and

Collapse of the north tower of the World Trade Center on September 11, 2001 ***(Reuters News Media, Inc./CORBIS)***

apply engineering lessons that can be learned from the event. Such was the case in the first attack on the World Trade Center in 1993 and the attack in Oklahoma City in 1995, both of which involved large bombs. Forensic engineers evaluate blast damage and can make recommendations for modifying structures to better withstand the extreme forces involved in events such as bombings, earthquakes, and other deliberate acts such as the second World Trade Center attack on September 11, 2001.

The collapse of the towers created the largest crime scene ever processed. The overriding forensic issue was identification of the remains of the 2,830 people who were killed. The catastrophic forces of impact, fire, and collapse meant that very few remains were recovered in any recognizable form. Thus, alternative modes of identification were needed including forensic ANTHROPOLOGISTS, ODONTOLOGISTS, FINGERPRINT examiners, and forensic biologists to perform DNA TYPING. The New York City Medical Examiner's office, working with private labs, took the lead in identification tasks. After initial recovery efforts were completed, rubble was collected and barged to a landfill site where it was meticulously sorted and studied for evidence and human remains. These efforts were completed during the summer of 2002. For many of the victims, no traces of their remains were ever located.

Forensic engineering can also be critically important in cases where crimes are not involved. A well-known recent example is the crash of TWA Flight 800 off Long Island in 1996, where a number of theories involving bombs and missiles were put forward. (See essay under MASS DISASTERS). It was eventually shown, through engineering and forensic analysis, that the crash was likely the result of an accidental fuel tank explosion.

Fire and explosion investigation is another aspect of forensic engineering, related to but not the same as arson investigation. Although arson cases are often widely publicized, less than 30 percent of structural fires are caused by arson. Fire investigators work alongside fire departments to identify the causes of fires, study how they spread, and determine what factors, technical or procedural, contributed to the fire. Among other things, fire investigators study electrical wiring, burn patterns, and equipment and appliances in an attempt to locate the point of origin of a fire. Computer modeling is becoming an important tool of the fire investigator, helping to study how a fire might have started and spread and how fire protection systems worked during the fire.

See also "Identification of the Dead" essay; PRODUCT LIABILITY.

Further Reading

Carper, K. L., ed. *Forensic Engineering*. 2d ed. Boca Raton, Fla.: CRC Press, 2000.

Noon, R. K. "Structural Failures." In *Forensic Science: An Introduction to Scientific and Investigative Techniques.*

Learning from Tragedy: Forensic Science and Terrorist Attacks

Recent tragedies in the United States have brought forensic engineers into the spotlight and have demonstrated that thorough investigations can improve building design and reduce future threats. While engineering and design changes can never completely protect building occupants from deliberate terrorists acts, they can go a long way toward reducing the toll they take. Engineering modifications can also help new buildings withstand other kinds of abrupt forces such as those experienced during earthquakes. Additionally, these tragedies, along with other large-scale crimes such as the anthrax mailings of October 2001, illustrate how forensic scientists from many disciplines work together to deal with and learn from mass disasters and other large-scale criminal acts.

Oklahoma City Bombing

Construction of the Alfred P. Murrah Federal Building in Oklahoma City was completed in 1976. The building met all codes in place at the time, and the city is located in an area not prone to earthquakes or other seismic events. On April 19, 1995, the building was bombed and destroyed. The attack came in the form of a huge ammonium nitrate and fuel oil bomb (ANFO) loaded into the back of a Ryder truck. The blast was roughly equivalent to that produced by 4,000 pounds (1814.3 kg) of TRINITROTOLUENE (TNT) and created a crater that was approximately 30 feet (9.1 m) across and 7 feet (2.1 m) deep. Explosives experts from agencies such as the FBI and ATF sifted through the debris and were able to quickly pin down the type of explosives used. Residues of this same material were recovered from suspect Timothy McVeigh's clothing. McVeigh and Terry Nichols, an accomplice, were later convicted of the crime, and McVeigh was executed.

The bombing also brought civil engineers to the site to study the disaster and to make recommendations based on their findings. The pressure wave produced by the explosion devastated the north side of the structure and caused the collapse of three critical columns. The failure of these columns led to a progressive collapse of much of that portion of the building and resulted in the deaths of 168 people. Several other nearby buildings were damaged. Detailed analysis by a Building Performance Assessment Team (BPAT) dispatched by the Federal Emergency Management Agency (FEMA) studied the blast effects and issued findings. Among their conclusions was that the damage spread beyond that done by the explosion due to construction techniques used, which did not incorporate the same level of resiliency as is now recommended. Much of these newer design criteria originally were meant to counter earthquake scenarios in which collapses are likely. The BPAT team noted that had compartmentalized construction, additional reinforcement, and/or seismically hardened construction techniques been used, the progressive collapse could have been avoided and many lives saved.

World Trade Center Attacks

The attack on the World Trade Center by two hijacked jetliners laden with fuel was a scenario unforeseen when engineers designed and built the "Twin Towers" in the 1960s and 1970s. Although the towers were built to withstand the impact of a Boeing 707, the consequences of the fuel load of such a plane were not taken into account. On September 11, 2001, the Twin Towers showed remarkable resiliency to the physical force of the impact but were doomed by fires that followed, which weakened the building's structural steel. Apparently, the jet fuel itself was not the principal cause of the failure; rather, it was the tons of office furniture and paper that kept the fires burning long enough and hot enough to destroy the integrity of the steel.

The towers were built using a redundant modular design that consisted of outer wall support columns surrounding an inner core that contained elevators, stairwells, and building support and fire suppression systems. Floor joists secured to the inner core and outer columns by angle clips or brackets supported the 110 floors in each tower. The construction technique is somewhat similar to that used to build a deck using the wall of the house as a support. Each floor was designed to support approximately 1,300 tons (1,179.3+) beyond its own weight. The steel within the support structures was coated with a fireproofing material designed to protect it from the deleterious effects of heat. To impart maximum fire safety, systems in tall buildings such as the Twin Towers were designed to maintain structural integrity (by protecting steel from heat) long enough to allow complete evacuation, usually a two to three hour task. Thus, while the towers were designed to resist impact or fire, the combination of both proved devastating.

The impact of the jets destroyed several of the perimeter columns and damaged the cores, notably the stairwells down which people would have to go to evacuate. The extent of the damage must have been substantial given that only four people who were above the impact zones were able to escape. Fire suppression systems were also damaged. However, structurally the towers remained strong, and, short of another catastrophic impact, earthquake, or windstorm, they would have stayed standing.

Each jet carried with it approximately 24,000 gallons (90,849.8 l) of highly flammable fuel that spilled out, ignited, and spread. Although intense, this fire burned out fairly quickly and burned at a relatively low temperature as indicated by the plumes of thick black smoke that poured out of the stricken buildings. The temperatures, although hot, were not sufficient to melt steel, which requires a temperature of 2,732°F (1500°C). Unfortunately, the fires produced enough heat to weaken the steel such that it lost half of its normal strength. Complicating this effect was that much of the fireproofing applied to the steel supports in the area of the impact was probably dislodged by the collision. Furthermore, the fires burned with different intensities in different locations, causing uneven distortions in the steel. It appears as if the first to fail were the floor joists in the areas that were the most damaged by impact and heat. The impacted flooring collapsed onto the floor below, exceeding the excess capacity and causing the angle brackets to fail. This led to an unstoppable "pancaking" effect of increasingly accelerated floor failures driven by more and more weight collapsing onto each floor. The rubble, when it struck the concrete, was moving at an estimated 120 mph (193 km). Surprisingly, the debris piles were only a few stories high. This was due to the construction techniques used—most of the volume (approximately 95 percent) of the towers was air.

Engineering studies will continue for years, but preliminary results are pointing the way to improving building design, both for new and existing skyscrapers. Building evacuation routes, including emergency lighting, will be reexamined, as will fire suppression systems. Of particular concern are widespread, catastrophic fires that spread in seconds rather than in minutes. Complete control of fires such as those created by the jet impacts is probably an unrealistic goal, but controlling them long enough to buy escape time should not be. Given that the Twin Towers displayed remarkable mechanical stability, it is likely that fire suppression and improved escape and safety systems are where the key lessons will be learned and applied.

Anthrax Mailing

In October 2001, another terrorist attack occurred that also challenged the forensic science community and law enforcement community in ways never anticipated. Deadly ANTHRAX bacteria in a highly concentrated powder were mailed in four envelopes from Trenton, New Jersey, to locations in Florida, New York City, and Washington, D.C. Congressional and other government offices were contaminated, as were postal facilities all along the East Coast. Five people died, and another 13 were hospitalized; hundreds were given preventative treatments. The FBI investigation spread over many states and became the second-largest investigation in the agency's history (September 11 being the largest). The investigative challenges were complicated by the forensic ones and led to an unprecedented multiagency effort involving the FBI, Postal

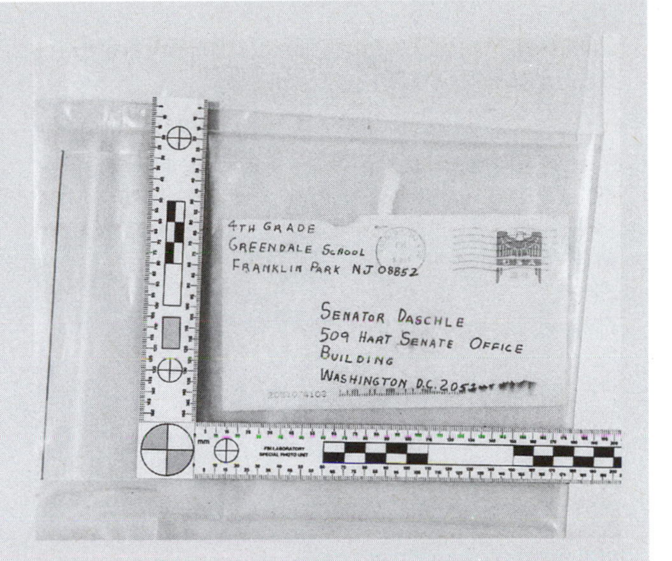

Envelope containing anthrax that was mailed to Senator Tom Daschle in October 2001 ***(Reuters NewsMedia Inc./ CORBIS)***

Service, Centers for Disease Control (CDC), EPA, and military facilities specializing in biological weapons. Since the anthrax was mailed in handwritten envelopes with handwritten letters enclosed, a variety of forensic specialties were needed, including QUESTIONED DOCUMENTS, LINGUISTICS, TRACE EVIDENCE, and DNA. One of the letters, sent to Senator Patrick Leahy of Vermont, contained enough of the powder to allow for extensive genetic analysis that confirmed that the bacteria were of the Ames strain. A radiocarbon dating test also showed that the powder was relatively fresh, a sign that it had been recently prepared.

Traditional forensic analysis of the evidence could not be undertaken until it was rendered safe to handle. This had to be accomplished while causing minimal or ideally no damage to the envelopes or their contents. Late in 2001, the FBI and the CDC, working with Army scientists at the U.S. Army's Medical Research Institute of Infectious Diseases (United States AMRIID) at Fort Detrick, Maryland, announced that a protocol had been established. The letter was decontaminated at Fort Detrick and transferred to the FBI laboratory for testing. There, the envelope was examined with a microscope to recover any trace or transfer evidence and examined for latent fingerprints. Cautiously the envelope was opened and the letter removed, allowing questioned document examiners, forensic linguists, and other specialists to analyze the envelope and the letter. Despite the scope of the investigation and all of the scientific and forensic work, as of spring 2007, no suspect has been arrested in the case, which has proven that disparate

(continues)

Learning from Tragedy: Forensic Science and Terrorist Attacks *(continued)*

agencies can work together on a wide-scale criminal and forensic investigation. Sadly, the case has also redefined the scope of forensic science well beyond the traditional disciplines. In the age of terrorists, any science can become forensic science.

Further Reading

Corle, W. G., et al. "The Oklahoma City Bombing: Summary and Recommendations for Multihazard Mitigation." *Journal of Performance of Constructed Facilities* 12, no. 3 (1998): 100.

Eagar, T. W., and C. Musso. "Why Did the World Trade Center Collapse? Science, Engineering, and Speculation." *JOM: The Journal of the Minerals, Metals, and Materials Society* 53, no. 12 (2001): 8.

Johnston, D., and W. J. Broad. "Anthrax in the Mail Was Newly Made, Investigators Say." *New York Times,* June 23, 2002.

Mlarkar, P. F., et al. "The Oklahoma City Bombing: Analysis of Blast Damage to the Murrah Building." *Journal of Performance of Constructed Facilities* 12, no. 3 (1998): 113.

Sozen, M. A. "The Oklahoma City Bombing: Structure and Mechanisms of the Murrah Building." *Journal of Performance of Constructed Facilities* 12, no. 3 (1998): 120.

The F.B.I. "The Anthrax Investigation." Available online. URL: http://www.fbi.gov/anthrax/amerithraxlinks.htm. Downloaded April 15, 2007.

2nd edition. Edited by S. H. James and J. J. Nordby. Boca Raton, Fla.: CRC Press, 2005.

Shepherd, R., and J. D. Frost, eds. *Failures in Civil Engineering: Structural, Foundation, and Geoenvironmental Case Studies.* Reston, Va.: American Society of Civil Engineers, 1995.

entomology, forensic The application of the study of arthropods such as flies and beetles to legal proceedings, both criminal and civil. Arthropods are animals with jointed legs and include insects, arachnids (spiders), centipedes, millipedes, and crustaceans. Forensic entomology, principally related to insects, has become a common tool in death investigation, in assigning a time since death occurred (POSTMORTEM INTERVAL [PMI]), and/or for determining a site of death. Death and criminal investigations are broadly categorized as medicocriminal/medicolegal entomology, while medical entomology is the subdiscipline that deals with insects that can affect human health and carry disease such as mosquitoes and ticks. Forensic entomologists can also be involved in legal proceedings classified as urban entomology (involving insects such as cockroaches), and stored product entomology, which includes infamous cases (or allegations) of insects and insect parts in fast food.

Flies and beetles are the insects used most often to determine time or place of death. The entomologist studies the types of insects present at a death scene, their number, and stages of the life cycles that are observed. Techniques known as insect succession are of particular value for determining the PMI, since different species will be attracted to the body depending on the stage of decomposition. The entomological analysis of a death scene involves studies and observations at the site as well as in the lab. At a death or CRIME SCENE, the entomologist must study the area around the body carefully, take measurements of critical variables such as temperature and sun exposure, document and photograph, and obtain samples with as little disturbance as possible. Normally, samples from the body itself are taken during AUTOPSY. In the lab, insects collected at the site are studied and often are raised through their life cycle. This allows for definitive identifications that can be difficult at earlier stages in the life cycle and can improve the estimate of the PMI. Rearing the insects also provides the entomologist with information and data that may be useful in future casework.

Temperature is critical in determining what types of insects will appear at a death scene and how quickly they will develop. Other factors are the season, how much sun exposure the scene gets, what time of day the body was placed at the scene (time of day determines which types of insects will be active), whether the body was buried and, if so, how deeply, how the victim is clothed or wrapped, and whether the scene is in an urban or rural environment. Scavengers also can be an issue. Conversely, insect damage to a body may look like trauma, injury, or burns, and the forensic entomologist can be invaluable in distinguishing this kind of postmortem damage from injuries that might have contributed to death. Finally, insects that feed on or move through BLOODSTAINS at a scene can create patterns that may be difficult to interpret. Insects that have fed on blood can leave droppings that often look like BLOODSTAIN PATTERNS.

Future trends in forensic entomology include the use of computer models to study insect succession and improve estimates of the PMI and the inclusion of DNA and toxicological methods to increase the amount of information available to the entomologist. DNA typing could be used to improve the identification of insect species but also for identification and associated purposes. For example, if an insect feeds on portions of a body and that material can be recovered and subjected to DNA typing, it could be possible to definitively link that particular insect to a particular victim. There is growing interest in the field of entomotoxicology, in which insects found at a death scene are chemically analyzed to determine what substances they ingested while feeding on a body. Preparation of insects for chemical analysis uses techniques similar to those developed for the analysis of hair and fingernails, since the chemical composition of insect shells is similar to these materials. Entomotoxicology is valuable in cases when a body is discovered so long after death that the blood and tissues needed for standard toxicological analyses have disappeared. In some cases, both drugs and toxins can be detected in the insects originating from the body that they fed on. However, these same compounds can poison the insects themselves. Ingestion of these toxins may influence the growth and development of the insects. These effects in turn can impact the PMI estimation that arises from any analysis of that insect population.

Further Reading

Byrd, J. H., and J. L Caster, ed. *Forensic Entomology: The Utility of Arthropods in Legal Investigations.* Boca Raton, Fla.: CRC Press, 2001.

entomotoxicology *See* ENTOMOLOGY, FORENSIC.

entrance wounds *See* BULLET WOUNDS.

environmental forensics The study of environmental data and information that are involved in legal proceedings. *International Journal of Environmental Forensics* is also the title of a journal in the field that began publishing in 2000. Often the goal of the litigation is to allocate responsibility for environmental contamination and to determine what parties will be responsible for the clean up or other remedial action. Environmental cases are not typically handled by forensic laboratories but rather by laboratories and private consultants working in the environmental field. The ENVIRONMENTAL PROTECTION AGENCY (EPA), along with various state and local agencies, are responsible for enforcement actions. Cases are often large and complex, involving experts from several disciplines including biology, chemistry, epidemiology, geography, and geology. Analyses are performed by government agencies or by private consultants.

The techniques used can be broadly categorized as noninvasive/nonintrusive or invasive/intrusive. Examples of noninvasive techniques include computer modeling, mapping, and photography, while examples of invasive techniques include sampling, drilling, testing, and laboratory analysis. Complex calculations and computer models are extensively used for describing and predicting the spread of pollutants through soil, air, water, and groundwater. Contaminant transport models are one example, and they are used to study the release of a pollutant or pollutants over space and time and can be employed in attempts to locate point of origin and when a spill occurred. Like most models, data that are used as input include physical or chemical parameters, mapping and geological data, and the results of testing. Such models normally produce a range of possible results or a result with an associated uncertainty.

Further Reading

Morris, R. D. *Environmental Forensics Principles and Applications.* Boca Raton, Fla.: CRC Press, 2000.
Morris, R. D., and B. L. Murphy, eds. *An Introduction to Environmental Forensics.* San Diego, Calif.: Academic Press, 2002.

EPA (Environmental Protection Agency) The agency charged with overseeing environmental affairs, regulation, and enforcement at the federal level. The EPA has investigative authority and as a result requires scientific analysis of environmental samples and evidence. Currently, the EPA's Office of Criminal Enforcement, Forensics, and Training (OCEFT) oversees criminal investigations, compliance issues, and forensic analyses. The National Enforcement Investigations Center (NEIC), housed under this office, is charged with providing support for litigation activities and includes a forensic COMPUTING team, while the Laboratory Branch maintains facilities and equipment for the analysis of evidence; increasingly the term used to describe these activities is ENVIRONMENTAL FORENSICS. The EPA maintains an extensive Web site at www.epa.gov.

ephedrine Along with pseudoephedrine, a common ingredient in over-the-counter (OTC) drugs used to treat colds and flu, specifically the symptom of congestion. Ephedrine and pseudoephedrine occur naturally in plants in the Ephedra family, and ephedrine is a central nervous system (CNS) stimulant. Tablets containing ephedrine and/or caffeine are occasionally sold on the street, purporting to be amphetamine. Ephedrine can also serve as a precursor in the clandestine synthesis of methamphetamine.

See also CLANDESTINE LABS.

epidemiology The study of disease patterns in large populations. Epidemiology is a particularly important part of environmental forensics, in which disease patterns related to pollution are a critical concern.

erasures *See* OBLITERATIONS; QUESTIONED DOCUMENTS.

ethics and accountability Several forensic science professional organizations have codes of ethics, but there are no uniform national standards. The AMERICAN ACADEMY OF FORENSIC SCIENCES (AAFS), the AMERICAN BOARD OF CRIMINALISTICS (ABC), and the CALIFORNIA ASSOCIATION OF CRIMINALISTS (CAC) all have ethical guidelines that vary in length and detail. Common themes in these ethical guidelines include prohibitions against misrepresenting or exaggerating one's scientific qualifications; use of the best available scientific methods of analysis; use of procedures to ensure the quality and reliability of data produced; and an obligation to report those that violate the guidelines. Ethical guidelines are important in any profession, but even more so in forensic science given the nature of the work and the integral relationship with the justice system. Ethical codes enforced by professionals working in the field ensure that the work done is trustworthy, reliable, and impartial.

evidence *See* PHYSICAL EVIDENCE.

evidence, digital *See* COMPUTING, FORENSIC.

evidence collection How physical evidence is collected varies depending on the type of evidence and the situation in which it is collected. A police officer or other law enforcement agent may collect evidence, but crime scene personnel or evidence collection tech-

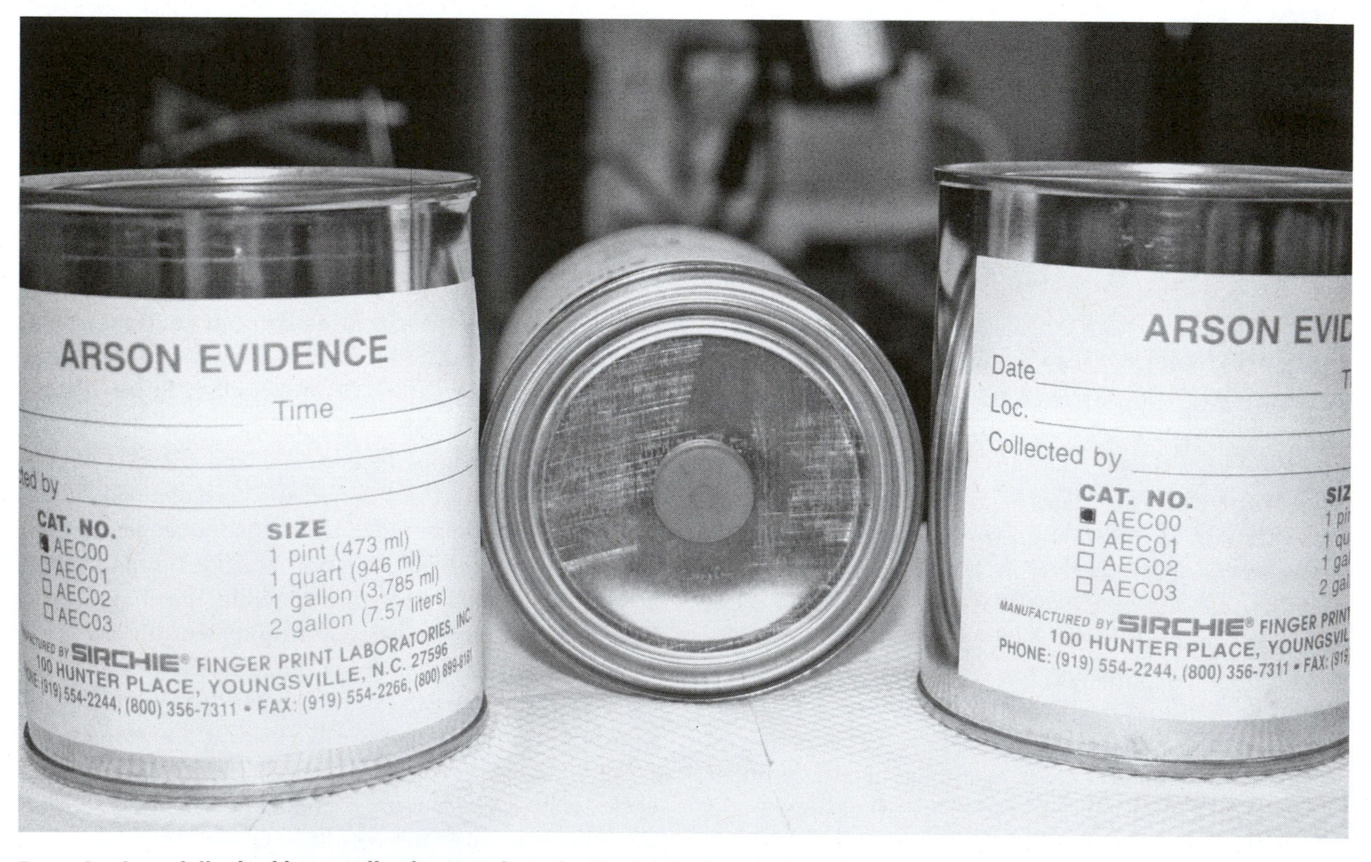

Example of specialized evidence collection containers for fire debris ***(Suzanne Bell)***

nicians may also. Bloodstains and other biological evidence are normally stored in paper to provide air flow, while other types of evidence such as cartridge cases can be stored in plastic bags. Regardless of the specifics of a particular situation, some general procedures and guidelines always apply. First, a CHAIN OF CUSTODY must be started and maintained by whoever initially collects the evidence. Complete documentation in the form of notes, photographs, and so on must also be recorded. Proper labeling is essential and must be done in such a way as to avoid altering the evidence. For example, CARTRIDGE CASES can be labeled directly, but care must be taken to ensure that the writing does not obscure or alter markings on the case that can be used in any subsequent analysis. In some cases, such evidence is not marked directly but rather is placed in a container or bag that is marked and sealed. Some types of evidence, such as biological evidence (BLOOD, BLOODSTAINS, and BODY FLUIDS) must be thoroughly dried before packaging and delivery and may require additional preservation such as refrigeration.

In many cases, additional control samples are needed for comparison analysis. For example, in an ARSON case if debris consisting of burned carpet is submitted, a control sample consisting of a portion of unburned carpet is essential (if it can be found). In the case of crime scenes, ELIMINATION PRINTS are also collected and submitted along with any other evidence that will be subject to fingerprint or impression evidence analysis.

evidentiary fact A fact ascertained from the evidence directly, but not one that speaks directly to the matter in question. For example, consider the case of a murder involving a gun. If a suspect is identified and forensic analysis is performed that shows that there was GUNSHOT RESIDUE (GSR) on his or her hands, that finding is an evidentiary fact. By itself, it does not prove that the suspect committed the murder, which is the matter in question. Additional information and evidence is required to establish whether the suspect is innocent or guilty of the crime.

exchange principle *See* LOCARD'S EXCHANGE PRINCIPLE.

exclusionary evidence Evidence that excludes or eliminates a person or disproves a possible scenario. For example, if semen involved in a rape case is found to be of a type that does not match that of a suspect, that is considered exclusionary evidence since it eliminates that person as a possible source.

exemplars *See* CONTROL SAMPLES; QUESTIONED DOCUMENTS.

exit wounds *See* BULLET WOUNDS.

expert witness A person who is accepted by the TRIER OF FACT as having specialized knowledge that is relevant to the case and beyond the expertise of the average person. A key component of any forensic scientist's job is to offer expert testimony in cases that involve evidence that they have analyzed in the lab. This communication to a court constitutes the "forensic" portion of their work, whereas the laboratory analysis and interpretation constitute the "science" portion. The expertise of a witness is judged based on education, training, and experience and must pass any challenges offered by the defense. There is no set standard for what constitutes acceptable credentials; rather, it is up to the court to decide on a case-by-case basis who qualifies as an expert in that particular case. Just because one is a forensic scientist does not automatically mean that he or she will be accepted as a witness in a given case. An expert in FINGERPRINTS could not testify as an expert in a drug case because his or her expertise is not relevant to the issue in question.

Ideally, the expert witness is not an advocate for either the prosecution or the defense but rather presents results impartially. While other witnesses are limited to testifying about things they have personal knowledge of, an expert witness is required to present results, interpret them to the court, and offer opinions based on these results. One of the most important jobs of an expert witness is to convey complex technical and scientific information in a way that is understandable to the average layperson. As forensic analyses become more complex (such as DNA TYPING), this is becoming an increasingly challenging task. In addition, the weight given to an expert's testimony often hinges on how the jury perceives the witness, regardless of the reliability and quality of the results presented. Finally, it is not uncommon for opposing expert witnesses to reach entirely different opinions based on the same evidence. Thus, each expert must explain and defend the methods used and conclusions reached.

See also COURTROOM PROCEDURES.

explosive A chemical compound or mixture that decomposes rapidly to produce heat and gas. Explosions are similar to COMBUSTION in that a solid or liquid is converted to gases in a reaction that is exothermic (heat-producing). A source of fuel is required (the material that decomposes during the explosion), as well as an oxidant. What distinguishes an explosion from combustion is the speed at which the reaction occurs. Explosions create huge amounts of gas that can travel at speeds of nearly 7,000 miles per hour (11,265 km), and this shock wave or "blast effect" can do enormous damage to anything in its path. A by-product of the blast effect is an extremely loud noise created by the pressure wave. In forensic science, the analysis of explosives is usually categorized with ARSON and is considered part of forensic chemistry.

Explosives can be divided into low and high explosives. Low explosives burn very quickly and must be kept in a confined space to actually explode. Accordingly, low explosives are occasionally referred to as burning explosives. Examples of low explosives include BLACK POWDER and SMOKELESS POWDER (used as PROPELLANTS in AMMUNITION), which are frequently used to make homemade explosives and pipe bombs. Another low explosive, made infamous by the Oklahoma City Bombing on April 19, 1995, is composed of ammonium nitrate and 6 percent fuel oil (ANFO). A similar mixture of urea nitrate and other materials was used in the first attack on the World Trade Center in 1993. The maximum burning speed of low explosives is around 3,281 feet per second (1,000 meters/second). Low explosives are sensitive to heat, friction, and sparks and are thus not very stable. The detonation of a low explosive generates what is referred to as pushing power, in which large objects are moved rather than shattered. Fragments of the container in which the explosive was placed are relatively large, and there are often significant amounts of residues remaining after the explosion.

High explosives are further divided into primary and secondary explosives. Primary high explosives are shock and/or heat sensitive and are often used as primers that ignite secondary high explosives. PRIMERS in ammunition and blasting caps contain primary high explosives. Secondary high explosives are much more stable and are usually detonated by the shock generated from a primary explosive. High explosives decompose at a much faster rate than do low explosives, and detonations generate shattering power, produce smaller, sharper fragments, and generally leave minimal

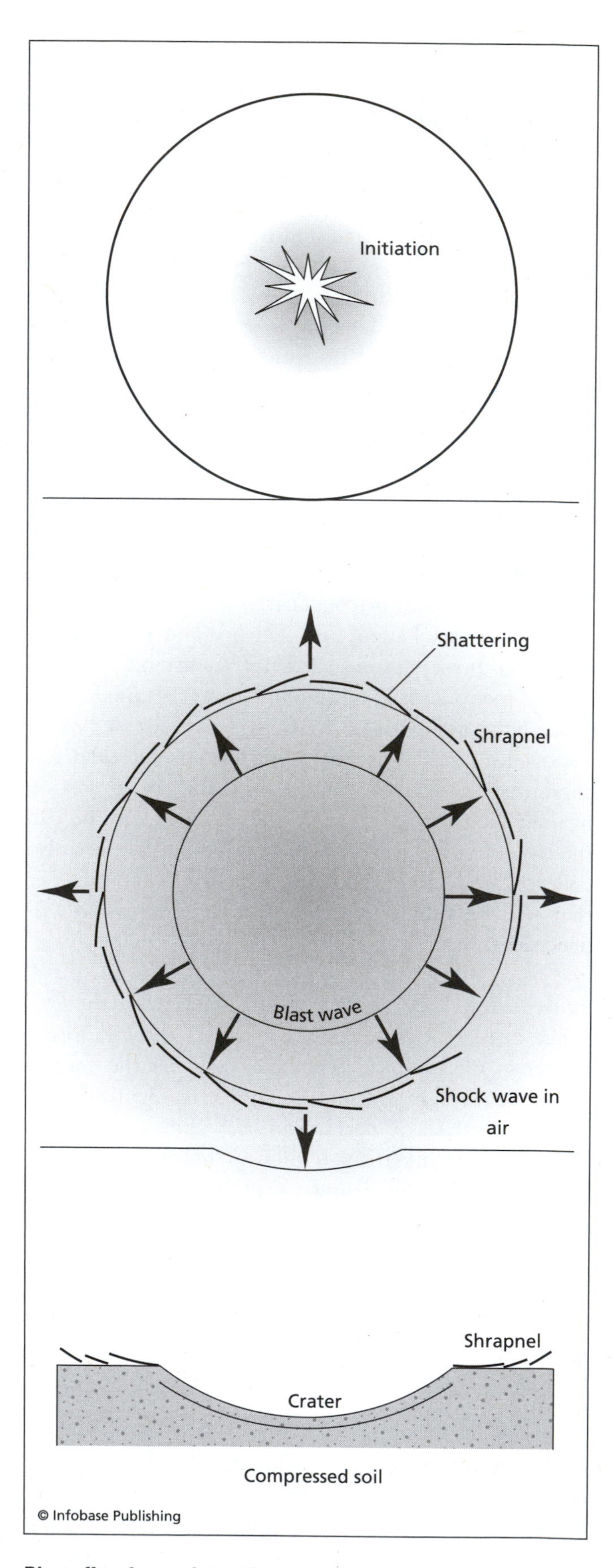

Blast effect from a detonation

residue. The accompanying figure shows the chemical structures of several high explosives.

High explosives include the famous NITROGLYCERIN ("nitro"), which was invented in 1847. Nitroglycerin alone is unstable and dangerous to handle, and it took refinements introduced by Alfred Nobel to package it in a stable form. By combining nitroglycerin with diatomaceous earth (DE), Nobel created what came to be known as dynamite. Nobel, who was from Sweden, was a prolific inventor and is perhaps better known for the Nobel Prize, which his fortune created. The design of dynamite continued to evolve, the primary goals being to reduce sensitivity to shock and to extend usefulness to damp and wet environments. The military also has a number of high explosives including trinitrotoluene (TNT), HMX, RDX, tetryl, and PETN. The terms "plastic explosive" or "plastique" usually refer to RDX mixtures that are moldable; "C4" is a complex that is 90 percent RDX.

Laboratory analysis of explosives and explosive residues utilizes a variety of techniques. When working with debris, preliminary microscopic examination can be helpful in identifying residues, but these are not always obvious among burned and blackened debris. For the tiniest residues, SCANNING ELECTRON MICROSCOPY (SEM) can be used. A number of chemical screening tests (PRESUMPTIVE TESTS) can be employed to identify components such as nitrates, nitrites, and chlorates, which are commonly found in explosives. Example tests include the GRIESS TEST, phenylamine, and alcoholic potassium hydroxide (KOH). There are also CRYSTAL TESTS available. Solvent EXTRACTIONS are frequently needed to isolate the explosives from debris. Explosive residues can also be collected using swabs and wipes taken from surfaces such as walls and floors.

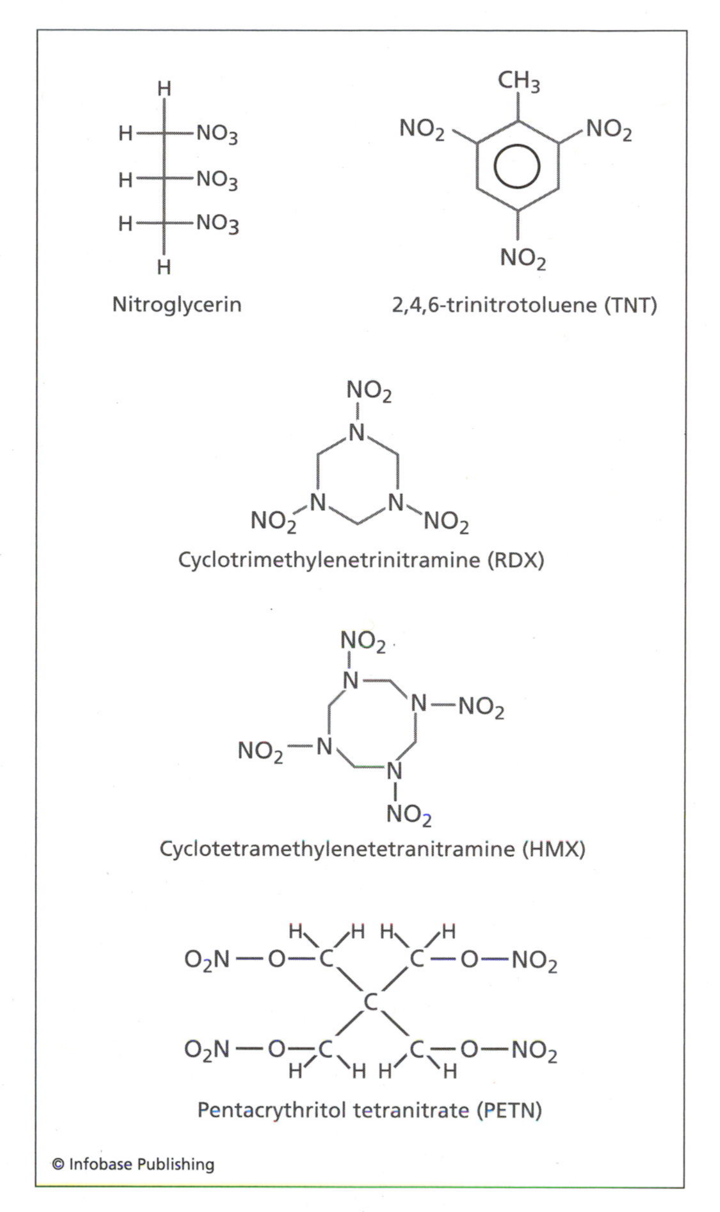

Chemical formulas of different high explosives

Other techniques used in explosive analysis include THIN LAYER CHROMATOGRAPHY (TLC), INFRARED SPECTROSCOPY (IR), GAS CHROMATOGRAPHY/MASS SPECTROMETRY (GC/MS), and HIGH PERFORMANCE LIQUID CHROMATOGRAPHY (HPLC). HPLC is particularly useful in explosives analysis since the sample is not heated as in gas chromatography. Inorganic components and residues such as potassium nitrate and chlorates can be identified using X-ray diffraction (XRD). ATOMIC ABSORPTION (AA) has also been used.

A number of screening and field-deployable devices have been used to identify explosive materials and devices. Such technologies are deployed at airports and at the scenes of mass disasters where explosives may have been involved. Common screening techniques include ION MOBILITY SPECTROMETRY (IMS) and portable gas chromatographs. Dogs can also be trained to sniff out explosives and are a common sight at airports and events that draw large crowds. The federal agencies and forensic laboratories that work with explosives are the FBI and BUREAU OF ALCOHOL, TOBACCO, FIREARMS, AND EXPLOSIVES (ATF).

See also EXPLOSIVE POWER; X-RAY TECHNIQUES.

explosive power The destruction associated with explosives depends on the pressure and speed at which the blast wave moves. High explosives have higher

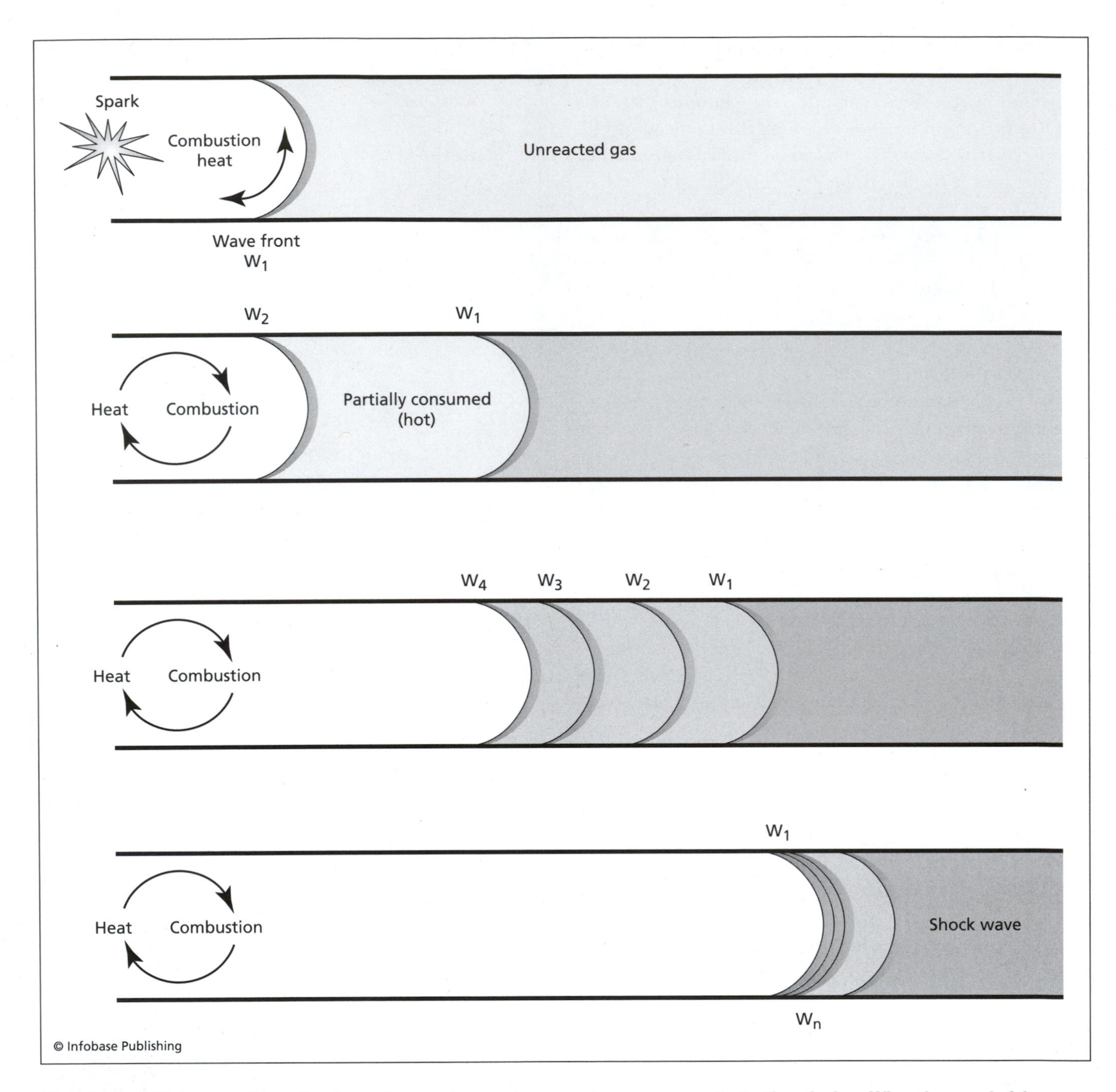

The process of detonation. Due to heating and expansion, each successive wave moves faster than the last. When the speed of the wave exceeds the speed of sound, detonation occurs.

shattering power, also called "brisance," compared to low explosives, which generate what is often called "pushing power." Consider a pipe bomb as an example, with explosive materials confined within. Detonation of a low explosive bomb would produce few fragments, and the blast wave might displace heavy objects nearby. A high explosive pipe bomb would shatter the pipe, create showers of sharp shrapnel, and shatter objects nearby. Shatter power depends on the pressure created by the shockwave. Behind the shockwave is the blast wave of hot compressed gaseous products of the combustion.

Explosive power fundamentally depends on the heat evolved (Q) and the volume of gaseous products produced (V) and is expressed as the product of these two (Q*V). The relative power of one explosive compared to another is obtained through the power index:

$$PI = \frac{QV_{explosive}}{QV_{picric\ acid}} * 100$$

The explosive index of common explosives is shown in the following table. Typically, the index is calculated based on 1.00 gram of material.

Further Reading

Akhavan, J. "Combustion, Deflagration, and Detonation." In *The Chemistry of Explosives*. Cambridge: Royal Society of Chemistry, 1998.

———. "Thermochemistry of Explosives." In *The Chemistry of Explosives*. Cambridge: Royal Society of Chemistry, 1998.

external reflectance A general class of techniques used in INFRARED SPECTROSCOPY (IR), with forensic applications in DRUG ANALYSIS and TRACE EVIDENCE analysis. Traditional spectroscopy is based on measuring the degree of absorbance of a sample. For example, an IR ABSORBANCE SPECTRUM of a drug is obtained by exposing a sample to successive wavelengths of energy in the infrared portion of the electromagnetic spectrum and plotting how much energy was absorbed by the sample at each wavelength. External reflectance techniques measure energy that is reflected by the sample rather than absorbed by it. Types of external reflectance techniques include ATTENUATED TOTAL REFLECTANCE (ATR) and DIFFUSE REFLECTANCE (DRIFTS).

The explosive shockwave is evident from the image of an atomic bomb. ***(U.S. Department of Energy/Photo Researchers, Inc.)***

Explosive	Structure	Power index
Low/primary		
Mercury fulminate	$Hg(C{\equiv}N{-}O)_2$	14
Lead styphnate	[benzene ring with NO_2, O^-, NO_2, O^-, NO_2] Pb^{2+}	21
Lead azide	$N_3{-}Pb{-}N_3$	13
High/secondary		
Nitroglycerin	O_2N, NO_2, NO_2	171
Picric acid	HO–(ring) with NO_2, NO_2, NO_2	100
TNT 2,4,6- trinitrotoluene	CH_3 ring with NO_2, NO_2, NO_2	331.2
HMX octogen; cyclotetramethylenetetranitramine	NO_2, $H_2C{-}N{-}CH_2$, $NO_2{-}N$, $N{-}NO_2$, $H_2C{-}N{-}CH_2$, NO_2	455
RDX hexogen; cyclotrimethylenetrinitramine	NO_2, N, H_2C, CH_2, $O_2N{-}N$, $N{-}NO_2$, CH_2	457
Nitrocellulose (guncotton)	CH_2ONO_2, H, ONO_2, ONO_2, H, ONO_2, CH_2ONO_2 (repeating unit)$_n$	variable
Tetryl 2,4,6- trinitrophenylmethylnitramine	H_3C, NO_2, N, O_2N, NO_2, NO_2	355
PETN Pentaerythritol tetranitrate	$O_2N{-}O{-}H_2C$, $CH_2{-}O{-}NO_2$, C, $O_2N{-}O{-}H_2C$, $CH_2{-}O{-}NO_2$	452
TATB 1,3,5-triamino-2,4,6- trinitrobenzene	NH_2, O_2N, NO_2, H_2N, NH_2, NO_2	273

(Source: Adapted from information found in Akhavan, J. Ch. 3: "Combustion, Deflagration, and Detonation." In *The Chemistry of Explosives.* Cambridge, U.K.: Royal Society of Chemistry, 1998.)

extraction In a chemical or biochemical analysis, the process of preparing the sample such that the compounds or components of interest are separated from any materials that may complicate or interfere with the analysis of those samples. Many forensic tests require that the samples be extracted one way or another. For example, to perform DNA TYPING on a bloodstain, it is necessary to extract the DNA from the blood and from the material it is deposited on. In DRUG ANALYSIS, extractions are used to separate the suspected drug from any cutting agents that might be present. Toxicological analysis of blood and other fluids often requires an extraction step to separate the drug from the complex blood matrix that can interfere with subsequent testing. Each type of analysis has a separate and specific extraction procedure. Once the sample has been extracted, it is then ready for further analysis as appropriate.

See also PARTITIONING.

extractor marks **(ejector marks)** Markings created on CARTRIDGE CASINGS that can be useful in FIREARMS analysis and identification. These impressions are created by the metal-to-metal contact between the cartridge case and the extractor and ejector mechanisms used in the weapon. The extractor mechanism removes a cartridge from the chamber and the ejector throws the cartridge away once it is extracted. REVOLVERS do not have ejectors, but automatic and semiautomatic weapons (pistols and rifles) do, and as a result the cartridge cases used in such weapons are designed differently from AMMUNITION used in revolvers. As shown in the figure, ammunition used in revolvers is essentially cylindrical, whereas cartridges used in automatics and semiautomatics are designed with an extractor groove that provides a place for the mechanism to grasp (and thus mark) the cartridge. Extractor and ejector markings are usually found on the bottom of the cartridge casing and can be useful in linking a cartridge case to a specific weapon (INDIVIDUALIZATION). Analysis can be complicated if the cartridge case has been reloaded and reused; in these cases there will be multiple sets of markings on the case, one set for each time the cartridge was run through the mechanism of a gun.

Two cartridge cases showing extractor grooves ***(Courtesy of Michael Bell, West Virginia University, Forensic and Investigative Sciences program)***

ex visitatione divina A term frequently used to assign a cause of death into the 19th century. With AUTOPSY rare and medical knowledge still primitive, many deaths investigated by CORONERS were attributed to "a visit from God." Barring obvious wounds or other visible symptoms, most deaths, including sudden or unexpected ones, were usually ascribed to a visit by the deity. This bias made it easier to get away with a well-executed poisoning. "Acts of God" were still the most commonly assigned cause of death in Britain well into the 1800s. Other interesting causes of death recorded over the centuries included evil, lice, lethargy, and French pox (syphilis).

F

fabric impressions/fabric prints A type of IMPRESSION EVIDENCE that can occur when fabric comes in contact with a surface capable of retaining the pattern. A fabric impression can be studied for class characteristics such as weave, and if the fabric has unique features such as tears, wrinkles, creases, or wear patterns, it is sometimes possible to INDIVIDUALIZE the impression and match it back to a COMMON SOURCE. For example, fabric impressions can result from hit-and-run accidents in which pedestrians are struck at high speed by a vehicle. Patterns in the clothing—including fabric weave, patches, or buttonholes—may be impressed into metal surfaces with sufficient detail to match the impression to a specific article of clothing. Another high speed impact that can produce a similar impression is the passage of a lead bullet through fabric, with the relatively soft surface of the bullet retaining the impression. Fabric impressions can also be left by contact of the cloth with a dusty or bloody surface, in which case the evidence is treated much like a latent FINGERPRINT. Glove prints are another example of a fabric impression that have the potential to be useful depending on the level of detail retained in the print. Leather is a form of skin with unique patterns, and that fact, coupled to unique wear patterns in the gloves, can sometimes allow an examiner to link a glove print to a single glove.

facial reconstruction A group of techniques used to assist in the identification of badly decomposed or skeletal remains. A two-dimensional reconstruction is simply a drawing (or computer generated drawing) of the face of the deceased, while a three-dimensional reconstruction is a sculpted likeness built upon a skull or portions of it. Facial reconstruction is usually a joint effort of forensic ANTHROPOLOGISTS and forensic ARTISTS. Forensic sculptors are forensic artists who specialize in three-dimensional reconstructions. The anthropologist, usually a specialist in physical anthropology or osteology, contributes knowledge about bony structures and how they dictate placement of features (such as nose and ears), while the artist re-creates the features and attempts to show what the person looked like while alive. For both two- and three-dimensional reconstructions, the starting point is the skull itself, or a casting made of it. If the skull is the starting point, it needs to be reasonably complete, although the forensic artist can estimate the dimensions of missing portions. Fragility is also a concern and can limit the amount of handling and manipulation that the skull undergoes. Since it is evidence, the artist strives to minimize any damage or permanent changes to the skull as a result of the reconstruction. Also common to all reconstructions is extensive photography of the skull from different angles and under different lighting conditions, which provides the artist with several perspectives from which to work. Tissue depths are estimated using tables of data and are usually marked on the skull for the photography.

Two-dimensional reconstructions (drawings) are more appropriate than are sculptures when the skull is badly damaged or unusually fragile. It is also less time consuming and labor intensive and so is less expensive. An additional advantage is the ability to create or scan drawings into computers, where specialized software is used to generate different versions of the reconstruction. For example, if no hair is found with the body, the artist has no guidance as to hair color or hairstyle. The computer can create several variations, one of which may be recognized by someone viewing it. Inclusion of eyeglasses, different ages, and different eye colors are examples of other features that can be easily varied in a two-dimensional reconstruction.

Three-dimensional reconstructions have been performed for more than a century and originated primarily as a tool in archaeology. These early efforts were quite subjective and did not gain much attention in forensic science until the late 1920s, when the first applications were noted in Britain. It was not until the 1940s that there was significant use of forensic sculpture in the United States, and even then it was another 20 or so years before it became an accepted part of forensic science. By that time, more reliable guidelines were available for tissue depths, which had originally been crudely estimated by sticking needles or knives into cadavers. This turned out to be misleading, given the dramatic changes in skin and tissue characteristics that take place after death. Since the goal of reconstruction is to produce a likeness of the deceased as a living person, use of newer and more reliable tissue depth tables in the latter part of the century greatly improved the quality and thus the acceptability of sculpture for facial reconstruction.

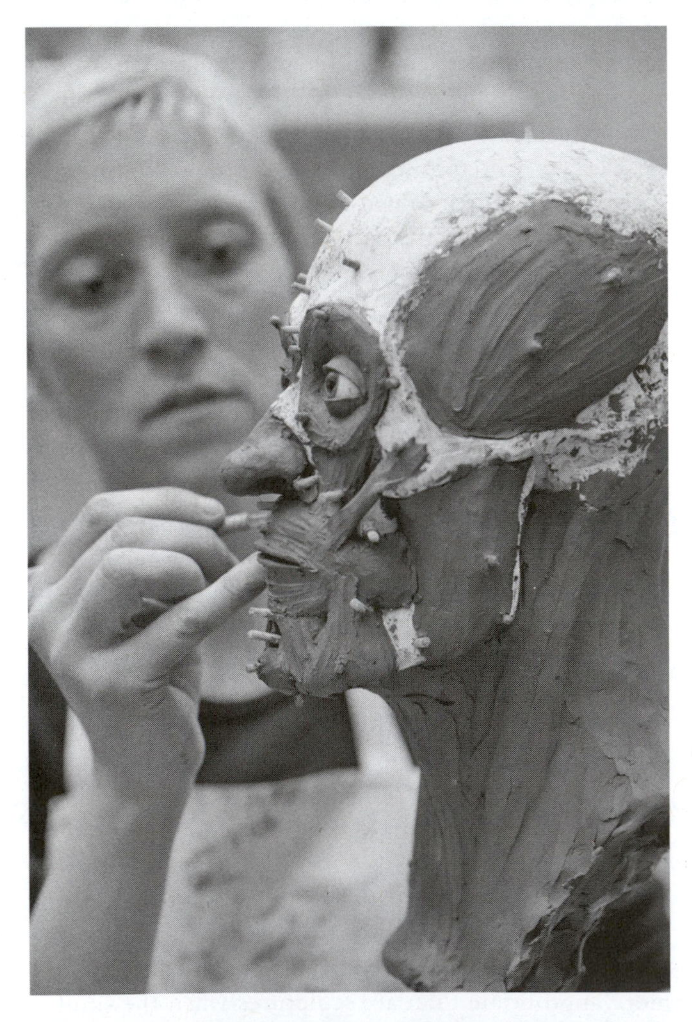

The process of facial reconstruction ***(Courtesy of Michael Donne, University of Manchester/Photo Researchers, Inc.)***

The sculpture starts with the same procedure as drawing, with the skull and tissue depth markers (clay or plastic) attached to the skull with dissolvable glue. This assures that the original skull can be retrieved with minimal permanent changes. Clay is then used to connect the tissue markers and to gradually cover all surfaces of the skull. Glass eyes are positioned early in the process and require the artist to make an assumption about eye color. Other decisions the artist makes include seemingly mundane choices like whether to have the sculpted person smiling or not, what the ears looked like, and, as in two-dimensional reconstructions, how to place and style the hair. Each choice can have a significant impact on the likeness produced; for example, a person who rarely smiled in life but is sculpted with a grin will not be as easily recognized. Similarly, ears can be a distinctive or unremarkable feature in life, and misinterpretation in the reconstruction can be significant. Unfortunately, since the visible portions of ears are made of cartilage and not bone, the artist again must make educated assumptions about size and shape, and, unlike a drawing, it is not a simple task to alter the likeness once it has been produced. However, features such as hair can be addressed using wigs. As a final touch, some sculptors use paint or cosmetics to try to re-create a lifelike skin tone.

Not surprisingly, recent advances in these techniques revolve around computer-assisted or computer-generated reconstructions using software that can display images in three dimensions. Archaeologists are also now using medical CAT scans of mummies and other old or ancient remains as a basis of reconstruction or even as an "electronic mold" from which a physical skull cast can be created.

Further Reading

Nafte, M. "Reconstructing Identity." In *Flesh and Bone: An Introduction to Forensic Anthropology.* Durham, N.C.: Carolina Academic Press, 2000.

Taylor, K. *Forensic Art and Illustration.* Boca Raton, Fla.: CRC Press, 2001.

Ubelaker, D., and H. Scammell. "The Face Is Familiar." In *Bones: A Forensic Detective's Casebook.* New York: HarperCollins, 1992.

facsimile (FAX) *See* QUESTIONED DOCUMENTS.

failure analysis In the forensic context, a technique used in forensic ENGINEERING as part of the investigation of building collapses, airplane crashes, or other events that involved the failure of some kind such as mechanical, electrical, or materials. For example, if a bridge collapses, various components used in construction such as bolts and cables would be subject to failure analysis as part of the investigation.

false negative A type of error that occurs when a test or analysis produces an incomplete or negative result when it should have been positive and/or definitive. False negatives as well as FALSE POSITIVES can occur in many kinds of forensic tests including DRUG ANALYSIS, TOXICOLOGY, and detection and analysis of BLOOD and BLOODSTAINS. As an example, consider a URINE sample submitted by someone who has recently ingested a drug. If the instrumentation used in the testing is not functioning properly and fails to detect the drug, a false negative result is obtained. Another type of false negative can occur when typing blood or in other attempts to determine if two pieces of evidence have a COMMON SOURCE. For example, if DNA TYPING fails to match a suspect to a bloodstain even if the blood really did come from that person, that is an example of a false negative. Problems with the samples (for example too old or subject to microbial degradation) or the laboratory techniques can produce false negatives. As a part of laboratory QUALITY ASSURANCE/QUALITY CONTROL (QA/QC), procedures are designed to minimize false negatives and to detect them when they occur. CONTROL SAMPLES are just one example of how false negatives can be detected and corrected. In the example above citing the urine analysis, if the laboratory routinely analyzed standards known to contain drugs alongside evidentiary samples, the analyst would have detected the error immediately since the control sample would have also been negative, a result the analyst would know could not be correct. Even the best QA/QC may not prevent all false negatives. In the bloodstain example, if the problem is degradation of the stain, perfect laboratory techniques will still produce a negative result. In such cases, analyst experience is critical in recognizing potential problems and in interpreting results.

false positive A type of error that occurs when a test or analysis produces a positive result or incorrect result when it should not have. False positives as well as FALSE NEGATIVES can occur in many kinds of forensic tests including DRUG ANALYSIS, TOXICOLOGY and detection and analysis of BLOOD and BLOODSTAINS. False positives are particularly common with presumptive tests for drugs and blood. For example, many of the reagents used to test stains for the presence of blood will give positive results with plant matter or chemical oxidants. The potential of these substances to cause false positives is the reason the tests are classified as presumptive rather than conclusive or definitive. Another type of false positive can occur when attempting to assign two pieces of evidence to a common source, such as when a bloodstain is subjected to DNA TYPING in an attempt to link it to a specific suspect. If the typing produces a match even when the suspect was not the true source of the stain, this is an example of a false positive. As with false negatives, the sample itself and/or laboratory procedures can produce false positives, and QUALITY ASSURANCE/QUALITY CONTROL (QA/QC) using CONTROL SAMPLES and other protocols help to reduce or eliminate them when they do occur. If, in the example above, the false positive was the result of contaminated equipment in the lab, analysis of a blood sample with a known type would produce the same false positive and alert the analyst to the problem.

falsifiability One of the hallmarks of scientific theories and a key consideration in separating science from PSEUDOSCIENCE. A scientific theory or explanation is purposely designed and stated so that there is a way to test it and thus to prove it false. Scientific theories are stated this way either by using a specific mathematical formula or by wording the theory in such a way that it can be verified by experimentation or observation. Standard physical relationships such as "for every action there is an equal and opposite reaction," or $F = ma$ (force equals mass times acceleration), are clearly stated and can be tested experimentally. Conversely, astrologists claim that "I can tell you about your future by reading the stars," but this statement is vague and cannot be experimentally tested by any objective method. The idea of falsifiability is central to forensic testing. For example, if an analyst performs a DNA analysis on a bloodstain and determines the specific type, that result (the identification of a specific type) can be tested by another analysts doing a separate and independent test. If the results can be objectively tested, they can also be proven to be wrong.

See also HYPOTHESIS AND THE SCIENTIFIC METHOD.

FBI laboratory scandal A series of controversial cases and practices that emerged in the 1990s and led to questions about the laboratory's practices in areas such as explosives characterization and bullet analysis. In the late 1990s, the FBI laboratory came under intense national scrutiny as a result of allegations made by one of its senior explosives chemists, Frederic Whitehurst, who joined the FBI in 1982 and moved to the laboratory in 1986. He began filing complaints soon after, charging that work was sloppy, testimony misleading, and that his complaints resulted in retaliation. His criticisms involved some of the biggest cases of the 1990s, including the first World Trade Center bombing in 1993 and the Oklahoma City bombing in 1995.

Because of the controversy, the DEPARTMENT OF JUSTICE, Office of the Inspector General, launched an investigation and issued a report in 1997. The report was critical of the laboratory personnel and procedures and initiated a wave of reform as well as increased media attention. Another scandal erupted in 2004 concerning FBI practices related to the analysis of bullets, using elemental analysis techniques. The National Academy of Sciences Press in 2004 published *Forensic Analysis: Weighing Bullet Lead Evidence*, critical of the statistical analysis methods used to interpret the results.

Although the laboratory has been subject to much public scrutiny, it is not surprising given the size and prominence of the facility. Since the 1997 report, the laboratory has obtained ASCLD/LAB accreditation and worked to correct problems it identified. In the case of the bullet lead analysis, the FBI requested the evaluation after a prominent metallurgist in the bureau raised questions about the practice.

See also FEDERAL BUREAU OF INVESTIGATION (FBI).

Further Reading

Smith, W. D. "The FBI Laboratory." *Analytical Chemistry* May 1, 2004: 175A–178A.

U.S. Department of Justice, Office of the Inspector General. "The FBI Laboratory: An Investigation into Laboratory Practices and Alleged Misconduct in Explosives-Related and Other Cases, April 1997." Available online. URL: http://www.usdoj.gov/oig/special/. Downloaded June 27, 2007.

feathers A widespread component of many consumer products and thus a form of TRACE EVIDENCE. Feathers and down, the fine underfeathers of birds such as geese and ducks, are used in bedding, sleeping bags, pillows, mattresses, coats, and vests, and it is not unusual for feathers or portions of them to escape from an article and to become TRANSFER EVIDENCE. Parts of feathers are also encountered as a component of DUST, one of the easiest materials to transfer. Chicken, duck, turkey, and pigeon feathers make up the majority of feathers encountered in forensic cases, but others are occasionally seen. Trained analysts, occasionally in consultation with ornithologists, can often identify the species or family of bird from which a feather came based on structural features identified under a MICROSCOPE. Light microscopes as well as POLARIZING LIGHT MICROSCOPY (PLM) and SCANNING ELECTRON MICROSCOPY (SEM) are the primary tools used in forensic applications.

feces Until the advent of DNA TYPING, feces were not considered to be a particularly valuable type of biological evidence. Feces are most commonly encountered in burglary, sodomy, and homosexual assault cases, and a test for the compound urobilin is used to confirm the presence of fecal material. Simple examination of the contents is sometimes useful to determine the origin of the material; the presence of hair and undigested food matter is suggestive of animal rather than human origin. Because fecal matter contains a great deal of cellular matter, DNA typing can be utilized in some cases.

Further Reading

Greenfield, A., and M. M. Sloan. "Identification and Biological Fluids and Stains." In *Forensic Science: An Introduction to Scientific and Investigative Techniques*. 2nd edition. Edited by S. H. James and J. J. Nordby. Boca Raton, Fla.: CRC Press, 2005.

Federal Bureau of Investigation (FBI) The investigative arm of the U.S. DEPARTMENT OF JUSTICE

(DOJ) and home to the largest forensic laboratory and forensic research center in the world. The bureau maintains an extensive Web site at www.fbi.gov. The bureau traces its origins to the early part of the last century, when the need for a permanent investigative unit within the DOJ was becoming evident. In 1908, Attorney General Charles Bonaparte appointed a core of special agents transferred from the SECRET SERVICE that were christened the Bureau of Investigation in 1909. Over the years, the bureau became responsible for an ever-increasing list of crimes, with major attention focused on espionage, smuggling, bootlegging, organized crime, and civil rights. In 1924, the bureau secured its most famous director, J. Edgar Hoover, who led the bureau until his death in 1972, 48 years later. In 1926, the bureau assumed the primary responsibility for fingerprint cards and identifications for law enforcement agencies throughout the country, a responsibility that still represents one of its largest single tasks.

The bureau was renamed the U.S. Bureau of Investigation in 1932, the same year that a technical laboratory was established. Early examinations focused on FIREARMS and QUESTIONED DOCUMENTS. The laboratory moved to the Department of Justice building in Washington, D.C. Although not the first forensic lab in the United States, the Technical Laboratory, as it was called then, was the largest and evolved into the single most influential source of forensic analysis and information in the world. The organization of the lab served as a model for many other state and local facilities created in subsequent years. Later in that same decade (1935), the bureau was renamed again to the Federal Bureau of Investigation. Also in 1935, the National Academy was opened and became a center for the training of law enforcement officers from all over the country and, eventually, all over the world. During the war years, the Technical Laboratory contributed in intelligence operations and was active in the field of cryptography (making and breaking codes). In 1950, the now famous "Ten Most Wanted" list was born. In 1974, the FBI moved, along with the laboratory, to the new J. Edgar Hoover Building in downtown Washington. Then the lab moved yet again in 2003 to Quantico, Virginia, also home to the FBI Academy. Currently, the laboratory provides forensic services in many diverse areas including chemistry, DNA TYPING, FIREARMS and TOOLMARKS, EXPLOSIVES, visual and audio recordings as well as IMAGE ENHANCEMENT, FINGERPRINTS, questioned documents, and TRACE EVIDENCE.

The laboratory grew along with the bureau, expanding into new areas. During the 1980s and early 1990s, attention turned to computer crimes and digital evidence, with a special response team formed to retrieve digital evidence. In 1981, the Forensic Science Research and Training Center was established at the FBI Academy in Quantico, Virginia, allowing forensic scientists from around the country to obtain training in many different areas. Research is conducted in many areas, and procedures updated and standardized. By the early 1990s the bureau had taken a leading role in developing DNA typing techniques.

In the late 1990s, problems surfaced at the laboratory, particularly in the area of explosives analysis. An inquiry by the inspector general in the Department of Justice began in 1995 after a chemist in the lab, Frederic Whitehurst, made several allegations concerning mishandling of evidence, improper reporting, poor scientific technique, and more serious charges. He had been making such charges since 1989, and the investigation affected several major cases, including the first World Trade Center bombing in 1993 and the Oklahoma City bombing in 1995. Problems were uncovered involving other cases dating back to the late 1980s, which in turn cast a shadow over ongoing investigations such as the crash of TWA FLIGHT 800. The inspector general's report, released in 1997, found no evidence of criminal misconduct, fabrication of evidence, perjury, or obstruction of justice, but it did point out many shortcomings that demanded action. A number of changes and reforms resulted, including the hiring of a new laboratory director.

One of the most important aspects of the FBI Laboratory Division is the numerous databases and reference collections maintained and coordinated there. This includes the Combined DNA Index System (CODIS), the Integrated Automated Fingerprint Identification System (IAFIS), and reference collections of glass, footwear tread patterns, tire tread patterns, explosives, paints, and numerous collections relating to questioned documents. Collation and exploitation of this type of database is emerging as one of the key components of the future of forensic science.

See also FBI LABORATORY SCANDAL.

Further Reading

Johnston, D. "Report Criticizes Scientific Testing at F.B.I. Crime Lab." *New York Times,* April 16, 1997.

McKenna, J. T. "Report Cites Flaws in FBI Bomb Analysis," *Aviation Week and Technology Review* 146, no. 18 (April 28, 1997): 44.

Federal Rules of Evidence A set of rules describing the admissibility of scientific evidence and expert witness testimony. The rules, embodied in Article VII, "Opinions and Expert Testimony," form the basis of admissibility of such evidence in many jurisdictions outside the federal system. Rule 702, "Testimony by Experts" states:

> *If scientific, technical, or other specialized knowledge will assist the* TRIER OF FACT *to understand the evidence or to determine a fact in issue, a witness qualified as an expert by knowledge, skill, experience, training, or education, may testify thereto in the form of an opinion or otherwise.*

In 1923, the *FRYE* DECISION laid down a standard of admissibility that could be summarized by the term "general acceptance," meaning that if the scientific technique in question was generally accepted in the scientific community, then it was worthy of consideration in the court. However, as scientific methods became more specialized, the limitations of this standard became apparent. In a crucial 1993 ruling (the *DAUBERT* DECISION), the Supreme Court decided that the standards set forth in the Rules of Evidence were more flexible and that general acceptance was not an absolute requirement. The Court assigned to the trial judge the role of determining what scientific evidence and expert testimony is relevant to a particular case and whether that evidence and expert testimony should be admitted.

fentanyl A powerful narcotic analgesic that is 100 times more powerful than MORPHINE, but shorter acting. It has been marketed under the name Sublimaze. Because it is such a potent drug, low dosages are taken, complicating the task of the forensic chemist or toxicologist who must identify a sample or detect its presence in URINE or BLOOD. The chemical structure of fentanyl has been the basis of several illegally synthesized analogs, or DESIGNER DRUGS. Fentanyl and its analogs have been available on the street since the late 1970s. Currently it is used and abused in the form of skin patches. These patches are designed to release low doses continually for treatment of chronic pain.

FEPAC The Forensic Education Program Accreditation Commission, which was formed to review and accredit forensic science education programs at colleges and universities. The commission was launched as a committee associated with the AMERICAN ACADEMY OF FORENSIC SCIENCES in 2001. FEPAC commissioners are either academicians or practitioners of forensic science who oversee the accreditation process. The process begins with the review of self-study documents completed by forensic science educational programs. A self-study document describes the program's curriculum, courses, faculty, staff, support, facilities, and equipment. If the application is accepted, an inspection team visits the educational site and reports its findings to the commission, which then decides if the program should be accredited. As of late 2007, there are 16 accredited forensic science education programs in the United States.

ferric chloride A PRESUMPTIVE TEST used in DRUG ANALYSIS. The reagent is prepared by dissolving ferric chloride ($FeCl_3$) at a concentration of 10 percent in water. It is used to detect the presence of MORPHINE, which will cause the solution to turn a blue-green color progressing to a green. It also reacts with other substances such as aspirin.

fibers Given that profuse varieties of textiles, fabrics, and cloth are mass produced, fibers are one of the most commonly encountered forms of trace evidence. As illustrated in the figure, fibers can be divided into several classes based on their origin, with the first division between natural and artificial fibers. Natural fibers include those of mineral origin such as glass and asbestos; vegetable origin (cotton and linen); and animal origin (wool). HAIR is a specialized fiber of animal origin and is addressed in a separate entry.

Cotton is the most common vegetable fiber and is composed of cellulose, as are all plant fibers. Cellulose is a POLYMER of glucose, meaning that it comprises long chains of interconnecting individual units of glucose. It is also a carbohydrate and is a component of the cell walls of plants. The individual units of a polymer such as cellulose are called monomers, and a chain of connected monomers is called a polymer. Most fibers are polymers of different sorts. The artificial or synthetic fibers encompass those that are derived from natural fibers and those that are completely synthetic. Rayon, the first widely used artificial fiber, is made by dissolving cellulose and reconstituting it after drawing it

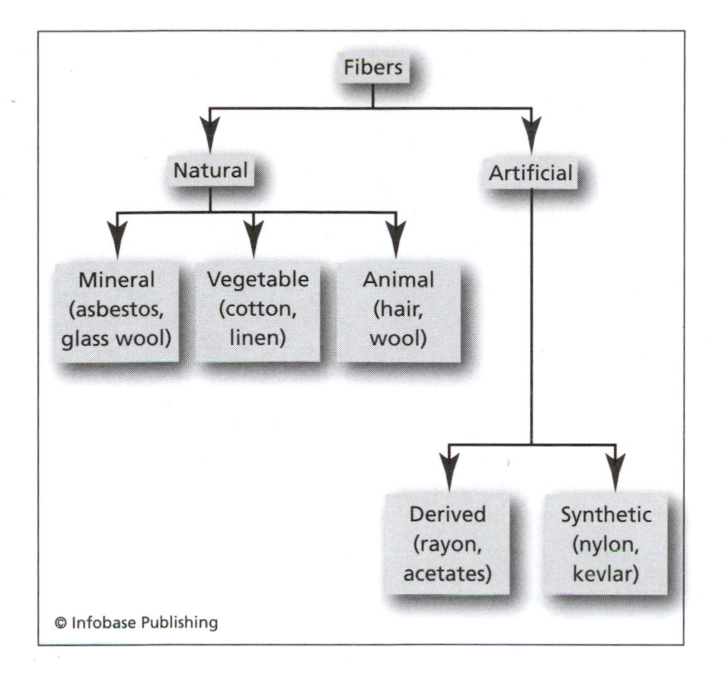

Types of fibers

through a narrow tube the diameter of the desired fiber. Cellulose acetates (often called "acetates") are another type of derived fiber. The largest subgroup of fibers are completely synthetic, all polymers. Nylon, the first such polymer (a polyamide) was developed in 1938 and soared to popularity the following year when it was used in women's stockings. From then on, these stockings were known as "nylons," and the market for synthetic fibers exploded to include such familiar groups as the polyesters and polypropylenes. Kevlar is related to nylon and is classified as an aramid, meaning it is a polymer composed of aromatic *amides*. Kevlar is well known for use in body armor and bulletproof vests, an amazing application for a fiber. The Federal Trade Commission (FTC) has developed a list of generic names for families and subclasses of synthetic fibers, and the different versions made by different manufacturers are identified by their trade names. For example, one of these families is spandex, and a specific brand of spandex is Lycra, which is used in sportswear and swimsuits.

Because fibers are mass-produced and used in such large quantities, it is not possible to individualize them and to conclusively link any two fibers to one COMMON SOURCE. Cotton fibers in particular are so common that they are rarely useful evidence. However, the value of fibers increases if more than one type is found in common, as was the case in the WAYNE WILLIAMS CASE in Atlanta, in which multiple types of fibers were found on victims and in Williams's house and car. Similarly, if fibers show distinctive patterns of wear or damage (such as burning), or similar types of surface contaminants, that increases the likelihood of a common origin but does not guarantee it. Thus, the job of the fiber examiner often is to characterize fiber samples in as much detail as possible to either prove that they did not come from the same source or determine that they could have.

Sources of fibers are as varied as types of fibers themselves. Clothing, furniture, car and building carpeting, building materials, paper, wood, sawdust, and so on are all potential sources of fiber. Given that fibers occur in every environment in which there are people, the chance of fiber transfer in any contact is high. However, since fibers are easily transferred, this presents a challenge to investigators. The majority of transferred fibers are lost after just a few hours, so unless a suspect is caught soon after a crime, few fibers relating to the contact in question will remain. As a result, the lack of common fibers between a suspect and a victim or crime scene is not conclusive evidence that contact never occurred. Also, since fibers move easily, the danger of contamination and SECONDARY TRANSFER is significant. Ideally, any piece of evidence that might have fiber evidence is delivered immediately

Fiber embedded in the surface of a bullet *(Courtesy of William M. Schneck, Microvision Northwest—Forensic Consulting, Inc.)*

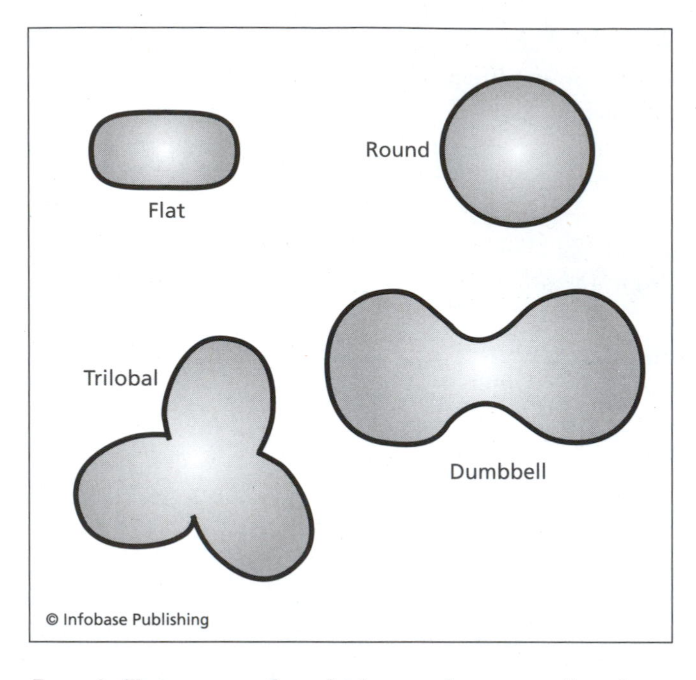

Example fiber cross sections. A microscopic cross section of several man-made fibers can be found in the color insert.

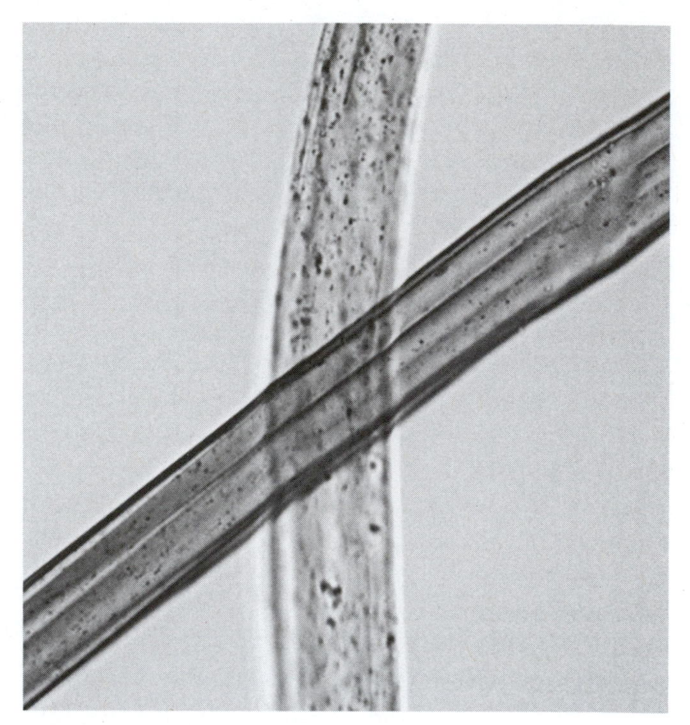

Two polyester fibers viewed under normal illumination ***(Courtesy of Rebecca Hanes, Bennett Department of Chemistry)***

to the laboratory for analysis. If this is not possible, fibers can be collected using TAPE LIFT techniques, by tweezers, by scraping the item over a clean piece of paper, or occasionally by vacuuming.

Analysis of fibers varies from case to case but almost always starts with a microscopic examination. The value of such examinations was understood as far back as the 1850s, when the microscope was first applied to the forensic analysis of fibers. In the hands of a skilled microscopist, many fibers are readily identifiable at this stage. Cotton, for example, has a distinctive twisted appearance that is easily recognizable to a trained expert. Structural and physical characteristics (the MORPHOLOGY) of the fiber are examined, looking at features such as diameter, color, any patterns internally or on the surface, surface treatments such as delustering agents (TiO_2 is common), and twisting or braiding. The cross section of a fiber can be helpful in identification, and some typical cross-sectional types are shown in the figure above. The shape of the orifice through which it is drawn during manufacture determines the cross-section of a synthetic fiber. Fiber analysts have access to collections of fibers that can assist in these comparisons and evaluations.

After initial microscopic evaluation, the examiner can turn to additional tests like melting behavior (using a HOT STAGE microscope), optical characterization of

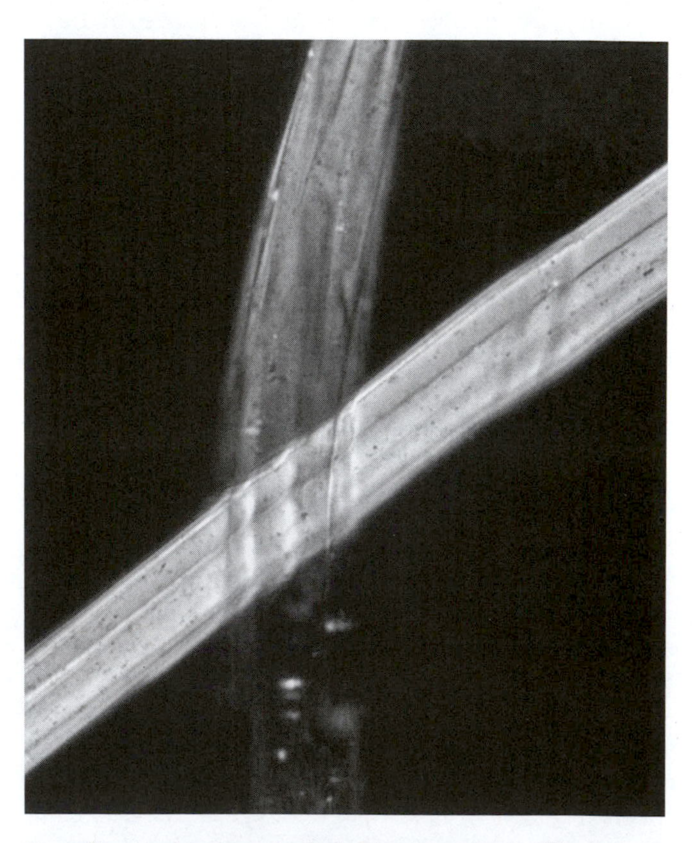

Same fibers under crossed polarization. Note that the fiber has nearly disappeared. ***(Courtesy of Rebecca Hanes, Bennett Department of Chemistry)***

relative refractive indexes (RIs), and solubility tests. The optical properties of fibers are also evaluated, looking at such things as fluorescence (or lack of it) and RI using the BECKE LINE test. Spectroscopic analysis of the fiber, particularly infrared microscopy, is valuable for identifying the subgroup to which the fiber belongs as well as identifying coatings and contaminants that might be present. Similarly, microspectrophotometry in the ultraviolet (UV/VIS) and infrared (IR) can be employed to analyze dyes in or on fibers. THIN LAYER CHROMATOGRAPHY (TLC) is also sometimes used for this purpose, although it is a destructive technique. Pyrolysis GAS CHROMATOGRAPHY/MASS SPECTROMETRY (GC/MS) has been explored for fiber characterization but is also destructive and not widely used. Several images of fibers are found in the color insert.

Further Reading

Gaudette, B. "The Forensic Aspects of Textile Fiber Examination." In *Forensic Science Handbook*. Vol. 2. Edited by R. Saferstein. Englewood Cliffs, N.J.: Regents/Prentice Hall, 1988.

Kubic, T. A., and N. Petraco. "Microanalysis and Examination of Trace Evidence." In *Forensic Science: An Introduction to Scientific and Investigative Techniques*. 2nd edition. Edited by S. H. James and J. J. Nordby. Boca Raton, Fla.: CRC Press, 2005.

Robertson, J., ed. *Forensic Examination of Fibers*. London: Ellis Horwood, 1992.

fingernails and fingernail scrapings Fingernails are a tissue that is chemically similar to hair and as such can prove useful for forensic analysis. In addition, scrapings taken from underneath the fingernails of the victims of such crimes as homicide and sexual assault often contain minute but valuable traces resulting from the assault including BLOOD, skin, HAIRS, and FIBERS. DNA TYPING techniques have been successfully applied to matter recovered from fingernail scrapings as well as the fingernails themselves. In addition, broken fingernails can serve as transfer evidence, and PHYSICAL MATCHING techniques can sometimes be used to link a fingernail fragment back to the nail it came from.

fingerprints As a mark of individuality, fingerprints have a long history. Ancient cultures such as the Babylonian and Chinese used them as a signature, although it is not known if the ancients recognized that fingerprints were unique to each individual. Modern interest in fingerprints as an aid to law enforcement (also called dactyloscopy) dates back to the middle of the 19th century. In 1858, the chief administrative officer of the East India Company, Sir John Herschel, was faced with problems of impersonations among native workers. To combat this, he adopted the ancient traditions and began to use handprints as a form of identifier on documents. He moved quickly to using only the fingerprints from the last joint to the tip, and by 1877 he was using the same procedure to assist in identifying prisoners. In 1880, a Scottish physician, Henry Fauld, published an article in *Nature,* a respected scientific journal, concerning the identification of criminals based on the uniqueness of fingerprints. This led to further advances in the new field, and in 1892, Sir Francis Galton, a cousin to Charles Darwin, published a book entitled *Finger Prints*. In 1894, Scotland Yard adopted fingerprints to go along with the BERTILLON system of body measurements (ANTHROPOMETRY) used at the time for criminal identification. Across the Atlantic, JUAN VUCETICH, an Argentinean police official, was maintaining a large catalog of fingerprints and, in 1892, recorded the first case in which a fingerprint was used to solve a crime. A mother had murdered her son and blamed a neighbor, but finding her bloody fingerprints on a doorpost was enough to elicit her confession. Vucetich would later develop a fingerprint classification system that is still used in much of South America.

The new century saw increasing interest in and acceptance of fingerprints in law enforcement. In 1900, Sir EDWARD RICHARD HENRY of the Metropolitan Police Force (London) published *Classification and Use of Finger Prints* based on the ideas put forth by Galton eight years earlier. The work formalized these ideas into a system of fingerprint classification that has come to be known as the Henry system, variations and extensions of which are used in the United States, Europe, and elsewhere. By 1902, the United States had begun to adopt fingerprinting for identification with the New York Civil Service, followed in 1903 by adoption at Sing Sing prison. The military soon followed. By 1915, the first professional society in the area of fingerprinting, the INTERNATIONAL ASSOCIATION FOR IDENTIFICATION (IAI), was formed. In 1923, the International Association of Chiefs of Police established the National Bureau of Criminal Identification (NBCI) at Leavenworth prison in Kansas. The next year, an Identification Division was established at the Bureau of Investigation, the precursor of the FBI. They merged the NBCI collection with theirs to create a national centralized repository for fingerprint information. This collection, now housed within the FBI, is the largest in

History of Fingerprints: Stay Tuned!

FINGERPRINTS have long been considered the mainstay of forensic identification, even into the DNA age. Since fingerprints are unique to the individual, it is assumed that a match between a suspect print and a comparison print is sufficient to INDIVIDUALIZE the evidence and link it to a COMMON SOURCE. However, the key words are "unique" and "assumed." Increasingly, courts are taking a harder look at fingerprint evidence and the body of knowledge underlying fingerprint identification. This reexamination has led to spirited debate within the forensic community over what is science versus what is technology and why such distinctions are important. The debate is spreading to other areas of forensic investigation such as FIREARMS, TOOLMARKS, BITE MARKS, SHOE PRINTS, and TIRE IMPRESSIONS, all areas that involve visual pattern matching. The outcome of these discussions will have a significant impact on forensic science in the years to come.

The courts have accepted fingerprint evidence for nearly a century, but there is an important distinction between judicial acceptance and scientific acceptance. Scientific acceptance follows the SCIENTIFIC METHOD and involves experimentation, observation, peer review, reproducibility, repeated testing, and refinement. This takes place within the scientific discipline to which the work belongs. For example, scientific acceptance of DNA TYPING occurred first in the domain of molecular biology, biochemistry, and related fields. On the other hand, judicial acceptance occurs in the court in an adversarial environment, not in a scientific one. Because fingerprints were first admitted before modern rules for judicial acceptance of scientific evidence were defined, this distinction becomes important.

The *DAUBERT* case, a U.S. Supreme Court decision in 1993, laid out new guidelines for the acceptance of scientific and other expert witness testimony. This decision assigned to judges the role of gatekeeper and provided a series of criteria that could be employed to determine if the evidence should be admitted. These criteria include general acceptance by the scientific community, testability (and thus FALSIFIABILITY), use of standard procedures, validation of the method, and determination of error rates. These last two criteria are central to recent challenges of fingerprint evidence. The advent of DNA TYPING in many respects set the standard for how the *Daubert* decision would be applied to scientific evidence. Admissibility hearings on DNA evidence (also referred to as Daubert hearings) subjected DNA evidence to rigorous judicial scrutiny and led to its routine acceptance by courts today. This model is now being applied to fingerprint evidence, and the results have caught many in the fingerprint community off-guard.

When blood or body fluid evidence is typed and compared to a standard, such as a suspect's or victim's blood, the results are not reported in absolute terms but rather based on probabilities. The forensic biologist uses databases of known frequencies of DNA types and obtains a number such as one in a trillion. This statement means that the odds of finding a random match in the population is extremely small, but not zero. In contrast, fingerprints are reported as matching or not matching. This absolutism is becoming a serious and contentious issue for fingerprint examiners.

There is good reason and sound science behind the premise that a person's fingerprints are unique. This assumption is based, however, on examination of prints from all 10 fingers. In many criminal cases, the examiner may have little more than a partial of one print or prints. This disparity does not mean that the original premise is invalid, but it does mean that more study of the statistical and probabilistic nature of matches based on partial prints will need to be explored. Although several models have been developed to define the probability of finding two identical prints

the world. In the late 1960s, the FBI began research into the use of digital and computer technologies to assist in fingerprint classification and identification, with the first operable reader available in 1972. This technology continued (and continues) to evolve to the current system of automated fingerprint identification (AFIS) and the Integrated AFIS or IAFIS system.

Characteristics of Fingerprints

Like most forms of evidence, fingerprints have class characteristics (such as loops, arches, and whorls) that can be used to divide fingerprint patterns into categories. However, fingerprints also contain unique patterns within them that allow for their INDIVIDUALIZATION. Galton estimated that fingerprints were unique to one in approximately 60 billion people. Even identical twins do not have the same fingerprints, making them, as the Chinese and Babylonians seemed to have appreciated, a distinctive mark of an individual. In addition, other aspects of fingerprints have lent them to identification purposes. First, fingerprints are formed *in utero* and do not change throughout the lifespan (except for expansion due to growth, scarring, or purposeful alteration). Because fingerprints can be or partial prints, none has been validated by an accepted, large-scale scientific study.

Another issue being raised in fingerprint examination is the lack of a set of standards for how many comparison points must match to obtain a positive identification. In a typical case, an examiner first submits a questioned print to an AFIS system, a database of prints. This submission process produces a set of potential matches, or an indication that no good potential matches were identified. The examiner then compares the questioned print with possible matches and determines if, indeed, a match has been made. During the comparison process, the examiner concentrates on the minutiae, the ridge detail of the prints, comparing features to see if they appear on both the questioned fingerprint and a potential match. However, in the United States, there is no set standard as to how many points must be found in common before a match is declared.

This lack of standardization can be worrisome given that many partial prints have very few identifiable points to begin with. Consequently, the final determination is based on the experience and skill of the examiner. Regardless of how good the examiner is, the process introduces subjectivity into fingerprint comparisons, which conflicts with the goal of scientific evaluation—maximum objectivity. In other words, when an examiner declares a match, this declaration is an opinion. Although this does not invalidate such a declaration, the courts are taking a harder look at the process of fingerprint matching and the broader area of pattern matching as practiced in firearm, toolmark, shoeprint, tire, and bite mark analysis.

Another emerging challenge to fingerprints is based on the concept of known error rates. In any type of analysis, error can be roughly broken down into two types. The first is error that arises from the technique itself; the second type of error arises from mistakes made by the person performing that technique. In DNA typing, a procedural error might arise if the instrument being used becomes contaminated. In fingerprint identification, such an error would arise from an incorrect assessment of a ridge detail, for example. The DNA analyst uses CONTROL SAMPLES and other aspects of QUALITY ASSURANCE AND QUALITY CONTROL (QA/QC) to detect and correct errors that arise from instrumentation or procedures. In fingerprinting, it is not as simple to address procedural errors, nor has a reliable procedural error rate been determined. Some claim that the method has no errors, since it is based on computer screening and simple comparisons. But intuitively, the idea of an errorless procedure is always suspect, and there have been no comprehensive studies to quantify procedural errors encountered in fingerprint comparisons. Similarly, in DNA typing, analyst error is addressed and minimized using specific QA/QC procedures, along with strict training and certification guidelines. Fingerprint examiners also undergo lengthy training and certification, but error arising from individual examiners is harder to specify or control. None of these concerns invalidates fingerprint evidence, but in light of how DNA procedures have been tested and accepted, it does raise new challenges for fingerprint examination. Some day, probably sooner rather than later, it may be possible to obtain a DNA type from a latent fingerprint. Until then, the courts will continue to wrestle with these new issues.

See also MAYFIELD CASE.

Further Reading

Cole, S. A. *Suspect Identities. A History of Fingerprinting and Criminal Identification.* Cambridge, Mass.: Harvard University Press, 2001.

Stoney, D. A. "Measurement of Fingerprint Individuality." In *Advances in Fingerprint Technology.* 2d ed. Edited by H. C. Lee and R. E. Gaensslen. Boca Raton, Fla.: CRC Press, 2001.

initially subdivided by class characteristics, that means it is possible to develop sophisticated and efficient systems for classifying, filing, storing, and identifying them. Finally, due to recent advances in digital imaging, computer databases, and software, fingerprints can be cataloged and searched electronically, providing the fingerprint examiner with a small collection of individuals who might have left the print wherever it was found.

Fingerprints are patterns of ridges (called friction ridges) that appear to have evolved to assist in gripping. The appearance of the ridges is the result of the pattern of subsurface structures called papillae that exist between the upper epidermis layer and the lower epidermis layer of the skin. Friction ridges are also found on the palms, toes, and soles of the feet. Areas of the skin that have friction ridges also have pores, which are the surface opening of sweat glands, but no oil glands. Any oil that is deposited with a fingerprint must come from oily areas such as the face or near the hair. The patterns of the pores on fingerprints is yet another unique characteristic and is the subject of study in poroscopy. However, it is the RIDGE CHARACTERISTICS (type, number, and relative position) that make

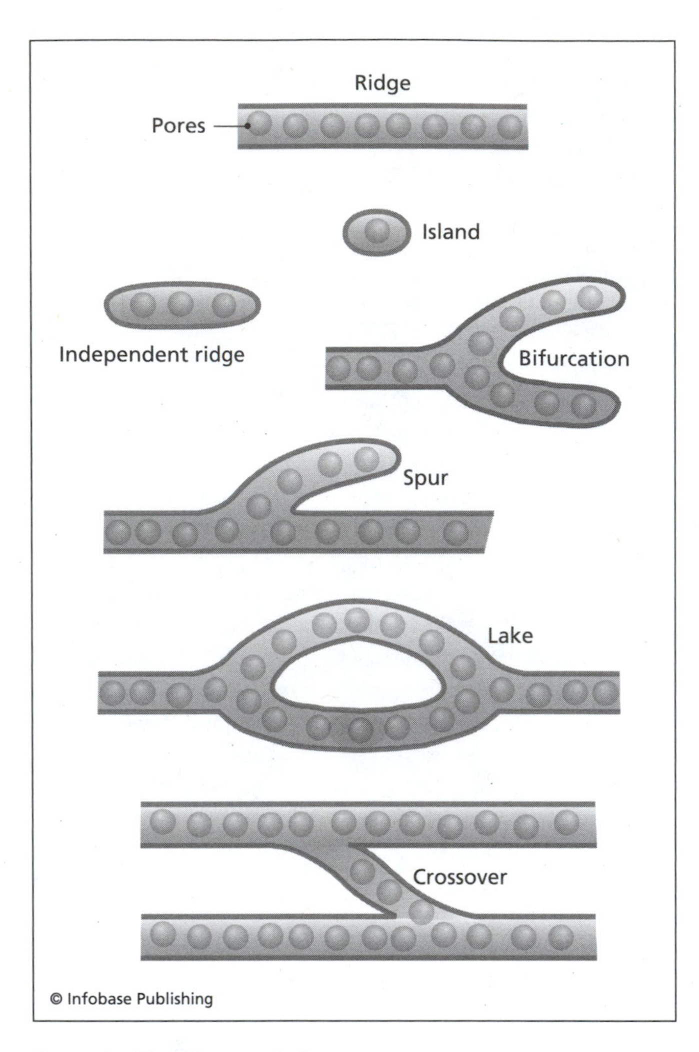

Example ridge characteristics

each fingerprint unique, and these features collectively are called the minutiae. Other examples of minutiae include bifurcations (a ridge that splits), a delta (named for its similarity to a river delta), ridge dots, ridge endings, and enclosures. Examples of some of these ridge characteristics are illustrated in the figure.

Classification

The patterns of fingerprints can be divided into four basic types: whorls, tents, arches, and loops, each of which can be further subdivided as shown in the accompanying figure. Systems of fingerprint classification are based on the presence or absence of these features and are used to categorize similar patterns into smaller groups that can then be searched and compared for the minutiae that are used to make individual identifications. The FBI system, an extension of the Henry system, is the most widely used in the United States and is called a 10-finger system since prints from all fingers are used. The system consists of levels of subdivision to categorize prints into ever-smaller groups. The first level, called the primary classification, divides all fingerprint patterns into 1,024 groups. For the primary classification, fingers are numbered from 1 to 5, starting with the right thumb. Thus, the right thumb is assigned the number 1 and the right little finger 5. The left thumb is 6 and the left little finger is 10. A numerical value is assigned to each finger based on the presence or absence of whorls. If a whorl is present in either the right thumb or right index, a value of 16 is assigned to each finger. If a whorl is found in the right middle or right ring finger, the value assigned is 8. The pattern of cutting the value assigned in half continues until the left ring and little fingers, which will have a value of 1 assigned if a whorl is present. The final step in the primary classification is to arrange these values in fractional form:

$$\frac{\text{Even finger values} + 1}{\text{Odd finger values} + 1}$$

The 1 is added to each to assure that there is never a value of zero in either the numerator (value on top) or the denominator (value on the bottom). This would occur if a person's fingerprints consisted only of arches and loops, with no whorls on any of the fingers. Such a person would be classified in the 1/1 group. In contrast, someone with whorls on all fingers would generate a primary classification of 32/32:

$$\frac{16 + 8 + 4 + 2 + 1 + 1 = 32}{16 + 8 + 4 + 2 + 1 + 1 = 32}$$

The other subdivisions that can be performed include a secondary classification, subsecondary, final, major, and key, also displayed in fractional format using both letters and numbers. The fraction is displayed in the order: Key Major Primary Secondary Subsecondary Final.

One of the limitations of the 10-finger classification system is that rarely do investigators recover anything but partial latent prints at a crime scene. Filing systems based on single fingerprints have been attempted, but until the advent of computerized database systems, these proved to be too large and cumbersome to be useful. AUTOMATED FINGERPRINT IDENTIFICATION SYSTEMS (AFIS) work by digitizing carefully collected images of a fingerprint, encoding details of the minutiae, and submitting the results to a database search. These systems do not identify a person uniquely but rather provide a list of potential matches for the fingerprint examiner to evaluate and make the final determination.

At a crime scene, three different types of fingerprints can be left. Latent prints are those that are nearly invisible, while visible prints are deposited in a material that makes them easy to locate such as blood or dust. Plastic prints are those embedded in a moldable material such as window putty or soft caulk. Since there are no oil glands on the ridged surfaces of fingers, a latent print is composed primarily of water (98.5 percent) from the sweat, with the remaining being dissolved solids such as ions (K^+, Na^+, Ca^{2+}, Cl^-, PO_4^{3-}, CO_3^{2-}, and SO_4^{2-}), fatty acids, sugars, and proteins. The total mass of transferred material is usually on the order of one microgram (μg, a millionth of a gram). Any oily material in a latent print originates from the fingers touching an oily part of the body such as the nose. Almost immediately, the components that make up the fingerprint are subject to degradation by chemical and microbial processes. Additionally, since so little solid material is deposited with a fingerprint, latent prints can be difficult to detect. Recent advances include the use of lasers and chemical treatments in conjunction with high intensity light sources. The latter, generally referred to as alternate light sources (ALS), exploit induced FLUORESCENCE in the components that make up fingerprints. ALS units can be made portable, and when combined with different chemical treatments, can be effective and sensitive methods to detect latent fingerprints.

Developing and Visualizing

Numerous methods exist for visualizing and preserving latent fingerprints, all based on chemical reactions with components deposited with the print. The process starts and ends with photographic documentation, but between these actions, there are many options available. The surface on which a print is deposited is critical in determining what approach is to be used. Visualizing a print left on a porous surface such as wood, leather, or skin is a challenge different from developing one left on a nonporous material such as glass or tile. Generally, prints left on porous materials are visualized using chemical methods, while those on nonporous surfaces are developed using powder or cyanoacrylate (Super Glue). Powders come in a variety of colors, and the choice of which to use depends on the background and what will produce the best contrast for photography. Powders come in different colors and can contain metallic, luminescent, or thermoplastic materials. Thermoplastics are the same types of components that are used in many copiers and laser printers and are set by heat. Luminescent powders will give off light either as a result of exposure to lasers or other light sources or as a result of chemical reactions. Regardless of the type, powders are generally applied using fine brushes and adhere to moisture and oils found in the latent print. For delicate prints, actual brush contact with the print can be avoided by using a MagnaBrush and magnetic powders, which adhere to the brush and can be sprinkled onto the print with a gentle movement. Once a print is "dusted," it is documented by photography and can be carefully transferred to adhesive surfaces using a tape lift technique. The lifted print is then placed on a background that provides maximum contrast.

Several different chemical developers have been used, with many more being studied and researched. One of the oldest techniques involved an iodine fuming in which the object to be printed is placed in a cabinet along with a solid piece of iodine (I_2). The iodine is then heated, and it quickly sublimes, meaning it is converted directly from a solid to a gas in the same manner that dry ice (frozen carbon dioxide) turns into a gas without going through a liquid stage.

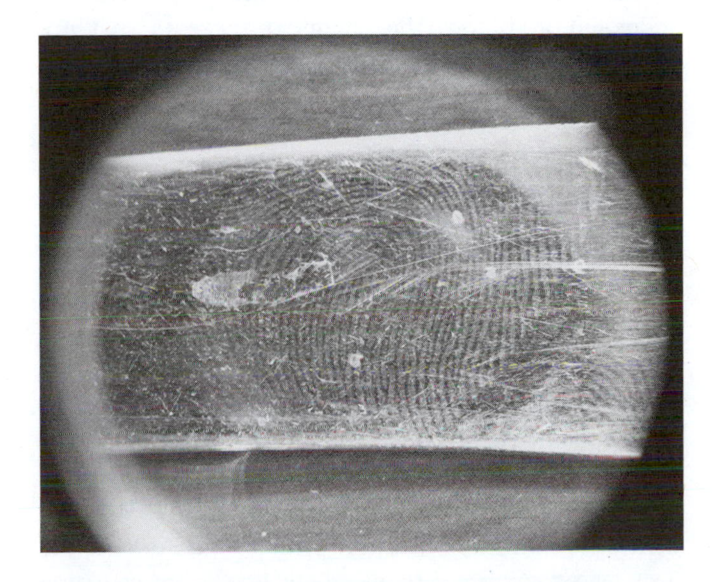

A fingerprint on a knife blade treated with cyanoacrylate (superglue) ***(Courtesy of Dr. Keith Morris, West Virginia University, Forensic and Investigative Sciences program)***

The iodine vapors react with components deposited in the fingerprint, resulting in a purplish visualization of the pattern. To set the pattern, additional ingredients such as starch are required; otherwise, the developed print quickly fades. A compound called Physical Developer (PD) is also used and is based on reactions between oils and silver nitrate ($AgNO_3$). Similar reactions of silver compounds are the basis of black and white photography. Currently, the most commonly used chemical developer is NINHYDRIN, which produces a bluish purple in a reaction with the AMINO ACIDS deposited in the latent print. Ninhydrin coupled with zinc chloride ($ZnCl_2$) followed by laser illumination is an example of the use of chemical development coupled to an ALS. Finally, a number of ninhydrin derivatives have been developed and used including 1,8-diazafluoren-9-one, or DFO. Often, latent print examiners will use a series of visualizing agents in a specific order such as DFO-ninhydrin-PD.

Arguably the most famous reagent used to develop latent fingerprints is Super Glue, a cyanoacrylate that reacts with components in the fingerprints to create a grayish-white pattern. Like iodine, fuming cabinets can be used, and the process is initiated by the addition of sodium hydroxide (lye, NaOH). Even fingerprints left on skin can sometimes be recovered by placing the limb or the body in a tent that is filled with the fumes. Wands have been developed for use at crime scenes, and entire enclosed spaces such as cars can be fumed using this reagent. For fingerprints deposited in blood, the chemical reagents that are used for PRESUMPTIVE TESTS for blood (such as phenolphthalein) can be used to visualize and develop the patterns.

This fingerprint was treated with fluorescent dye to make it brighter. See the color insert for the true-color image. ***(Courtesy of Dr. Keith Morris, West Virginia University, Forensic and Investigative Sciences program)***

Fingerprint evidence can be collected, preserved, and delivered in several ways, the choice of which depending on the circumstances. When objects such as guns or paper are found at a scene and are to be fingerprinted, the object is normally delivered to the fingerprint examiner rather than being developed in the field. Similarly, prints deposited on removable surfaces can be excised and delivered rather than processed in the field. In any case and as with all evidence, documentation by complete and detailed photography is essential, particularly in cases in which it is unlikely that the print can be developed, transferred, or lifted from where it was found. Tape lift techniques (described previously) are frequently used after photography to preserve the print for later evaluation. Digital imaging and image enhancement techniques are becoming common and can be used to improve contrast between the print and the background or to improve the quality of otherwise poor quality prints. Digitalization is also the first step for preparing a print for computerized storage and searching.

For the identification of a fingerprint, the process often starts with submission to AFIS, which, contrary to the common misconception, does not produce the name of an individual to which the fingerprint belongs. Rather, a list of potential sources is provided from which the fingerprint examiner works to make the final identification. However, even recov-

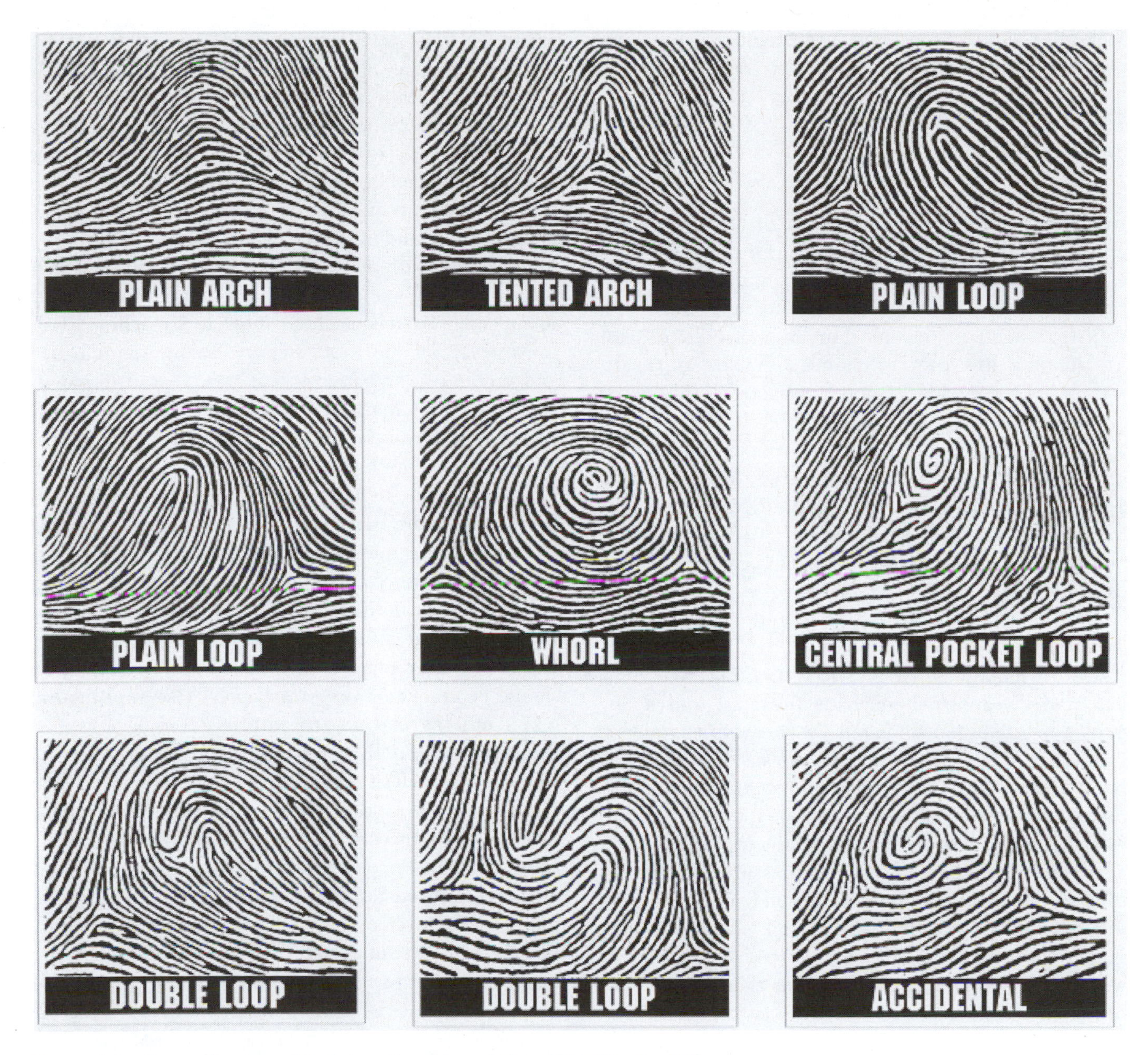

Fingerprint patterns ***(Courtesy of the International Association of Identification, IAI).***

ering a perfect fingerprint is not a guarantee that identification can be made. If the person has never been fingerprinted or fingerprints are not entered into a centralized database, then there is nothing to compare to and thus nothing that would allow for definitive identification. However, assuming that the examiner does have known fingerprints to compare to, he or she will proceed to compare the minutiae to determine if two prints match. Although there is no set standard in the United States as to specifically how many matched points must be found, the range is usually 6 to 12 points. In making a final identification, both the quality of the latent print and the quantity of matched points is taken into account by the latent print examiner.

See also MAYFIELD CASE.

Further Reading

Gaensslen, R. E., and H. Lee, eds. *Advances in Fingerprint Technology.* 2d ed. Boca Raton, Fla.: CRC Press, 2001.

Moler, E., et al. "Fingerprint Identification Using Image Enhancement Techniques." *Journal of Forensic Sciences* 43, no. 3 (1998): 689.

U.S. Department of Justice. *Fingerprint Identification.* Washington, D.C.: U.S. Department of Justice, Federal Bureau of Investigation, 1991.

firearms Forensic examination of firearms was established in the United States and Europe during the early years of the 20th century. In many laboratories, firearms examiners are also involved in TOOLMARKS, GUNSHOT RESIDUE, chemical analysis of bullets, SERIAL NUMBER RESTORATION, and DISTANCE DETERMINATIONS of how far a shooter was from the victim or target. Although the term "forensic ballistics" is sometimes used to describe this work, it is not a correct description. Ballistics is the study of the motion and trajectory of projectiles, while firearms analysis focuses on the study of BULLETS, CARTRIDGE CASES, and other materials associated with firearms as physical evidence.

Types of Firearms

Pistols or handguns, rifles, SHOTGUNS, machine guns, automatic weapons, homemade firearms, and hobby guns can all be encountered by firearms examiners. Pistols are smaller guns designed to be fired with one hand and include the revolver and semiautomatic pistols, often erroneously called "automatics." True automatics fire continuously as long as the trigger is pulled. Semiautomatic guns exploit gas pressure and springs to eject the spent cartridge, load a new one, and cock the weapon for the next shot, but a separate trigger pull is required to fire the next cartridge. Homemade weapons are usually small handguns that carry few or one shot and have been referred to as "zip guns" and Saturday Night Specials. Hobbyists such as those that reenact Civil War battles use smooth bore guns with round projectiles (ball shot) and black powder, and so a forensic firearms examiner must be familiar with such weapons.

Analysis

Tasks of firearms investigators are similar to those of other forensic analysts in that they seek CLASS CHARACTERISTICS and characteristics that will INDIVIDUALIZE evidence by linking a specific weapon to a specific bullet or cartridge. Modern guns all have rifled barrels, meaning that a series of grooves (LANDS AND GROOVES) are machined into the barrel in a spiraling pattern. When the bullet is forced over these lands and grooves, spin is imparted to the bullet, which stabilizes the trajectory and greatly increases accuracy over smooth bore weapons. This contact creates distinctive STRIATION patterns on the bullet that can later be matched to the weapon. The striations are usually unique to one gun based on the way the gun is manufactured and used. To machine lands and grooves into a barrel, different tools can be used, all of which involve metal-to-metal contact. Although a tool is used to cut many barrels, each barrel that is machined will cause wear on the tool, creating small irregularities and random wear patterns that can be transferred to the next barrel machined. This next barrel will also create irregularities and so on, meaning that the markings on each new barrel cut will be slightly different and thus unique. In addition, once a gun is purchased, it will have a different usage pattern from every other gun and thus a unique wear pattern on the lands and grooves. Thus, slight differences created during milling combined with individualized wear patterns lead to a unique striation pattern for most guns. As a result, the marks each gun imprints on bullets and cartridge cases should be unique or very nearly so. These imprints are a kind of IMPRESSION EVIDENCE, as are toolmarks. For this reason, many firearms examiners are also responsible for the analysis of toolmark evidence.

The classifications associated with firearms evidence include the CALIBER of a weapon or AMMUNITION, the number of lands and grooves, the direction and degree of twist in the rifling pattern, and the make and manufacturer of a weapon or ammunition. Thus, when evidence is submitted such as a bullet or case, the examiner will attempt to identify as many characteristics as possible. The same approach would be used if a weapon were submitted. When a weapon as well as bullets and/or casings are available, then the examiner can attempt to prove or disprove the theory that the suspect weapon fired the bullet and casing. The primary tool used to do so is the COMPARISON MICROSCOPE, which allows for a side-by-side comparison of a bullet or casing submitted as evidence (for example, if a bullet was recovered at autopsy and a casing was found at a crime scene) with bullets or casings obtained from test firings of the weapon. To obtain these standards, the firearms examiner discharges the suspect weapon into a trap of some kind (a water tank or boxes filled with cotton or oil-soaked cotton) using the same type of ammunition as the submitted evidence. By firing into the trap, all of the striations on the bullet are preserved and can be used for comparison to the submitted bullets. If the striations on the test bullet or casing match up with those of the evidence, then the same weapon was used to fire both.

The handling of firearms evidence is driven first by safety concerns. If a weapon is recovered at a scene, it is normally unloaded, but only after complete and careful documentation. Information such as how many bullets were fired, which chamber was under the hammer (in the case of revolvers), and general condition of the weapon can be crucial. Any marking of guns, bullets, or casings is done carefully to avoid damaging the striations. If bullets are imbedded in walls, floors, or other materials, it is usually preferable for entire sections of the material containing the bullet to be removed and transported to the lab, where the examiner can carefully remove the projectile without damaging it. Conversely, the examiner will study the evidence carefully, looking for latent fingerprints and other trace evidence such as hair, blood, and tissue on or in the barrel or impressions such as textile prints on bullets. All must be carefully documented and preserved before the actual firearms examination can begin.

History

Firearms examination had become widely accepted by the 1930s. The first reported use of a comparison microscope came in 1922, with increasing use of bullets and cartridge cases as evidence. In the early 1920s, Charles Waite became involved as an investigator in a case in which Charles Stielow was convicted of the murder of his landlord and housekeeper based primarily on testimony related to firearms. An expert at the trial stated that Stielow's pistol fired the fatal shots, and Stielow was sentenced to die. Shortly thereafter, another man confessed to the crime, prompting a reexamination. No evidence of a conclusive match of the bullet to Stielow's pistol was found. As a result of his involvement in the case, Waite collected extensive information about the rifling characteristics of firearms in the United States and Europe. In 1924, he started a private laboratory called the Bureau of Forensic Ballistics staffed by pioneers such as CALVIN GODDARD, John Fisher, and Philip Gravelle. Waite left shortly after that, while Gravelle became a strong advocate of the use of the comparison microscope. Fisher developed an instrument called a helixometer used to measure the angle of twist of the rifling, and Goddard wrote extensively about firearms examination. Goddard became involved in a pivotal early firearms case, the Sacco-Vanzetti trial that had taken place in 1922 and resulted in convictions and sentences of death for the defendants. Goddard joined in a review of the case in 1927.

On April 20, 1920, a robbery took place in South Braintree, Massachusetts, in which five men robbed a paymaster, killing him and the guard. A month later,

View looking down the barrel of semi-automatic pistol. The lands, grooves, and twist are seen. *(Courtesy of Michael Bell, West Virginia University, Forensic and Investigative Sciences program)*

two men, Sacco and Vanzetti, were charged with the crime. The case took a political twist since both men were of foreign descent and were anarchists, a movement that was spreading across the United States and Europe at the time. A .38 pistol was recovered from Vanzetti, but it could not be conclusively tied to any of the evidence recovered. From the bodies, .32 ACP bullets were recovered, and Sacco had a .32 pistol in his possession when arrested. Many experts testified about the firearms evidence while riots and protests broke out. The men were convicted, and, during his review, Goddard was able to link one cartridge case and a fatal bullet to Sacco's gun. The executions took place seven years after the crime, on August 23, 1927. Although controversial at the time, later firearms examination supported Goddard's findings.

In 1929, Goddard became involved in the famous St. Valentine's Day Massacre and was able to show that the murdered men had been killed by two Thompson submachine guns. His work in the case was well received, and shortly thereafter Goddard left the Bureau of Forensic Ballistics to direct the new Scientific Crime Detection facility in Chicago that would eventually become the Chicago Police Department Laboratory. During this period, developments in Europe were paralleling those in the United States, led by men such as VICTOR BALTHAZARD. Balthazard was the first in Europe to identify FABRIC IMPRESSIONS of textiles on bullets and to use FIRING PIN impressions, BREECHBLOCK MARKS, and EXTRACTOR/EJECTOR MARKS. The study of such impressions employing a comparison microscope is still the primary approach used.

Matching striations on two bullets as seen through a comparison microscope ***(Courtesy of Michael Bell, West Virginia University, Forensic and Investigative Sciences program)***

Other tasks undertaken by firearms examiners include the restoration of obliterated serial numbers from guns or other metal objects, distance determinations, reconstruction of shooting events, and, occasionally, the chemical characterization of bullet fragments when the damage prevents the standard microscopic approach. The most well-known example of this was conducted on the bullets involved in the assassination of President JOHN F. KENNEDY in 1963. In that case, NEUTRON ACTIVATION ANALYSIS (NAA) was applied to bullets and fragments in an attempt to determine how many bullets were represented by the recovered pieces. As in all areas of forensic science, computers and computer databases are becoming increasingly important. The National Integrated Ballistics Information Network (NIBIN) is coordinated by the federal BUREAU OF ALCOHOL, TOBACCO, FIREARMS AND EXPLOSIVES.

See also BALLISTIC FINGERPRINTING; D.C. SNIPERS.

Further Reading

Rowe, W. F. "Firearm and Toolmark Examination." In *Forensic Science: An Introduction to Scientific and Investigative Techniques*. 2nd ed. Edited by S. H. James and J. J. Nordby. Boca Raton, Fla.: CRC Press, 2005.

fire investigation Similar to the investigation of ARSON and occasionally used synonymously with the term *arson investigation*. While arson and fire investigator often work side-by-side, the primary difference between the two jobs arises from the ultimate cause of the fire. If arson is involved, a criminal act has occurred, and that would be the responsibility of the arson investigator. However, most fires (more than 85 percent) are not the result of arson. The investigation of these fires is the subject of fire investigation. Fire investigation is often considered to be part of or closely related to forensic ENGINEERING, with similar goals. Fire investigators seek to find the causes and contributing factors to a fire (both physical and human) and to make recommendations on how to avoid a similar incident in the future.

Further Reading

Redsicker, D. R. "Basic Fire and Explosion Investigation." In *Forensic Science: An Introduction to Scientific and Investigative Techniques*. 2nd ed. Edited by S. H. James and J. J. Nordby. Boca Raton, Fla.: CRC Press, 2005.

Microscopic comparison of two firing pin impressions ***(Courtesy of Michael Bell, West Virginia University, Forensic and Investigative Sciences program)***

firing pin impressions (or **indentations**) Small marks created in the PRIMER by the impact of the firing pin. The action of pulling the trigger of a firearm causes the firing pin to strike the primer, which consists of a tiny amount of a shock sensitive explosive. Ignition of the primer in turn ignites the PROPELLANT, causing a rapid buildup of gases behind the bullet. The pressure drives the bullet down the barrel and out toward the target. The surface of the primer struck by the firing pin is relatively soft metal, which can pick up the pattern on the surface of the firing pin. Either as a result of the original machining or through use and wear, the markings on the surface of the firing pin will become unique. FIREARMS examiners can use firing pin impressions to link cartridges to specific weapons. Unlike cartridge cases, which can be reloaded and reused, primers can be used only once, so any fired cartridge (not yet reloaded) will show only one set of firing pin impressions.

See also AMMUNITION.

FISH An acronym with two different meanings in the forensic context. A Forensic Information System for Handwriting is a database of handwriting samples used in QUESTIONED DOCUMENT analysis. Letters or other writings involved in cases (bank robbery notes and so on) are scanned into a database for future reference and comparison. If a new document is submitted and shows similarities to others already in the database, a document examiner can be alerted so that further study can be made. The FISH system was developed by the United States SECRET SERVICE. The other meaning of FISH is fluorescent in-situ hybridization sex determination, a method that can be used to determine the sex of cells.

Further Reading

Geradts, Z. "Use of Computers in Forensic Science." In *Forensic Science: An Introduction to Scientific and Investigative Techniques*. 2nd edition. Edited by S. H. James and J. J. Nordby. Boca Raton, Fla.: CRC Press, 2005.

Fish and Wildlife Service (**wildlife forensics**) The federal law enforcement agency responsible for protection of wildlife and prosecution of crimes such as poaching. In 1989, the agency opened the first and currently the only forensic laboratory dedicated to the investigation of crimes related to wildlife, a specialty referred to as wildlife forensics. The laboratory is located in Ashland, Oregon, and serves the national and international communities. The lab specializes in the identification of animals by species and subspecies based on bones, biological fluids, or parts of the animal submitted as evidence. Cause of death is determined to the extent possible, and PHYSICAL EVIDENCE such as FINGERPRINTS and FIREARMS (BULLETS, CARTRIDGE CASES, and so on) are tested. The goals of the analysis are the same as in a traditional forensic laboratory: to provide investigative leads and to link individuals to a crime or crime scene. Assistance with crime scenes is available, and analysts testify in court as needed. Different groups within the lab work in the areas of CRIMINALISTICS, PATHOLOGY (cause of death investigation), MORPHOLOGY (identification of species by structures such as bones and feathers), and genetics. Toxicological techniques are used to detect and identify poisons as well. The lab employs experts in mammals, birds, and reptiles. The FWS maintains a Web site at www.fws.gov.

flame ionization detector (**FID**) A type of detector used in GAS CHROMATOGRAPHY (GC), which is frequently used in the analysis of ARSON evidence. An FID is sensitive to compounds containing bonds between carbon and hydrogen, making it an ideal detector for the analysis of hydrocarbons such as gasoline and other common ACCELERANTS. As shown in the figure on the next page, gas emerging from the outlet of the GC instrument is mixed with flammable hydrogen gas and ignited so that the flame burns constantly whenever the detector is on.

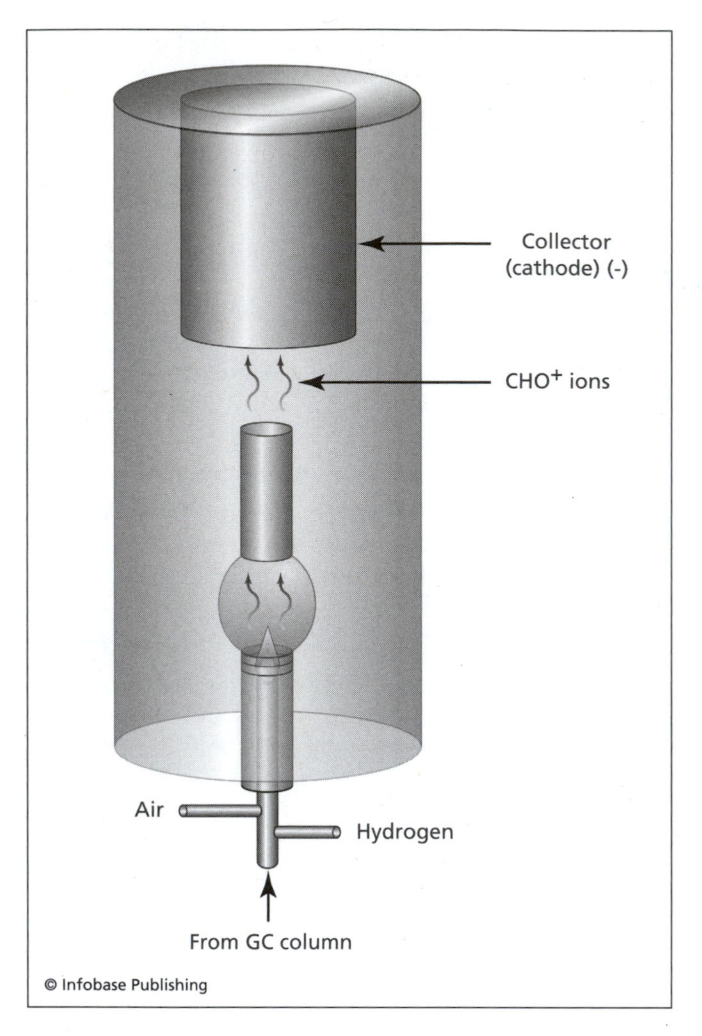

Simplified schematic of a flame ionization detector (FID)

Compounds are separated in the GC and enter into the flame, where the hydrocarbon molecules burn to form CHO+ ions. Because ions carry a positive electrical charge, they will be drawn toward the collector, which has a negative charge. The greater the number of positive ions collected, the greater the detector response. When an arson sample is analyzed using a GC/FID, the resulting pattern produced by the detector can be used to identify the type of accelerant used, such as gasoline, diesel, or kerosene.

flammable liquids *See* ACCELERANT.

flash point *See* ACCELERANT.

fluorescence A property of atoms and molecules that can be exploited in the forensic examination of materials such as GLASS, FIBERS, DYES, rubber, FINGERPRINTS, oils, grease, and TAPE. When atoms or molecules absorb ELECTROMAGNETIC RADIATION (EMR) of the appropriate wavelength, the energy that is absorbed can be used to promote the electrons into what is called an excited state. When the electrons decay back to their original ground state, energy is emitted. In the case of fluorescence, the energy is emitted in the form of electromagnetic energy that is the same wavelength or longer (lower energy) than the excitation energy. This emission of energy is called fluorescence, and it ceases as soon as the excitation energy source is turned off. If the emission continues after the source is turned off, the phenomenon is called phosphorescence, familiar to many in the form of glow-in-the-dark watches.

Fluorescence can be used to classify and identify many materials of interest to the forensic scientist and can be performed using techniques called SPECTROFLUOROMETRY (SPF) or using fluorescence microscopes. In most applications, the incident light used to excite the molecules is in the ultraviolet (UV) range, undetectable to the human eye, while the fluorescent energy is often in the visible or, in some cases, the infrared range of the electromagnetic spectrum. Unlike the straight-through design of absorption spectrophotometers, fluorescence radiation is detected at a 90° angle, keeping it separate from the incident radiation. The incident radiation is the light that is initially shined on the sample and travels in a straight line through it. One advantage of fluorescent detection is sensitivity, which is greater than for traditional ABSORPTION techniques; single molecules have been detected based on fluorescence. A photo of a fingerprint treated with fluorescent agents is found in the color insert.

flurazepam A member of the BENZODIAZEPINE family of drugs and marketed under the name of Dalmane. It is prescribed to treat sleeping disorders and insomnia.

footprints (footwear impressions) *See* IMPRESSION EVIDENCE; SHOE PRINTS.

forensic In its simplest and oldest meaning, forensic relates to public debates (in a "forum"). The term is still used in this context to describe debating clubs and societies found in some high schools and universities. However, the common modern use of the term is related to the legal system, a formalized public debate

The Future of Forensic Science

The pace of progress in forensic science is accelerating just as the pace of science and technology in our society is. Although advancements in DNA TYPING are the first to come to mind, it is certainly not the only area that will see dramatic changes in the coming years. Predictions are always a gamble, but the following can be reasonably expected:

Miniaturization of DNA Typing Apparatus and Expansion of Databases

Practitioners in the field can envision a day when miniaturized DNA typing apparatus will be available, so-called labs on chips that may allow typing to be performed at a crime scene. Turnaround time for analysis could drop from days and weeks to hours or even minutes. Although this seems astonishing, little more than 15 years has passed since DNA typing supplanted traditional serological techniques such as ABO blood group and isoenzyme typing. With the completion of the sequencing of the human genome in 2001, movement into proteomics (the study of how genes make proteins), and astounding advances in biotechnology, even these predictions may be conservative.

Along with improved DNA typing will come larger databases of DNA types, following a pattern set by the rapid expansion of fingerprint databases seen in the last century. These improved techniques, along with increased hiring of DNA analysts, will spur testing of evidence from older cases. In turn, this will allow investigators to link serial crimes in a way never before possible. Already these "cold hits" are making headlines, as in the case of Seattle's "Green River Killer." In 2001, a suspect was finally arrested for the crimes, a series of murders that took place in the Seattle area in the 1980s. The evidence facilitating the arrest was DNA typing of a saliva sample on gauze collected from the suspect by having him chew on it. The gauze was preserved from collection in 1984 until analysis in 2001.

Finally, as DNA techniques improve, smaller samples will be needed. Consequently, it may be possible to type the DNA found in fingerprints using these improved methods. If so, LOCARD'S EXCHANGE PRINCIPLE, that every contact leaves a trace, will take on new meaning. Everything a criminal touches could be used to convict him or her. However, if this does occur, more evidence will require increased automation and, inevitably, more forensic scientists to conduct the tests. Indeed, the success of DNA technology may be the crucial innovation that moves forensic science to one of the top government spending priorities.

Increasing Reliance on Computer Modeling

Computer modeling has become an indispensable tool in all aspects of science, engineering, and technology and as a result is making rapid inroads into forensic science. Applications in accident reconstruction, crime scene reconstruction, and fire and arson investigation are already common. Accidents ranging from traffic accidents to major disasters such as airplane crashes can be modeled and studied using computer programs designed to re-create the sequence of events leading up to the incident. For airplane accidents, data comes from the "black boxes" that record information such as control settings and instrument readings for the last few minutes of a flight. This data can be fed directly into a modeling program to visualize the movement of the plane during the incident. Such reconstructions were used in the analysis of the crash of TWA FLIGHT 800, which crashed off the coast of Long Island in 1996.

For traffic accidents, information obtained from physical evidence such as skid marks, from witnesses, and from data recorders found in some cars is used to achieve the same goal—a computer animation that depicts the likely course of events leading up to the accident. Fire investigators can turn to models to simulate the origin and spread of blazes. Other areas that may benefit from increased use of computer modeling include FIREARMS and bullet trajectories and BLOODSTAIN PATTERN interpretation. As more data is collected as part of an investigation, the potential for accurate modeling of the events in question increases as well.

New Techniques in Forensic Deception Analysis

Technology is finally making inroads to assist investigators in answering a fundamental question of investigation: Is this person telling the truth or a lie? Psychiatrists and psychologists use a battery of interview and testing tools, while POLYGRAPH operators use the lie detector to measure bodily responses to emotion. Neither is completely trustworthy. However, recent research gives hope that greatly improved methods will soon become part of the investigative arsenal. In one approach, called brain fingerprinting, electrical impulses in the brain are measured in response to images. Theoretically, by showing someone a picture of a crime scene, it would be possible to determine if they were familiar with it or not. Another technique is using medical MRI images (based on nuclear magnetic resonance [NMR]) to image the brain while a question is being answered. Preliminary results indicate that different parts of the brain are activated when deception is involved. Given human

(continues)

The Future of Forensic Science *(continued)*

nature, it is unlikely that there will ever be a method that is 100 percent effective in detecting lies, but the next few years or decades should see dramatic improvements.

Increasing Sensitivity and Decreasing Size of Chemical Instrumentation

Each year brings greater improvements in the capability, automation, and portability of chemical instrumentation. These advances trickle down to forensic chemistry and forensic biology, increasing throughput (number of samples that can be run in a day) and decreasing turnaround time. Unfortunately, soaring caseloads at laboratories often overwhelm these improvements. Yet, without these improvements, crime laboratories would be quickly and hopelessly inundated. Autosamplers, which allow instruments to run unattended, have been a boon for drug analysis and toxicology, in which much of the analysis requires instrumental data. Robotics systems are now capable of automating many of the tasks of sample preparation and transfer. While instruments and accessories cannot replace a forensic chemist, toxicologist, or biologist, automation maximizes the productivity of each, allowing the scientist to concentrate on analysis and interpretation.

New instrumentation and improved design of older instruments is also influencing forensic analysis. Much of this can be traced to decreasing costs as technologies mature. Many labs now use HIGH PRESSURE LIQUID CHROMATOGRAPHY (HPLC) and HPLC-mass spectrometry (HPLC-MS), techniques that were rarely seen in forensic labs until the 1990s, both due to cost and method development considerations. A workhorse instrument in forensic chemistry, the infrared spectrophotometer, has seen advances in capability that will likely continue unabated. The driving force of much of this improvement lies with advancement in lasers and electronics. For inorganic analysis, instruments such as inductively coupled plasma mass spectrometry (ICP-MS) and inductively coupled plasma atomic emission spectroscopy (ICP-AES) are expanding into forensic labs and will likely become commonplace in the next 10 to 20 years. Likewise, surface analysis instruments such as X-RAY FLUORESCENCE (XRF) and SCANNING ELECTRON MICROSCOPES (SEM) are becoming affordable to more forensic labs, increasing capabilities and enlarging the scope of analyses available to the forensic examiner.

Databasing, Statistical Analysis, Networking, and Pattern Matching

Forensic science is becoming computerized, automated, and networked. As this trend continues, labs will be able to share more data, and distributed databases of many different kinds of evidence will become available. Currently, most labs can access databases of FIREARM evidence, FINGERPRINTS, SHOE PRINTS and TIRE PRINTS, and DNA types. With increasing connectivity, more labs will join the networks and will be able to contribute their case results as well. As these databases grow, so do their utility and value. However, along with this expansion comes greater need for security and prevention of unauthorized access. For the CODIS DNA database, this is a particularly important concern given the sensitive nature of the data it contains.

Driven by DNA typing, statistics and related disciplines are becoming an integral part of forensic science. The *DAUBERT* DECISION will also continue to push disciplines such as fingerprinting to incorporate more statistical information and justification for their practices. Advances in computer hardware and software will likely bring commercial "data mining" techniques to forensic science and will facilitate more effective utilization of the ever-increasing amounts of data collected.

epitomized in the courtroom. Thus, FORENSIC SCIENCE can be broadly considered to be the application of science to a legal context.

forensic disciplines For example, to find "forensic art," look under ART, FORENSIC. The table on page 163 provides a sampling of forensic disciplines, but it is not complete.

See also individual entries.

Forensic Files A television show broadcast by the truTV network. Each episode highlights investigations in which FORENSIC SCIENCE played a significant role. Along with other television shows such as *CSI,* the program has helped to popularize forensic science.

forensic science In the broadest sense, the application of the techniques of science to legal matters, both criminal and civil. Forensic science includes a number of disciplines and subdisciplines such as forensic anthropology, engineering, and medicine. There is some debate as to how the terms CRIMINALISTICS and forensic science relate to each other, but usually criminalistics is considered to be the largest subdivision of forensic science that encompasses the analysis of physical evidence. In this scheme, criminalistics includes the

Forensic surveying	Forensic deception analysis	Forensic neuropsychology
Forensic accounting	Forensic dentistry	Forensic nursing
Forensic anthropology	Forensic ecology	Forensic odontology palynology (pollen)
Forensic archaeology	Forensic engineering	Forensic pathology
Forensic art	Forensic entomology	Forensic psychiatry/ psychology
Forensic biology	Forensic geology	Forensic psychophysiology
Forensic biomechanics	Forensic limnology (fish)	Forensic radiology
Forensic botany	Forensic linguistics	Forensic seismology
Forensic chemistry	Forensic mathematics	Forensic sculpture
	Forensic medicine	

traditional divisions found in forensic science laboratories such as SEROLOGY and DNA, CHEMISTRY, and TRACE EVIDENCE.

Forensic Science International *See* JOURNALS AND PERIODICALS.

forgery As defined by *Black's Law Dictionary,* forgery is the "false making or material altering of a document with intent to defraud. . . . A signature of a person that is made without the person's consent and without the person authorizing it." At its core, forgery is an attempt to fraudulently imitate an original, be it a signature, lottery ticket, will, painting, sculpture, historical document, or ancient artifact. In forensic science, evidence associated with forgery usually falls within the purview of QUESTIONED DOCUMENTS, but there are many cases of forgery that involve things such as art and archaeological artifacts. In such cases, forensic work is involved, but usually forensic scientists do not perform it. Rather, experts in the specific fields do it.

As an example, specialized techniques of INSTRUMENTAL ANALYSIS are often used to detect art forgeries or alterations. One of the critical considerations in these applications is that the technique should be nondestructive to preserve the work as much as possible. When destruction is inevitable, then microscopic techniques such as INFRARED MICROSCOPY or SCANNING ELECTRON MICROSCOPY (SEM) are favored since such small samples are required. Chemical analysis can be useful in determining what types of paints and varnishes are used and whether the materials found on a painting are consistent with the materials that would have been available at the time the painting was made, assuming it is authentic. One of the more interesting applications of chemical and microscopic analyses in a suspected forgery was the recent investigation of the Shroud of Turin, in which the shroud, purported to be the burial cloth of Jesus, was found to have been created several hundred years later.

Further Reading

Kaye, B. "Forgery and Fraud." In *Science and the Detective.* Weinheim, Germany: VCH Press, 1995.

Shearer, J., et al. "FTIR in the Service of Art Conservation." *Analytical Chemistry,* July 1983, 874A.

Fourier Transform Infrared Spectroscopy (FTIR) *See* INFRARED SPECTROSCOPY (IR).

fraud *See* ACCOUNTING, FORENSIC.

freebase A term that usually refers to the basic form of DRUGS, specifically COCAINE, giving rise to the term "free basing." Many drugs can exist in both an acidic form (acidic salt) and a basic form, with the appearance and physiological effects being quite different. Cocaine hydrochloride is a sparkling white powder that is usually snorted or dissolved in water and injected. Cocaine freebase is a sticky, resinous material that can be smoked, causing an almost instantaneous effect. Conversion of a drug from one form to the other requires only simple chemistry and can be accomplished using reagents as simple as baking powder or sodium hydroxide (lye, NaOH).

frequency estimates Estimates of how many people in a specific population have a given genetic characteristic. For example, in the ABO BLOOD GROUP SYSTEM, approximately 42 percent of the Caucasian population are Type A, and this figure is a frequency estimate. The term is now used primarily relating to DNA TYPING, but the idea can be extended to any kind of evidence that exists within known classes or subdivisions.

See also DISCRIMINATION POWER; POPULATION GENETICS.

***Frye* decision** (1923) A court ruling in 1923 that has greatly influenced how scientific evidence and expert witness testimony is admitted and used. The ruling was a rejection of the validity of the POLYGRAPH (LIE DETECTOR) test, which stated in part:

> *Just when a scientific principle or discovery crosses the line between the experimental and demonstrable stages is difficult to define. Somewhere in the twilight zone the evidential force of the principle must be recognized, and while courts will go a long way in admitting expert testimony deduced from a well-recognized scientific principle or discovery, the thing from which the deduction is made must be sufficiently established to have gained general acceptance in the particular field in which it belongs.*

This ruling led to a criterion referred to as "general acceptance" that governed the admissibility of scientific evidence in many jurisdictions. However, the ruling became more problematic as scientific advances continued and scientific disciplines became more specialized and compartmentalized. As this occurred, the idea of general acceptance within a particular field became more difficult to obtain or define. Furthermore, the criteria set forth in *Frye* were seen as restrictive of innovative techniques that might be known and accepted by only a small portion of scientists within any given discipline. Some of these limitations were addressed in the *DAUBERT* DECISION in 1993, which broadened the acceptance criteria based on the FEDERAL RULES OF EVIDENCE and assigned to the judge the primary responsibility for determining the admissibility of scientific evidence and expert testimony. Although the rules set forth in *Daubert* were widely accepted, some jurisdictions still use *Frye* to guide the admissibility of scientific evidence and expert testimony.

fuel/air ratio An important quantity used in fire and arson investigation to determine a particular combination of fuel, such as gasoline and air. For combustion to take place, there must be fuel present, a source of oxygen (the oxidant, typically air in ARSON cases), and a source of ignition. As long as the fuel/air mixture is in the combustible range, a flame will be self-sustaining. The events that bracket the flame "event" are initiation (ignition) and quenching or suppression. The range of combustibility is referred to as the flammable range and is defined as the ranges of fuel/oxidant ratios that permit steady flame propagation. The lower end of the scale is called the lower flammability limit (LFL) or the lean limit, whereas the upper range is the upper flammability limit (UFL) or rich limit. The terms *lower* and *upper explosive limit* (LEL and UEL) are also used.

G

Galton, Sir Francis (1822–1911) English *Researcher/Fingerprint Pioneer* Galton, aside from being the cousin of Charles Darwin, is known as an important figure in the history of fingerprinting. A researcher in heredity, he started collecting thumbprints in 1888, then prints of all fingers in 1890. He is credited with developing the first classification system for FINGERPRINTS, which was adopted by the British government as an adjunct to the BERTILLON system of body measurements and photographs that was then the primary method of criminal identification. In 1892, he published the influential book *Finger Prints,* which helped bring fingerprinting to the forefront of criminal identification. It is still considered to be one of the primary references in the field. The book was notable for stating the fundamental principles that fingerprints are unique and unchanging. Galton also was the first proponent of classification using the basic patterns of the loop, arch, and whorl. In the United States, the terms *Galton ridge* and *Galton points* are used to describe features found in fingerprints.

Sir Francis Galton, cousin to Charles Darwin and a key figure in the history of fingerprinting ***(National Library of Medicine, National Institutes of Health)***

gamma ray spectroscopy *See* NEUTRON ACTIVATION ANALYSIS.

gas chromatography (GC; GC/MS) An instrumental technique used forensically in DRUG ANALYSIS, ARSON, TOXICOLOGY, and the analyses of other ORGANIC COMPOUNDS. GC exploits the fundamental properties common to all types of CHROMATOGRAPHY, separation based on selective partitioning of

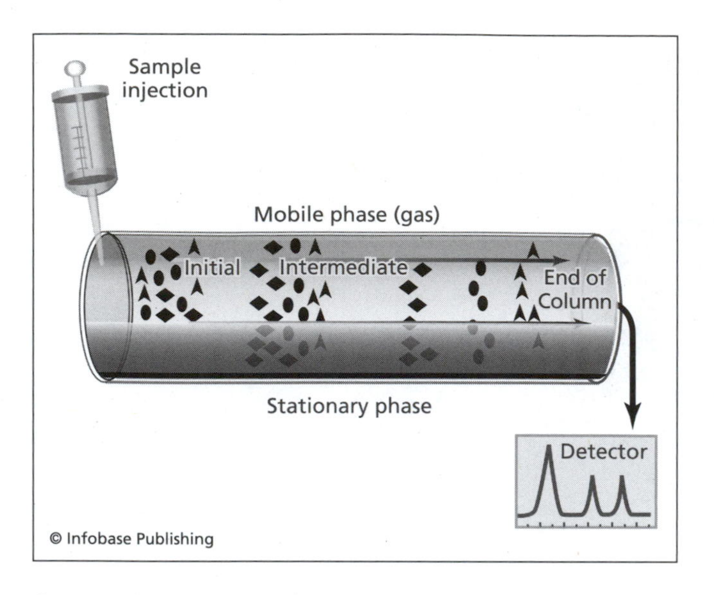

Process of separation in gas chromatography. See text for details.

compounds between different phases of materials. Here, one phase is an inert gas—helium (He), hydrogen (H_2), or nitrogen (N_2)—that is referred to as the mobile phase (or carrier gas), and the other is a waxy material (called the stationary phase) that is coated on a solid support material found within the chromatographic column. In older GC systems, the stationary phase was coated on tiny beads and packed into glass columns with diameters about the same as a pencil and lengths of 6 to 12 feet (1.8–3.7m), wound into a coil. The heated gas flowed over the beads, allowing contact between sample molecules in the gaseous mobile phase and the stationary phase. Called "packed column chromatographs," these instruments were widely used for drug, toxicology, and arson analysis. Around the mid-1980s, column chromatography began to give way to capillary column GC, in which the liquid phase is coated onto the inner walls of a thin capillary tube (about the diameter of a thin spaghetti noodle) that can be anywhere from 49 to 328 feet (15 to 100 m) long, also wound into a coil. Capillary column chromatography represented a significant advance in the field and greatly improved the ability of columns to separate the multiple components found in complex drug and arson samples. However, a few applications still require packed columns.

The purpose of the gas chromatograph is to separate mixtures into individual components that can be detected and measured one at a time. A plot of the detector output is called a CHROMATOGRAM, which charts the detector's response as a function of time, showing the separate components. The separation occurs based on differences in affinities for the two phases. As shown in the figure, the sample is introduced into the GC column by way of a heated injector, which volatilizes all three components and introduces them into the gas flowing over the stationary phase. In this example, the compound represented by the arrowhead (▲) has the least affinity for the stationary phase. As a result, it moves ahead of the other two components and will reach the detector first. The compound symbolized by the diamond (◆) has the greatest affinity for the stationary phase and spends the most time associated with it. As a result, this compound will be the last to reach the detector. Separation has been achieved based on the different affinities of the three types of molecules found in the sample. In reality, complex mixtures cannot always be completely separated, with some compounds emerging from the column simultaneously. This is called coelution, which can often be overcome using detectors such as MASS SPECTROMETERS (MS). The second figure shows an actual chromatogram of a cocaine sample along with the mass spectrum of one of the components, labeled "cocaine artifact."

In most forensic applications of GC, a sample is prepared by dissolving it in a solvent, and the solu-

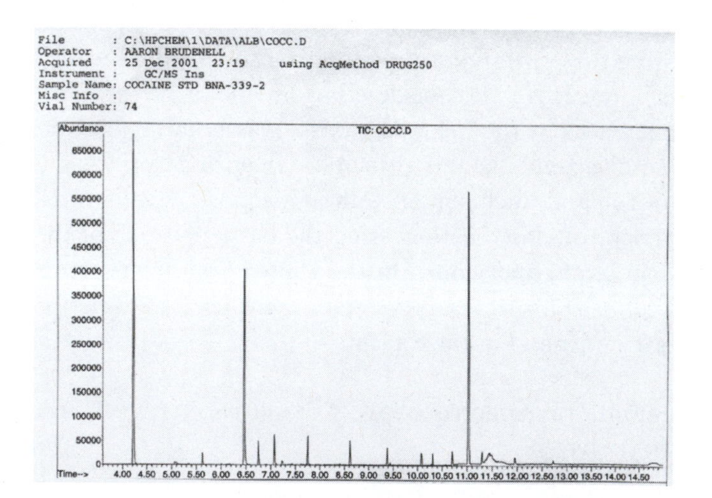

Gas chromatograph of a cocaine sample. This plot was produced by a gas chromatograph coupled to a mass spectrometer detector system, or GC/MS. The peaks are very narrow, typical of modern capillary GC instruments. Each peak corresponds to a component in the sample, and the mass spectrum in the lower frame corresponds to the peak at 6.472 minutes, third largest in the chromatogram. ***(Courtesy of Aaron Brudenelle, Idaho State Police)***

tion is injected into the instrument using a syringe. For example, to analyze a white powder suspected of being cocaine, a small portion is weighed out and dissolved in a solvent such as methylene chloride, methanol, or chloroform. A tiny portion of the sample is then drawn up into a syringe and injected into the heated injector port of the instrument. The mobile phase gas (called the carrier gas) also enters the injector port, picking up the volatilized sample and introducing it into the column where the separation process occurs. If the sample contains cocaine, it will emerge from the column at a given time (known as the retention time) that can be compared to the retention time of a known standard sample of cocaine. The retention time in conjunction with information obtained from the detector is used to positively identify the compound as cocaine if indeed it is present. Another method of sample introduction for GC is called PYROLYSIS, in which a solid sample such as a fiber or paint chip is heated in a special sample holder to extreme temperatures, causing the sample to decompose into gaseous components that can then be introduced into the GC. Pyrolysis is used when the sample is not readily soluble in common GC solvents.

A number of different detectors are available for use in gas chromatography. In forensic applications, the most commonly used are mass spectrometry (often abbreviated as MSD for mass selective detector), FLAME IONIZATION (FID), and NITROGEN-PHOSPHORUS (NPD). The MSD is the most common of the three, principally because it can provide definitive identification of compounds (in almost all cases) along with quantitative information. The FID is used in arson cases because of its sensitivity to hydrocarbons, the primary ingredient in most ACCELERANTS. The NPD is used in drug analysis and toxicology.

One of the advantages of chromatographic systems such as GC is the ability to provide both QUALITATIVE information (identification of individual components) and QUANTITATIVE information (concentrations of individual components). If the instrument is properly calibrated, it can be used to determine quantities of materials present in samples, and this is commonly done in drug analysis. For example, the purity of a drug sample seized as evidence can provide important information and may be used as part of the prosecution. Similarly, when a sample of plastic or a fiber is analyzed by pyrolysis GC, the relative abundance of the individual components can be useful in creating a chemical signature or fingerprint of that particular sample.

gasoline *See* ACCELERANT.

GC *See* GAS CHROMATOGRAPHY.

GC/MS *See* GAS CHROMATOGRAPHY.

gender determination *See* SEX DETERMINATION.

General Electric v. Joiner *See* *DAUBERT* TRILOGY.

General Knowledge Examination (GKE) A certification examination offered by the AMERICAN BOARD OF CRIMINALISTICS (ABC, www.criminalistics.com) first introduced in 1993 and modified in 2007. The current version of the initial written test begins with a general examination in one of five areas: drug analysis, fire debris analysis, molecular biology (DNA), trace evidence, and general criminalistics. Each exam contains a mix of general FORENSIC SCIENCE topics along with more specialized questions related to the topic area. Upon passing this examination, the person is awarded "Diplomate" status designated as D-ABC. To advance to "Fellow" status (F-ABC), the analyst must take and pass laboratory proficiency tests on an annual basis.

generalist v. specialist The issue of whether a forensic scientist should be a generalist well versed in most aspects of the field or a dedicated specialist in one area has been debated since the earliest days of FORENSIC SCIENCE. On one extreme are those who argue that the discipline has advanced to the stage that specialization is not only the norm, but required given the depth and technical complexity of each area (firearms, questioned documents, and so on). The opposing viewpoint holds that forensic scientists should be knowledgeable across a broad spectrum of areas, even if they do not work in those areas on a daily basis. Pioneers such as HANS GROSS and PAUL KIRK held this opinion, and it was Kirk who expressed the view that a CRIMINALIST is in fact an expert in comparison and INDIVIDUALIZATION rather than a specialist in any one forensic area. In this framework, the work that a firearms expert performs to link a bullet to a specific gun (a COMMON SOURCE) is conceptually similar to the work that a forensic biologist does in linking a BLOODSTAIN to a specific person or that a trace evidence analyst does to link hairs and fibers to a common source. In addition, the analyst working on a case may see all kinds of different evidence and needs to have some generalized knowledge to be effective. For example, a firearms analyst may find a fiber on a gun that could provide crucial information, and he or she needs to recognize it as such. A generalist, in this view, takes a more holistic approach and works within the context of the entire case, not just in one isolated aspect.

On the other hand, analysts working in areas such as drug analysis or DNA TYPING often argue that their areas are specialized and that it is not necessary for them to have extensive knowledge outside their own field, nor is it realistic to expect them to obtain and maintain updated knowledge beyond the scope of their daily responsibilities. Such work, in this viewpoint, is fairly isolated from other aspects of the investigation, and a breadth of knowledge across disciplines is not essential. Furthermore, the scope of forensic science is expanding rapidly into new territory such as forensic videography, forensic computing, forensic botany, forensic archaeology, and even forensic seismology. Thus, keeping up with developments, even at a superficial level, is becoming more difficult.

It is probably fair to say that most forensic scientists and criminalists take a middle view and believe that a basic foundational understanding of the core aspects of the field are beneficial, even if the analyst works in only one area. Anyone involved in forensic work, for example, should understand LOCARD'S EXCHANGE PRINCIPLE, the concept of the CHAIN OF CUSTODY, and the basics of legal proceedings and testimony. This notion of a uniform core of information is reflected in the only national analyst certification test, the GENERAL KNOWLEDGE EXAMINATION (GKE). The first phase of the program tests, as the name suggests, general knowledge in forensic science, with the option of testing in specialty areas. However, as areas such as DNA typing become more sophisticated, more automated, and applied to more cases, the pressure toward specialization is likely to increase.

genetic marker systems Prior to the advent of DNA TYPING, genetic marker systems were used to analyze BLOOD and BODY FLUIDS. Genetic markers are inherited characteristics that show variation within a population. Such a variation is called a POLYMORPHISM ("many forms") and can be illustrated by the most well-known genetic marker system in blood, the ABO BLOOD GROUP SYSTEM. A person's ABO blood type is inherited from the parents and can be Type A, B, AB, or O. Thus, ABO is a genetic marker system that is polymorphic, with four types in the population. Although a person's ABO blood type is not unique, by typing more genetic marker systems, the combination of types can narrow the range of possible donors substantially. Genetic marker systems exist in blood and can also be typed, under certain circumstances, in body fluids such as semen or saliva. To be useful in forensic work, a genetic marker system should have several variants that subdivide the population into several smaller groups, the more the better. Furthermore, the system must be robust and must resist degradation long enough to be typed in bloodstains that may be old and/or in very poor condition. This last requirement limited the widespread typing and use of many genetic marker systems. Although fairly easy to type in fresh whole blood, many of the systems are fragile and could not be routinely and reliably typed in stains.

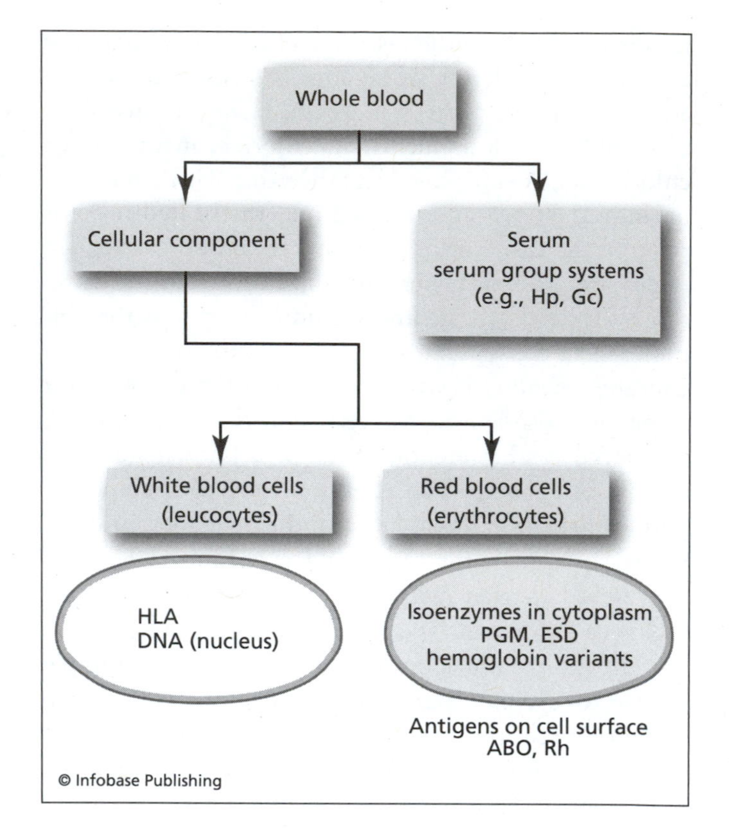

Genetic marker systems in blood. Prior to the adoption of DNA typing, these systems were widely used in forensic serology.

As shown in the figure, blood can be subdivided into different fractions, all of which contain genetic marker systems. After centrifuging, blood separates into serum and the cellular components. Serum, a yellowish liquid, contains serum blood group systems such as haptoglobin (Hp) and group-specific component (Gc) that are polymorphic. Within the cellular component, white blood cells (leucocytes) contain the HUMAN LEUCOCYTE ANTIGEN system (HLA) that contains many different factors and types. Both the serum blood group systems and HLA system were difficult to type in stains and were not routinely used in forensic casework.

Unlike red blood cells, the white blood cells have a nucleus, which is the source of DNA used in DNA typing. The 13 loci that are usually typed in current practice can also be loosely classified as genetic markers since they are inherited and polymorphic. They are not genes per se; rather they are loci on chromosomes that do not appear to code for anything. Red blood cells are the richest source of non-DNA genetic marker systems that were once widely used in forensic SEROLOGY. These cells (erythrocytes) have on their surface the antigens that make up blood group systems such as ABO and Rh. Within the cell are found the ISOENZYME SYSTEMS such as PHOSPHOGLUCOMUTASE (PGM) and ESTERASE D (ESD) as well as variations of the hemoglobin molecule. The ABO blood group and isoenzymes were the most used in casework.

See also DISCRIMINATION INDEX; SECRETORS.

Further Reading

Spalding, R. P. "Identification and Characterization of Blood and Bloodstains." In *Forensic Science: An Introduction to Scientific and Investigative Techniques*. 2nd edition. Edited by S. H. James and J. J. Nordby. Boca Raton, Fla.: CRC Press, 2005.

genetics In forensic science, fundamental genetics underlies concepts of DNA TYPING and in the past was crucial in the typing of other GENETIC MARKER SYSTEMS in blood and BODY FLUIDS. Genetics is the study of heredity, the fundamental unit of which is the gene. Genes are encoded in DNA, and the place in which a particular gene is found is called its locus. Multiple sites are called loci. Different forms of the same gene are called alleles. Human genes are found on 46 chromosomes organized into 23 pairs. Within those pairs, one chromosome originated from the mother, the other from the father. Many of the genetic characteristics of interest in forensic science are codominant, meaning that the allele from both the mother and the father is expressed. For example, the ABO BLOOD GROUP SYSTEM is a genetic marker system with the Types A, B, AB, and O. The A type is a result of AA or AO alleles. The observable, measurable way in which any gene is expressed is called the phenotype, and so the phenotype of both AA and AO is A in this example (disregarding subtypes). Another example is found in the ISOENZYME SYSTEM called PGM (PHOSPHOGLUCOMUTASE) in which types of 1 (1-1), 2 (2-2), and 2-1 are observed. When variations occur at a given locus, such as in these examples, it is referred to as a POLYMORPHISM ("many forms") since variants exist within the population. Population genetics, the study of the relative frequencies of different types or variations within a given population (such as Caucasian, African American, and so on) is of critical importance in interpreting the results of forensic DNA or genetic marker analysis.

genetics, forensic Broadly speaking, the application of GENETICS to forensic science. Most often this term refers to POPULATION GENETICS, FREQUENCY ESTIMATES, and issues related to DNA TYPING.

geology, forensic Primarily the study of soils, which are a common type of TRANSFER EVIDENCE. HANS GROSS and EDMOND LOCARD, pioneers of early forensic science, were among the first to recognize the value of soil evidence in linking an individual to a particular place. Early work and explorations of forensic geology were reported in the middle of the 19th century, but the first documented cases are generally credited to Georg Popp of Frankfurt, Germany. In the early 1900s, Popp used SOIL analysis, and particularly evidence found in layers of soil, to link suspects to crime scenes. This is still one of the primary roles of the forensic geologist, and the use of geological information and soil characterizations in investigations continues to expand. Today, forensic geology can play a role in both criminal and civil cases.

Geochemistry

Soil is a complex mixture of minerals, organic material, botanical and animal products and debris, and anthropogenic (humanmade) materials that are naturally found on the earth's surface. Although thousands of minerals have been identified, most soil samples contain only a few, typically 20 or less. Examples of common minerals are dolomite (chemical formula $CaMg(CO_3)_2$) and talc (chemical formula $Mg_3(Si_4O_{10})(OH)_2$) The chemical and physical characteristics of soil vary greatly depending on location, both along the surface (horizontally or laterally) and below it (vertically). This variability offers both good news and bad news to the forensic analyst. Because soils vary so much, this increases their uniqueness and offers a better chance at linking a questioned soil to a specific location. But given this variability, soil samples taken six inches (15.24 cm) or a foot (.3 m) apart may show significant differences, as can two soils taken from the same location but different depths. Thus, the soil samples must be carefully collected and must include several control samples taken within close proximity. What defines "close" will vary from situation to situation. However, the necessity of collecting multiple samples for soil analysis is inescapable.

In cases in which soil comparisons are to be performed, the questioned soil is that which is collected at a crime scene or is somehow directly related to a criminal or other disputed act. The CONTROL SAMPLES are collected from places that could be possible sources. For example, if murder is committed and a body is found in a wooded location, soil from that location will be found on the body. Some portion of that soil will also inevitably be transferred to the murderer. A suspect is arrested and in the tread of his shoes, soil is found and recovered for analysis. This is the questioned soil, while controls would be gathered from the crime scene. If the suspect claims to have been at another location, say at a beach picnic, the geologist would also collect samples from the beach, and such samples would be referred to as controls or "alibi samples." Comparisons of the questioned soils to the controls would be used in an attempt to determine which story is true. In fact, the analysis of alibi samples is an important aspect of most forensic geology. Geologists also make use of reference materials such as "Soil Survey" data, published by the U.S. Geological Survey (USGS) to help locate possible origins of a soil when no comparison samples are available. Other tools that are used are geological and topographical maps and aerial photographs.

Analysis and Comparison

The analysis of soil is similar to the analysis of glass in that it relies heavily on the measurement of physical properties. In most soil comparisons, soil color can be used to distinguish between samples. Care must be taken to ensure the soils are dry, as the color of moist soil differs from that of dry soil. Moisture content is sometimes used, although this will obviously vary based on environmental conditions. The pH of soil (how acid or alkaline it is) can be important as well, although many factors can influence soil pH. Organic content of the soil (wood, leaves, and so on) can be measured by subjecting the soil to very high temperatures, which cause the organic material to burn off as vapors. The resulting weight change results from the loss of organic material only, since at the temperatures used, none of the minerals or inorganic material present will melt or vaporize.

If enough of the sample is available, particle size distribution and microscopic examination under polarized light (POLARIZED LIGHT MICROSCOPY, PLM) are among the most common forensic tests applied to soils. Density distributions, once a mainstay of soil analysis, are now rarely done. Particle size distribution can be determined using a manual sieving procedure or more recently using automated microscopic methods. Density gradients can be constructed in long narrow tubes in which the density varies down the length of the tube. Soil particles introduced at the top of the tube will settle to the point at which their density is the same as that of the liquid. The pattern of densities in similar soils should be comparable. Another useful property is the fluorescence of the sample.

Perhaps the most important tool for the forensic geologist is microscopy using STEREOMICROSCOPES, polarizing light microscopes (PLM), and scanning electron microscopy/energy dispersive X-ray instruments (SEM/EDX). Using the microscope, the analyst can sort through samples, study individual particles, identify minerals, and perform further testing. Chemical tests that can be applied to soils are varied, with the most common being EDX for ELEMENTAL ANALYSIS, X-ray diffraction (XRD) to identify crystal structures, X-ray fluorescence (XRF), and inductively coupled plasma (ICP) techniques. Organic components of the soil can be studied using GAS CHROMATOGRAPHY/MASS SPECTROMETRY (GC/MS) and HIGH PERFORMANCE LIQUID AND CHROMATOGRAPHY (HPLC). INFRARED SPECTROSCOPY (IR) can be used to identify both organic and inorganic components of soil.

One of the fascinating aspects of soil evidence is that it can collect in layers, and the characteristics and patterns of those layers can be critically important in deciphering a person's movements or reconstructing a crime. Consider your shoes, which most likely have a tread pattern of some type. Those treads can trap and retain soil or other material from wherever you walk, and those materials will accumulate in layers, with the top layer reflecting your most recent movements. The same thing can occur with tire treads or layers of dirt and grime that accumulate on clothing. Not surprisingly, in these cases, collection techniques are critical, and it is important that the layering structure be retained and preserved.

Finally, forensic geology can be valuable in providing investigative leads. If the analyst can look at traces of soil found at a scene or on a tool or implement, he or she can suggest to the investigators possible sources of these materials. For example, identification of a specific substance such as coarse-grained granite, pollen, or some other soil component found in a suspect's vehicle could be used to help locate a CLANDESTINE GRAVE.

See also X-RAY TECHNIQUES.

Further Reading

Murray, R., and L. Solebollo. "Forensic Examination of Soil." In *Forensic Science Handbook*. Vol. 1, 2nd edition. Edited by R. Saferstein. Upper Saddle River, N.J.: Prentice Hall, 2002.

Murray, R., and J. Tedrow. *Forensic Geology*. Englewood Cliffs, N.J.: Prentice Hall, 1992.

geophysical methods Instruments and techniques used to measure the characteristics and properties of the earth, principally beneath the surface. An example of a forensic use of a geophysical technique would be employing ground-penetrating radar to search for a CLANDESTINE GRAVE. Other geophysical methods include metal detectors, seismographs, and magnetometers.

Gettler, Alexander O. (1883–1968) American *Toxicologist* Dr. Gettler organized a chemical analysis laboratory within the Office of the Medical Examiner of New York City. Gettler's lab became the nexus of TOXICOLOGY in the United States under his direction, which lasted until his retirement in 1959. He testified hundreds of times and contributed in many areas. One example came in cases of suspected drowning, where he noted that "years ago, medical examiners and chemists only guessed when they gave 'drowning' as a cause of death. I wanted to find some positive way of telling and finally evolved a way precise and beyond contradiction." He did so via a comprehensive study of both sides of the hearts of drowning victims. Specifically, he measured the amount of chloride in the serum in the separate chambers.

Freshwater rivers and the sea surround New York City, so a determination of drowning as a cause of death tells half a tale. If a person drowns in salt water, they inhale it, along with the high concentration of sodium chloride it contains. Blood leaving the right side of the heart travels to the lungs, where it picks up oxygen. This blood then travels to the left side of the heart and on out to the body, where oxygen is consumed. If there is seawater in the lungs, the chloride ion moves into the blood that ends up in the left side of the heart. Since the heart stops beating relatively quickly after immersion, this high chloride ion concentration does not have time to disperse and even out. The concentration of chloride ion in the blood in the left side of the heart will be higher than that of blood in the right side of the heart. Gettler's work provided a definitive and quantitative method of determining drowning and in what type of water. He also improved testing methods for detection of poisons and worked with testing of blood and body fluid stains. The many toxicologists he trained spread his practices and ideas to other cities and jurisdictions.

GHB *See* DATE RAPE DRUGS.

glass As a ubiquitous material of the modern world, glass is also one of the most common forms of physical evidence. Glass can be involved in automobile accidents including hit-and-runs, burglaries, assaults, and many other types of crimes. Because glass is so common and is so easily broken into small pieces, it can be a rich source of TRANSFER EVIDENCE that can be used to link a person to a scene. Glass has been manufactured for centuries, but only recently has the process become automated, leading to huge quantities of mass-produced glass products such as windshields, windows, mirrors, bottles, eyeglasses, and so on. Unfortunately (at least from the forensic perspective), with mass production comes greater uniformity and less ability to discriminate. However, glass can still provide critical evidence and, under certain circumstances, can be INDIVIDUALIZED or linked to a COMMON SOURCE.

Characteristics and Chemistry

Glass is an amorphous solid, meaning that it lacks the rigid, ordered crystal structure of materials such as table salt. Glass is also considered in the strictest sense to be a viscous solid or a super-cooled liquid rather than a true solid. Because of this lack of order

Bullet holes in glass ***(Courtesy of Michael Bell, West Virginia University, Forensic and Investigative Sciences program)***

and pattern at the molecular level, glass breaks in random patterns and has other unique properties that make it such a valuable and widely used material. The most common ingredient in most glass is quartz sand, SiO_2. Soda lime glass, of which windows are usually made, has additives of Na_2O, CaO, MgO, and Al_2O_3, all metal oxides. Different additives are used to impart different physical characteristics to the glass. Colored glasses are made by adding oxides from the transition metal family. Red glass is made by adding cadmium or selenium, while adding cobalt compounds can make blue. Leaded glass used in fine crystal contains high concentrations of lead oxide (PbO). Tempered or safety glass is created by subjecting the glass to thermal cycling during the manufacturing process, resulting in a glass that breaks into small squares rather than into sharp shards. Windshields are yet another type of specialty glass made by sandwiching glass on both sides of a plastic sheeting material. Most glass made today is called float glass because of the way it is manufactured. Earlier processes used blowers or rollers to mold molten glass into flat sheets. Float glass is made by pouring the molten material onto a bath of melted tin, resulting in a very smooth surface that needs little or no additional polishing.

Analysis

Broken glass panes in windows or windshields are a frequently encountered type of evidence in burglaries and violent crimes. The process of breaking glass creates a shower of particles both in the forward and reverse direction, which may be transferred to the person breaking the glass if they are standing close enough. In many (but not all) cases, it is possible to determine from what direction the glass was broken. As can be seen in the photo, an impact in glass produces two kinds of fractures, radials (radiating out from the point of impact) and concentric (forming a circle around the impact). The patterns found in the cross-sections of these fractures can indicate from which direction the force was applied. The lines created in the glass, called ridges or Wallner lines, show distinctive patterns depending on whether the fractures from which they are collected are concentric or radial. The most reliable results are obtained closest to the original impact, and consistency among several comparisons is highly desirable for the most reliable analysis. Impacts of bullets or other small projectiles such as pebbles also produce characteristic patterns, usually a crater with the largest portion on the face of the glass opposite from the impact. In other words, the exit hole is larger than the entrance hole. Lower velocity impacts may not penetrate the glass but leave only a pit or crater on one side. If a gun is involved at close range, GUNSHOT RESIDUE (GSR) may be transferred to the glass and can aid in interpretations. Also, if there are multiple impacts in the same pane of glass, the sequence of events can be deduced from the fracture patterns. Existing fractures from the first impact act as barriers to fractures created by the second impact, and these abrupt terminations are easily identifiable. Shattering of glass by heat creates a distinctive and different fracture pattern characterized by wavy smooth cracks.

Currently, the only way to link two fragments of glass to a common source is by the process of PHYSICAL MATCHING. Since glass breaks randomly, each breakage is unique. As a result, if a fragment of glass such as a piece of a broken window or headlight can be fitted back into the original like a puzzle piece, that fit individualizes the glass and proves that it could come only from that source. In some cases, unique patterns or textures in the glass such as dimpling or striations can assist in physical matching procedures. Short of being able to perform a physical match, the forensic examiner still can conduct several tests to show whether glass might have come or definitely did not come from a common source. After a substance is identified as glass (using microscopic evaluation), the next step is to obtain measurements and descriptions of simple physical properties such as color, thickness, shape, and texture. In many cases, these measurements will be enough to exclude a common source, meaning that two glass samples could not have come from the same place (EXCLUSIONARY EVIDENCE). Next, the analyst can perform more detailed analysis of the glass such as evaluating fluorescence and surface flatness.

The two most common physical parameters used to characterize glass samples are measurements of the REFRACTIVE INDEX and DENSITY. The refractive index is usually measured using a microscope equipped with a hot stage. The glass is immersed in an oily material with a known refractive index. The material is slowly heated, changing the refractive index of the oil. When the glass is no longer visible in the oil, its refractive index is the same as that of the oil, and the value is then recorded. Density is usually determined using a floatation technique in which the composition of a liquid (and/or its temperature) is varied until the glass sample remains suspended in it. At that point, the density of the glass and the liquid are the same, and the value can again be recorded. Even if two glass samples have identical refractive indexes and densities, that does not mean that they came from the same source, only that they might have. The FBI maintains a database of several hundred glasses showing the frequency of refractive indexes and densities of all those catalogued, and these frequencies can be used to estimate how common a certain type of glass is. However, this information does not individualize a glass. On the other hand, if two glasses have different values, they could not have come from the same source. Often, this type of exclusionary evidence is critical. Additionally, in some cases a glass sample will have distinctive backing materials such as paint, glue, or tape that can be used in further testing.

Chemical characterization of glasses is not widely used in forensic science for a variety of reasons. Glass samples often display significant internal variations, meaning that chemical composition, particularly at the trace level, can vary within the same sample of glass. Instrumentation needed is often expensive and requires specialized training and a great deal of time, further limiting usefulness. Finally, many methods of chemical analysis require that the sample be dissolved in acids, and such destructive analyses are often not feasible or appropriate. Instrumental techniques that have been used for chemical characterization of glass include NEUTRON ACTIVATION ANALYSIS (NAA), ATOMIC ABSORPTION (AA), SCANNING ELECTRON MICROSCOPE/energy dispersive X-ray (SEM-EDX), X-ray fluorescence (XRF), and X-ray diffraction spectroscopy (XRD).

See also X-RAY TECHNIQUES.

Further Reading

Koons, R., et al. "Forensic Glass Comparisons." In *Forensic Science Handbook*. Vol. 1, 2nd edition. Edited by R. Saferstein. Upper Saddle River, N.J.: Prentice Hall, 2002.

glove prints *See* FABRIC IMPRESSIONS.

glyoxalase I (GLO I) *See* ISOENZYME SYSTEMS.

Goddard, Calvin (1891–1955) American *Physician, Firearms Examiner* Dr. Calvin Goddard is often credited with establishing scientific examination of FIREARMS evidence in the United States. During the early part of the 20th century, the examination of firearms evidence was neither systematic nor reliable, and many of the experts called to testify were self-proclaimed. Testimony of such experts was, at best, misleading and, at worst, completely wrong. Dr. Goddard, along with other early workers in the field, helped to change that by developing the tools and procedures necessary to make firearms examination an accepted part of forensic science. Goddard was a retired army physician and professed gun enthusiast who had risen to the directorship of the Johns Hopkins Hospital in Baltimore, Maryland. In 1925, he joined the Bureau of Forensic Ballistics, a private organization founded by Charles E. Waite. A few years earlier, Waite had been involved in the Stielow case in which firearms evidence was used to reverse the conviction of an innocent man. Goddard worked with others at the bureau and quickly became a vocal advocate of the use of the forensic COMPARISON MICROSCOPE in firearms casework. Waite died in 1926, Goddard became the director of the bureau, and during this time he wrote several articles on firearms examination, including an article that appeared in *Popular Science Monthly* in 1927. Also in 1927, Goddard became involved in the Sacco-Vanzetti trial in which his examination of bullets and guns conclusively linked Sacco's gun to a fatal bullet, a finding that was upheld during a reexamination several years after Goddard's death.

Calvin Goddard ***(Courtesy Northwestern University Archives)***

As a result of his growing reputation, Goddard was called to examine evidence of the infamous St. Valentine's Day Massacre in Chicago in 1929. During the coroner's grand jury inquest, he was able to show that all the murdered men had been killed by two Thompson submachine guns. Some of the jurors in that case later raised money to establish a forensic laboratory at Northwestern University in Evanston, Illinois, called the Scientific Crime Detection Laboratory. Goddard was appointed the first director and stayed until 1932. The laboratory moved to Chicago and became the Chicago Police Department laboratory in 1938. Goddard also assisted the FBI in establishing a firearms analysis capability at their new lab, inaugurated in 1932.

gold bromide and gold chloride Reagents used in DRUG ANALYSIS. Gold bromide (AuBr) and gold chloride (AuCl) along with the platinum analogs (PtBr and PtCl) are used in CRYSTAL TESTS to make microcrystals that are characteristic of a given drug. However, obtaining a crystal is not always considered definitive, and further, more specific instrumental methods of analysis are required to confirm the identification. Gold chloride is used for COCAINE, and a gold chloride volatility test is used to form crystals with the AMPHETAMINES. In this variation, the suspected amphetamine sample is dissolved in a basic (alkaline) solution, where the FREEBASE form is volatile and evaporates slowly. A drop of gold chloride solution is held above the evaporating sample, and if an amphetamine is present and in the vapor state, it will dissolve in the drop and form the characteristic crystal. Since most of the materials commonly used to dilute street drugs (CUTTING AGENTS) such as sugars will not volatilize, these materials cannot interfere with the crystal test. A crystal test can also be used as part of the analysis of heroin using platinic chloride.

Gonzales, Thomas A. (1878–1956) American *Forensic Pathologist and Medical Examiner* Dr. Thomas A. Gonzales succeeded Dr. CHARLES NORRIS as the chief MEDICAL EXAMINER of New York City, who had been with the Office of the Chief Medical Examiner (OCME) since its formation in 1918. Having begun his career at the New York office as an assistant to Dr. Norris, Gonzales became the acting chief upon Norris's death in 1935. Soon after, he became the Chief Medical Examiner by outperforming all other candidates on written examinations testing the applicants.

In addition to providing evidence that convicted hundreds of murderers and saved many innocent people accused of crimes never committed, Gonzales is noted for being the coauthor of the 1937 textbook *Legal Medicine and Toxicology*. He directed the ME's office during World War II, a time when scientific developments accelerated and modern analytical instrumentation began to appear in forensic laboratories. One of his key contributions was hiring Dr. Alexander Wiener, a scientist who had worked with KARL LANDSTEINER, the Nobel Prize winner who discovered the ABO and Rh blood groups. Wiener took charge of serological evidence such as blood and BODY FLUID stains, bringing the New York office into the forefront of American forensic laboratories. One of the assistant MEs under Gonzales was Dr. MILTON HELPERN, who later became chief and one of the first of the celebrity forensic scientists of the 20th century.

Gonzales was a key figure in the foundation of the AMERICAN ACADEMY OF FORENSIC SCIENCES in 1947, the first forensic professional society to encompass many forensic specialties. One of the more noteworthy cases during his time was the 1945 crash of an Army B-25 bomber into the Empire State Building. Gonzales retired with distinction in 1953 and passed away a few years later in 1956.

good laboratory practice (GLP) A series of guidelines for laboratory analysis designed to ensure the accuracy and reliability of the results produced. Originally designed for use in environmental, food, and pharmaceutical analyses, the concept of GLP has expanded to cover any type of laboratory analysis. Although specific GLP guidelines vary among disciplines and laboratories, they generally cover aspects such as personnel training and testing, laboratory procedures and protocols, documentation, facilities and equipment, calibration and standardization, and QUALITY ASSURANCE/QUALITY CONTROL (QA/QC). GLP for forensic science laboratories is addressed through analyst and laboratory CERTIFICATION and ACCREDITATION procedures.

grand jury An inquisitional judicial proceeding used at both the state and federal levels. The job of the grand jury is to hear evidence and decide if it is sufficient to indict someone. If so, the grand jury returns an indictment, and criminal charges can be brought against the person in question. At the state level, a grand jury indictment is often required for capital crimes and felonies, but the standards vary among jurisdictions. An indictment does not mean that the person is guilty, but only that there is sufficient evidence to prove that a crime did occur and that the person might have committed that crime. Forensic scientists can be called to testify before grand juries just as they can in any other court proceeding. The grand jury traces its origins to English common law.

graphology The analysis of handwriting for the purpose of identifying personality traits and other psychological factors associated with the writer. Graphology is a PSEUDOSCIENCE and is not to be confused with forensic handwriting analysis conducted by a QUESTIONED DOCUMENTS examiner. In such cases, the examiner is evaluating writing samples for purposes such as detecting forgeries or determining if the same person was responsible for different writing samples. This type of handwriting examination is systematic and based on sound principles and procedures subject to constant review and improvement. Graphology is not part of forensic document examination and is not considered admissible or reliable in courts. It is to forensic document examination what astrology is to astronomy.

Griess test A PRESUMPTIVE TEST used to detect the residuals of GUNPOWDER or EXPLOSIVES. The test detects the nitrite ion (NO^{2-}) and can be modified to also detect the nitrate ion (NO^{3-}). Reagents used in the test are napthylamine in methanol, sulfanilic acid in acetic acid, and zinc metal. Nitrates and nitrites are produced whenever a gun is fired, and nitrate is a breakdown product of nitroglycerin, an ingredient found in many explosives and in SMOKELESS POWDER used as a PROPELLANT in AMMUNITION. Thus, this test was at one time widely used in an attempt to determine if a suspect had recently fired a gun or had handled explosives containing nitroglycerin. The test is also used in DISTANCE DETERMINATION in which GUNSHOT RESIDUE (GSR) can be visualized using the test. Finally, the reagents can be used as a developer for THIN LAYER CHROMATOGRAPHY (TLC) applied to explosives. As with all presumptive tests, FALSE POSITIVES are common with the Griess test, and a positive result is not conclusive.

Gross, Hans (1847–1915) Austrian *Lawyer, Forensic Scientist* Gross, an examining magistrate in Graz, Austria, is the man credited with coining the term CRIMINALISTICS to describe what was then an infant science. Gross viewed forensic science holistically and believed that experts from diverse fields would contribute to the analysis of physical evidence and solution of crimes. He understood the value of biological evidence, SOIL, DUST, and many other types of transfer and TRACE EVIDENCE. In 1893 he published the first textbook in forensic science, which was translated into English under the title of *Criminal Investigation,* and he started a journal called *Kriminologie,* which is still published. Gross was a pioneer in the field, and he exerted a strong influence on his contemporaries, including other pioneers in the field such as EDMOND LOCARD.

ground-penetrating radar (GPR) A technique used to search for CLANDESTINE GRAVES. Other terms describing this technique are ground-probing radar, subsurface radar, and surface-penetrating radar. The word *radar* is an acronym for *ra*dio *d*irection *a*nd *r*anging developed during World War II to locate aircraft and ships. GPR works by sending radar waves (in the microwave region of the electromagnetic spectrum) into the ground and examining the echoed waves that reflect off materials in its path. Different materials, such as SOIL, BONES, and human remains, interact with the energy in different ways; some absorb more, while others reflect more. These interactions are revealed in the way the initial pulse is reflected and returns to the instrument. Geologists, engineers, and archaeologists also make use of GPR.

group-specific component (Gc) A GENETIC MARKER SYSTEM found in the serum of BLOOD. Gc is produced by the liver and shows variations in the populations with the primary types of 1, 2-1, and 2. Type 1 is the most common, with about 51 percent of the population showing it. Type 2-1 is found in about 41 percent, with the remainder being type 2. Typing can be accomplished using ELECTROPHORESIS and ISOELECTRIC FOCUSING. With the ascendance of DNA TYPING techniques, typing of Gc in bloodstains is rarely if ever needed or used.

guaiacum test A PRESUMPTIVE TEST for blood, one of the first developed but no longer used. It relied on guaiacum (a resin isolated from trees) in combination with hydrogen peroxide. If a stain turned blue when treated with these reagents, it was considered a positive result indicative of blood. However, as with all presumptive tests, the results are not conclusive, nor do they prove that the blood is human.

gunpowder A term most often used to describe the PROPELLANTS used in AMMUNITION. There are two kinds of powder available, both classified as a low explosive, and the term *gunpowder* has been applied to both. BLACK POWDER, the original gunpowder, consists of 75 percent potassium nitrate (KNO_3 or saltpeter), 10 percent charcoal, and 15 percent sulfur. Because of the smoke produced, use in firearms ended in the 1800s, but black powder is still available for hobby guns and is used in applications such as fuses. Smokeless powder is used in modern firearms and consists of nitrated cotton or NITROCELLULOSE (single base) or nitroglycerin combined with nitrocellulose (double base). *Gunpowder* usually refers to smokeless powder because it is the modern standard, but the term can and often is used to describe black powder, particularly in a historical context.

guns *See* FIREARMS.

gunshot residue (GSR) The residue that escapes from a gun when it is fired. GSR is considered a type of TRANSFER EVIDENCE and can be detected using chemical tests and instrumental analyses. The residue comes mostly from the primer, which is part of the CARTRIDGE containing the BULLET and PROPELLANT in modern AMMUNITION. Particles of unburned powder are also part of the residue. The most common elements found in GSR are lead (Pb), antimony (Sb), and barium (Ba), and this combination of elements is telling. When a particle is found to contain these three elements, it is almost certainly gunshot residue. Unquestionably lead, antimony, and barium exist in other materials, but the combination of all three in a single particle is usually considered to be conclusive of GSR, since this type of particle is formed when a cartridge is fired. Other elements that can be found include copper (Cu), aluminum (Al), iron (Fe), and zinc (Zn). Chemicals in the propellant (usually SMOKELESS POWDER) that can be found in GSR are nitrate ions (NO^{3-}), nitrite ions (NO^{2-}), NITROCELLULOSE, and nitroglycerin. During the firing process, other compounds can form between metal elements and atoms such as carbon, nitrogen, oxygen, and hydrogen, which are produced as a result of the combustion. Thus, GSR is chemically complex, particularly at the trace level. The size of the GSR particles and their chemical makeup will vary depending on the type of weapon, powder, primers, and projectiles used. The particles themselves as well as the individual elements and the chemical compounds in GSR can be detected using chemical, microscopic, and instrumental methods of analysis.

GSR can be found on the hands, clothing, or any other surface that is close to a gun when the weapon is fired. The appearance of the pattern will vary with distance and can be used in DISTANCE DETERMINATION studies. Samples can be taken from the hands of someone who is suspected to have recently fired a gun, and these samples can be tested for the presence of GSR. A negative result does not mean that a gun was not fired since the time that the residue will remain depends on many factors, including the frequency and thoroughness of any hand washing. Typically, residue will persist for a few hours unless the hands are washed. When samples are collected from the hands, they are usually done with plastic swabs dipped in dilute nitric acid or a similar procedure. The location from which the swabs are taken from the hands is important, and separate swabs are taken for different parts of each hand, along with CONTROL SAMPLES. This could provide critical information such as which hand was used to fire the gun, if indeed the person fired one at all.

Many of the older chemical tests (PRESUMPTIVE TESTS) for GSR targeted nitrite and nitrate ions. These include the GRIESS TEST, the DIPHENYLAMINE TEST, and the WALKER TEST. Limitations of these tests include the lack of specificity and the widespread use of nitrates and related compounds in many products. Nitrates are found in fertilizers, cigarette smoke, and cosmetics, to name just a few sources, so finding nitrates on a person or an object is not conclusive evidence of GSR. A variation of these tests, the DERMAL NITRATE TEST, used a paraffin cast of the hands that was then tested for the presence of nitrates. Due to the number of FALSE POSITIVES encountered, this test has been largely abandoned. Another presumptive test exists for lead using sodium rhodizonate. A photo of a sodium rhodizonate test is found in the color insert.

Since GSR contains particulates, one of the most valuable tools for identifying it is the microscope. Particles that appear to be GSR can be separated and subjected to specific chemical and instrumental tests to confirm the identification. A scanning electron microscope coupled to an X-ray device (SEM/XRD) is particularly valuable since chemical analysis can be conducted on the particle as it is being observed. Samples that are to be analyzed using SEM are collected on an aluminum disk with an adhesive attached, and particles can be lifted this way from clothing or other surfaces. Other instrumental analysis techniques that have been used for GSR analysis are ATOMIC ABSORPTION (AA), NEUTRON ACTIVATION ANALYSIS (NAA), HIGH PRESSURE LIQUID CHROMATOGRAPHY (HPLC), and CAPILLARY ELECTROPHORESIS (CE).

See also BULLET WOUNDS.

Further Reading

Schwoeble, A. J., and D. Exline. *Current Methods in Forensic Gunshot Residue Analysis.* Boca Raton, Fla.: CRC Press, 2000.

H

hairs Along with fibers, one of the most common types of physical evidence encountered. Hair analysis is closely related to FIBER analysis, leading to the common usage of the phrase "hairs and fibers" to describe the forensic analysis done in this area. Hair is an animal fiber (a mammalian pithelial structure) characterized by a scaly cuticle that is easily recognizable when viewed under a microscope. An experienced forensic examiner can distinguish animal hair from human, but it is rarely possible to definitively link a questioned hair to an individual unless DNA TYPING can be performed. Thus, for hair evidence, the primary method (and often the sole method) of analysis is visual examination. The examiner can conclude that the hair could have, could not have, or cannot be determined to have come from the same source as a known hair sample. Rarely if ever can the analyst state with certainty that it was unique to that source.

Structure

As shown in the illustration, hair is produced below the surface of the skin in the follicle, and the base of the hair is comprised of living tissue. As the hair grows and reaches the surface, the cells dry and become keratinized. Keratin is a strong protein that is found in high concentrations in the fingernails and imparts to the hair tremendous durability and resistance to degradation and chemical attack. Inside the follicle is the root, where the living portions are found, along with the bulb and sheath tissue. The follicle is associated with an oil gland called a sebaceous gland.

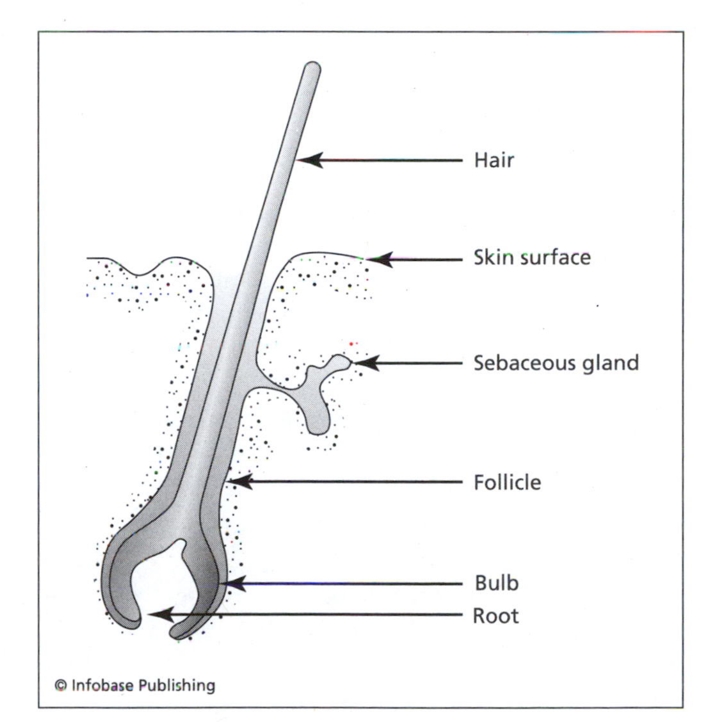

Features of a strand of hair in the skin

As shown in the second figure, hair possesses scales that protrude from the outer cuticle, with the unattached "flap" of the scale pointing toward the tip of the hair. On the surface, the scales form an irregular pattern most often compared to shingles on a roof. The "shine" of hair is related to the position of the cuticles and how loose they are on the cuticle. Inside the cuticle (the cortex), hair consists of fibrils of protein arranged

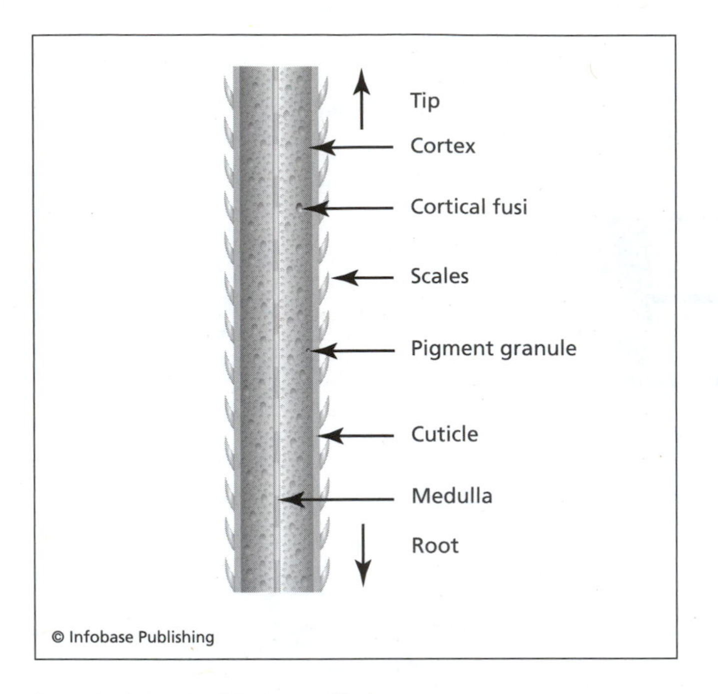

Important structural features of hair

in a stacked arrangement of an individual protein strand enclosed in a protofibril, which in turn is encased in a microfibril and then a macrofibril. A tubular structure called the medulla runs down the middle of the hair, and it may be thick, thin, continuous, or discontinuous. Pigment granules are responsible for color and are scattered throughout the cortex. There are three colors of pigments found in human hair: yellow, black, and brown; only gray hairs have no pigment. The cortical fusi are void spaces within the cortex that are concentrated nearer the root end.

Individual hair growth occurs in stages independent of neighboring follicles. The first phase, growth, is called the anagen phase and can last as long as six years. During this period, the follicle is producing hair at a normal rate. During the catagen phase, a transition occurs that can last several weeks, leading to the final telogen phase in which the follicle is dormant and growth ceases. This phase can last up to six months and will end when the hair is naturally shed and the cycle begins again. Since the pigment cells also go dormant, the last portion of the hair is white. The structure and appearance of the root (its MORPHOLOGY) can be used to determine at what phase a hair was lost. Likewise, if a hair is forcibly removed, which is often the case in violent crimes, portions of the follicular tissue will adhere to it and can be identified microscopically.

Analysis

The forensic importance of hair as trace evidence has been appreciated for more than a century, with cases involving the microscopic examination of hair being reported in the mid-1800s. Pioneers in the field such as VICTOR BALTHAZARD and PAUL KIRK understood the importance of hair evidence and wrote about it. However, since hair varies not only from person to person but also within hair samples from the same person, multiple hairs are necessary for any examination. By studying a range of hairs from the same source, the examiner obtains a sense of the range of variation to expect. CONTROL SAMPLES are also essential, particularly when some of the hairs recovered might have come from investigators or first responders to a scene. Microscopy is the primary tool of the hair examiner, who typically uses a STEREOMICROSCOPE, COMPOUND MICROSCOPE, POLARIZING LIGHT MICROSCOPE (PLM), and COMPARISON MICROSCOPE for casework. The examiner first works

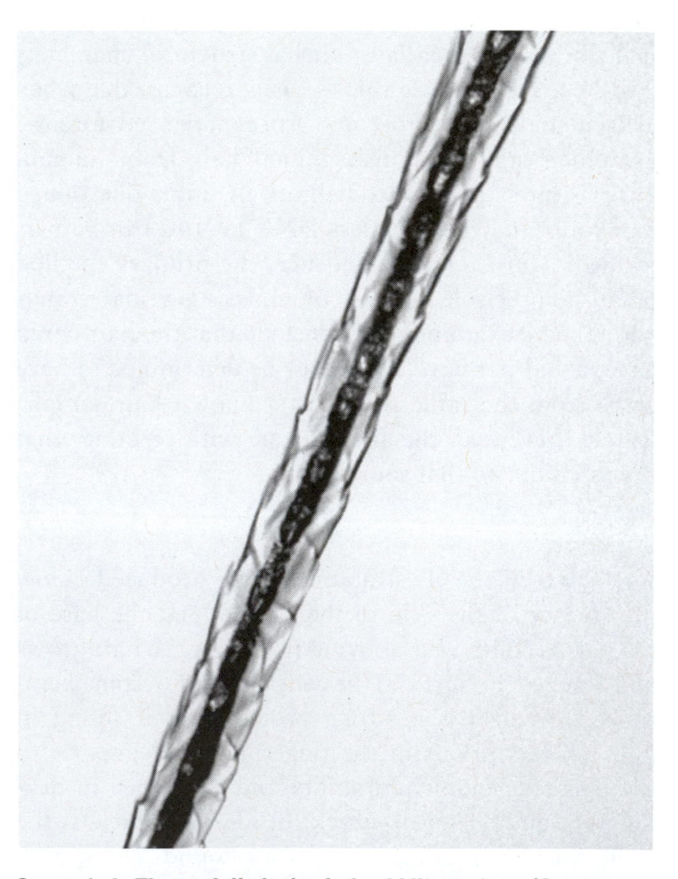

Coyote hair. The medulla is the dark middle portion. ***(Courtesy of Rebecca Hanes, Bennett Department of Chemistry, West Virginia University). A color image of hair can be found in the color insert.***

to confirm that an unknown fiber is indeed a hair and, if so, if it is human or animal. If the hair is animal, the species can provide important information. A valuable measurement obtainable through the microscope is the medullary index, which is the ratio of the diameter of the medulla divided by the diameter of the hair. In animals, the medulla tends to be wide, and the medullary index of animal hair can be diagnostic of species.

Hair differs in appearance depending on what part of the body it is from, an important consideration since hair comparisons must be made based on hairs that come from the same area. Hair has different characteristics depending on if it is from the legs, arms, chest, pubic area, anal area, face, head (scalp), eyebrows, or eyelashes. After microscopic examination, experienced forensic hair analysts can usually assign a region of origin to a hair. Age is impossible to determine reliably from a hair except in the case of infants, which have extremely fine hair that is easily recognizable. Certain diseases can create patterns in hair, as can chemical treatment such as bleaching or dyeing. It is also possible to determine if a hair has been cut and if the cutting was done with scissors or a razor. Other useful characteristics of hair are the diameter, general texture, cross-sectional appearance, shape of the medulla (if present), and appearance of the cuticle.

The pattern made by the scales on the cuticle is often evaluated by preparing a hard cast of the pattern in clear nail polish. The hair is pressed into the wet polish and then carefully removed, leaving an impression that can be observed microscopically. Features within the cortex such as pigment distribution can also be studied and used to compare between questioned hairs and known or control samples. Due to wide variations, the appearance of the medulla is not always of value, and many people lack a medulla or exhibit discontinuous patterns. Although other features may help identify ancestry, these determinations are estimates and may not correlate with how the person self-identifies. Because of all these factors, a comparison microscope often proves valuable for direct comparison of two hairs simultaneously, somewhat similar to the way in which questioned and known bullets are compared using the same instrument. For hair comparisons, transmitted light is used, while for firearms examination, reflected light is employed. However, the conclusions drawn in hair comparisons are not as definitive as is possible for bullets.

Researchers have studied methods to INDIVIDUALIZE hair with limited success, at least until recently. If a hair is forcibly removed or retains follicular tissue, the hair could sometimes be typed for the ABO blood group of the donor. However, this kind of typing was not applicable to hairs that had been shed naturally (the telogen phase). Unfortunately, most hairs encountered have shed naturally, and so typing was not often feasible. These older grouping procedures have been superceded by DNA typing techniques, which offer the potential for individualizing at least some hair evidence in the same way DNA typing allows for the individualization of blood and bloodstain evidence. If root and follicular tissue remain on a hair (one that was forcibly removed), this tissue can be subjected to PCR techniques in which small amounts of DNA are amplified and typed. Still, some follicular tissue must be present for this to be successful. On the horizon, typing of mitochondrial DNA (mtDNA) appears promising for hair analysis and in particular for typing hairs without follicular tissue. The current disadvantage to mtDNA typing is that it is time consuming, expensive, and not widely performed.

Finally, chemical analysis of the hair can be used in some cases to detect poisoning. For example, hair analysis has led to speculation that NAPOLEON BONAPARTE died not of stomach cancer but of arsenic poisoning. Other heavy metals can be deposited in hair, and since hair grows at a regular rate, it provides a record of what was ingested when. However, attempts to use chemical signatures of hair as a means to individualize it have not met with success.

Further Reading

Bisbing, R. "The Forensic Identification and Association of Human Hair." In *Forensic Science Handbook*. Vol. 1, 2nd edition. Edited by R. Saferstein. Upper Saddle River, N.J.: Prentice Hall, 2002.

Kubic, T. A., and N. Petraco. "Microanalysis and Examination of Trace Evidence." In *Forensic Science: An Introduction to Scientific and Investigative Techniques*. 2nd edition. Edited by S. H. James and J. J. Nordby. Boca Raton, Fla.: CRC Press, 2005.

half-life A term that applies to concepts of radiochemistry and TOXICOLOGY and more broadly, to chemical reactions. The half-life of a process is the time that is required for the process to reach a halfway point ($t_{1/2}$) starting from an initial time symbolized as $t = 0$. In radioactive decay, it is the time

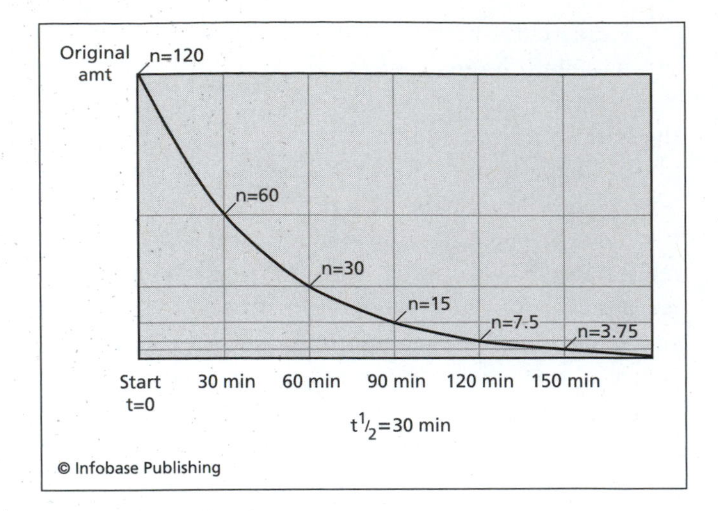

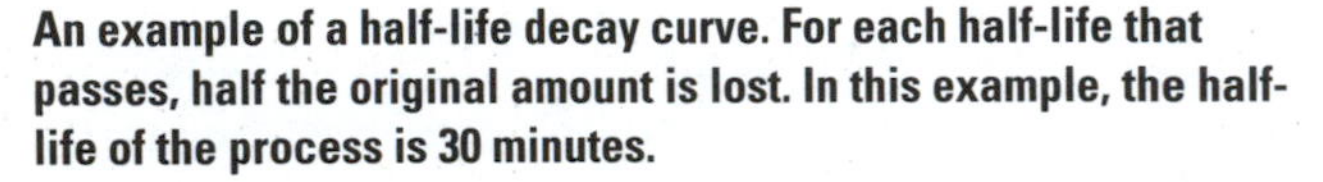
An example of a half-life decay curve. For each half-life that passes, half the original amount is lost. In this example, the half-life of the process is 30 minutes.

required for half the original amount of a substance to undergo radioactive decay. In toxicology, the half-life of a substance is the time required for the original concentration to decrease to half of its concentration at t = 0. For example, if a person ingests heroin, the drug is absorbed and the concentration of heroin in the blood reaches its maximum concentration within a few minutes. Metabolic processes begin degrading the heroin into monoacetylmorphine, and when half the original heroin has been converted, that is one half-life. The monoacetylmorphine itself is metabolized into morphine.

See also ADME.

half-truth A concept of law that first appeared in English law during the Middle Ages. Calling something a half-truth was not derogatory but rather indicated what value should be placed on that evidence. Alone, a half-truth was not enough to prove or disprove an issue, but it did have value in supporting one version of it over another. A half-truth in medieval law did not imply a purposeful omission of key information; rather it weighted the evidence as somewhere between definitive and useless. Half-truths arose in part because of the use of ORDEALS and torture to validate testimonial evidence.

The church, which used torture frequently to obtain evidence, eventually grew uncomfortable with the concept of the ordeal. By the 13th century, ordeals were uncommon, and the courts returned to the ancient practice of relying on sworn oral testimony as the best evidence. There was some reluctance to do so given that the church portrayed and taught that humans are innately sinful and untrustworthy. Partially in recognition of this tension, half-truths appeared. Admission of supporting evidence and half-truths demonstrates a trend that would influence scientific and supporting evidence as much as it did oral testimony. By definition, the role of supporting evidence is to make a stronger case. With each piece of consistent supporting evidence, one version of events becomes the most trustworthy. The tipping point occurs when the probability of truth is sufficient. The concept of sufficiency underlies both science and law and by extension, FORENSIC SCIENCE.

hallucinogens *See* DRUG CLASSIFICATION.

handguns A classification of FIREARMS that includes REVOLVERS, pistols, and homemade guns. The term handgun describes how they are used and contrasts these kinds of guns with what are called shoulder guns such as RIFLES and shotguns. Of all firearms, handguns are the ones most often used in crimes.

handwriting *See* QUESTIONED DOCUMENTS.

hanging Death by means of ASPHYXIA (loss of oxygen to the brain) caused by a noose tightened by the body weight of the victim. The pressure of the noose quickly cuts off the blood flow to the brain, leading to the rapid onset of unconsciousness. The airway also closes, leading to death in a few minutes. This effect can be achieved without complete suspension, so people can die of hanging while kneeling or otherwise still being in partial contact with the ground. Nooses can be made of rope, a belt, a sheet, a towel, or anything that is long enough to go around the neck and be secured to a stationary surface. Hanging is rare as a means of murder but is seen frequently as a means of suicide. Accidental hangings can occur in children or in cases of SEXUAL ASPHYXIA. Hanging produces a hanging groove in the neck that is the deepest opposite the point of suspension. Thus, if the noose is tied in the back of the neck, the hanging groove will be most prominent on the front of the neck. The wider the noose material, the less deeply the groove will be cut, so the groove from a towel will be different from that produced by narrow cording such as a clothesline.

A sink in which bloody evidence has been rinsed *(Courtesy of Amy Aylor, Bennett Department of Chemistry, West Virginia University)*

Same sink treated with luminal *(Courtesy of Amy Aylor, Bennett Department of Chemistry, West Virginia University)*

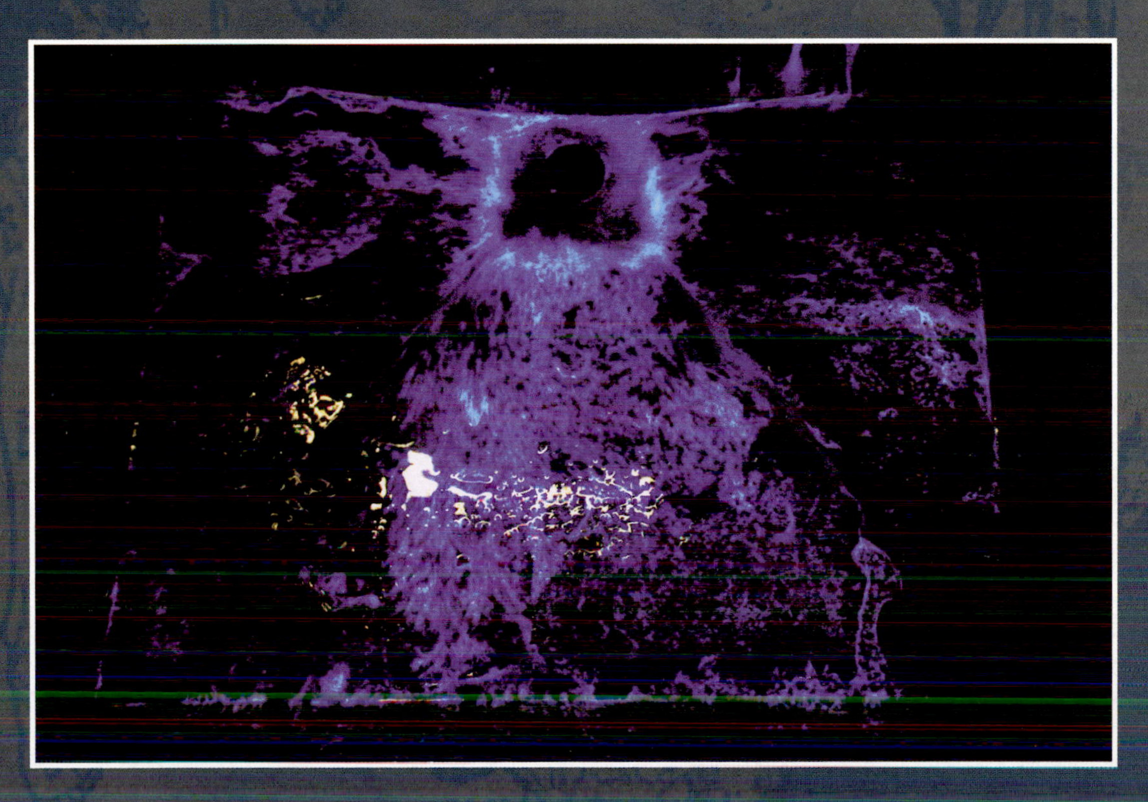

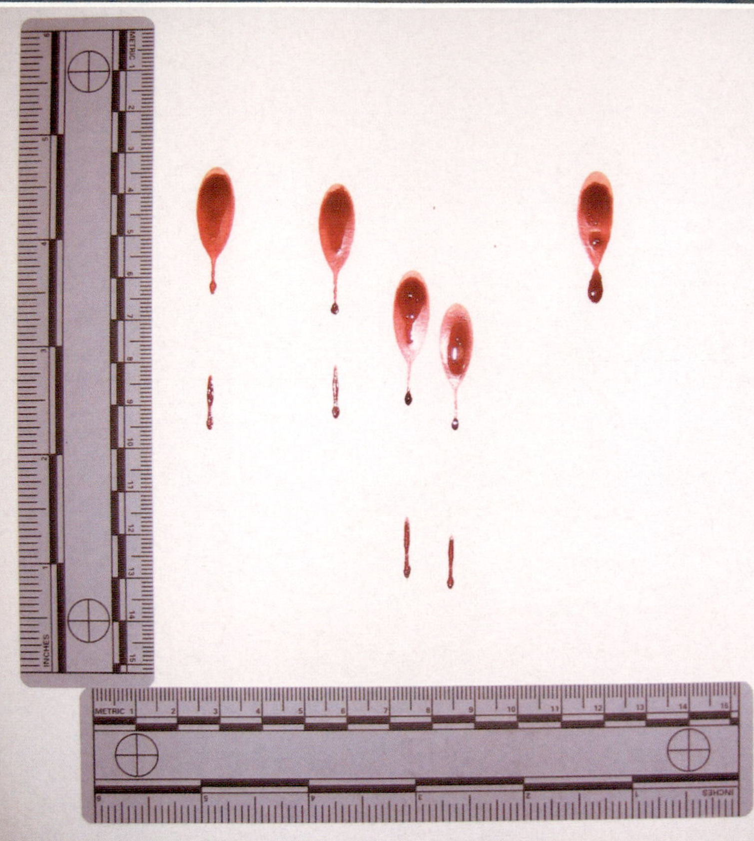

Blood drops deposited at an angle of approximately 30 degrees *(Courtesy of Michael Bell, West Virginia University, Forensic and Investigative Sciences program)*

Shoe prints treated with fluorescent reagents *(Courtesy of Michael Bell, West Virginia University, Forensic and Investigative Sciences program)*

Fingerprint treated with fluorescent chemicals *(Courtesy of Michael Bell, West Virginia University, Forensic and Investigative Sciences program)*

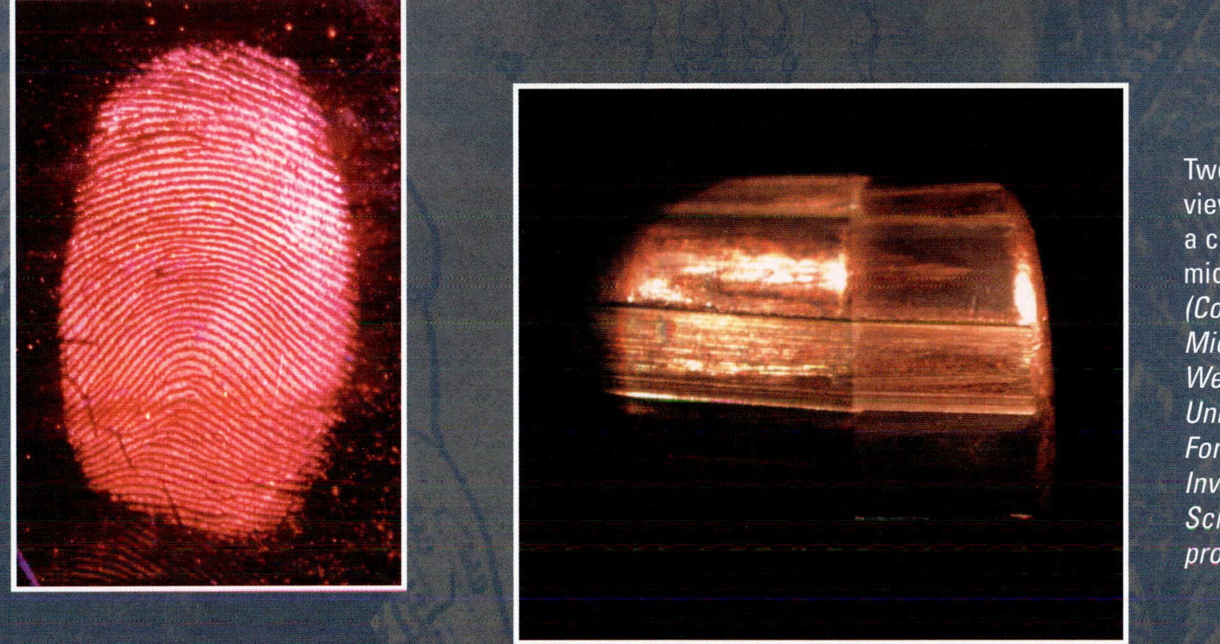

Two bullets viewed under a comparison microscope *(Courtesy of Michael Bell, West Virginia University, Forensic and Investigative Sciences program)*

An outdoor crime scene illuminated using painting with light techniques. The ghost images are the result of different exposures captured on the same image *(Courtesy of Michael Bell, West Virginia University, Forensic and Investigative Sciences program)*

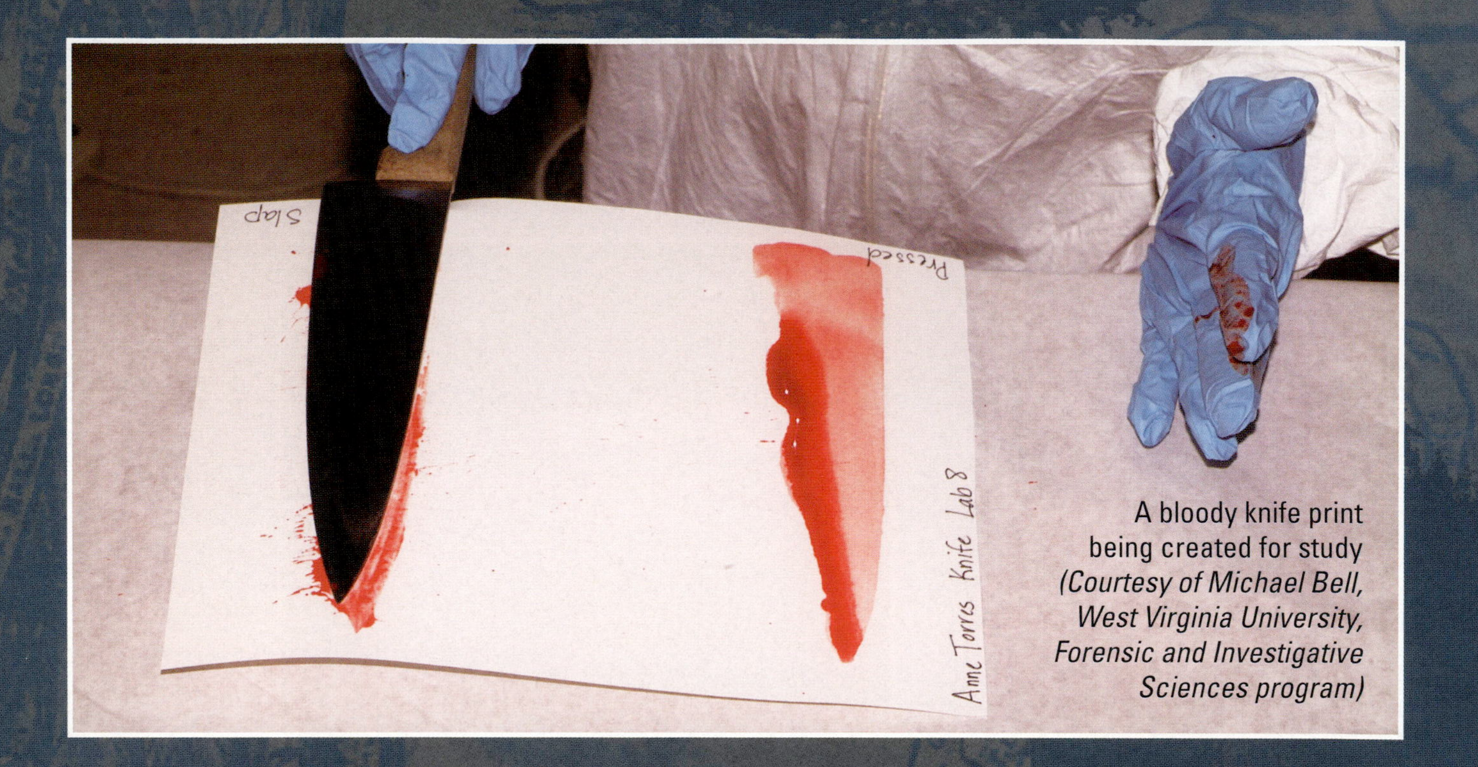

A bloody knife print being created for study *(Courtesy of Michael Bell, West Virginia University, Forensic and Investigative Sciences program)*

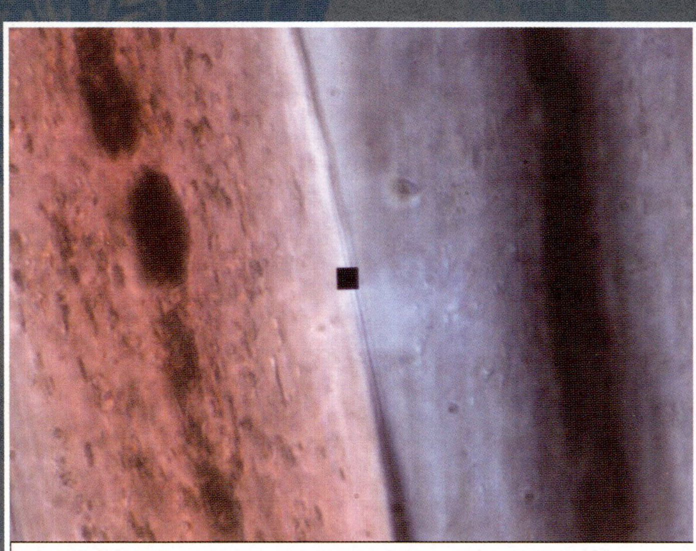

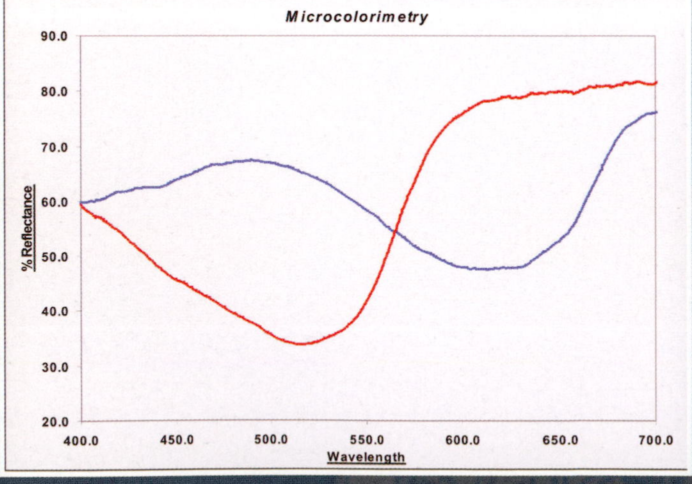

Two colored cotton fibers (top) and their color spectra (below) *(Suzanne Bell)*

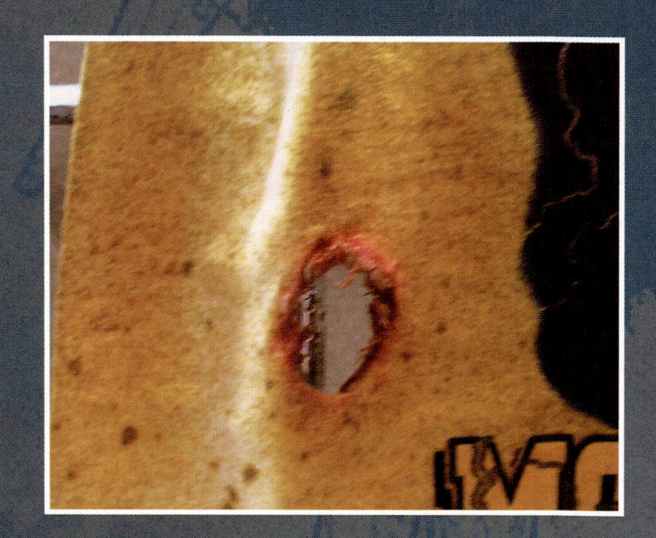

A bullet hole treated with sodium rhodizonate, which reacts with lead from the primer *(Suzanne Bell)*

A cotton fiber under normal illumination *(Courtesy of Rebecca Hanes, Bennett Department of Chemistry, West Virginia University)*

The fiber under polarized light with the filters 45 degrees to each other *(Courtesy of Rebecca Hanes, Bennett Department of Chemistry, West Virginia University)*

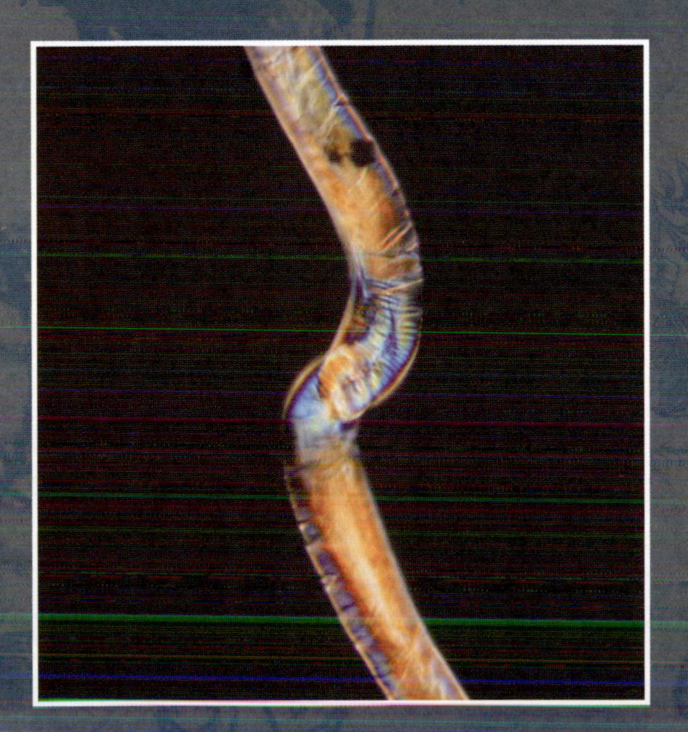

The fiber under crossed polars *(Courtesy of Rebecca Hanes, Bennett Department of Chemistry, West Virginia University)*

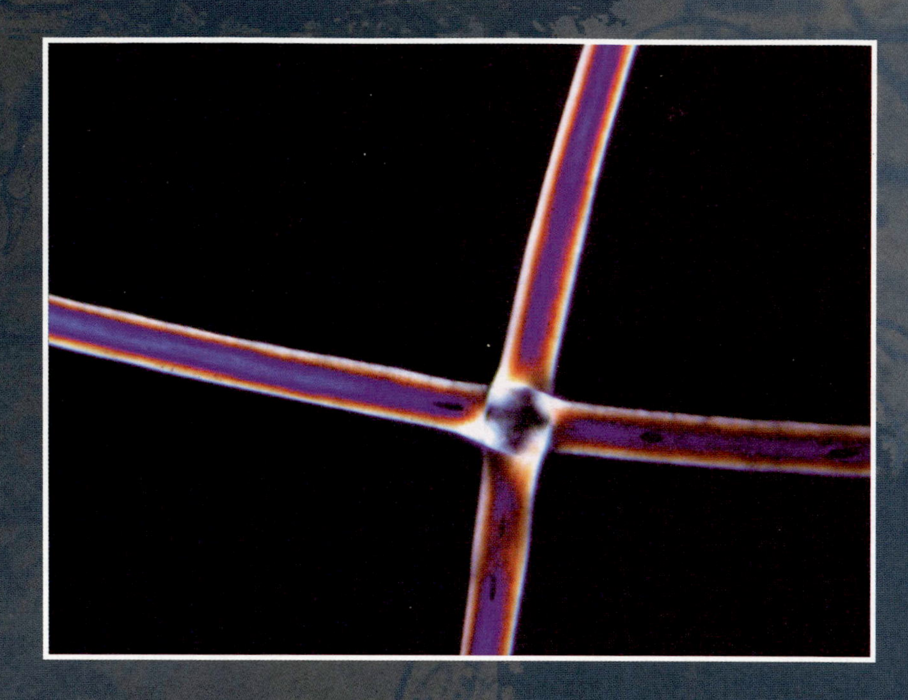

Two nylon fibers under crossed polars. Note the interference pattern where the fibers cross. *(Suzanne Bell)*

Training for bloodstain pattern analysis *(Courtesy of Michael Bell, West Virginia University, Forensic and Investigative Sciences program)*

Color-based presumptive tests used in drug analysis *(Suzanne Bell)*

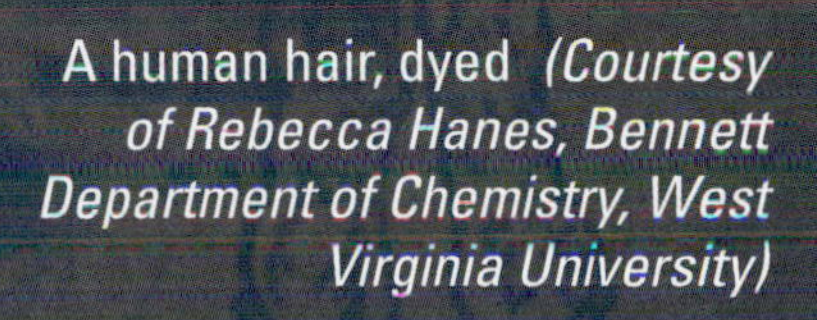

A human hair, dyed *(Courtesy of Rebecca Hanes, Bennett Department of Chemistry, West Virginia University)*

The sequence of the Duquenois-Levine test for the presence of marijuana, plant extract only *(Suzanne Bell)*

Extract, reagent, and hydrochloric acid *(Suzanne Bell)*

Chloroform added forms a layer at the bottom of the test tube. *(Suzanne Bell)*

The dye formed by the reaction is soluble in the chloroform and drops into the chloroform layer, a positive result. *(Suzanne Bell)*

Although murder by hanging is uncommon, it is not unheard of, and such cases present a challenge to investigators. At AUTOPSY, a finding that death was caused by hanging is not enough by itself to determine if the hanging was accidental, suicidal, or murder. Signs of struggle are expected if it was murder unless the victim was already incapacitated. Occasionally, murderers will hang a body after death in an attempt to conceal the crime, but this can be detected at autopsy. Some suicides show signs of earlier failed attempts that can at first appear to be the wounds associated with a struggle. Conversely, people bent on killing themselves sometimes will bind their own hands, further complicating the investigation. Fiber evidence can assist in determining how a rope was pulled and from what direction a person was suspended, while careful examination of the scene can provide important clues about the circumstances. Trace evidence may be trapped in the knot as well.

haptoglobin (Hp) A serum blood group system and GENETIC MARKER SYSTEM. Prior to the ascendance of DNA TYPING, this was one of the systems that could be typed in bloodstains using electrophoresis. There are three types of haptoglobin with 2-1 being the most common, found in approximately 40–50 percent of the population depending on race. Type 1 is shown by about 20–30 percent and type 2 by 20–35 percent.

Hardy Weinberg law A relationship used by forensic biologists to estimate the FREQUENCIES of alleles (variants of a gene) and the resulting distribution of types within a population. It provides a statistical method to test frequencies and to provide estimates of the DISCRIMINATING POWER of a given type. The following examples show how the calculations are performed using simple situations.

For the first case, consider two consecutive flips of a coin, which mimic the combination of two genes (one from a mother and one from a father) to create a variant at the site of the gene (its locus). For any one toss, the probability of obtaining a heads (H) is 0.50, meaning that 50 percent of the time, heads should appear. This is assigned to a variable labeled p. The variable q, the frequency of tails (T) is also 0.50, and p + q must always equal one. In terms of the Hardy Weinberg law, this is expressed as $(p + q)^2$, and then by using a technique called a binomial expansion, the expression $p^2 + 2pq + q^2$ is obtained. This is the equation that can be used to estimate frequencies. If a coin is tossed twice, there are three possible outcomes: two heads (HH), one of each (HT), and two tails (TT). The equation can be used to estimate how often each combination is expected to occur. HH would correspond to p^2, or $50 \times .50 = 0.25$, or 25 percent. This means that 25 percent of the time, when a coin is flipped twice, a combination of HH should occur, and the same for TT. For one of each, the predicted frequency is $2 \times 0.50 \times 0.50$ (2pq) or 0.50, meaning that half the time, two flips will result in one of each. These are the frequencies expected assuming the coin tosses are completely random and the coin is not altered to favor heads or tails.

This analogy can be extended to a hypothetical genetic marker system in which three types are possible: 1-1, 2-1, and 2-2, a common scenario. The types are combinations of the gene variant (the allele) from the mother and the father. However, in this case the forensic scientist would have actual data on types in the population, and these frequencies would be used to estimate p and q. Assume that 350 people have been typed for this system with the following results: 100 of them are found to be 1-1 (28.6 percent of the total), 200 are found to be 2-1 (57.1 percent), and 50 are found to be 2-2 (14.3 percent). The value of p is calculated from this data. First, the occurrence of the allele 1 is = 100 (the 1-1 types) + $\frac{1}{2} \times 200$ (half the 2-1) = 200. To obtain the frequency (p), 200 is divided by 350 (the total) to get 0.571. Similarly, for q, the value is 0.429, and $0.571 + 0.429 = 1.000$, as is required. Now these values can be substituted into the expression used above to obtain the following expected frequencies:

- 1-1 = $0.571^2 = 0.326$, or 32.6%
- 2-1 = 2(0.571)(0.429) = 0.490, or 49%
- 2-2 = $0.429^2 = 0.184$, or 18.4%.

Note that these percentages add up to 100 percent.

Using statistical techniques, it is possible to compare the actual frequencies with the observed ones to determine if the Hardy Weinberg relationship applies. For example, the predicted occurrence here of type 2-1 was 49.0 percent, and the actual data showed this percentage to be 57.1 percent. A test called the Chi-squared test would be used to determine if this difference is significant. If it is not, then the frequencies estimated can be considered reliable. If the difference is significant, this suggests that something else is affecting the types. It could be that the initial database of

350 people is too small, or it may mean that one of the types is favored in nature and that natural selection is influencing frequency. The Hardy Weinberg relationship and statistical calculations will not reveal why a difference is significant, only that it is.

Further Reading

Melvin, J., et al. "Paternity Testing." In *Forensic Science Handbook*. Vol. 1. 2d ed. Edited by R. Saferstein. Upper Saddle River, N.J.: Prentice Hall, 2002.

Harrison Narcotics Tax Act This federal statute, signed in 1914 by President Woodrow Wilson, was the first federal legislation aimed at controlling drugs of abuse. It was directed at MORPHINE, which by the turn of the 20th century had become a significant problem. The act was related to taxing and control of the supply rather than banning or controlling substances as does the CONTROLLED SUBSTANCES ACT. Under the Harrison Act, materials classified as narcotics had to have the official tax stamp to indicate that they had been legally imported and distributed.

hashish A potent derivative of marijuana (*Cannabis sativa* L) that has a high concentration of the psychoactive ingredient Δ^1-*trans*-tetrahydrocannabinol (THC). Hashish is a tarry substance that varies in color from dark green to almost black. It is the resin of the marijuana plant and is often prepared by extracting the flowering tops with an alcohol. The resulting concentration of THC can approach 60 percent, far above the concentration found in the leaves of the plant. For this reason, hashish is extremely potent, and there are reports that users of this form can develop at the least a strong psychological dependence on the drug. Like marijuana, hashish is classified as a hallucinogen and is listed on Schedule I of the CONTROLLED SUBSTANCES ACT (CSA). It is not, contrary to common misunderstanding, a narcotic. Hashish is often found as blocks or chunks of resin mixed with plant matter.

In the lab, the forensic drug analyst will perform a series of tests to determine if a substance is hashish. The first is a microscopic examination in which the analyst searches for the presence of distinctive features of the marijuana plant such as the seeds and CYSTOLITHIC HAIRS (also called bear claws). Depending on the procedure used to prepare the hashish, these may or may not be present. Other analytical tests are the same as are those used for marijuana. The modified DUQUENOIS-LEVINE test is a PRESUMPTIVE TEST for the presence of THC, and a purple color in a layer of chloroform is considered a positive result. A final confirmation is obtained using a chromatographic technique, usually THIN LAYER CHROMATOGRAPHY (TLC).

Further Reading

Huestis, M. "Marijuana." In *Principles of Forensic Toxicology*. 2nd edition. Edited by B. Levine. Washington, D.C.: American Association of Clinical Chemistry, 2003.

headlamps and headlights A form of physical evidence (whole or more often in fragments) that can be important, particularly in the investigation of traffic accidents and hit-and-run cases. Headlights as well as taillights traditionally were made of GLASS, but increasingly plastic components are seen. During accidents, the glass or plastic can shatter and become a form of TRANSFER EVIDENCE in which fragments are transferred to another vehicle or a person. Larger fragments can sometimes be matched to a make or model of car or truck since headlight or taillight designs are often customized. Alternatively, a PHYSICAL MATCH may be possible when fragments of the lamp can be exactly fitted back to the original. Lacking this, glass fragments can be analyzed, with DENSITY and REFRACTIVE INDEX (RI) being the properties usually tested. Often forensic examiners are requested to determine if lights were on at the time of an impact, something that is usually easy to do. Inside most types of lights are filaments through which electric current is run to generate light. Since they carry current, these wire filaments get very hot. If the sealing glass or plastic that encloses it is broken while the filament is on and hot, a coating of oxidation can be observed (visually or microscopically) on the filament. If the light is off and the filament is cold, no such oxidation will be seen.

headspace analysis A technique used in the analysis of ARSON evidence, forensic TOXICOLOGY, and other specialized chemical analyses. The headspace is the air or other gas that sits over a sample (solid or liquid) that is enclosed in a sealed container. Fire debris, for example, is usually collected in paint cans, filling a portion but not all of the can's volume. The lid is sealed, creating an enclosed air space above the fire debris. Any materials that are volatile enough (evaporate at a low enough temperature) will collect in this headspace. By withdrawing a sample of it, volatile components from the debris can be identified using different analytical techniques. In an arson case, the components of interest are any ACCELERANTS such as gasoline used to start and sustain a fire. Headspaces can be cold, or the container may be heated to increase volatilization. The headspace can also be dynamic, meaning that it is periodically swept out of the container and directed over a material that can trap the components contained in it. This technique is also known as purge-and-trap. In the analysis of fire debris, several variations of headspace sampling have been or are used including cold headspace, heated headspace with collection on a charcoal trap, collection of headspace vapors on a charcoal strip, and dynamic headspace with a charcoal trap. In forensic toxicology, headspace analysis can be employed in the analysis of CARBON MONOXIDE (CO) in blood. A sample is sealed into a small vial along with a compound called a releasing agent that drives the carbon monoxide out of the blood and into the headspace above it. The headspace can then be sampled and the presence of CO confirmed using GAS CHROMATOGRAPHY (GC).

hearsay Testimonial evidence from a witness that relates to events that the witness does not have personal knowledge of. Rather, the witness is testifying about something he or she heard another person say while that other person was not under oath. In a courtroom, if a witness is on the stand and asked a question such as: "What did Mr. Jones say about that?" or "What did you hear Mr. Jones say about that?" the response the witness gives would be hearsay evidence. Hearsay evidence is usually inadmissible, but there are exceptions to the hearsay rule.

heavy metals In forensic science, this term usually refers to metals that are also POISONS. In this context, the metals of interest include ARSENIC (As), LEAD (Pb), MERCURY (Hg), cadmium (Cd), bismuth (Bi), antimony (Sb), and thallium (Tl). Of these, arsenic is the best known as a poison—accidental, suicidal, and homicidal. Strictly speaking, it is classified as a metalloid or semimetal in terms of chemical properties and behavior. Antimony is also found in gunshot residue, and thallium is used in the electronics industry.

Detecting heavy metals is the responsibility of forensic toxicologists, who analyze blood and other body fluids and tissues for the presence of drugs and poison. To detect metals and to determine their concentrations, elemental analysis is required using ATOMIC ABSORPTION (AA) and inductively coupled plasma techniques (ICP-AES and ICP-MS). Sample preparation is crucial, and steps must be taken to prevent contamination. Since some of these metals are found naturally in the body, any contamination can make results difficult to interpret. The analysis of mercury is particularly challenging since it is a relatively volatile metal that can evaporate during sample heating.

Further Reading

Saady, J. "Metals." In *Principles of Forensic Toxicology*. 2nd edition. Edited by B. Levine. Washington, D.C.: American Association of Clinical Chemistry, 2003.

Helpern, Milton (1902–1977) American *Physician and Medical Examiner* Dr. Milton Helpern became the chief of the New York City Office of the Medical Examiner in 1953 after being with the office for nearly 20 years. He became one of first MEDICAL EXAMINERS known by name to many in the public outside of New York City. He participated in many high-profile and controversial cases and founded a museum associated with the ME's office. A famous example of Helpern's views came in the trial related to the murder of Malcolm X. The defense attorney stated that Helpern did not know who killed Malcom X, to which Helpern replied "I don't know and I don't care; I'm interested in what did it, not who done it." Helpern's death was tinged with irony given that his last illness came on during the annual meeting of the AMERICAN ACADEMY OF FORENSIC SCIENCES being held in San Diego in 1977.

Helpern was a prodigious writer in scientific journals and coauthor of a text with THOMAS A. GONZALES and Dr. Benjamin Vance entitled *Legal Medicine and Toxicology*. His public recognition came in part from books he wrote for general audiences, the most famous of which was modestly called *Autopsy: The Memoirs of Milton Helpern, The World's Greatest Medical Detective*. Although Helpern was not the first to write a popular book, he started a trend joined by other MEs such as Dr. Michael Baden, who later held the same position as Helpern. Baden took the trend a step further by overseeing a television series on HBO called *Autopsy* and by frequent appearances on cable television networks.

Further Reading

Sturner, W. Q., "The Wit and Wisdom of Milton Helpern: A Glimpse in Time." *The American Journal of Forensic Medicine and Pathology* 1998, 19, (3). (Source of quote, p. 289)

hemagglutination Also called agglutination, this is a clumping reaction that occurs when red blood cells are mixed with ANTISERA to the ANTIGENS found on their surface. An example is the agglutination reaction used to determine a person's ABO BLOOD GROUP SYSTEM type.

See also ABSORPTION-ELUTION AND ADSORPTION INHIBITION TESTS.

hematin test (Teichman test) A confirmatory test for blood based on the formation of distinctive hematin crystals that are viewed under a microscope. The test was developed in 1853 by Ludwig Teichmann and is identified with both the Teichman and Teichmann spellings. The Teichman test is one of a group of confirmatory tests for blood referred to as CRYSTAL, microcrystal, or microcrystalline tests. Crystal tests require a larger sample than do the PRESUMPTIVE TESTS for blood and are susceptible to interferences. Thus, for very small stains, the crystal test is not used since it will consume material most likely needed for DNA TYPING.

Further Reading

Shaler, R. "Modern Forensic Biology." In *Forensic Science Handbook*. Vol. 1. 2d ed. Edited by R. Saferstein. Upper Saddle River, N.J.: Prentice Hall, 2002.

Spalding, P. R. "Identification and Characterization of Blood and Bloodstains." In *Forensic Science: An Introduction to Scientific and Investigative Techniques*. 2nd edition. Edited by S. H. James and J. J. Nordby. Boca Raton, Fla.: CRC Press, 2005.

hemochromogen test (Takayama test) A confirmatory test for blood based on the formation of distinctive hemochromogen crystals that are viewed under a microscope. The test was developed in 1912 by Masao Takayama in Japan and has evolved into the most commonly used of the so-called microcrystalline (microcrystal or CRYSTAL is also used) tests for blood. The test requires a larger sample than do the PRESUMPTIVE TESTS for blood and is susceptible to interferences. Thus, for very small stains, the crystal test is not used since it will consume material most likely needed for DNA TYPING.

Further Reading

Shaler, R. "Modern Forensic Biology." In *Forensic Science Handbook*. Vol. 1. 2nd edition. Edited by R. Saferstein. Upper Saddle River, N.J.: Prentice Hall, 2002.

Spalding, P. R. "Identification and Characterization of Blood and Bloodstains." In *Forensic Science: An Introduction to Scientific and Investigative Techniques*. 2nd edition. Edited by S. H. James and J. J. Nordby. Boca Raton, Fla.: CRC Press, 2005.

hemoglobin (Hb) The predominant protein in red BLOOD cells that is used to transport oxygen from the lungs to the tissues. The hemoglobin molecule is made up of four protein subunits (the globin portion), two alphas (α) and two betas (β), and for this reason, hemoglobin is classified as a tetramer. Each of the four subunits possesses a heme unit with an iron ion (Fe^{2+}) at the center, and it is the heme units that bind to the oxygen, four per hemoglobin molecule. PRESUMPTIVE TESTS and microcrystalline confirmatory tests for blood used in forensic science rely on reactions with the hemoglobin molecule. There are also variants of hemoglobin that can be typed using ELECTROPHORESIS and ISOELECTRIC FOCUSING techniques.

Presumptive tests for blood have been around for more than a century, and all are based on a type of chemical reaction known as an oxidation. In these reactions, hemoglobin acts as a catalyst, resulting in a color change that is observable. Examples of presumptive tests are BENZIDINE, PHENOLPHTHALEIN, and LUMINOL. The last reaction is different in that the chemical reaction results in the emission of light, a phenomenon called chemoluminescence. The luminol test is valuable when suspected blood is deposited on a dark background where color changes would not be visible or when an attempt has been made to clean up bloodstains. These chemical presumptive tests are quite sensitive and can detect hemoglobin in trace amounts. Another procedure used to detect hemoglobin is based on UV/VIS SPECTROSCOPY in which the light absorption pattern of hemoglobin is obtained. An added advantage of this technique is that it can distinguish hemoglobins that are bound to molecules other than oxygen (the oxyhemoglobin form). For example, the mechanism of CARBON MONOXIDE poisoning is based on the ability of that molecule to bind to hemoglobin even more strongly than does oxygen, forming carboxyhemoglobin. The difference between these forms is easily seen using spectroscopic analysis.

Hemoglobin is also the target molecule in confirmatory tests in which chemical reagents react with it to form distinctive crystals viewed under a microscope. The TEICHMAN test, also known as the hematin test for the type of crystal formed, was developed in the 1850s, and the TAKAYAMA test (hemochromogen test) was developed in the early 1900s. A third type, the Wagenaar test (ACETONE CHLOR-HEMIN) has also been used, but the Takayama test is the most common. Crystal tests are less sensitive than are the presumptive tests and thus require larger samples. Neither the presumptive tests nor the microcrystalline tests prove that a blood sample is human, nor are they foolproof. FALSE POSITIVES occur with presumptive tests, while interferences can be a problem with the crystal tests.

There are variants of hemoglobin aside from the normal adult form (Hb A). The sickle-cell trait found in some black and Hispanic populations is caused by the difference of one AMINO ACID in the protein chains that make up the molecule. This type of hemoglobin is signified as Hb S, and if a person shows the hemoglobin type of Hb SS, that individual will have the disease sickle-cell anemia. Interestingly, a person with the type Hb SA carries the trait but actually has increased protection from the disease malaria, suggesting why the Hb S variant has survived. Fetal hemoglobin is also a distinct type and is labeled Hb F (fetal hemoglobin). Another variant is Hb C, and these variants can be separated and identified using electrophoresis or related techniques.

Further Reading

Shaler, R. "Modern Forensic Biology." In *Forensic Science Handbook*. Vol. 1. 2d ed. Edited by R. Saferstein. Upper Saddle River, N.J.: Prentice Hall, 2002.

Spalding, P. R. "Identification and Characterization of Blood and Bloodstains." In *Forensic Science: An Introduction to Scientific and Investigative Techniques*. 2nd edition. Edited by S. H. James and J. J. Nordby. Boca Raton, Fla.: CRC Press, 2005.

Henry, Sir Edward (1851–1931) English *Police Officer* Henry, like Sir FRANCIS GALTON and JUAN VUCETICH, was a key figure in the early use of FINGERPRINTS in criminal investigations. He first employed them for identification purposes while he was the inspector-general of police in Bengal, India. Later, he became commissioner of the Metropolitan Police in London (Scotland Yard), and in 1901, his fingerprint classification system was adopted. Modifications of the Henry system are still used throughout Europe and in the United States.

Henry's law An expression that describes the relationship between the solubility of a gas in a liquid as a function of pressure. The equation is written as:

$$K_H = [A]/P_A$$

in which K_H is called the Henry's law constant, [A] represents the concentration of gas A dissolved in a liquid, and P_A represents the pressure of that gas over the liquid, in question. At a set temperature, the value of K must remain the same, which means that if the pressure of a gas increases over a liquid (P_A increases), then the concentration of that gas dissolved in the liquid ([A]) must also increase. This can be illustrated by the everyday situation of carbonation (CO_2 gas) in beverages such as soda pop, sparkling wines, and champagne. These beverages are bottled under elevated pressure, allowing more carbon dioxide to dissolve in them. When the cap is opened, the familiar hissing sound indicates that the pressure is equalizing, and pressure over the liquid is falling back to normal atmospheric levels. In terms of the equation, P_A is decreasing. At the same time, bubbles of carbon dioxide form inside the liquid and escape, since at the lower pressure it is less soluble ([A] is also decreasing). Since both the pressure and the dissolved concentration decrease, the value of K_H remains a constant, as the law states it must. If the bottle is left open, it will shortly go "flat," with all excess carbon dioxide removed. Henry's law is important in forensic science in the area of BLOOD ALCOHOL CONCENTRATION (BAC), BREATH ALCOHOL, and in some types of specialized chemistry analyses.

heroin A derivative of MORPHINE that is highly addictive and widely abused. Also called diacetylmorphine, diamorphine, and acetomorphine, heroin is easily synthesized from morphine by the addition of acetyl chloride or acetic anhydride. The morphine is obtained from OPIUM poppies, and so heroin is classified as one of the opiate ALKALOIDS. It is most commonly found in the form of a hydrochloride salt with a white or off-white color; however, a brownish black resinous form known as "Black tar" originating in Mexico is also seen. Heroin is water soluble and so is usually injected, although snorting and smoking are possible. Because of its potency, heroin is both physically and psychologically addictive and has no acceptable medical use. It is listed on Schedule I of the CONTROLLED SUBSTANCES ACT (CSA).

Heroin obtained on the street is rarely pure but rather is cut with various CUTTING AGENTS (diluents) such as quinine, which shares a similar bitter taste with heroin. Sugars, powdered milk, procaine, lidocaine, and, in the case of darker colored heroin, cocoa may also be used. Heroin use increased during the 1990s, but there are signs that it may be leveling off. Sources of heroin are Mexico, South America, Asia, and southwest Asia (Turkey, Afghanistan, and the neighboring regions).

The analysis of heroin begins with the use of PRESUMPTIVE TESTS in which a chemical reagent is added to the suspected heroin sample. These tests are not definitive but are useful in guiding further analysis. For heroin, the MARQUIS and FROEHDE reagents are used. Additional tests follow and may include any combination of CRYSTAL TEST, THIN LAYER CHROMATOGRAPHY (TLC), INFRARED SPECTROSCOPY (IR), GAS CHROMATOGRAPHY/MASS SPECTROMETRY (GC/MS), and HIGH PERFORMANCE LIQUID CHROMATOGRAPHY (HPLC). Heroin forms distinctive crystals with several reagents such as mercuric iodide, sodium acetate, platinic chloride, and gold chloride. These crystals are observed under a microscope, hence the name microcrystalline tests. Thin layer chromatography performed alongside a heroin standard is also useful but like the presumptive and crystal tests is not definitive. However, analysis by GC/MS or IR is definitive.

Further Reading

Drug Enforcement Agency (DEA) Web site. "Heroin." Available online. URL: www.usdoj.gov/dea/concern/heroin.html. Downloaded May 4, 2007.

high explosives *See* EXPLOSIVE.

high performance liquid chromatography (HPLC) An instrumental system based on chromatography that is widely used in forensic science. The "HP" portion of the acronym is sometimes assigned to the words *h*igh *p*ressure (versus *h*igh *p*erformance), but it refers to the same analytical system. HPLC is used in DRUG ANALYSIS, TOXICOLOGY, EXPLOSIVES analysis, INK analysis, FIBERS, and PLASTICS to name a few forensic applications.

Like all chromatography, HPLC is based on selective partitioning of the molecules of interest between two different phases. Here, the mobile phase is a solvent or solvent mix that flows under high pressure over beads coated with the solid stationary phase. While traveling through the column, molecules in the sample partition selectively between the mobile phase and the stationary phase. Those that interact more with the stationary phase will lag behind those molecules that partition preferentially with the mobile phase. As a result, the sample introduced at the front of the column will emerge in separate bands (called peaks), with the bands emerging first being the components that interacted least with the stationary phase and as a result moved quicker through the column. The components that emerge last will be the ones that interacted most with the stationary phase and thus moved the slowest through the column. A detector is placed at the end of the column to identify the components that elute. Occasionally, the eluting solvent is collected at specific times correlating to specific components. This provides a pure or nearly pure sample of the component of interest. This technique is sometimes referred to as preparative chromatography.

Many different types of detectors are available for HPLC. The simplest and least expensive is the REFRACTIVE INDEX detector (RI). Although this detector is a universal detector, meaning it will respond to any compound that elutes, it does not respond well to very low concentrations and as a result is not widely used. On the other hand, detectors based on the absorption of light in the ultraviolet and visible ranges (UV/VIS

An HPLC system in use ***(Courtesy of Jennifer Mercer, Bennett Department of Chemistry, West Virginia University)***

detectors and UV/VIS SPECTROPHOTOMETERS) are the most commonly used, responding to a wide variety of compounds of forensic interest with good to excellent sensitivity. The photodiode array detector (PDA) is especially useful since it can produce not only a peak-based output (a CHROMATOGRAM) but also a UV/VIS scan of every component. In many ways, the ideal detector for HPLC is a MASS SPECTROMETER (MS), which provides both quantitative information and in most cases a definitive identification of each component (qualitative information). LC-MS systems are becoming more common in forensic TOXICOLOGY laboratories. Other detectors that are sometimes used include fluorescence detectors (which are very sensitive) and electrochemical detectors.

Unlike in GAS CHROMATOGRAPHY (GC) in which the mobile phase is an inert gas, the mobile phase in HPLC can be one of many different solvents or combinations of solvents. This imparts to HPLC a greater flexibility and range of application than has GC. Because the sample does not have to be converted to the gas phase, compounds such as explosives that break down at high temperatures are much more amenable to HPLC than GC. For HPLC, all that is required is that the sample be soluble in the solvents selected for the analysis. In addition, there are several types of HPLC defined by the type of mobile phase and stationary phase that is used. For forensic applications, one of the most commonly used types of HPLC is referred to as "reversed phase." In this type of HPLC, the mobile phase is a solvent or mix of solvents that are polar, meaning that different parts of the individual solvent molecules carry a partial positive or negative charge. Water, methanol (methyl alcohol), ethanol (ethyl alcohol), and acetone are examples of polar solvents. The stationary phase in reverse phase HPLC is a nonpolar material such as a long chain hydrocarbon molecule. In this type of HPLC, components in the sample will partition and separate based on their degree of interaction with the stationary phase relative to the mobile phase. In other words, the separation is based primarily on the relative polarity of the sample molecules. Reverse phase HPLC is used in drug analysis (LSD for example), analysis of CUTTING AGENTS such as sugars, explosives, and GUNSHOT RESIDUE (GSR), and forensic toxicology.

Normal phase HPLC uses a polar stationary phase and a nonpolar mobile phase, but this is not widely used in forensic applications. Size exclusion chromatography (SEC) separates compounds based on relative sizes. The stationary phase in SEC is composed of a gel with different sizes of microscopic pores through it. The larger the molecule, the longer it takes for it to navigate through the pores and reach the detector. SEC is useful for the analysis of large molecules that come in a range of sizes such as plastic POLYMERS, PROTEINS, and NITROCELLULOSE, a component of GSR. Chiral chromatography, a relatively recent development, is making inroads into forensic science since it is capable of separating ENANTIOMERS, molecules that are mirror images of each other. This capability is particularly valuable in forensic toxicology and drug analysis. Finally, ion exchange chromatography is available for detection species such as nitrate (NO_{3-}) and other ions.

Further Reading

Northrup, D. "Forensic Applications of High-Performance Liquid Chromatography and Capillary Electrophoresis." In *Forensic Science Handbook*. Vol. 1. 2d ed. Edited by R. Saferstein. Upper Saddle River, N.J.: Prentice Hall, 2002.

HLA *See* HUMAN LEUKOCYTE ANTIGEN (HLA).

HLA-DQA1 *See* DNA TYPING.

HMX *See* EXPLOSIVE.

hollow point bullets *See* AMMUNITION.

Holmes, Sherlock The famous fictional detective created by SIR ARTHUR CONAN DOYLE. Over the course of 40 years (1887–1927), Doyle published 56 short stories and four novellas starring Holmes and his good-humored assistant Watson as they solved crimes using logic and science. In his career, Holmes delved into many areas of forensic science including forensic BIOLOGY, TRACE and TRANSFER EVIDENCE, FIREARMS, and QUESTIONED DOCUMENTS using fictional techniques that in many cases accurately predicted later developments in the field. Doyle's stories influenced and inspired many of the pioneers of forensic science including HANS GROSS and EDMOND LOCARD. Throughout the century-plus that has passed since the first story, *A Study in Scarlet*, was published in 1897, many readers have been enticed by Sherlock Holmes into pursuing a career in forensic science.

Holmes is best described as a forensic chemist, although his scientific interests and skills were broad

Fiction and Forensic Science

Forensic science began to emerge as a distinct discipline in the late 1800s. Coincidently, this was also a time when detective fiction, led by SHERLOCK HOLMES, was capturing the popular imagination. Like any good writer of science fiction, Sir ARTHUR CONAN DOYLE correctly predicted many developments in forensic science several years before they were actually used. HANS GROSS and EDMOND LOCARD, pioneers of forensic science, were fond of Doyle's stories and recommended them to their students. Even now, the bug still bites future forensic scientists when they read their first Sherlock Holmes story.

As forensic science evolved into an accepted part of science and law enforcement, it also became a more common plot device in fiction. Since the 1990s a body of fiction has emerged in which forensic detection takes center stage. Examples are series penned by Patricia Cornwell (forensic PATHOLOGY, PROFILING, and other forensic disciplines) and Kathy Reichs (forensic ANTHROPOLOGY). Many of the books written by these authors have made the *New York Times* best-seller list. The public loves a good scientific detective yarn as much now as it did a century ago.

Within the last few decades, the public's exposure to fictional forensic science has expanded beyond novels to visual media. Films such as *Silence of the Lambs* (1991, based on the novel by Thomas Harris) and *The Bone Collector* (1998, based on the novel by Jeffrey Deaver) are just two examples incorporating aspects of forensic science. *Silence of the Lambs,* starring Jodie Foster, won several Oscars and was notable for aspects of PROFILING and forensic ENTOMOLOGY. *The Bone Collector* revolved around a hero who had been an ace forensic investigator who has become paralyzed after an accident at a crime scene.

Television has become the medium through which forensic science garners the most public exposure by way of news shows, documentaries, and crime drama. Networks including Court TV, HBO, and the Discovery Channel all have broadcast series devoted to real-life forensic science. As in novels, forensic science has taken the lead in several drama shows as well. The 1970/80s show *Quincy,* starring Jack Klugman, centered on a medical examiner and was inspirational for the first generation of forensic scientists born and raised in the television age. *C.S.I.* (*Crime Scene Investigation*) was a top-rated drama on network television in 2001 and led to a spin-off, *C.S.I. Miami,* launched in 2002. With a former crime scene investigator as one of the writers, *C.S.I.* concentrates on evidence and analysis rather than police work. The surprising popularity of the series has kindled popular fascination with forensic science to a new audience. As a result, more young people are becoming interested in a career in forensic science. In turn, this raises the public's awareness of forensic science and increases general interest in science as a result. This certainly is a good thing, but is there also a downside to this popularization?

As forensic science becomes integrated into modern crime fiction, more poetic license is taken. For example, many a book incorporates a "perfect match" between a

and deep. In *A Study in Scarlet,* Holmes announces the discovery of the "Sherlock Holmes" test, a chemical reagent that detected not just blood, but specifically human blood. This was in 1887, 14 years before an immunological test for human blood was announced in 1901. It was not until 1904 that the venerable BENZIDINE test, a PRESUMPTIVE TEST for blood that reacts with hemoglobin was developed. As Doyle wrote:

> *"I've found it. I've found it," he shouted to my companion, running towards us with a test tube in his hand. "I've found a reagent which is precipitated by hemoglobin and by nothing else . . . The old guaiacum test is very clumsy and uncertain. So is the microscopic examination for blood corpuscles. The latter is valueless if the stains are a few hours old . . . Are they blood stains, or rust stains or fruit stains, or what are they? This is a question which has puzzled many an expert and why? Because there was no reliable test. Now we have the Sherlock Holmes test, and there will no longer be any difficulty."*

The GUAIACUM test that Doyle referred to was the oldest presumptive test for blood, and like all presumptive tests was subject to FALSE POSITIVES, meaning that the reagent would react with materials other than blood. The other option he mentions, namely, looking for red blood cells (corpuscles), only works on fresh stains that are still wet. Once a stain dries out, the cells lyse (break open) and are no longer recognizable under a microscope. As happens in fiction, a hero from literature manages to accurately predict developments years ahead of his or her time. Holmes is still considered by many to be the greatest fictional detective ever created.

See also "Fiction and Forensic Science" essay.

hair or fiber found at a crime scene and one recovered from a suspect. This plot contrivance ignores the fact that unless DNA is recovered, it is nearly impossible to find such a "perfect" situation. Additionally, fictional portrayals tend to glamorize the profession and extend the purview of forensic scientists into realms they would never enter in real life. No doubt forensic science is a fascinating career, but it rarely comes close to matching the dazzling world created by screenwriters and novelists.

For example, very few forensic scientists carry guns as part of their job, and few ever become involved in investigative tasks such as interviewing suspects. A major crimes squad often handles crime scenes, with crime scene technicians an integral part of the team. However, their principal job is to process and document the scene and to deliver evidence to the lab. They rarely are involved in scientific testing that occurs after evidence is transferred. In other jurisdictions, scenes are handled by police officers who may call in forensic laboratory analysts for assistance as needed. For these scientists, crime scenes are an adjunct to their laboratory work. In any case, the investigation of a crime is a team effort, and as such it succeeds only when everyone stays in their assigned positions. Forensic scientists must protect their objectivity, since it is objective analysis the court needs to help ascertain truth. If an analyst tests a piece of evidence and then talks to the suspect that might be convicted by it, objectivity is lost. The analysis cannot be admitted regardless of how valuable and reliable it is.

Obviously, the principal objective of these fictional books, shows, and movies is to entertain; it is called "fiction" for a reason. However, if the same liberties are repeatedly taken, people unfamiliar with forensic science may develop false ideas. Young people in particular watch television shows or movies and tend to assume that what they see is what a career in the field is like. As a result, they begin pursuing such a career based on false assumptions. The essays entitled "Careers in Forensic Science: A Reality Check" and "Myths of Forensic Science" provide additional information of interest, and a little homework can save months or years of frustration pursuing a job that does not exist. However, with the proper perspective, forensic science and detective fiction make a wonderfully entertaining combination, as they have since the days of Sherlock Holmes.

Further Reading

Gerber, S. M. "Forensic Science in Detective Fiction." In *More Chemistry in Crime, From Marsh Arsenic Test to DNA Profile.* Edited by S. M. Gerber and R. Saferstein. Washington, D.C.: American Chemical Society, 1997.

Ramsland, K. *The Science of CSI.* New York: Berkley Boulevard Books, 2001.

Thomas, R. R. *Detective Fiction and the Rise of Forensic Science.* Cambridge: Cambridge University Press, 1999.

Thomashoff, C. "The Science of Success: It's All in the Details." *The New York Times,* March 3, 2002.

Weintraub, B. "Gory Details Cheerfully Supplied; Trading Police Forensics for Writer's Life in *'C.S.I.'*" *New York Times,* December 24, 2001.

Further Reading

Doyle, Sir Arthur Conan. *The Complete Sherlock Holmes: All 4 Novels and 56 Short Stories.* New York: Bantam Press, 1998.

Gerber, S. "A Study in Scarlet: Blood Identification in 1875." In *Chemistry and Crime: From Sherlock Holmes to Today's Courtroom.* Edited by S. M. Gerber. Washington, D.C.: American Chemical Society, 1983.

homicide A purposeful killing undertaken with the intention of causing the death of the victim. The investigation of homicides represents an appreciable portion of the workload in forensic science, but typically not as much as the public assumes. Although each case is unique, in general homicides generate a plethora of PHYSICAL EVIDENCE across a wide range of types that include (but are not limited to) BLOOD and BODY FLUIDS, BLOODSTAIN PATTERNS, weapons, FINGERPRINTS, HAIRS, FIBERS, and TRANSFER EVIDENCE. The starting point is the body, although there have been homicide cases pursued even when a body was never found.

The first responder to the CRIME SCENE must determine if the victim is already dead or if medical help is necessary. Once the fact of a death is established, then many forensic specialties will be called upon. The medical examiner or forensic pathologist will assume control of the body and will perform an autopsy to determine the cause of DEATH, whether accidental, suicidal, or a homicide. Any physical evidence that can be obtained from the body will also be collected, documented, and photographed. Forensic toxicologists assist in analyzing blood and body fluids for the presence of drugs or poisons. Analysis of evidence from the scene becomes the responsibility of the appropriate forensic analysts such as fingerprint experts, firearms examiners, and trace evidence analysts. Chemists become involved if arson was part of the crime or if drugs were associated with

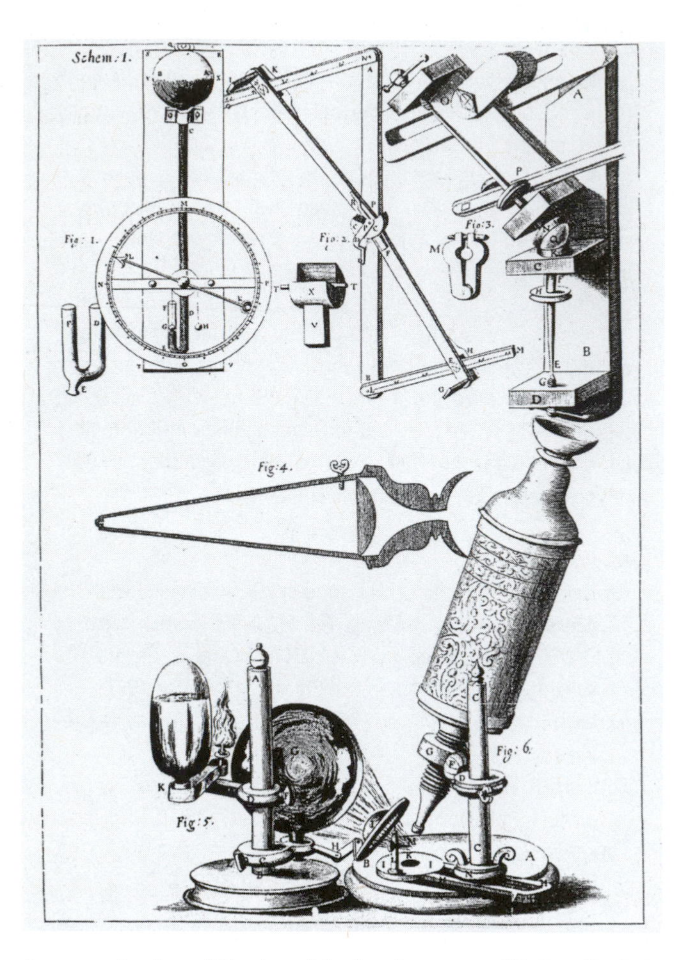

An engraving from *Micrographia* showing some of Robert Hooke's equipment. Above, left to right, are his barometer, refractometer for measuring refractive power of liquids, and lens-grinding machine. *(HIP/Art Resource, NY)*

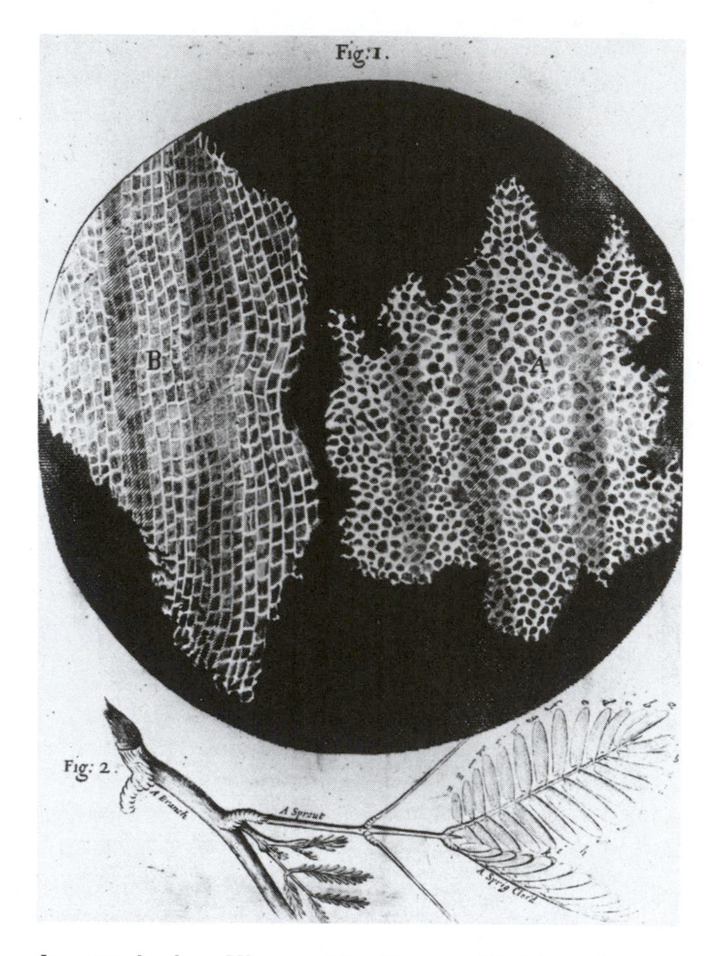

An engraving from *Micrographia* with a magnified view of cork tissue showing cellular structure and a cork tree branch *(National Library of Medicine)*

the crime. As information is obtained, investigators coordinate the information and use it to develop theories and suspects. Other experts may attempt crime scene reconstructions. Finally, if suspects are developed, additional tests and comparisons may be performed such as DNA TYPING of blood from suspects and comparison to blood found on the victim or at the scene. Thus, a major homicide case typically calls on many of the disciplines within forensic science.

See also MURDER.

Further Reading

Fisher, B. A. F. "Homicide Investigation." In *Techniques of Crime Scene Investigation.* 6th ed. Boca Raton, Fla.: CRC Press, 2000.

Hooke, Robert (1635–1703) English *Scientist* Robert Hooke, along with ROBERT BOYLE, was one of the most important early chemists because his contributions would prove crucial to the birth of modern FORENSIC SCIENCE. Among many other achievements, Hooke did significant development work with microscopes, the first forensic instrument. Hooke was more of a generalist than Boyle, having dabbled in mechanics, physics, and BIOLOGY, as well as CHEMISTRY. The microscope served him well in all areas.

Hooke built on the work of many earlier scientists to produce the first true compound microscope (one that included a lamp as a light source). In addition he published a book entitled *Micrographia* in 1665 and was the first to use the word *cell* to describe that fundamental biological unit. (Antoni van Leeuwenhoek had referred to them as "animalcules.") Hooke was also the first to describe sperm cells, a task that echoed in modern forensic laboratories where the microscope is still considered the only reliable method for identification of these cells in RAPE KITS and other SEXUAL ASSAULT evidence. From the middle of the 1600s on, the microscope's basic design has not changed, although some of the early ones are as much art as instrumentation.

Hoover, J. Edgar *See* FEDERAL BUREAU OF INVESTIGATION (FBI).

hot stage An accessory used with a microscope for the analysis of FIBERS and GLASS. In glass analysis, a piece of glass is immersed in oil with a known REFRACTIVE INDEX (RI). The hot stage gradually heats the oil, changing the refractive index while the analyst observes a feature known as the BECKE LINE. The Becke line is a halo of light that surrounds the particle but that vanishes when the refractive index of the liquid matches that of the glass. This method allows the examiner to determine the refractive index of the glass. For the analysis of fibers, the hot stage heats the fiber while the analyst observes the behavior and records observations such as fiber swelling, curling, contraction, and burning. The melting point of the fiber (if there is one that can be reached with the hot stage) is also useful in identifying the type of fiber, as is a study of changes in the optical properties of the fiber as heat is applied. Hot stages can control temperature very accurately and can increase it by tenths of a degree per minute. Hot stage analyses can also be automated to some extent, simplifying tasks such as determining the refractive index of many glass fragments.

human hair *See* HAIRS.

human leukocyte antigen (HLA) A GENETIC MARKER SYSTEM widely used in paternity testing and organ transplant. Although the subject of research in the 1980s, an effective and reliable method of typing HLA in BLOODSTAINS never emerged, and forensic applications were limited. With the advent of DNA TYPING, the need for developing such methods was all but eliminated. The term HLA is still seen, however, used in conjunction with DNA typing in the form of the HLA-DQA1 system. This was the first system targeted by PCR DNA typing techniques.

HLA are human leucocyte ANTIGENS, meaning that they are found associated with white blood cells (leucocytes) and in tissues. However, unlike other blood group and ISOENZYME systems, HLA is not found associated with the red blood cells (erythrocytes). The HLA system is part of the histocompatibility complex called MHC for major histocompatibility complex and consists of four different sites or loci on chromosome number 6. Since there are so many different factors that can be typed within the HLA system, it has a high DISCRIMINATION POWER, on the order of 90 percent. This means that successful typing can eliminate approximately 90 percent of all possible donors. However, the cost of typing combined with the technical challenge of typing HLA antigens in bloodstains prevented any significant application to forensic casework.

Further Reading

Melvin, J., et al. "Paternity Testing." In *Forensic Science Handbook*. Vol. 2. Edited by R. Saferstein. Englewood Cliffs, N.J.: Regents/Prentice Hall, 1988.

hydrocarbons *See* ACCELERANT.

hydrocodone A synthetic drug classified as an opiate and used for treatment of pain and severe coughing, much as codeine is. Depending on what the drug is combined with, it is found on Schedule II of the CONTROLLED SUBSTANCES ACT (CSA).

hypergeometric methods A probability-based approach used in forensic applications such as sample selection. Consider a case in which 10,000 unknown white pills are submitted to a forensic laboratory for testing. It may not be realistic or necessary to test all of these, as long as a representative and reliable sample is selected. Much of sample selection is dictated by legal requirements such as weights or purity, but in many cases, the key information required is an identification of drug or CONTROLLED SUBSTANCE that is present and in what quantites.

The hypergeometric method is appropriate in situations where samples are to be taken from a large population and not replaced. A common example of such a process would be drawing cards sequentially from a deck of cards, which is a finite population of samples, 52 in all. If 10 cards are drawn at random and not replaced in the deck, a hypergeometric distribution could be employed to calculate the probability of drawing all four queens as part of the 10 cards drawn. The hypergeometric concept is related to the BINOMIAL DISTRIBUTION used in forensic DNA analysis to calculate gene frequencies. It applies to cases where the outcome can be only one of two options.

Coin flips are often used as an example. To calculate the probability of getting three heads in a row, the individual probabilities are combined by multiplication as 0.5 × 0.5 × 0.5 = 0.125. This can be expressed as a ~13 percent chance that three coin flips will produce three heads. The probability of getting 10 in a row is much smaller, 0.001, which can be stated as 1 chance in 1024. In the case example above of 10,000 unknown white pills, selection of any one tablet can be considered an either/or option—either it will contain an illegal substance or it will not. The hypergeometric method allows an analyst to select a defensible number of samples based on probabilities derived from knowledge of a few samples.

hypothesis and the scientific method The procedures and framework in which science is practiced and the way in which forensic scientists approach their analyses. A hypothesis is often referred to as an educated guess, meaning that it is an idea offered to explain an observation or the result of an experiment. A hypothesis is not a theory or a natural law but rather a starting point by which scientists can work their way to a verified truth. As shown in the figure, the first step in the process is the collection of initial data either by experiments (empirical data) or by observation. This data is then analyzed, and, based on this, a hypothesis is offered. If no reasonable hypothesis can be made, then the cycle jumps back to the beginning for the collection of additional information. A key element of any hypothesis is that it must be testable. A specialized type of hypothesis often used in forensic science is what is called the null hypothesis. In this approach, the forensic analyst starts with a hypothesis that can be disproved. For example, if he or she is testing a blood sample, the null hypothesis might be that "this blood came from the suspect." It is stated as a hypothesis, and by using comparative DNA TYPING, the analyst will have the opportunity to disprove the null hypothesis.

During testing, a hypothesis may be modified and retested several times before it moves to the next stage, sometimes called a theory. However, many texts consider a hypothesis and a theory to be interchangeable, and the difference is not significant here. The theory is also subject to rigorous testing that attempts to disprove it. Like the earlier phases, several iterations and modifications can take place. What eventually emerges is the final answer, either a proven theory (or natural law) or, in the case of forensic analysis, a final answer that is the best that can be obtained with the evidence that was available. Using the previous example, the final result would be that either the blood sample did come from the suspect or it did not, within a reasonable level of scientific certainty. In reality, however, many times the best answer is more ambiguous and less specific, such as "this glass may have come from that window" or "this fiber is consistent with the carpet in that house." The forensic scientist uses the scientific method to obtain the best answer possible, but it may not be a definitive answer as far as the case is concerned.

Some critical characteristics of the scientific method are that it is objective (versus subjective), is based on experiment, observation, and fact, includes testable ideas, and is quantitative. Science is also fluid in that current theories are constantly open to study, revision, and review. It is also governed by a process called peer review in which scientific findings are reviewed by scientists knowledgeable in the field before publication. When findings are published, other researchers usually attempt to reproduce the experiments and confirm the

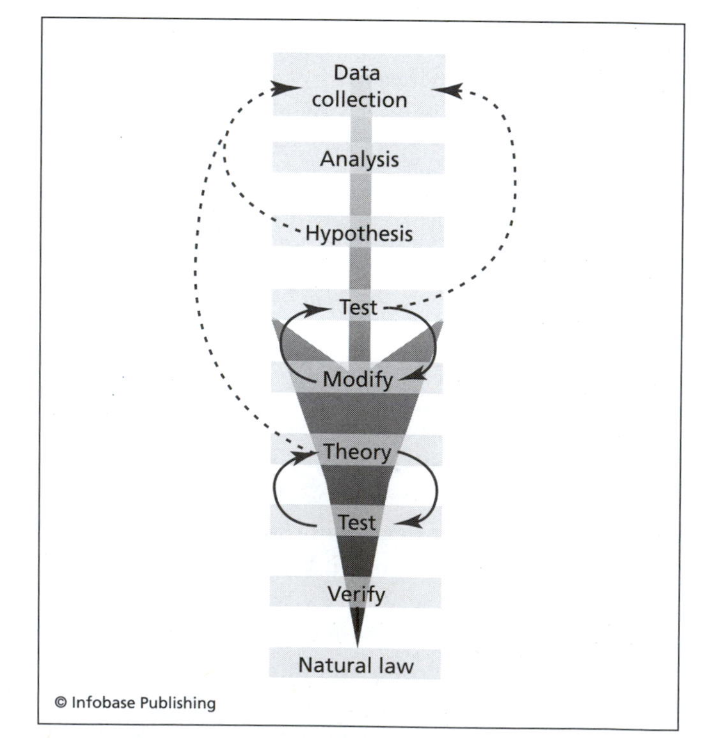

Overview of the process of the scientific method

findings. In this way, the scientific method is designed to be self-correcting and to constantly move in a forward direction, albeit at a sometimes methodically slow pace. In the context of a forensic analysis, the scientific method is used to ensure that the best possible results are obtained.

See also BAYESIAN STATISTICS.

Further Reading

Lee, J. A. *The Scientific Endeavor: A Primer on Scientific Principles and Practice.* San Francisco: Addison Wesley Longman, 2000.

hypothetical question A type of question that can be asked of a forensic scientist during courtroom testimony. Often the court needs to hear the opinion of the expert witness, but the need has to be tempered with the realization that the expert may not have all the information necessary to answer the question. For example, a fingerprint expert may have linked a latent print to a specific person, but that fact alone does not mean that the person was responsible for the crime. Asking for an opinion based on that identification could easily become a hypothetical question based on how it was phrased. Thus, the TRIER OF FACT must be very careful when considering whether to allow or disallow an expert to answer a hypothetical question. The court must decide if the information sought is relevant and if it relies too heavily on information the expert is not aware of.

Further Reading

Siegel, J. "The Forensic Expert Witness in the Courtroom: History and Development." In *More Chemistry and Crime, From the Marsh Test to DNA Profile.* Edited by S. M. Gerber and R. Saferstein. Washington, D.C.: American Chemical Society, 1997.

IABPA (International Association of Bloodstain Pattern Analysis) *See* BLOODSTAIN PATTERNS.

IAFIS (Integrated Automated Fingerprint Identification System) *See* AUTOMATIC FINGERPRINT IDENTIFICATION SYSTEMS (AFIS).

IBIS system An automated system used to assist in the analysis of BULLETS and CARTRIDGE CASES. IBIS was developed by the BUREAU OF ALCOHOL, TOBACCO, FIREARMS, AND EXPLOSIVES (ATF) along with Forensic Technology, a Canadian company. IBIS equipment is supplied to participating laboratories through ATF's National Integrated Ballistic Information Network (NIBIN) program. Much like automated fingerprint database software, IBIS is designed as a screening tool but does not replace the firearms examiner for making final identifications. IBIS has two computational components, one for data acquisition and one for analysis, and an analysis produces a list of candidate matches from the central databases. IBIS has been particularly valuable in linking seemingly unrelated crimes by showing that the same weapon was used in both or all instances.

See also BALLISTIC FINGERPRINTING; D.C. SNIPERS.

Further Reading

Bureau of Alcohol, Tobacco, and Firearms. "ATF's NIBIN Program." Available online. URL: www.nibin.gov. Downloaded May 4, 2007.

Geradts, Z. "Use of Computers in Forensic Science." In *Forensic Science: An Introduction to Scientific and Investigative Techniques*. 2nd edition. Edited by S. H. James and J. J. Nordby. Boca Raton, Fla.: CRC Press, 2005.

Tontarski, R., and R. Thompson. "Automated Firearms Evidence Comparison: A Forensic Tool for Firearms Identification—An Update." *Journal of Forensic Science* 43, no. 3 (1998): 641.

ICP techniques Instrumental techniques used for inorganic and ELEMENTAL ANALYSIS. ICP stands for *i*nductively *c*oupled *p*lasma and refers to the method used to convert a sample to its constituent atoms or ions. In an ICP torch, gaseous argon (Ar) is ionized by a Tesla coil to form Ar^+ and free electrons. These ions are accelerated and confined in a stable magnetic field. The resulting high energy collisions generate tremendous heat in the range of ~18,500°F (10,300°C). Under these extreme conditions, most chemical compounds are broken apart, forming free atoms and ions in this plasma "flame," although it is not fire in the same sense as combustion. Plasma is considered to be a separate state of matter in addition to solid, liquid, and gas. It consists of free electrons (electrons not associated with a specific atom), atoms, and ions and is characterized by an intense glow reminiscent of a flame. However, plasmas emit harmful ultraviolet radiation, so the user or viewer only sees the plasma through a thick layer of protective glass.

There are two ICP instrumental techniques used in forensic science. ICP-AES is a form of EMISSION SPECTROSCOPY (*a*tomic *e*mission *s*pectroscopy) in which the heat of the plasma is not only sufficient to atomize the sample but also to place many of the atoms

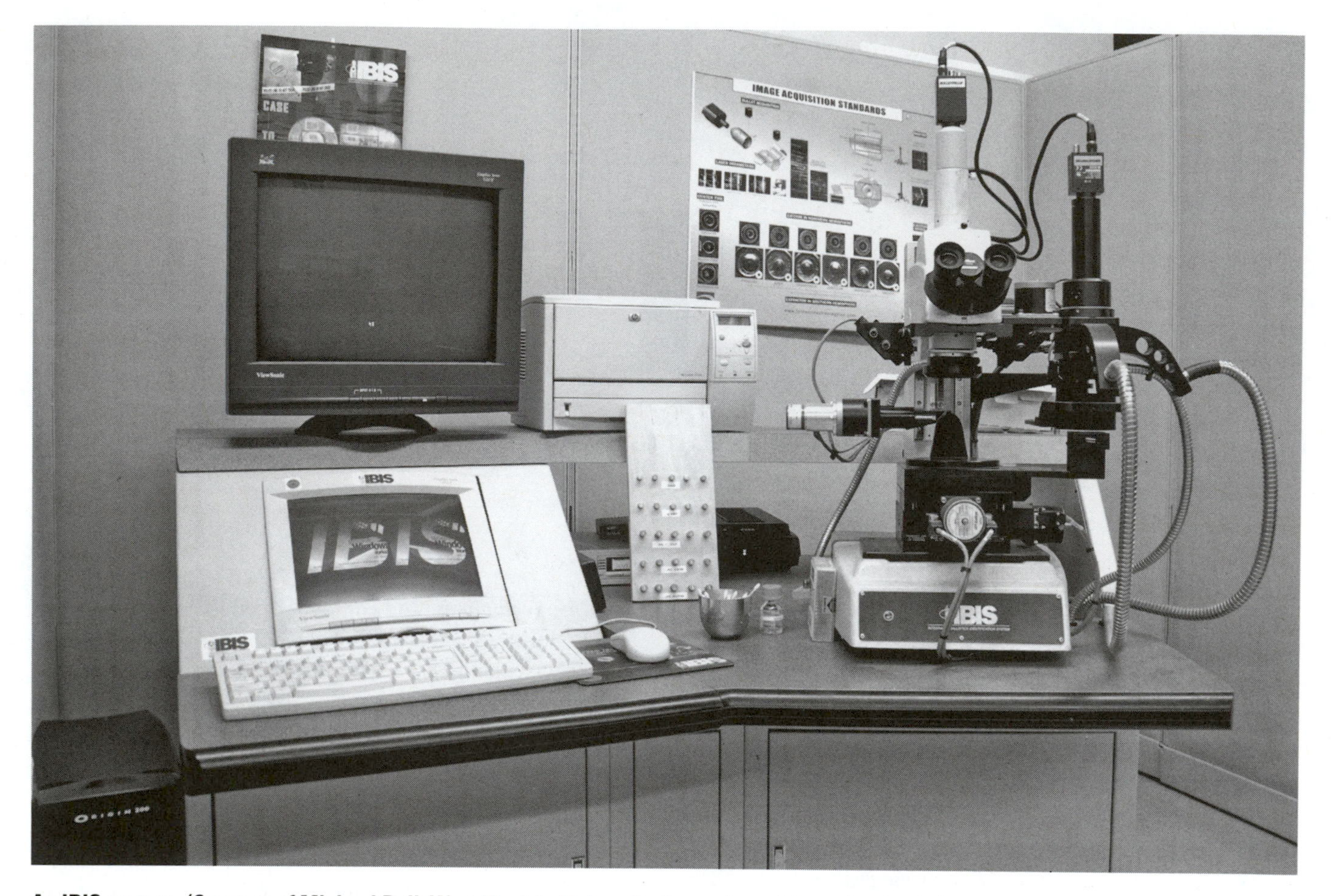

An IBIS system ***(Courtesy of Michael Bell, West Virginia University, Forensic and Investigative Sciences program)***

into the "excited state." In such a state, an atom emits characteristic wavelengths of light in the visible and ultraviolet ranges that can be used to identify specific elements in the sample and also to determine their concentrations. ICP-AES has many advantages over competing techniques such as ATOMIC ABSORPTION (AA) for elemental analysis, increased sensitivity, and the ability to analyze for several elements simultaneously. Like AA, ICP-AES has been used for the characterization of glass, gunshot residue, toxicology (heavy metal poisons such as arsenic), and soils. The disadvantage of the technique includes increased cost and complexity and potentially complicated sample preparation.

The second forensic application of ICP is in MASS SPECTROMETRY (ICP-MS) in which the plasma is the source of elemental ions. These ions are directed into the mass spectrometer where they are separated, detected, and can be quantitated based on their masses. ICP-MS can detect isotopes of elements (such as carbon-14, an isotope used in archaeological dating) and also multiple elements simultaneously. The instrument can also be used to analyze solids directly, avoiding the need for complex acid digestions and sample preparation. In a technique called laser ablation, lasers are used to vaporize small portions of a sample surface, and the vapors are then directed into the plasma.

identification In the analysis of physical evidence, determination of the chemical substance or substances of a material. Unfortunately, this term is often misunderstood or misused in the forensic context. For example, a statement such as "the blood was identified," meaning that it was shown to come from a specific person, is technically incorrect. What was meant is that the stain was INDIVIDUALIZED or proven to come from a COMMON SOURCE. Similarly, saying that "the fingerprints were identified" is also incorrect. The difference between identification, classification, and individualization as defined in forensic science is illustrated in the figure. Here, the physical evidence is a reddish stain found at a crime scene. Presumptive and confirmatory tests that react with hemoglobin allow the stain to be *identified* as blood. The stain is further analyzed using more specific and sophisticated tests that allow the blood to be *classified* as human, Type A. However, to determine if the stain originated from a specific person, it is necessary to perform DNA TYPING not only on the stain but also on a sample obtained from the person who is thought to be the common source. This is the key distinction between identification and individualization. To individualize evidence, a comparison is always required, but comparison is not needed for identification. Quite often, identification is required before classification and individualization, as shown here. It makes no sense to classify or perform DNA typing on a ketchup stain, so the identification of the stain as blood is part of the overall forensic analysis.

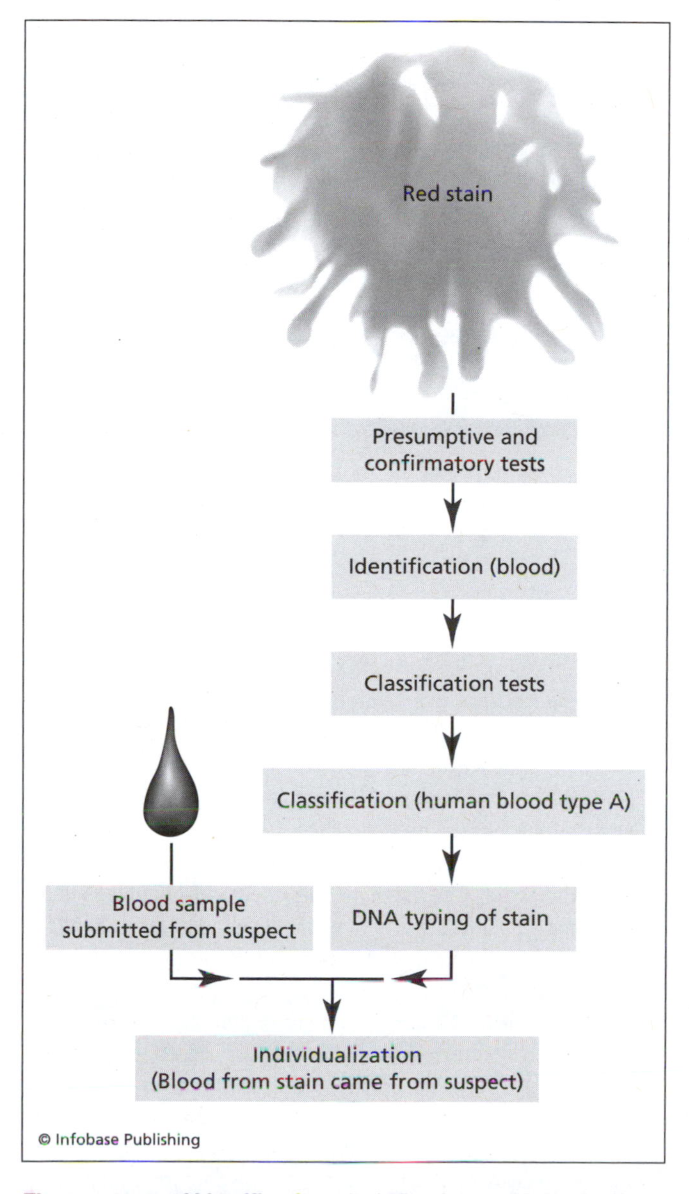

The processes of identification, classification, and individualization as applied to a red stain suspected of being blood

Identification typically requires some combination of visual examination and chemical testing, and the clearest example of identification in forensic science is seen in the area of drug analysis. Using a collection of different tests, the forensic chemist is usually able to identify what a suspected powder is and/or exactly what it contains. For example, if a white powder is submitted as evidence, the chemist can perform presumptive tests along with a microscopic examination and instrumental tests to determine that the powder contains cocaine, caffeine, and wheat flour. This is identification, not individualization. However, both of these elements could be part of a drug analysis under some circumstances. Investigators might have collected drug samples from different parts of the city and may need to know if they came from the same large batch. Similarly, during the analysis of the powder, the chemist may perform simple tests to determine what classification a drug falls into. Thus, identification, individualization, and classification are separate but closely related ideas in the context of forensic science.

Further Reading

Inman, K., and N. Rudin. "Classification, Identification, and Individualization—Inference of Source." In *Principles and Practices of Criminalistics, The Profession of Forensic Science*. Boca Raton, Fla.: CRC Press, 2001.

IED An abbreviation for improvised explosive device. The war in Iraq brought the term to the public's attention, but IEDs are nothing new. An IED is an explosive device created outside typical commercial or military manufacturing processes. IEDs may contain military explosives such as old artillery shells or plastic explosives, but the devices are assembled by methods that lie outside conventional manufacturing. A pipe bomb is an IED, as is a letter bomb.

See also EXPLOSIVES.

illumination The way an object is lighted, an important consideration in forensic microscopy and photography. In microscopy, there are two basic types of illumination, incident and transmitted. Incident radiation is the light that is reflected off an opaque object such as a bullet or paint chip. Transmitted light, as the

Identification of the Dead: Forensic Science and Mass Disasters

For many Americans, the year 2005 is associated with one word—*Katrina,* and the year 2001 with one day—September 11. Between came the D.C. snipers, a Christmas tsunami in South Asia (2004), and the London and Madrid terrorist bombings (2005 and 2004 respectively). A year later came one near miss—a disrupted plot to blow up airliners over the Atlantic in late 2006, near the five-year anniversary of September 11. The plotters planned to smuggle the ingredients of EXPLOSIVES aboard planes and synthesize and detonate them en route. For the forensic community, these dates were mileposts rather than turning points. These crimes and disasters culminated decades of slow evolution of microtragedy that was always the forensic scientist's stock and trade. Before September 11, a mass disaster was a plane crash or railroad accident with casualties in the hundreds. Agencies and first responders prepared for worst-case scenarios involving 500 deaths. After the attacks, CRIME SCENES were measured in miles and tons of debris; casualties were numbered in the thousands or tens of thousands. The public face of FORENSIC SCIENCE became less associated with CRIMES and EVIDENCE and more closely associated with recovery and identification of the dead. Forensic scientists did what they have always done. In the new century, they have emphasized scaling up over fundamental change.

People flee the World Trade Center in New York, September 11, 2001. ***(AP/Wide World Photos)***

Scale and Scope

September 11, 2001, was a day of cheerless milestones. It marked the largest deployment to date of the Federal Emergency Management Agency's (FEMA) Disaster Mortuary Response Team (DMORT), to the largest single crime scene ever created. Thousands of pieces of scattered and shattered human remains awaited recovery and identification. Multiple federal, state, and local agencies learned on the fly how to coordinate efforts on an unprecedented scale. A few weeks after the air assault, a still unidentified person sent ANTHRAX bacteria through the mail killing five, infecting 17, and forcing 30,000 people to take antibiotics. The dispersed but deadly trail of evidence stretched up the East Coast from Florida to Connecticut. In less than two months, these two terrorist incidents upended the very definition of the term *crime scene.* What was once limited to a room, house, or a body dumpsite now crossed state boundaries. Still, the dead held center stage.

The National Funeral Directors Association conceived DMORT in the 1980s. At the time, there were no uniform processes for handling mass disasters, nor was there a mechanism for local authorities to request assistance. The beginnings of DMORT were portable morgue units and teams of professionals trained to manage mass casualty incidents. The identification tools available to DMORT include visual identification of a body by relatives, use of personal effects like driver's licenses, tattoos and other markings, comparison of antemortem (before death) and postmortem X-rays, dental work, FINGERPRINTS, and DNA analysis. The combination of techniques proved invaluable when what little remained of a person or people was damaged, separated, burned, commingled with others, and unrecognizable as human, let alone as something that was once living.

The DMORT AUTOPSY emphasizes identification rather than cause of death, since the latter is usually obvious. The first DMORT was deployed in 1993, and the system now operates under the National Disaster Medical System (NDMS), a part of the Federal Emergency Management Agency (FEMA). Each team consists of doctors, dentists, X-ray technicians, anthropologists, and support staff, and each team brings with it a complete portable morgue, including refrigerated storage units. Four DMORT teams responded to New York City on September 11, the largest deployment in its history until August 2005, when all 10 DMORT teams responded to hurricane Katrina.

Identification of remains from the September 11 attacks in New York and Washington relied principally on DNA technology because the remains were fragmented, badly damaged, and removed from any useful context such as an airplane seat number. The obliterated planes fused with the structures, and remains of passengers mingled with those of building occupants. The collapse of the towers destroyed any sense of relative location (floor of the building or office number) that could have been helpful in associating a bone chip or tissue fragment with a small pool of likely victims. Rescue efforts yielded 287 intact bodies out of nearly 3,000 victims. Four years after the attacks, recovery efforts yielded approximately 20,000 separate samples from the site, which have been documented, cataloged, stored, and analyzed to the extent current technology allows. This number does not include comparison samples collected from relatives and family members. As of November 2005, DNA techniques have successfully identified 850 victims from the World Trade Center attacks out of the 2,749 listed as missing. When combined with other means such as fingerprints and dental records, nearly 1,600 individuals have been identified.

DNA typing methods, similar or identical to those used in forensic labs, produced the majority of the identifications. Some new technology and software emerged to address problems associated with damaged and degraded samples. The greatest challenge to forensic scientists was not scientific; it was organizational—developing a system to catalog each item, labeling and storing it, keeping track of results as they became available, and linking similar samples. All samples, linked or individual, had to be linked to one of the missing to provide conclusive identification. The comparison samples came from families and included toothbrushes and hairbrushes used by the victims as well as cheek swab samples obtained from relatives. Unlike most forensic cases, many of the matches were made indirectly using DNA types of relatives rather than by simple comparison of

(continues)

A corpse tied to a tree in floodwaters from Hurricane Katrina, New Orleans, September 15, 2005 ***(AP/Wide World Photos)***

Identification of the Dead: Forensic Science and Mass Disasters *(continued)*

a known and questioned sample. Development of this NEXT-OF-KIN SOFTWARE was indispensable to the identification efforts. Collection kits and methods developed in the wake of September 11 were distributed to other agencies and countries, as well as the lessons already applied in the 2004 tsunami and the 2005 hurricane.

The 972 deaths so far associated with Katrina presented a different identification challenge. Most bodies were intact and found in context such as in a home or hospital bed. Trauma from fire and collapse was absent, but severe decomposition arising from prolonged immersion in warm, swampy, and animal-infested water was typical. Much to the surprise of the forensic community and state officials, knowledge gained from September 11 was of limited use in the Gulf Coast. Despite a smaller death toll and intact bodies, identifications were proving to be problematic.

In the immediate aftermath of the storm, efforts focused on rescue before turning to recovery. In the hot and humid conditions, this delay allowed rapid decomposition that quickly made bodies unrecognizable, even to the victim's closest family and friends. A week or so in the water causes skin shedding and the loss of useable fingerprints. The storm's fury exhumed the long-buried dead, adding several hundred additional bodies to those killed by the hurricane. Soon after the storm, when dire predictions of thousands of deaths were the norm, the Federal Emergency Management Agency (FEMA), part of the new DEPARTMENT OF HOMELAND SECURITY formed in the aftermath of September 11, began planning construction and deployment of a morgue facility that could handle 150 bodies a week.

Utilization of postmortem records such as dental X-rays for identification requires the existence of antemortem data. Unfortunately, the generalized destruction destroyed dental and medical offices and storage sites as well as contaminated sample sources such as toothbrushes and hairbrushes that were so useful after September 11. Many victims were poor and had few medical or dental records to begin with, adding to the challenge. With personal possessions destroyed and relatives dispersed, collection of comparison samples was difficult.

The scope of the disaster magnified problems. On September 11, the three attack sites were each located within a single local jurisdictional area. The hurricane, however, struck across city, county, and state lines. Death investigation in most areas is the responsibility of local CORONERS rather than forensic pathologists and MEDICAL EXAMINERS, the latter often being too expensive for poorer jurisdictions.

Victims of the 2004 tsunami being recovered in Thailand's Phi Phi Island, December 28, 2004 ***(Luis Ascui/Reuters/Landov)***

Whether this had a direct effect on the recovery and identification process is not yet clear, but certainly it did not help. A few weeks after the storms, the projected death toll fell from the estimated tens of thousands to 1,000 to 2,000, and much of the initial work was completed. By February 2006, more than 900 bodies had been identified, leaving fewer than 100 still in need of identification. Most of the autopsies were completed before the FEMA morgue opened. The facility closed in February 2006, shortly after the last body was recovered.

As bad as Hurricane Katrina was, the worst mass disaster of the young century occurred in December 2004, when a massive tsunami inundated thousands of miles of southeastern Asia. Estimates of the dead range from 150,000 to nearly 220,000, but the final toll remains elusive. The disaster struck isolated islands and coastlines in an area far poorer and less developed than the U.S. Gulf Coast. The wave devastated tourist areas, killing hundreds of Europeans, Americans, and other visitors. The wave swept thousands irretrievably out to sea, never to return. Others washed ashore a mile away or in different countries. Survivors buried countless bodies to stave off disease. As would occur with Katrina, heat and humidity accelerated decomposition, and there was little if any access to refrigerated storage areas.

Because so many foreigners died in tourist areas, the government of Thailand coordinated with Interpol to form the Thai Tsunami Victim Identification (TTVI) operation. Forensic professionals from Europe, Canada, the United States, and other countries arrived within days of the disaster and began to assist in recovery and identification. Eventually, forensic practitioners from 29 countries were involved in the tsunami identification efforts. Compared to identification work after September 11, DNA was not used extensively. As of late 2005, dental identification accounted for about 48 percent of the identifications, fingerprints 34 percent, and DNA about 17 percent%. Approximately 3,000 victims had been identified, leaving about 750 unidentified. Tens of thousands remain missing.

Bracketed by September 11 and Katrina, Europe endured two bombing attacks, one in Madrid, Spain, on March 11, 2004, and one in London on July 7, 2005. The attack in Madrid targeted four commuter trains and killed 191 people. Two thousand were injured. The bombs used stolen commercial explosives detonated by cell phones, and the perpetrators blew themselves up to avoid capture. The London attack targeted the Underground subway system and involved three bombs that went off deep in the tunnels within less than a minute. A fourth bomb went off nearly an hour later, perhaps by mistake or by mishandling. In contrast to Madrid, the London bombings were suicide attacks. Some of the recovery workers who endured the heat of the Underground tunnels were veterans of tsunami recovery efforts. The Madrid attacks spawned an international controversy over a mistaken fingerprint match and wrongful arrest of a Muslim lawyer in Portland, Oregon, (*see* MAYFIELD CASE).

Recovery and identification of the dead has become the most visible forensic response and responsibility. Also new are the scope of these events and the resulting globalization of forensic science. Teams moved from New York to Madrid to Thailand to London, carrying new knowledge and hard-learned lessons to each new site only to find that

(continues)

Rescue workers cover up bodies at bombed train in Madrid, Spain, March 11, 2004. ***(AP/Wide World Photos)***

Aftermath of the July 2005 bombings in London ***(AP/Wide World Photos)***

what worked before was not entirely suitable for what was happening now. The definition of a crime scene, both the crime and scene elements, has become fuzzy and indistinct. Was September 11 an act of war or a crime? From the forensic perspective, the line has become vague and irrelevant. New Orleans and South Asia were not crime scenes, but they were definitely forensic scenes. In the new century, crime is no longer a forensic prerequisite.

name implies, travels through a transparent or semitransparent object such as a hair or fiber. Occasionally, both kinds of lighting are used in the study of physical evidence.

Four important characteristics of illumination are intensity, color, temperature, and angle. The intensity refers to how bright or strong the light is, while the angle refers to the angle at which the light intersects the surface being observed. The appearance of features, particularly three-dimensional features, can be altered as the angle of illumination (also called angle of incidence) is changed. The temperature of light does not refer directly to heat, but rather to the theoretical concept of an ideal black body emitter. A black body will emit characteristic wavelengths of light depending on what temperature it is heated to. At a temperature of about 10,292°F (5,700°C), a black body will emit light equivalent to natural daylight, which is a mixture of all colors of the rainbow at equal intensities. The color of light refers to the temperature that such a black body emitter would have to be to emit the same mixture of colors and intensities as the light source in question. The temperature is thus considered to be a measure of the "whiteness" of light.

In forensic photography, the same types of light are important (reflected and transmitted), and the angle of incidence is likewise critical. Traditional photographs that record patterns of visible light can be supplemented by pictures taken using ultraviolet light as well as infrared light. Photography can either be taken with the light present (called the available light) or supplemented by artificial lighting, which can make some features easier to see. Typically, to avoid problems that would be presented by any one illumination scheme, multiple photos are taken using different types of lighting and different angles.

Further Reading

DeForest, P. R. "Foundations of Forensic Microscopy." In *Forensic Science Handbook*. Vol. 1. 2d ed. Edited by R. Saferstein. Upper Saddle River, N.J.: Prentice Hall, 2002.

Nomenclature Staff, Royal Microscopy Society. *RMS Dictionary of Light Microscopy.* Royal Microscopy Society—Microscopy Handbook # 15. Oxford: Oxford University Press, 1989.

image enhancement/digital image enhancement Techniques that are becoming increasingly useful and important in the areas of questioned documents, fingerprints, and photographic identification. Images are obtained either from a traditional camera (or video camera) using film or from a digital camera. These cameras may be mounted on special devices such as microscopes. The images are digitized and then manipulated using specialized image enhancement software, usually with the goal of highlighting or making more visible the portions of the image that are of interest. This is often referred to as minimizing or removing the "noise" in the image. Care must be taken to document the steps used and to ensure that the image is merely enhanced, not materially altered to change the underlying feature. Image enhancement techniques are used in fingerprint identification, bite analysis, firearms and toolmarks, and facial identification, to name just a few applications.

Further Reading

Gerhardts, Z. "Use of Computers in Forensic Science." In *Forensic Science: An Introduction to Scientific and Investigative Techniques*. 2nd edition. Edited by S. H. James and J. J. Nordby. Boca Raton, Fla.: CRC Press, 2005.

immunoassay A group of techniques used in forensic TOXICOLOGY for the detection of drugs in urine, blood, and other body fluids. The term *immunosorbent assay* is also used, and these techniques yield both QUANTITATIVE and QUALITATIVE information. Several variations exist, based on the general technique shown in the figure. Immunoassay relies on an ANTIGEN-ANTIBODY reaction between the drug being tested and an antibody specific for it. The antibody is attached to a solid surface such as the bottom of a plastic or glass well. A complex that consists of the drug and a label is added, and the reaction occurs. As a result, the labeled drug is bound to the antibody. A sample that may contain the drug, such as urine, is added to the plastic well. If there is no drug or very little drug present, the labeled drug–antibody complex will remain undisturbed. However, if there is a large concentration

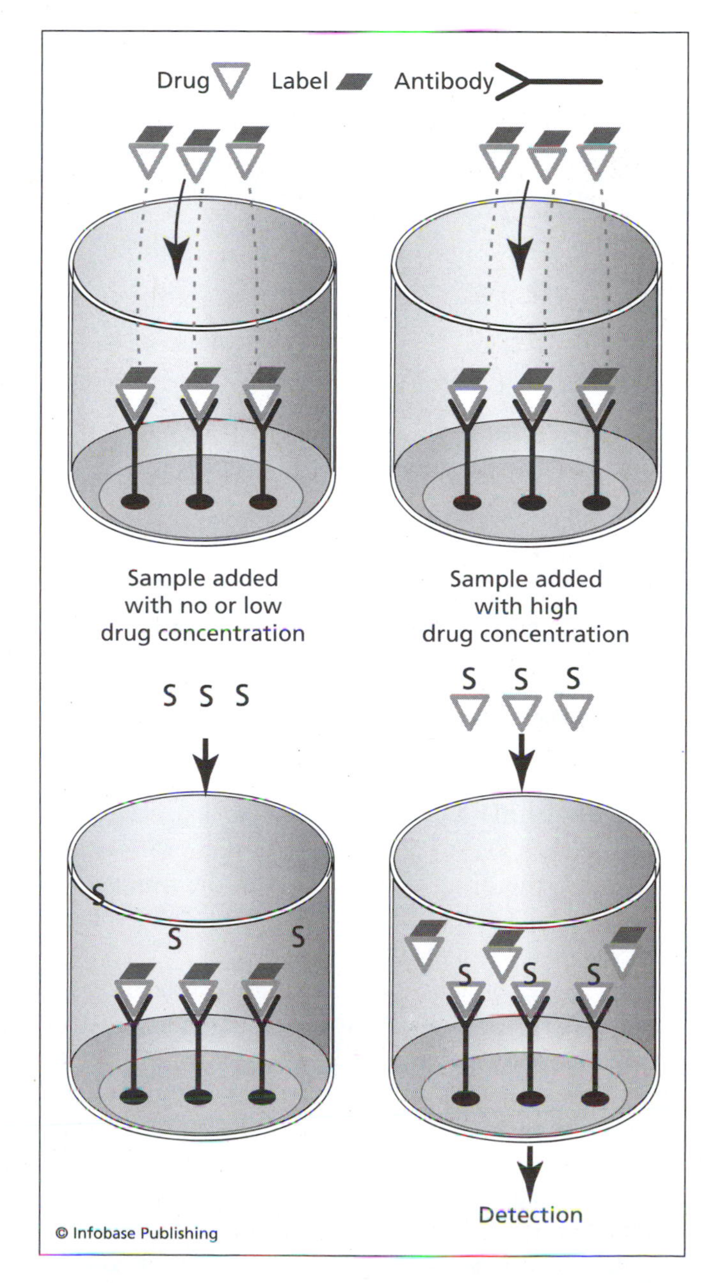

Overview of an immunoassay analysis. Refer to the text for details.

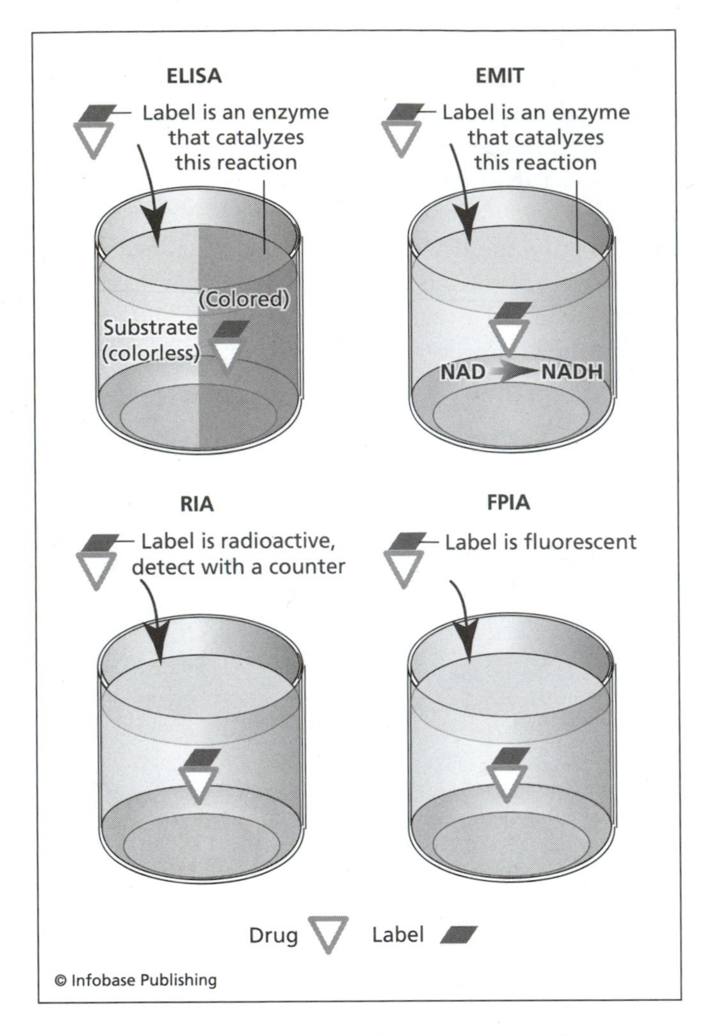

Different detection techniques used in immunoassay. The labeled drug or other target analyte is displaced and detected based on the appearance of color, catalytic action, radiochemical methods, or fluorescence.

of the drug, this will displace the labeled drug from the antibodies, releasing the labeled drug into solution. The higher the drug concentration in the sample, the more will be displaced. The amount of the displaced labeled drug is then measured.

The type of label used determines the type of immunoassay technique, and four common variations are illustrated in the second figure. In ELISA (enzyme linked immunoassay), a chemical compound called a substrate is added to the solution in the well. The label is an enzyme that catalyzes a reaction in which the substrate is changed, forming a colored solution. The deeper the color, the greater the concentration of the enzyme label present, which in turn implies a greater concentration of drug in the original sample. The EMIT (enzyme multiplied immunoassay technique) is similar; the label catalyzes a common biological reaction, the conversion of NAD (nicotinamide adenine dinucleotide) to NADH. The more NADH detected, the greater the concentration of drug in the sample. Radioimmunoassay (RIA) uses a radioactive label that can be detected by counting equipment. Finally, FPIA (fluorescent polarization immunoassay) uses a fluorescent label that interacts with POLARIZED LIGHT in different ways depending on whether it is bound to the antibody or free in solution.

immunodiffusion IMMUNOLOGICAL TECHNIQUES used in forensic SEROLOGY to determine the species of a blood sample. Immunodiffusion is a precipitin test,

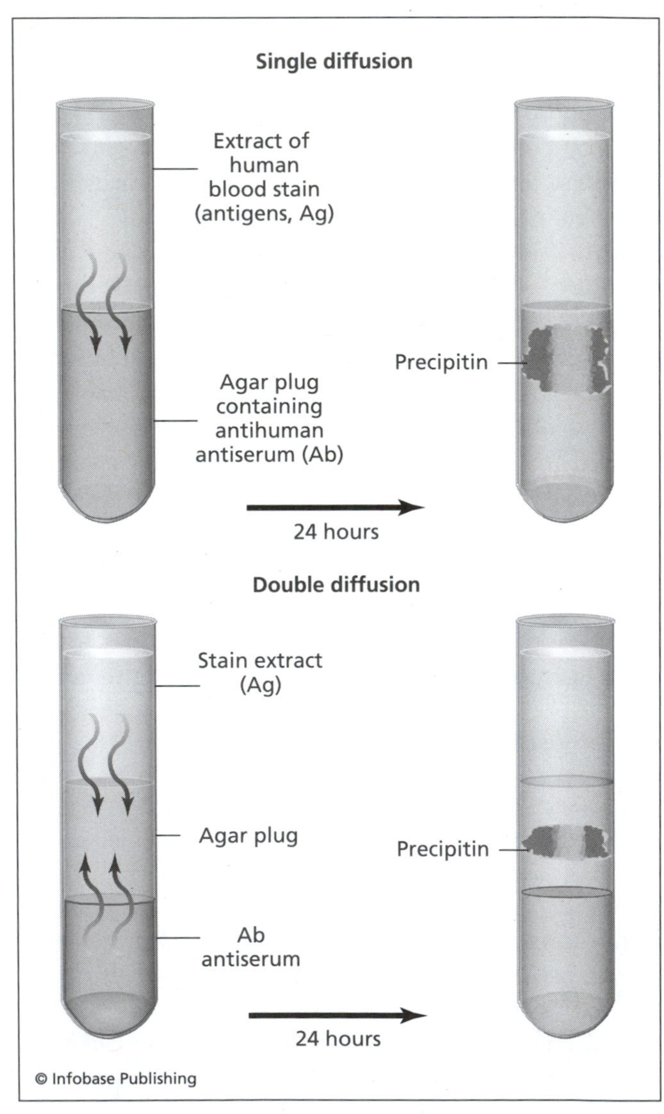

One-, or single-, dimension immunodiffusion. Appearance of the precipitin is a positive result.

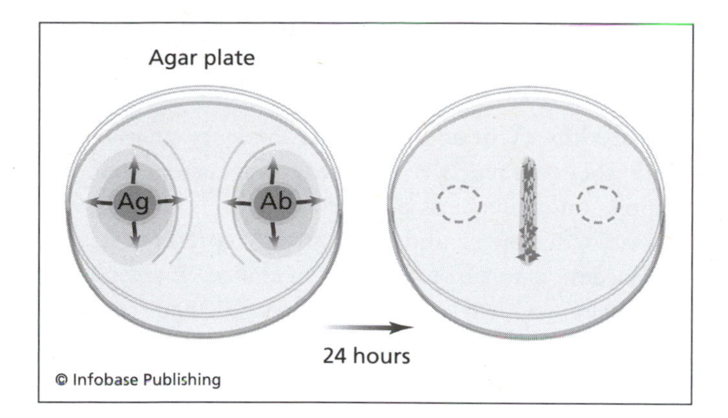

Two-, or double-, dimension immunodiffusion

meaning that a positive result is evidenced by the formation of a precipitate or solid (also called an immunoprecipitate) that can be easily seen. In this case, the solid forms as a result of an immunological reaction between ANTIGENS found in the blood sample and ANTIBODIES found in purified antisera applied in the test. Immunodiffusion tests rely on the process of diffusion, the natural spreading of a concentrated material or reagent into the surroundings. This phenomenon is seen when a drop of food coloring is placed in a glass of still water. The concentrated coloring diffuses throughout the glass until it reaches a homogeneous dilute concentration. One of the advantages of immunodiffusion tests is that a positive result confirms any PRESUMPTIVE TESTS that were performed to determine if a stain in question was blood. Thus, a positive result proves not only that a stain is blood but also shows what species it came from.

The first figure illustrates some common immunodiffusion techniques. The names reflect how the antigens or antibodies move through a gel medium. In one-dimensional techniques, the antigen or antibody moves only one way. Similarly, single diffusion means that only one component moves, whereas in double diffusion both the antigen and antibodies are mobile. In one dimension single diffusion, an extract from a stain containing the antigens (Ag) is layered on top of a gel layer in which the antibodies (Ab) are immobilized. The molecular structure of the gel allows water and solutions, such as those containing antigens and antibodies, to slowly permeate through it by the process of diffusion. In this case, the antigens in the stain diffuse into the gel, and a precipitin will form if the antibodies are specific to the species of animal from which the blood came. A variation of this is single dimension double diffusion, in which antiserum is placed in a small test tube. The gel is layered on top of that with nothing added to it. On top of the set gel, a small amount of the stain extract is placed, and the samples are allowed to sit for a period of about 24 hours. Antigens diffuse downward into the gel, while antibodies diffuse upward (double diffusion), eventually encountering each other in the gel. A precipitin band forming in the gel constitutes a positive result.

In most cases, the goal of immunodiffusion is to determine if a blood is human or not. Thus, the test tube will contain antihuman antiserum, and a positive reaction shows that the blood is human blood. Multiple tubes can be created to test the stain against other species such as cat, dog, and horse. In double diffusion (also called an Ouchterlony test), such multiple tests can be performed in a shallow Petri dish filled with agar. Small plugs are cut out to make wells into which the antiserum and stain extracts are placed. The diffusion occurs in a halo (radial diffusion) around the well moving out in all directions, with the same outcome as in single diffusion. A visible precipitin band indicates a positive reaction, while the lack of precipitin indicates that the antiserum and the antibody are not specific for each other.

In practice, stain extracts are usually tested against several antisera, not just one. For example, if blood is found in a home where pets have been kept, the stain extract might be tested against antidog and anticat sera. In wildlife forensics, the choice of antisera may be even broader, such as antielk and antibear. The third figure shows how a multiple double diffusion test might be set up to determine if a blood stain originated from a dog, horse, or cat. In this example, the stain was cat blood. The patterns of the precipitin lines can also be a rich source of information. The antigen molecules differ in

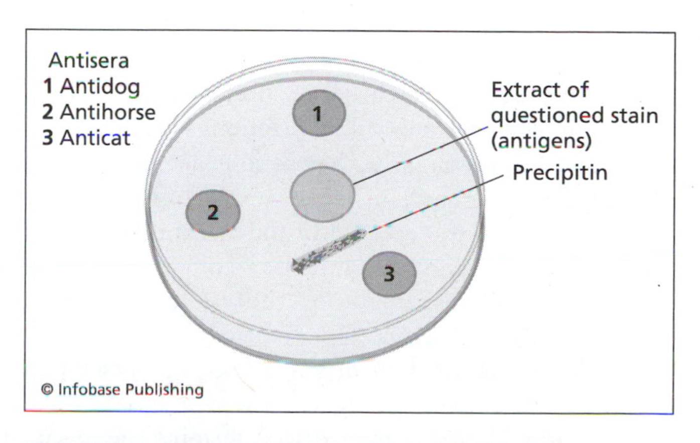

Method for species testing using immunodiffusion. A positive result is seen for cat blood.

structure from the antibody molecules, so they will diffuse at different rates through the gel. If the difference is significant, the relative diffusion rates can influence the appearance of the precipitin bands. Difference in the relative concentrations of antigens in the stain and the antibody in the serum can further complicate the patterns. Finally, the antisera of closely related species may interfere (cross-react) with each other. While not usually an issue in most cases, in wildlife forensics this could be a pivotal concern. For example, the ability to distinguish blood from a dog, wolf, or coyote might be crucial, but since these species are so closely related, some cross reactivity could be expected.

immunoelectrophoresis IMMUNOLOGICAL TECHNIQUES used in forensic serology to determine the species of origin of a blood sample or blood stain. These techniques combine ELECTROPHORESIS with the process of IMMUNODIFFUSION to detect ANTIGENS in a stain that can react with ANTIBODIES in an antiserum to produce a visible precipitate called a precipitin. Techniques that have been applied to forensic work include Rocket immunoelectrophoresis (the Laurell Technique), two-dimensional immunoelectrophoresis, and crossed-over electrophoresis. The latter technique works by placing antigens and antibodies opposite each other in small wells punched in a gel medium. An electrical field is applied, and the antigens and antibodies move toward each other, meeting at some location between the two wells. If the antiserum used is specific to the antigens found in the stain, a visible precipitin band, which is a milky white color, will form in the gel. For example, if one well is filled with the extract of bloodstain and the other well is filled with antihuman antiserum, a precipitin band will form if the blood is human blood. If the stain is dog blood, for example, no precipitin line will form.

immunological techniques (immunology) Tests and analyses that are based on reactions between ANTIGENS and ANTIBODIES. Such reactions are the basis of the immune response, in which an organism synthesizes antibodies to react with and neutralize foreign substances. If a person contracts a cold, the immune response includes synthesis of antibodies that attack the invading cold virus and eventually overcome it. Examples of forensic uses of immunological techniques include ABO BLOOD GROUP SYSTEM typing, IMMUNODIFFUSION and precipitin tests to determine the species of bloodstains, and IMMUNOASSAY techniques used in forensic TOXICOLOGY.

impression evidence Physical evidence that results from the contact between two objects or surfaces. Impression evidence is also referred to as imprint evidence or markings and as a group represents one of the largest classes of forensic evidence. Examples of impression evidence include markings made on cartridge cases and BULLETS; BITE MARKS; TOOLMARKS; FABRIC IMPRESSIONS; SHOE PRINTS; TIRE PRINTS; and in many cases, FINGERPRINTS. Impression evidence can be divided into groups depending on how it was made. Scraping produces impressions called STRIATIONS, examples of which include bullet markings and many toolmarks. For example, when a screwdriver scrapes across a metal surface, the scratching that results contains striations. In many cases, striation patterns must be studied under a microscope, as they are for bullet comparisons using a COMPARISON MICROSCOPE. The other way impressions can be made is by compression of one surface under the weight of another, such as when shoe impressions are made in mud or a fabric impression is transferred to a bullet passing through clothing. In general, an imprint is considered to be a flat impression (thin, as a fingerprint on glass), whereas an indentation also has depth, such as the muddy shoeprint. In other words, imprints are two dimensional, while indentations are three dimensional. In any case, the original is preferred for laboratory analysis,

The treads on tires can leave distinctive impressions in soft surfaces. ***(Courtesy of Michael Bell, West Virginia University, Forensic and Investigative Sciences program)***

and thus, if possible, the impression and the substrate are transported to the laboratory intact.

If it is not possible to transport the original impression, other alternatives are employed. Two-dimensional impressions can be recovered using lifting techniques such as TAPE LIFTS, coupled to chemical or other forms of visualization. Electrostatic lifting is another technique that is particularly useful for recovering dry impressions such as dusty shoe prints on floors. A film is placed over the impression, and a charge of static electricity is applied that attracts the dust to the film, in effect transferring the original to the film. For indentations, CASTING is the preferred method of studying and preserving the evidence if the original cannot be preserved and transported to the laboratory. Dental stone, plaster, Snow Print Wax, and sulfur have all been used to create casts of impression evidence. Casting is not routinely used for striation impressions.

Further Reading

Bodziak, W. J. "Lifting Two-Dimensional Footwear Impressions." In *Footwear Impression Evidence.* New York: Elsevier, 1990.

improvised explosive device *See* IED.

incendiary devices Devices that are used to ignite an ACCELERANT in an arson fire. They can range from simple to sophisticated and include items such as highway flares, fireworks, cigarettes, matches, black or smokeless powders (GUNPOWDER), and homemade mixtures such as sugar and chlorates. As in a combustion reaction or explosion, an incendiary device requires a fuel and an oxidant, be that oxygen in the air or a chemical source. For example, in the sugar-chlorate mixture, sugar is the fuel and the chlorate ($KClO_4$) is a compound that contains oxygen. A Molotov cocktail, typically a bottle with gasoline and a rag that is ignited, is another example of a device that can be used to start arson fires (also occasionally referred to as incendiary fires). Mechanical or electrical components may be part of it as well, with items such as flashbulbs used to ignite an accelerant. The incendiary device is found at the point of origin, and thus it or the portions of it that remain are critical components of the physical evidence associated with arson.

inclusionary evidence Evidence, including the results of a forensic analysis, that does not exclude a given possibility or disprove a given hypothesis. Inclusionary evidence is the opposite of exclusionary evidence and is best illustrated by an example. Assume a bloodstain is found on the clothes of a man suspected to have been involved in a violent rape. Blood is collected from the man as well as from the victim, and both are subjected to DNA TYPING, along with the stain. If the types of the stain and the suspect match, that result is inclusionary evidence. If, on the other hand, the types of the stain and the suspect do not match, that is exclusionary evidence, since the suspect has been excluded as a possible donor. Inclusionary evidence is not necessarily the same thing as conclusive evidence, however. For example, if the suspect's fingerprints are found in the victim's home, that is inclusionary evidence, but if the victim and the suspect knew each other before the crime, such evidence is not at all conclusive.

inconclusive result A result that is not useful or fails to answer investigative questions. Inconclusive results can occur in many situations such as when there is too small a sample (a tiny spot of blood so small that it cannot be reliably typed), there is fragmentary evidence (a fragment of a bullet), or because of limitations of procedures or instrumentation. Such problems can lead to incomplete, contradictory, or inconsistent results that are impossible to interpret with any confidence. Another instance that can lead to an inconclusive result is when two analysts examine the same evidence and come to different conclusions. If the two interpretations cannot be resolved, the result is considered to be inconclusive.

indented writing Writing that is transferred by the pressure of the writing instrument (pen, pencil, and so on) to the paper or other material that is beneath it. Indented writing is often undetectable to the naked eye and as such can be overlooked by a criminal. Accordingly, indented writing is a common form of QUESTIONED DOCUMENT evidence.

The common media portrayal of discovering indented writing by rubbing a pencil over it is a myth; by doing so, the evidence is irrevocably altered and possibly destroyed. The simplest and most common method of visualizing indented writing is with oblique (angled) lighting followed by photography. Other techniques employed include various types of infrared illumination and photography, infrared luminescence, fluorescent powders that cling to the indentations, and iodine fuming. An ELECTROSTATIC DETECTION APPARATUS

(ESDA) exploits the same principles used in laser printers and copiers to visualize the indented writing without damaging the surface it is on. Recently, digital IMAGE ENHANCEMENT has taken an increasingly important role in the detection and deciphering of indented writing evidence.

individualization (individuation) The process of linking physical evidence to a common source. Individualization is a process that starts with IDENTIFICATION, progresses through CLASSIFICATION, and leads, if possible, to assigning a unique source for a given piece of physical evidence. The term *individualization* is often mistaken for identification, which in the forensic context does not have the same meaning. In forensic science, to identify something means exactly that—to determine that a red stain is blood or that a white powder is cocaine. However, a statement such as "The fingerprint was identified as belonging to John Doe" is incorrect in the strict sense. Rather, the fingerprint has been individualized and linked to John Doe as the one and only possible source. The idea of a common source comes in when one considers how such a fingerprint might be used. If John Doe left a fingerprint on a gun, that latent print could be developed and subjected to a database search, which would, in this example, lead to the identification of several people whose prints were in the database and who might have contributed the print. A fingerprint examiner would compare the print found on the gun to the prints on record and determine that the evidence print matched that of Mr. Doe. The two prints, the one on the gun and the one in the database, have a common source—John Doe. The print on the gun is thus individualized.

Aside from fingerprints, the other kinds of evidence that can potentially be individualized in a similar manner are blood (via DNA TYPING) and IMPRESSION EVIDENCE, such as markings made on bullets and TOOLMARKS. Less obvious but extremely important in forensic analysis is individualization by way of a PHYSICAL MATCH. It is not that the evidence itself is unique but rather the way in which it was separated and pieced back together that allows for linkage to a common source. As an example, consider a case in which a body is wrapped in a torn white sheet. The sheet is recovered and preserved, ragged torn edges intact. Meanwhile, a suspect is developed and a search of her house leads to the discovery of a portion of a white sheet. If the torn edges of the burial sheet and the sheet found at the house match, that is an example of a physical match, which leaves no doubt that the two pieces were once part of the same whole. Sheets themselves are not chemically or physically unique in the same sense that blood or fingerprints are, but the random tearing pattern is.

inductively coupled plasma *See* ICP TECHNIQUES.

inductive reasoning (inferential reasoning) A mode or process of thinking that is part of the scientific method and complements deductive reasoning and logic. Inductive reasoning starts with a large body of evidence or data obtained by experiment or observation and extrapolates it to new situations. By the process of induction or inference, predictions about new situations are inferred or induced from the existing body of knowledge. In other words, an inference is a generalization, but one made in a logical and scientifically defensible manner. Since inductive reasoning is part of the scientific method, it is also ingrained in forensic science.

See also HYPOTHESIS AND THE SCIENTIFIC METHOD.

infrared microscopy *See* INFRARED SPECTROSCOPY; MICROSCOPY.

infrared (IR) photography A technique used in the analysis of QUESTIONED DOCUMENTS to reveal indented writing or other obscured or otherwise invisible writing. Illumination of documents by infrared light can increase contrast and reveal otherwise invisible, damaged, charred, obliterated, or obscured writing. Inks can also be studied using IR luminescence, in which inks are exposed to visible light and then give off energy that is in the infrared portion of the electromagnetic spectrum. Infrared photography works the same way as does conventional photography except that IR sensitive film is used.

infrared spectroscopy and microscopy (IR, FTIR) As a group, one of the most versatile and widely used techniques of instrumental analysis applied in forensic science. Infrared spectroscopy is a form of absorption spectroscopy in which electromagnetic radiation in the infrared range is absorbed by molecules. The pattern of absorption across the infrared range is unique for each different molecule and, as a result, an IR spectrum provides specific identification for compounds. However, definitive identification is possible only if

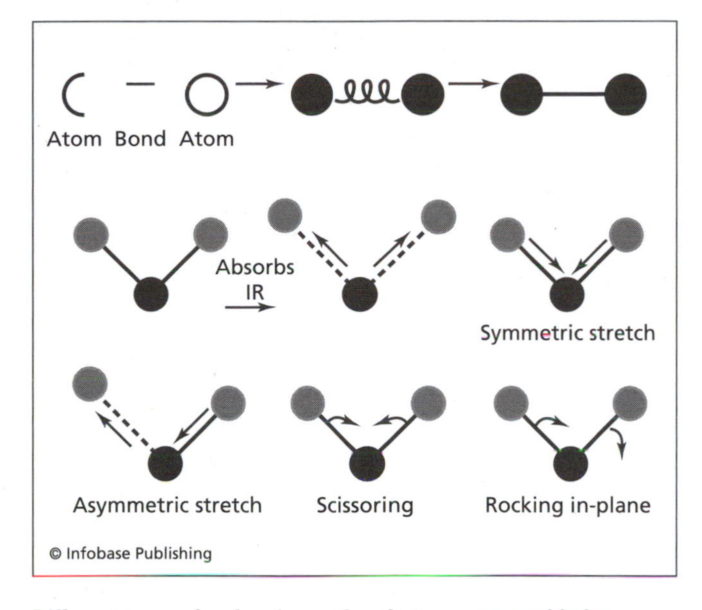

Different types of molecular motion that can occur with the absorption of infrared radiation. In this simplified depiction, two atoms linked by a chemical bond can be thought of as two rigid spheres connected by a spring. Absorbance or IR radiation induces vibrations.

the sample being studied is pure. IR methods are most widely used for the analysis of organic compounds such as drugs, synthetic fibers, and plastics, but the technique can also be used for many inorganic materials such as might be found in soil or paints.

Like all spectrophotometric techniques, IR depends upon molecules absorbing ELECTROMAGNETIC RADIATION (EMR). IR energy is not sufficient to break a molecule apart or to promote electrons to higher energy states. Rather, when a molecule absorbs IR radiation, the energy is converted to vibrational motion within the molecule. To visualize this effect, atoms within a molecule can be thought of as tiny steel marbles and the chemical bonds connecting them as springs. When IR radiation is absorbed, it causes the spring to flex and bend, but the energy is not sufficient to actually break the spring. Since the motion is three dimensional, there are many different types of motion that can occur, as illustrated in the first figure. In a symmetric stretch, the bonds expand, both in the same direction. In an asymmetric stretch, one bond stretches in one direction while the other compresses in the opposite direction. These motions can also be up and down and appear to be coming up out of the paper or going down farther into it. Scissoring, wagging, and twisting motions can also occur. Thus, because each bond between two different atoms has many different possible motions and because molecules are composed of many such atoms and bonds, the IR absorption pattern for any one molecule is unique.

The second figure shows how, in a very simplified way, an IR spectrum is recorded for a molecule such as water, which consists of three atoms and two bonds. Infrared radiation of different wavelengths (here symbolized as) is shined through water molecules. Some of these wavelengths will be absorbed, causing different kinds of vibrations in the molecules, while other wavelengths will not and will pass through the water. A detector records the results, and a plot is made of the degree of absorbance at each wavelength. The resulting plot is the IR spectrum of that compound. The third figure shows actual spectra of closely related compounds. MORPHINE and CODEINE are both found in OPIUM, whereas HEROIN is made from morphine. However, even though these compounds are closely related, the spectra are distinctive. The region of the spectra from 1,500 to 500 wavenumbers (the inverse of the wavelength) is called

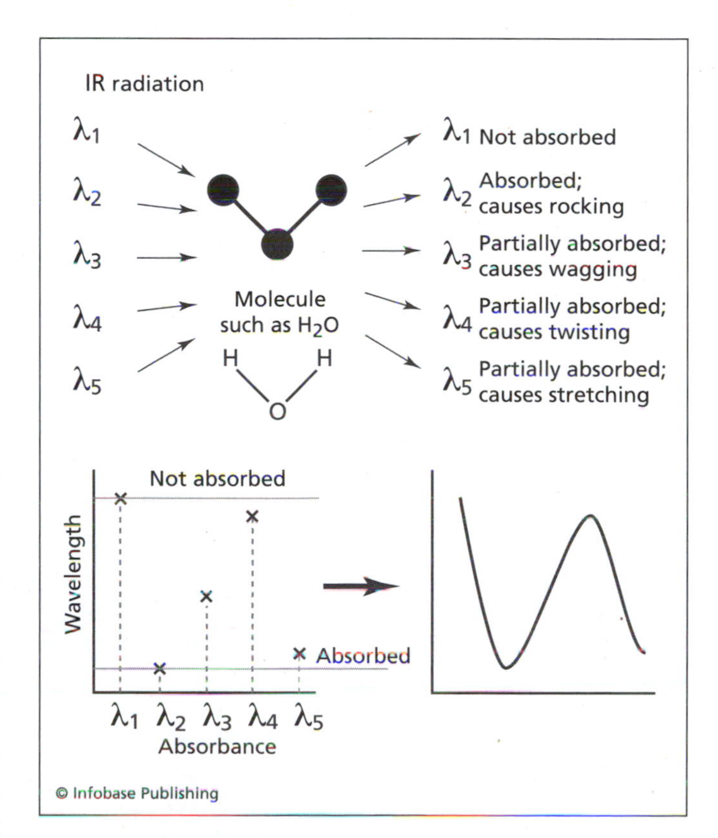

Creation of an infrared absorbance spectrum

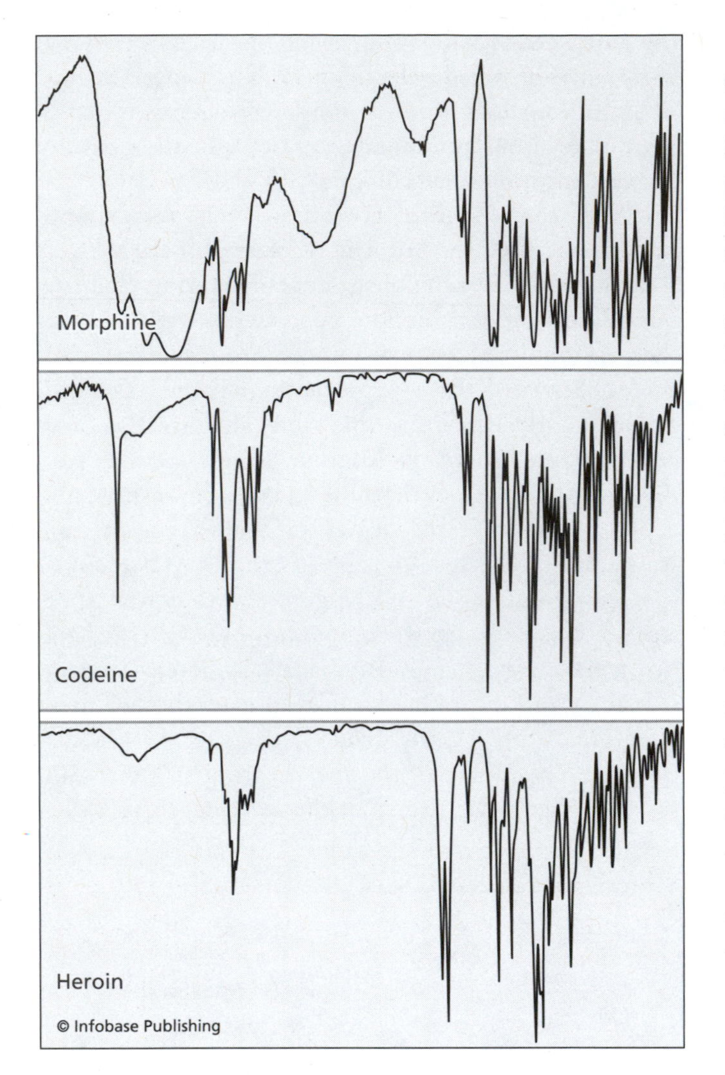

Example IR spectra of drugs

the "fingerprint region" and is distinctive enough to identify specific compounds.

Most IR spectrophotometers sold since 1990 use a device called a MICHELSON INTERFEROMETER coupled to a mathematical operation called a Fourier transform to obtain spectra. In this approach, all wavelengths of light are presented to the sample simultaneously rather than sequentially (one at a time). The resulting interference pattern (somewhat like a hologram) is then mathematically analyzed to yield the spectrum. This type of instrument and technique is referred to as FTIR. This instrumental design allows for many variations of IR applicable in forensic science and forensic chemistry. These include ATTENUATED TOTAL REFLECTANCE (ATR), internal reflectance, and microspectrophotometry (micro-FTIR), in which the instrument is coupled to a microscope.

Applications of IR in forensic science are widespread and numerous. It is the technique of choice for most drug analyses since identification is definitive. IR is also used in toxicology and in the analysis of fibers, plastics, paints, polymers, and soils. Micro-FTIR is becoming increasingly important for examination of trace evidence such as individual fibers and paint layers. In questioned documents, IR techniques, both microscopic and photographic, are widely employed. However, a critical element in any IR examination is sample purity. Although an IR spectrum is considered definitive identification, this is true only when the sample is pure. A mixed sample will produce a mixed spectrum that may be difficult or impossible to interpret. Finally, IR spectroscopy is not well suited to QUANTITATIVE ANALYSIS and thus is often supplemented with other instrumental techniques when such information is required. For example, if a white powder is found to contain cocaine, an extraction procedure is used to isolate the cocaine, followed by IR analysis to provide definitive identification. However, to determine what percentage of the sample is cocaine, a technique such as GAS CHROMATOGRAPHY/MASS SPECTROMETRY (GC/MS) is necessary.

Further Reading

Andrasko, J. "Microreflectance FTIR Techniques Applied to Materials Encountered in Forensic Examination of Documents." *Journal of Forensic Science* 41, no. 5 (1996): 812.

inhalants A class of psychoactive components ingested by inhalation. Also called "glue sniffing" or "huffing," ingestion of inhalants became significant in the 1960s and remains most popular among teenagers and young adults. Inhalants are volatile (evaporate easily), and most are gases at normal room temperatures and pressures. Abusers typically fill a plastic bag or soak a rag with the substance to dose themselves. Unlike drugs of abuse, inhalants are components of common products such as paint or cleaning solvents and are not regulated under the CONTROLLED SUBSTANCES ACT (CSA). Many are propellants used in spray cans containing paint or other similar materials. Toluene and butane are examples of hydrocarbon inhalants that are easily available in many consumer products. Another large class of inhalants is halocarbons, meaning the molecules contain chlorine, fluorine, or bromine. Halocarbons include CFCs (chlorofluorocarbons) and chlorohydrofluorocarbons, which replaced most uses of ozone-depleting CFC's. Another term used to refer to this class of compound and inhalant is "Freons," and they are available as coolants, fire suppressants, and propellants. Gaseous anesthetics are yet another group of halogenated compounds. One of the first anesthetics was chloroform, a volatile liquid that is often incorrectly depicted in movies and TV shows as a compound that can cause nearly instantaneous unconsciousness when someone is forced to inhale it off a saturated rag. Nitrous oxide ("laughing gas") can also be abused, as can amyl nitrate, an inhalant often associated with homosexual activity.

As a class, inhalants are considered to be central nervous system depressants with effects similar to alcohol, although they act much faster. Abuse of inhalants carries high risks, particularly for kidney, liver, and brain damage. Forensic identification of inhalants is a challenge since often there is no residual to test. The compounds are not detectable in urine, although some of the metabolites are. Blood tests can be used, often employing HEADSPACE analysis coupled to GAS CHROMATOGRAPHY (GC).

Further Reading

Broussard, L. "Inhalants." In *Principles of Forensic Toxicology*. 2nd edition. Edited by B. Levine. Washington, D.C.: American Association of Clinical Chemistry, 2003.

Drug Enforcement Agency (DEA) Web site. "Inhalants." Available online. URL: http://www.usaoj.dea.gov/concern/inhalants.htm. Downloaded May 4, 2007.

inks The type, age, and patterns of inks can provide important evidence in the examination of QUESTIONED DOCUMENTS. The composition of inks is much like that of paints—a solvent base (water or other), coloring materials (either INORGANIC pigments or ORGANIC dyes, natural and synthetic), and other additives that control thickness and final appearance. The oldest ink, called India ink but originally used in China, consists of carbon black (ground charcoal) suspended in water and containing adhesive gums and varnishing components. Iron gallotannate inks contain inorganic colorants along with tannic acid and are used in fountain pens. Inks used in ballpoint pens contain synthetic dyes dissolved in organic solvents and additives to maintain a thick consistency. Newer gel pens contain synthetic dyes impregnated into gel. The forensic analysis of inks focuses on analyses of the colorants and dyes.

The variety of writing instruments used to deliver ink to paper is nearly as rich as the variety of inks themselves. Feathers and fountain pens have been used with India ink, while modern writing instruments run the gamut from ballpoints to felt tips to gel pens. The ballpoint pen was developed in 1939 and made its way in significant quantities to the United States in 1945. Ballpoints work by delivering ink to a small ball valve in the tip of the pen that rotates and spreads the ink. Felt tips, introduced in the early 1960s, use a wicking action to deliver the ink, resulting in fairly quick wearing and widening of the contact surface. Recently introduced gel pens come in a wide variety of colors that are highly resistant to water and solvents. While this is desirable for creating permanent writing, this property complicates forensic analysis, since it is often necessary to extract the ink first using a solvent. Inks are also components of items such as typewriter and dot matrix printer ribbons, inkjet printers, copiers, and laser printers. Thus, ink analysis can be a part of any questioned document case.

In 1968, the BUREAU OF ALCOHOL, TOBACCO, FIREARMS, AND EXPLOSIVES (ATF) created a national repository for inks. The collection was transferred to the United States SECRET SERVICE forensic laboratory in 1985, where it currently resides and holds several thousand ink samples. The primary use of the collection is for dating and related determinations. A voluntary taggant program was also initiated in the 1970s in which manufacturers would add different "tagging" compounds to ink every year. If an ink sample is found to contain a taggant, the examiner knows what year the ink was produced.

Other techniques can be used to estimate the dates that ink was applied to a document. Inks, like paint, must dry, and this process can take years. The degree of dryness can be estimated by extraction of the ink into different solvents. The drier the ink, the more difficult it will be to extract. Although reasonably successful, dryness tests require that the examiner have a fresh sample of the ink that has been spread on a similar medium and stored in a similar fashion as has the questioned sample. Another test used to estimate age is the chloride migration test. As an ink dries, chloride ions (Cl-) contained within it begin to diffuse away from the initial pen stroke. The older the writing, the greater the dispersal and spread of the ions. Dating of inks is particularly important in fraud and forgery cases in which an original document created on a given date is altered at a later date using a different writing instrument and ink. If the analyst can prove the older document has been altered by the addition of newer ink, that is strong evidence of fraud. A simple example would be a case of a document purportedly written in 1910 but found to contain ballpoint pen ink. Since the ballpoint pen was not invented until the middle of the century, the document is either a forgery or has been altered.

Inks can be studied using nondestructive methods such as micro-FTIR (microscopic INFRARED SPECTROSCOPY) and INFRARED PHOTOGRAPHY. Examination under ultraviolet light (black light) and an evaluation of fluorescence can also be useful, particularly for differentiating inks and determining if two different inks have been used on the same document. These kinds of evaluations can reveal otherwise hidden or invisible writing or writing that has been obliterated or damaged by fire, for example. The most common method for evaluation of inks remains THIN LAYER CHROMATOGRAPHY (TLC), in which the dyes in ink are separated on a thin plate and visualized. Other instrumental analysis techniques that have been used for ink analysis include DIFFUSE REFLECTANCE INFRARED SPECTROSCOPY (DRIFTS), GAS CHROMATOGRAPHY/MASS SPECTROMETRY (GC/MS), HIGH PRESSURE LIQUID CHROMATOGRAPHY (HPLC), CAPILLARY ZONE ELECTROPHORESIS (CZE), and X-ray fluorescence (XRF).

See also X-RAY TECHNIQUES.

Further Reading

Brunelle, R. L. "Ink Dating—The State of the Art." *Journal of Forensic Science* 37, no. 1 (1992): 113.

inorganic compounds and inorganic analysis Chemical compounds can be broadly classified as organic or inorganic based on their composition. Organic compounds contain carbon atoms that are bonded to other carbons or atoms that are nonmetals such as hydrogen or oxygen. All other compounds are inorganic. There are some exceptions to these rules such as the cyanide ion (CN^-) and carbonate ion (CO_3^{2-}) that are considered to be inorganic even though they fit the definition of an organic compound. The use of the term "organic" as it applies to chemicals is often confused or misrepresented in the media and in advertising, where "organic" is used to mean "natural." This definition is separate from the chemical one.

In forensic science, the distinction is important because the methods of analysis will vary depending on the class of a compound. Drugs such as COCAINE and HEROIN are organic compounds, and they are analyzed using techniques such as INFRARED SPECTROSCOPY (IR). On the other hand, poisons such as ARSENIC and cyanide are inorganic and are analyzed using ATOMIC ABSORPTION (AA) or INDUCTIVELY COUPLED PLASMA (ICP) techniques. Many types of physical evidence contain both types of compounds. Paint contains organic solvents, binders, and metals, such as lead and zinc, which are inorganic. GUNSHOT RESIDUE (GSR) contains mostly inorganic materials such as nitrates and lead, but some of the components of gunpowder are organic. Both portions of such evidence can be analyzed, but using different sample preparation and testing procedures.

insects *See* ENTOMOLOGY, FORENSIC.

instrumental analysis In analytical chemistry, methods of analysis (qualitative and/or quantitative) that utilize instrumentation. Older methods of analysis are classified as traditional or "wet chemistry" and rely on simple equipment such as glassware and analytical scales (analytical balances). One such traditional method used in forensic science is THIN LAYER CHROMATOGRAPHY (TLC). Instrumental analysis is widely used in forensic science in drug analysis, toxicology, arson, and trace evidence to name a few. In drug analysis, instruments such as GAS CHROMATOGRAPHS/MASS SPECTROMETERS (GC/MS) and INFRARED SPECTROPHOTOMETERS (IR) are indispensable. Trace evidence examiners use SCANNING ELECTRON MICROSCOPES (SEM) coupled to X-ray diffraction (XRD) instruments. GUNSHOT RESIDUE (GSR) can be analyzed using atomic absorption or inductively coupled plasma instruments. Thus, instrumental analysis has become indispensable across many forensic disciplines.

See also X-RAY TECHNIQUES.

instrumental neutron activation analysis *See* NEUTRON ACTIVATION ANALYSIS (NAA).

interlaboratory variation A term used to describe the small variations in results that can occur when different laboratories (forensic or other) analyze the same sample. Some analyses should not produce any differences, such as DNA TYPING or FINGERPRINT analysis. However, areas such as drug analysis may. For example, if a single sample is submitted to 50 different laboratories to determine its purity, a range of values could result. One lab may find a value of 50.0 percent, while another might find 50.5 percent and yet another 49.8 percent. A determination of interlaboratory variation in such cases helps to set a baseline of what variation is normal and natural, the result of small random discrepancies. Measurement of variation is an important part of QUALITY ASSURANCE/QUALITY CONTROL (QA/QC).

internal standard A method or compound used in a chemical quantitation. Internal standard quantitation is highly accurate and provides a method for correction of problems associated with loss or suppression of an analyte during sample preparation and analysis. The most common forensic application is in chemistry (DRUG ANALYSIS and TOXICOLOGY). An internal standard is a compound or element that is different from the analyte of interest but still closely related to it. For example, LIDOCAINE or PROCAINE would be good choices as internal standards for the analysis and quantitation of cocaine, as long as neither compound is found in the sample itself. In MASS SPECTROMETRY (MS), isotopes can be incorporated into internal standards in a variation of the internal standard technique called isotope dilution. However, the cost and complexity of this approach has limited use in forensic applications.

An internal standard analysis begins by establishing a known ratio of response of the analytical system to both the analyte and the internal standard at known concentrations. Continuing with the above example, cocaine is quantitated using GAS CHROMATOGRAPHY (GC). The chemist would first determine the "response factor" of the GC detector to cocaine and procaine by analyzing solutions containing each compound at known concentrations. By adding the same amount of internal standard to an unknown sample, the response factor can be used to calculate the concentration of cocaine.

International Association for Identification (IAI) An international forensic science society initially focused on fingerprinting but that has expanded to include several forensic disciplines. The IAI was formed in 1914 by Harry Caldwell of the Oakland City Police Department at a time when scientific methods of identification were undergoing a transition from body measurements (ANTHROPOMETRY, or Bertillonage) to fingerprints. The society currently has more than 5,000 members and offers professional certifications in BLOODSTAIN PATTERNS, CRIME SCENE PROCESSING, footwear and TIRE IMPRESSIONS, forensic ART, forensic PHOTOGRAPHY, and latent fingerprints. The *Journal of Forensic Identification (JFI)* is a bimonthly scientific publication directed toward members and other forensic science professionals. The society is divided by discipline and includes the areas in which certifications are offered plus FIREARMS and TOOLMARKS, POLYGRAPH, QUESTIONED DOCUMENTS, voice identification, laboratory analysis, innovative and general techniques, and ODONTOLOGY. The organization maintains a Web site at www.theiai.org.

International Organization for Standardization An international standard-setting body formed in 1947 as a voluntary organization consisting of representatives from various member groups. The American National Standards Institute (ANSI) is the U.S. representative in ISO. ISO standardards are not regulations but voluntarily accepted standards that organizations, companies, or laboratories may adopt to improve quality. ISO standards are not specific to a given industry or profession. The AMERICAN SOCIETY OF CRIME LABORATORY DIRECTORS Laboratory Accreditation Board (ASCLD-LAB) has an international program based on ISO standards.

ion chromatography (IC) An instrumental technique that can be used to detect anions (negatively charged atoms or molecules such as Cl^-) and cations (positively charged species such as Na^+). IC has been applied in forensic science for the analysis of GUNSHOT RESIDUE (GSR) and EXPLOSIVES. The ions of interest include ammonium (NH_4^+), nitrate (NO_3^-), and chlorate (ClO_4^-), species that are often detected using color change or PRESUMPTIVE TESTS. The advantages of IC in these cases include specificity (presumptive tests are subject to FALSE POSITIVES and FALSE NEGATIVES) and increased sensitivity, down to the part-per-billion (ppb) range. A part per billion is 1 microgram (µg) per liter of water, and a microgram is 1/1,000,000 of a gram.

IC instrumentation works on the basis of ion exchange, the same principle used in water softeners. To perform an ion exchange, the sample is forced through a column bed that contains an exchange resin. In a water softener, the goal of ion exchange is to remedy hard water problems by removing the calcium and magnesium ions (Ca^{2+}, Mg^{2+}). To do this, the water supply is directed over a bed charged with sodium ions (Na^+). In the column, the calcium and magnesium will displace the sodiums at a 1:2 ratio to maintain charge balance. For every magnesium or calcium ion that is removed by the resin, two Na+ cations will be released into solution. Water hardness is removed by trapping calcium and magnesium in the bed, but as a result, soft water has a higher concentration of sodium, which can present problems to people with high blood pressure. Thus, ion exchange does not remove ions from solution, it only replaces one ion with others of the same overall charge.

There are two modes of ion exchange used in ion chromatographs, single ion exchange and suppression ion exchange. Single ion exchange works very much like a water softener, using resins that are specific for different cations, depending on what is of interest to the analyst. Suppression ion systems are used to reduce the ion concentration of the solution using an additional step. An example of ion suppression would be to create a column that replaces cations (positively charged ions) with the H^+ ion and anions (negatively charged ions) with the OH^- ion. When these two ions combine, they form water (H_2O), which is not ionic. Ion exchange still occurs, but by careful selection of resins, ions combine to form covalently bonded molecules (also called molecular compounds) that do not form ions in water. Regardless of the type of ion exchange column used, the detection system is most often a conductivity detector that relates concentrations of ions to the conductivity of the solution.

ion mobility spectrometry (IMS) A portable instrument used in forensic science to detect drugs, explosives, tear gas, and chemical warfare agents. Originally called plasma chromatography, IMS works similarly to ELECTROPHORESIS, except the charged species are separated in the gaseous state rather than in a gel. IMS works at atmospheric pressure, making it ideally suited for use as a portable monitoring system, and the militaries in several nations use IMS routinely for battlefield detection of chemical weapons. As shown in the figure, an air sample is drawn into the instrument and directed through an ionization region, where the radiation emitted by radioactive 63-nickle causes ionization and formation of ion/molecule clusters such as $H(H_2O)_3^+$. In this example, the ion is H^+, and the water is the molecule. An electronic pulse of a wire shutter (which looks like a screen door) admits clusters into the drift region where they move against a flow of a drift gas such as air or nitrogen. The clusters are separated based on the ratio of their size to their charge, much as separation is accomplished in gel electrophoresis, with the smaller clusters moving ahead of the larger ones. The clusters arrive at the detector, and the pulses are recorded as peaks, the heights of which are proportional to concentration. IMS has been used to detect drugs in closed shipping containers and also is being explored for a wider role in explosives detection at airports.

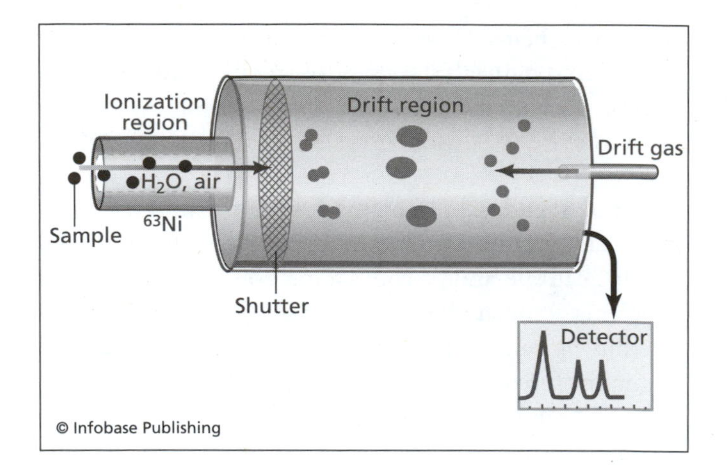

Operation of an ion mobility spectrometer

Further Reading

Eiceman, G. A., and Z. Karpas. *Ion Mobility Spectrometry.* 2nd edition. Boca Raton, Fla.: CRC Press, 2005.

ISO *See* INTERNATIONAL ORGANIZATION FOR STANDARDIZATION (ISO).

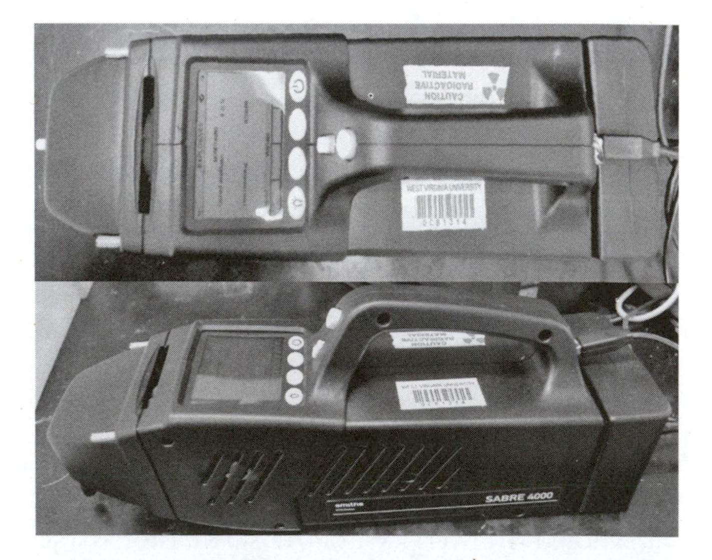

Two views of a portable ion mobility spectrometer ***(Courtesy of Michael Bell, West Virginia University, Forensic and Investigative Sciences program)***

isoelectric focusing (**IEF**) A technique used in forensic SEROLOGY to type ISOENZYME systems. Prior to DNA TYPING, the analysis of isoenzyme systems was used to type samples using ELECTROPHORESIS to separate the protein molecules on a gel. In gel electrophoresis, protein molecules, including those that make up the isoenzymes, migrate under the influence of the electrical field. The ratio of the size of a protein molecule to its charge determines how fast it moves, and the charge of the protein molecule depends on how many acidic sites are found within it. An acidic site is a location on a molecule that has an acidic hydrogen (also called a proton) attached to it. Acids, such as hydrochloric acid (HCl), are highly acidic because the proton (the H^+ ion) is easily removed from the molecule. In the same way, protein molecules have acidic sites at which acidic H^+ ions can be removed. Low concentrations of H^+ ions in the surrounding solution encourage deprotonation of the acidic sites, whereas high concentrations of H^+ discourage deprotonation. In proteins, acidic sites are usually associated with a chemical group called a carboxylic acid in which the hydrogen bonds to an oxygen atom as OH. If the proton is removed, the O remains behind with a net negative charge, indicated by O^-. Note that if the H^+ is present, the +1 and -1 charge neutralize, creating a net neutral group.

The concentration of the H^+ ion in an aqueous (water) solution is expressed by the pH scale, which varies from 0 to 14. A pH of 7.0 is neutral, whereas anything below 7.0 is acidic (high concentrations of H^+), and anything above 7.0 is basic or alkaline (low concentrations of H^+). In turn, the charge of a protein molecule depends on the pH. In a solution with low pH (an excess of H^+ ions available), the protein molecule will be fully protonated and will have a net zero, or neutral, charge. Conversely, at high pH values (low concentration of H^+), the proteins will tend to have a more negative charge since most of the OH groups will deprotonate, leaving O^-. Proteins are large, complex molecules, and different proteins will have different pH values at which their net negative charge is zero. Isoelectric focusing exploits these differences to yield very finely separated isoenzyme patterns.

In traditional gel electrophoresis, the separation is performed at a constant pH, such as 7.4 (slightly basic). In isoelectric focusing, a pH gradient is created in the gel, meaning that the pH changes with gel position. The figure gives a simplified overview of the IEF process. At the starting point, or origin, of the IEF gel, a negative charge is applied and the pH is high, meaning that H^+ ions are relatively scarce. This low concentration encourages deprotonation of the protein molecules (P). Since they are negatively charged, they will move away from the negative pole (like charges repel, unlike charges attract) toward the opposite positive side. However, the pH of the gel becomes more acidic (pH decreasing, H^+ abundance increasing) toward the positive terminal. At some pH, the protein molecule will become neutral and will cease to move since it is no longer charged. This point is called the isoelectric point, and proteins with different structures can be distinguished based on where they stop moving in the gel.

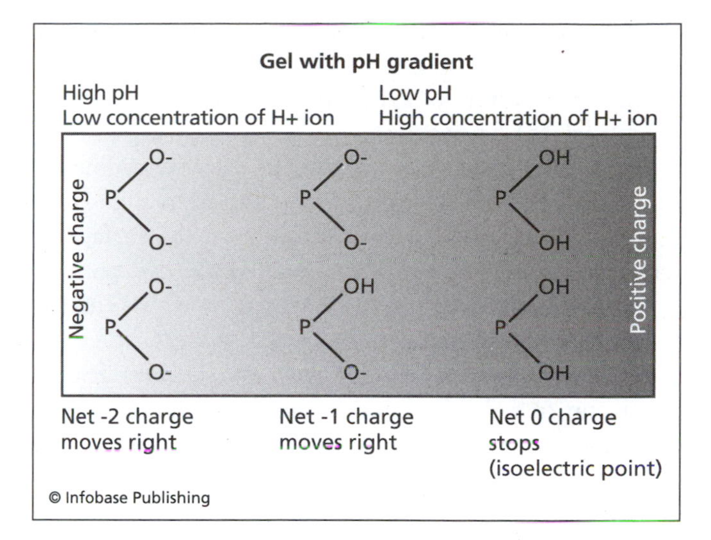

A gel with a pH gradient used for isoelectric focusing. The pH determines how many H^+ ions are present, which can protonate acidic sites such as P-O^-. A protonated site will not carry a charge, and the relative number of charged sites to uncharged sites will determine the overall charge and thus mobility of large ions in the gel.

IEF proved to have a higher resolving power than traditional gel electrophoresis and thus could distinguish more types within some of the isoenzyme systems. For example, the PGM isoenzyme system shows three types when analyzed by electrophoresis: 1-1, 2-1, and 2-2. Application of IEF to the same system yields 10 types, such as 2+ 1-. The difficulty of an IEF analysis is comparable to that of gel electrophoresis. However, the reagents are more expensive, and this cost prevented the routine use of IEF in forensic serology. With developments in DNA technology, isoenzyme typing is rarely performed anymore. As with electrophoresis, IEF is now performed in capillary tubes, a technique called capillary isoelectric focusing (CIEF).

isoelectric point *See* ISOELECTRIC FOCUSING.

isoenzyme systems Prior to the widespread acceptance of DNA TYPING, isoenzymes were used in conjunction with ABO BLOOD GROUP SYSTEM typing to categorize blood and body fluid evidence to the extent possible. These isoenzymes are found on the surface of red blood cells and are thus sometimes called red cell isoenzymes. Enzymes are biological catalysts necessary to speed up reactions that would otherwise be far too slow in an organism. The term *isoenzyme* refers to the group of red cell enzymes that are POLYMORPHIC, meaning that they exist in multiple different forms. The form that a person has is determined genetically. Since heredity (genes) determines the form, the isoenzymes are considered to be GENETIC MARKER SYSTEMS. The analysis and typing of isoenzymes was accomplished using gel ELECTROPHORESIS or closely related techniques.

Although a number of isoenzyme systems have been identified, not all were routinely analyzed in forensic applications. The six common systems typed were: phosphoglucomutase (PGM); adenylate kinase (AK); acid phosphatase (ACP or EAP); glyoxalase I (GLO); esterase D (ESD); and adenosine deaminase (ADA). The common ending-*ase* signifies that the compound is an enzyme. The usefulness of each isoenzyme in discriminating blood and eliminating possible sources varies from system to system and between racial groups. AK, one of the most robust of the six, has three types: 1, 2-1, and 2. Since it is robust, it tends to persist longest in stained materials and thus constitutes one of the more useful systems for testing older stains. Unfortunately, more than 90 percent of the population is type 1, so the DISCRIMINATING POWER of that system alone is low. The value of isoenzymes typically arose from typing several systems (if feasible) simultaneously, along with a determination of the ABO type of the blood. By combining probabilities, the discriminating power increased significantly. For example, if a stain was found to be ABO type O (~42 percent), AK 1 (~90 percent), PGM type 1 (~59 percent), and GLO 2-1 (~53 percent), the approximate percentage of the population that could be a source of that blood drops dramatically to (.42)(.90)(.59)(.53) = .12, or 12 percent. Typing the isoenzymes in addition to the ABO system adds significant discriminating power. However, this still does not individualize the blood.

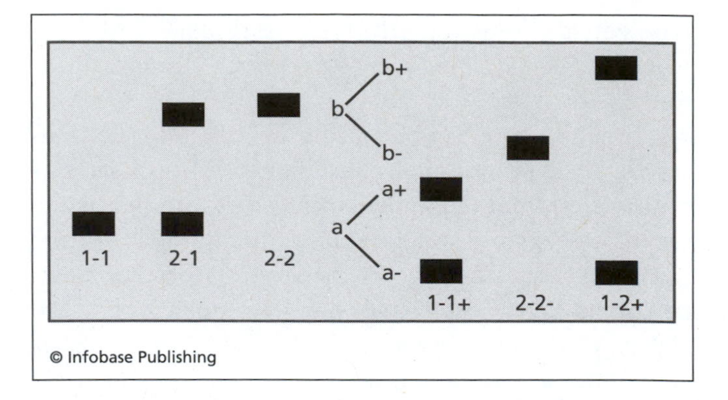

Comparison of the patterns of PGM isoenzymes that can be separated by electrophoresis and isoelectric focusing. To the left is the band pattern that would be seen using electrophoresis, and to the right additional bands separated by isoelectric focusing.

Gel electrophoresis was used to determine isoenzyme types, supplemented by a technique called ISOELECTRIC FOCUSING (IEF), which allowed for a finer division of the types. Whether reading the results of electrophoresis or IEF, the pattern obtained is a series of bands that appear in the gel after various developers are applied. The pattern is often referred to as a "bar code" that is specific for each type. The figure illustrates a representation of a portion of the electrophoresis pattern obtained for the three common PGM types (1, 2-1, and 2). When IEF is used for the same system, the "a" and "b" bands are further subdivided, leading to 10 subtypes, such as 1-1+. Thus, IEF increased the discriminating power of the PGM system, but the cost of the reagents needed to prepare the gels limited the widespread and routine use of IEF. The need for ever more isoenzyme systems possessing even greater discrimination power was eliminated with the advent of DNA typing.

See also STARCH WARS.

Further Reading

De Forest, P. R., R. E. Gaensslen, and H. C. Lee. "Blood." In *Forensic Science. An Introduction to Criminalistics.* New York: McGraw-Hill, 1983.

Spalding, P. R. "Identification and Characterization of Blood and Bloodstains." In *Forensic Science: An Introduction to Scientific and Investigative Techniques.* 2nd edition. Edited by S. H. James and J. J. Nordby. Boca Raton, Fla.: CRC Press, 2005.

J

Jabir (ca. 715–820) Arab *Alchemist* Jabir was an alchemist who played an important foundational role for FORENSIC SCIENCE and forensic TOXICOLOGY, although he never actually practiced either discipline. Also known as Jabir ibn-Hayyan or Geber, legend puts his birth around 720 C.E. and his death by murder in 815. However, to live so long in that age would have been remarkable. He was a member of the Hashashins, a radical religious group that took part in political murders, reportedly under the influence of hashish. Hashish is a potent hallucinogen derived from MARIJUANA. Jabir's cult and its practices may be the root of the word *assassin*. Jabir has been credited with hundreds of publications and that, combined with his purported age at death, has led historians to question if such a man actually existed or if the name represents collected works. Jabir's many noteworthy accomplishments include those related to DYES, INKS, GLASS, METALLURGY, and MEDICINES. He was probably the first alchemist to isolate arsenic metal from arsenic mineral, an important step toward detecting ARSENIC as a poison. Other Arab alchemists worked with acids such as vinegar and were the first to recognize that acids were a distinctive class of compounds. They also understood that some diseases were infectious, an idea well ahead of its time.

See also ALCHEMY.

Engraving of Jabir in a laboratory with chemistry utensils *(National Library of Medicine)*

Further Reading

Aromaticico, A. *Alchemy: The Great Secret.* New York: Harry N. Abrams, Inc., 2000.

Moran, B. T. *Distilling Knowledge: Alchemy, Chemistry, and the Scientific Revolution.* Edited by M. C. Jacob, S. R. Weart and H. J. Cook, *New Histories of Science, Technology, and Medicine.* Cambridge, Mass.: Harvard University Press, 2005.

Morris, R. *The Last Sorcerers: The Path from Alchemy to the Periodic Table.* Washington, D.C.: Joseph Henry Press, 2003.

Jack the Ripper The name given to the infamous London murderer who stalked the streets during the late summer and fall of 1888, known to have killed at least five women. Although the Ripper has spawned conspiracy theories, endless conjecture, and countless fictional spin-offs in books, television, and movies, the investigations made little use of forensic science. In the late 1800s, it was an infant science with few tools available to investigators at Scotland Yard and to the London City Police who were also involved in the investigation. The collection and use of FINGERPRINTS and BLOOD evidence were many years away, so forensic science was not a factor in the Ripper case. However, officers did make some CRIME SCENE sketches and took photos of victims, one at the scene. After one night in which two women were murdered (September 30), a bloody apron apparently used to wipe off the murder weapon and some graffiti chalked onto an outside wall were found. No tools existed to analyze the blood evidence, and so it was of no use. The graffiti contained a racial slur against Jews, so the decision was made to remove it before it could be photographed in the morning (flash photography had yet to be invented). Thus, the Ripper was one of the last serial killers to escape the scrutiny of forensic science.

Further Reading

Cornwell, P. *Portrait of a Killer: Jack the Ripper—Case Closed.* New York: Putnam, 2002.

journals and periodicals The field of forensic science has several professional journals, many of which are directed toward the many subdisciplines in the field. In the United States, the AMERICAN ACADEMY OF FORENSIC SCIENCES (AAFS) publishes the *Journal of Forensic Science* (commonly abbreviated JFS), which is widely read by a large cross section of forensic scientists. *Forensic Science International* is a similar publication with a largely European audience, and together these two publications are the widest read in the field. The *Journal of Forensic Identification* is published by the INTERNATIONAL ASSOCIATION FOR IDENTIFICATION (IAI) and focuses on fingerprinting, while the Forensic Science Society publishes *Science and Justice.* Examples of more specialized publications include the *Journal of Environmental Forensics* (also known as the *International Journal of Environmental Forensics*), *Journal of the American Society of Questioned Document Examiners, Forensic Linguistics,* and the *Journal of Analytical Toxicology.* Recently, several on-line journals have appeared, such as *Scientific Testimony* and *Forensic-Evidence.com.* In addition, many print journals are accessible on-line via subscription or through libraries. The following table lists a sampling of general and specialized forensic journals and periodicals:

ASCLD Newsletter (American Society of Crime Laboratory Directors)
Forensic Linguistics
Forensic Science Abstracts
Forensic Science Communications
Forensic Science International
Forensic Science Review
Forensic Science Abstracts
International Journal of Drug Testing
International Journal of Environmental Forensics
International Journal of Forensic Document Examiners
Journal of Analytical Toxicology
Journal of Canadian Society of Forensic Science
Journal of Clinical Forensic Medicine
Journal of Forensic Document Examination
Journal of Forensic Identification
Journal of Forensic Odontology and Stomatology
Journal of Forensic Sciences
Journal of Psychoactive Plants and Compounds
Journal of the American Society of Questioned Document Examiners
Medicolegal Update Journal
Microgram (published by DEA)
NIJ Newsletter (National Institute of Justice)
RCMP Gazette (Royal Canadian Mounted Police)
Science and Justice: The Journal of the Forensic Science Society
The American Journal of Forensic Medicine and Pathology
The Australian Journal of Forensic Science
The Biometric Digest
Wound Ballistics Review

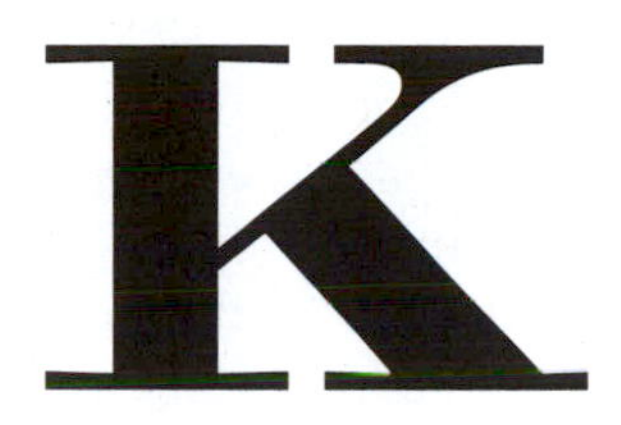

Kastle-Meyer color test (**Kastle Meyer color test**) A PRESUMPTIVE TEST used to identify BLOOD. The chemical substance used is phenolphthalein, which is made by boiling phenolphthalin in an alkaline solution containing potassium hydroxide (KOH). The test is also referred to as the K-M or KM test, or as the phenolphthalein test, even though the starting compound is phenolphthalin. Until recently, phenolphthalein was used as the active ingredient in some laxative products and is still commonly used in chemical analyses (titrations) of acids and bases. Despite some concerns about possible carcinogenic properties, the KM reagent is considered to be relatively safe and certainly safer than BENZIDINE, which was once widely used to identify blood but abandoned due to known carcinogenic properties. Phenolphthalein reacts with the hemoglobin in blood in the presence of hydrogen peroxide (H_2O_2) to cause a pink color to form. Like other presumptive tests, the KM test is not specific for blood and can produce FALSE POSITIVES with substances such as horseradish.

Katrina (hurricane) *See* IDENTIFICATION OF THE DEAD.

Kennedy assassination In modern times, no single murder has generated more speculation, conspiracy theories, conjecture, and fictionalization than the death of President John F. Kennedy on November 22, 1963, in Dallas, Texas. Kennedy was hit by two of three bullets fired from above and behind from the Texas School Book Depository building, and Lee Harvey Oswald was arrested within hours of the killing. Originally suspected of the murder of police officer J. D. Tippit, Oswald quickly became the prime suspect in the assassination. However, Oswald was himself killed by Jack Ruby before his interrogation was complete. Up to the time of his death, he continued to deny killing the president, and his murder so soon after the killing was one of many factors that would eventually spawn rumors of conspiracy that blossomed into an entire subculture of books, movies, and unending speculation that continue unabated even after two exhaustive federal investigations and forensic inquiries that have generally supported the conclusions originally drawn by the Warren Commission in 1964. This commission, led by the chief justice of the U.S. Supreme Court, Earl Warren, concluded that Oswald alone killed the president by firing three shots at Kennedy's limousine as it slowly made its way through downtown Dallas. The commission also concluded that the bullet that wounded Texas governor John Connally, who was sitting in front of Kennedy, was the same bullet that passed through the president's throat. This "single bullet" conclusion proved to be one of the more controversial issues to come out of the Warren Commission report.

Many facts laid the groundwork for the postulation that Kennedy was shot by one rifle by one man in one position. However, legitimate questions have been raised, all of which have been addressed by various experts, including forensic scientists and pathologists. For example, many people find it hard to believe that Oswald could have used a relatively cheap rifle (an Italian Mannlicher-Carcano 6.5 mm) to fire three shots, two accurately, in the space of a few seconds.

However, Oswald was a former Marine sniper who had trained to do exactly such a task. The distance between Oswald and the limousine was not very far (~270 feet) for such a weapon, and re-creations even by untrained marksmen were able to shoot a similar rifle accurately at that distance with roughly five seconds between shots. In addition, the vehicle was moving slowly, and for someone with Oswald's skills, would not have proven a difficult target. However, PRESUMPTIVE TESTS for GUNSHOT RESIDUE (GSR) on his skin were inconclusive. A paraffin cast/DERMAL NITRATE test was performed on Oswald's hands but was too contaminated to be useful. This test has since been abandoned as too prone to FALSE POSITIVES.

More vexing was the question raised by the infamous Zapruder film, which shows frame by frame the shooting, including the grisly head explosion that followed the second hit. The president's head clearly travels backward, toward Oswald's location behind him. Many believe that this motion is evidence of a second gunman (or gunmen) firing from a position in front of the limousines, from the equally infamous "grassy knoll." However, medical evaluation, WOUND BALLISTICS, and BIOMETRIC analysis show that such a movement is, in fact, consistent with a head shot coming from behind. In such a case, the bullet impacts, slows, fragments, and causes tremendous pressures to build, leading to a violent rupture, all within fractions of a second after impact. The violent rupture likely forced Kennedy's head to snap backward, and this interpretation is consistent with data obtained at the AUTOPSY. Although some questions were raised about occurrences at the autopsy, no credible evidence of any alterations to photographs or roentgenograms (X-ray pictures) has surfaced. Arguments for additional shooters in front of the motorcade were also based on the small size of the wound in the front of the president's neck, suggesting to some that this was, in fact, an entrance and not an exit wound. However, contrary to popular belief, it is not always easy to distinguish between entry and exit wounds, particularly based on size. In the case of the president, the location of his shirt collar easily explains the small exit hole.

The Magic Bullet

The "one bullet" conclusion is most often noted by conspiracy theorists as being the most unbelievable. They point to the fact that the bullet showed little deformity beyond bending and compression, with no fragmentation. How, they argue, could a bullet pass through Kennedy's spinal cord and throat, through Connally's chest, then through his wrist, and finally come to rest lodged beneath the skin of his thigh, where it fell out and was found on his stretcher? However, forensic chemical analysis of the bullets, coupled to an understanding of the bullet's approximate velocity and path, are completely consistent with this explanation. Although unusual, the one bullet theory stands up to scientific scrutiny far better than any of the numerous unsubstantiated alternatives.

Firearms analysis of the bullets, cartridge casings, and the Mannlicher-Carcano rifle found in the book depository indisputably proved that the bullets and casings had been fired from that weapon. Oswald's palm print was found on the rifle, linking him to it. The "one bullet," the second one fired and the first to strike the president, penetrated primarily soft tissue, not bone, until it reached Connally's wrist. It appeared to have exited the president's neck while tumbling, passing in a circuitous route through Connally's body, scraping along a rib and still moving fast enough to pass through wrist bones and finally come to a stop just under the surface of the skin on his thigh. Careful analysis of the film shows that Connally was clearly struck from behind a fraction of a second after the president and that they both reacted nearly simultaneously. However, the analysis that eliminated the second gunman theory was a chemical analysis of the bullets and recovered fragments first performed in 1963 and repeated in 1977 using NEUTRON ACTIVATION ANALYSIS (NAA). NAA utilizes a nuclear reactor to determine the elemental composition of samples and had been used to obtain a so-called chemical signature of bullets when other techniques had failed. The NAA undertaken by the FBI in 1963 was inconclusive. In 1977, a Select Committee on Assassinations formed by the House of Representatives commissioned a new NAA analysis using more sensitive detectors. The results showed conclusively that two and only two bullets were involved in the killing and that the bullet recovered from Connally's stretcher was the same one that deposited fragments in his wrist but was different from the fragments recovered from the president's brain tissue. Thus, although conspiracy theories thrive, they do so not because of forensic evidence, but despite it.

Further Reading

Artwhol, R. R. "JFK's Assassination, Conspiracy, Forensic Science, and Common Sense." *Journal of the American Medical Association* 269, no. 12 (1993): 1540.

Guinn, V. P. "JFK Assassination: Bullet Analyses." *Analytical Chemistry* 51, no. 4 (1979): 484A.

Lattimer, J. K. "Additional Data on the Shooting of President Kennedy." *Journal of the American Medical Association* 269, no. 12 (1993): 1544.

Petty, C. P. "JFK—An Allonge." *Journal of the American Medical Association* 269, no. 12 (1993): 1552.

kerosene *See* ACCELERANT.

ketamine A drug that is increasing in popularity particularly among juveniles and young adults taking part in parties or "raves." Ketamine is used as a veterinary anesthetic, and the only illicit source of the drug is through theft, mostly from veterinary clinics. It can also be obtained from foreign countries, particularly Mexico. As a veterinary drug, it is supplied as a liquid or a soluble powder that can be injected, sprinkled on other material and smoked, added to drinks, or snorted. The effects of ketamine have been compared to LSD and PCP, and the use of the drug is rapidly increasing. Street names for the drug include "Special K." It is listed on Schedule II of the CONTROLLED SUBSTANCES ACT (CSA).

Further Reading

Drug Enforcement Agency (DEA) Web site. "Ketamine." Available online. URL: http://www.usdoj.dea.gov/concern/ketamine.htm. Downloaded May 4, 2007.

Kirk, Paul (1902–1970) American *Forensic Scientist* Dr. Kirk is considered to be the father of modern American CRIMINALISTICS, having headed the first criminalistics program in the country. Kirk became involved in criminalistics as a result of collaboration between the Berkeley Police Department and the University of California at Berkeley encouraged by AUGUST VOLLMER. Kirk established the criminalistics program in 1937, and by 1948 it was a department under the university's School of Criminology. Kirk was active in research in many areas of evidence, including TRACE, HAIR, and FIBERS, as well as in teaching and casework. He authored a pioneering textbook, *Crime Investigation,* in 1953 with a second edition in 1974. The 1950s were among Kirk's most active years, and in 1955 he analyzed crime scene evidence from the SAM SHEPPARD case, in which Sheppard had been accused of murdering his wife, Marilyn. The famous case would later become the basis of the television show and the 1994 movie both entitled *The Fugitive.* Kirk's analysis, still controversial, concluded that Sheppard was not guilty of the crime. In many respects, the Sheppard case was to the 1950s what the O. J. SIMPSON case was to the 1990s, widely publicized and discussed, with forensic science playing a central role.

Paul Kirk in his laboratory ***(Bettman/CORBIS)***

Kirk's philosophy of criminalistics was generalist in the sense that he believed forensic scientists should have a broad scientific education and knowledge of many aspects of physical evidence. He considered the primary skill and distinction of forensic science as that of INDIVIDUALIZATION. Criminalists, in Kirk's view, were tasked with finding common sources and linking evidence to its source. Identification of evidence is secondary to individualization. Thus, the important question is not "Is this hair?" but rather "Did this hair come from this person?" The criminalistics program at Berkeley was disbanded in 1955, but by then many of the next generation of leaders in the field had been trained under Kirk, and his legacy continues to be central to forensic science in the United States.

knots A type of physical evidence important in cases of STRANGULATION, HANGING, or when victims are restrained using rope or cordage. One of the central questions in such cases is distinguishing between murder, suicide, and accidental death. The form and complexity of a knot can be a critical element in making that assignment. At the crime scene, the knot must be preserved to the extent possible, so if it is necessary to loosen the rope, cutting it rather than undoing the knot or knots is strongly preferred. Only if there is a chance of saving a victim are knots purposely undone or cut, and in cases of death, the ropes are usually not removed except as part of the AUTOPSY. Careful documentation (pictures and drawings) is essential in determining if the victim could have made the knot and positioned it where it was found. Similarly, in the case of restraints or homicide, the types of knots used can be critical in linking crimes or establishing a trademark of a certain individual or individuals.

Further Reading

Budworth, G. "Identification of Knots." *Journal of the Forensic Science Society* 22, no. 4, (1982): 327.

known samples (knowns) *See* CONTROL SAMPLES (CONTROLS).

Köhler, August (1866–1948) German *Microscopist* August Köhler is best known in the forensic community for developing a method of optimizing illumination in a microscope to produce the best possible photographs. While this does not sound difficult, Köhler developed the method in 1893, a time when microscope light sources, photography, and photographic film were primitive by modern standards. He optimized an alignment procedure that yields even and intense lighting without glare. In 1900, he went to work for the Zeiss company, which still produces microscopes today.

Further Reading

Davidson, M. W. *Molecular Expressions, Science, Optics and You: Pioneers in Optics.* Available online. URL: http://www.micro.magnet.fsu.edu/optics/timeline/people/index.html. Accessed November 14, 2007.

Köhler illumination German AUGUST KÖHLER developed the illumination pathway (pronounced "curler" illumination) that is standard today in forensic microscopy. He worked extensively in the field of photomicrography, an infant science in the late 1800s. Köhler used the method of illumination now named after him to obtain full, even, and bright lighting of specimens

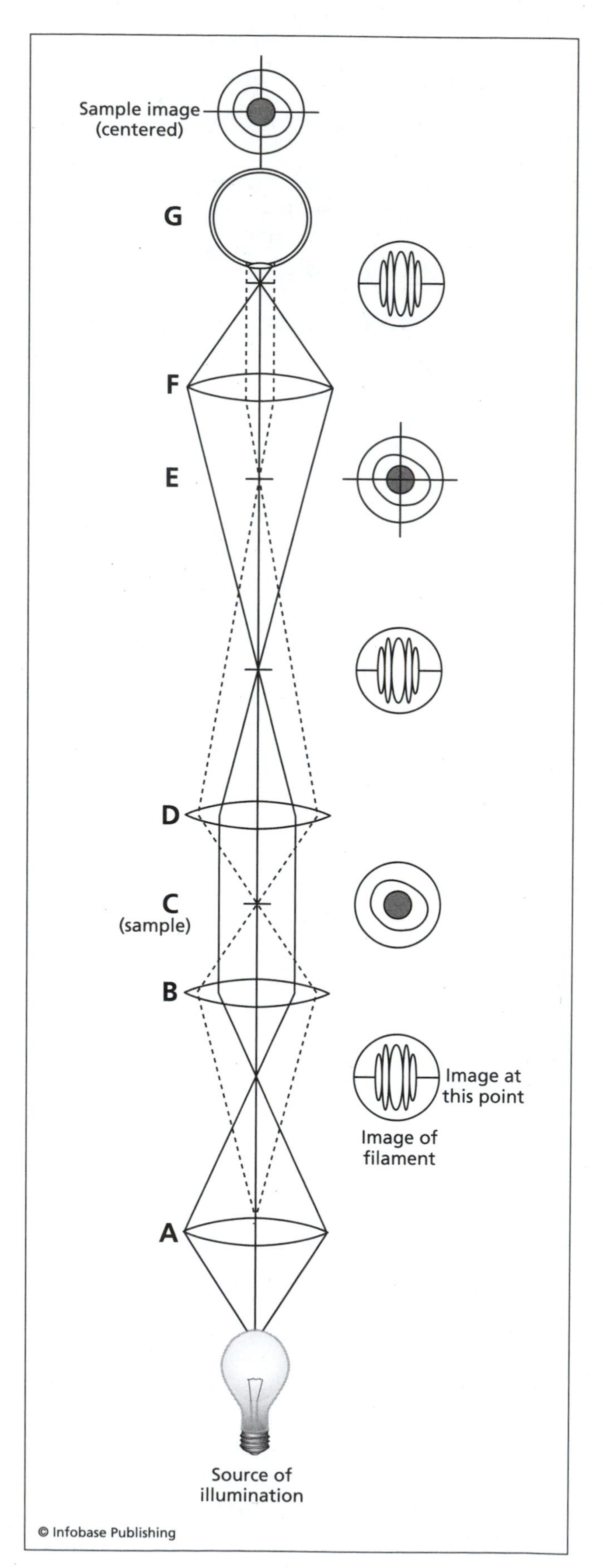

The microscopist uses the image of the filament to align the optics to ensure that the sample has maximum, glare-free, and even illumination.

that was essential for early photography. To establish Köhler illumination, the image of the filament is used to align the lenses and to set apertures such that the field of view (the portion of the sample being examined) experiences maximum illumination. The user adjusts the cone of light that passes through the sample to illuminate the field of view only and nothing more. The image of the filament is centered and focused using the Bertrand lens, which is inserted either inside the microscope body or near the ocular. Establishing Köhler illumination is analogous to aligning the optics in a spectrometer and setting slit widths.

See also MICROSCOPY.

Koppanyi test *See* DILLE-KOPPANYI TEST.

***Kumho Tire Company, Ltd. v. Carmichael* (526 U.S. 137 1999)** A decision by the U.S. Supreme Court that extended the "gatekeeper" role of trial judges in determining if the testimony of an expert should be admissible. In the *DAUBERT* DECISION of 1993, the Court assigned to judges the responsibility for deciding if the testimony of a scientific expert was admissible in the case in question. The *Carmichael* decision extended the gatekeeper role to include the testimony of any expert in any field, not just in a scientific or technical discipline.

L

laboratory accreditation *See* CERTIFICATION.

laboratory organization There is no standard for how a forensic science laboratory is organized or what sections and specialties it may contain, nor is there a standard way in which the laboratory fits into the criminal justice system. Laboratories can be part of federal agencies such as the FBI laboratory in Washington D.C. (the largest lab in the world) or the laboratories that are part of the BUREAU OF ALCOHOL, TOBACCO, FIREARMS AND EXPLOSIVES (ATF) and DRUG ENFORCEMENT ADMINISTRATION (DEA). The largest numbers of labs are associated with the states, cities, and counties under various police agencies such as state patrols, state police, county sheriffs, and city police. Many states have multiple laboratories in different regions coordinated under the state patrol, department of public safety, or similar governmental/law enforcement agency. Other labs are associated with entities such as the MEDICAL EXAMINER (ME)'s office. Even within the federal system, agencies such as ATF have several laboratories located in cities across the country.

A typical "full-service" crime laboratory will have analysts and sections that deal with DRUGS and ARSON (forensic chemistry), blood and body fluids (forensic SEROLOGY/forensic BIOLOGY), trace evidence and MICROSCOPY, FIREARMS and TOOLMARKS, and QUESTIONED DOCUMENTS. Other functions that may be present include a polygraph unit, photography and imaging, digital evidence analysis, TOXICOLOGY, and FINGERPRINTS. Some laboratories have crime scene units while others have evidence collection teams that process scenes, document them, and return evidence to the lab. On the other hand, some labs, particularly smaller branch facilities, may work strictly with drug analysis, which represents the largest workload in the vast majority of laboratory systems.

Lacassagne, Alexander (1843–1924) French *Physician and criminalist* Alexandre Lacassagne joined the faculty at the University of Lyon in 1880 as the first to chair the newly created department of forensic MEDICINE. He investigated several aspects of forensic medicine and death investigation including RIGOR MORTIS, the stiffening of the body that occurs after death. Fiction writers frequently use *stiff* as a synonym for a dead body, but that is true only for the freshly dead; rigor is usually completely released within three to four days. Lacassagne studied rigor as well, determining that the process actually appeared to begin with the heart. One of the first textbook mentions of rigor came in the 1882 edition of *Legal Medicine* by C. M. Tidy. At Lyon, Lacassagne trained many forensic scientists including EDMOND LOCARD.

lands and grooves Structures that are cut into the barrel of a firearm that cause the BULLET to spin. A spinning bullet does not tumble or wobble and as a result is much more accurate than a projectile fired from a smooth bore weapon such as a musket. Lands and grooves, collectively called rifling, are important in forensic FIREARMS examination. As shown in

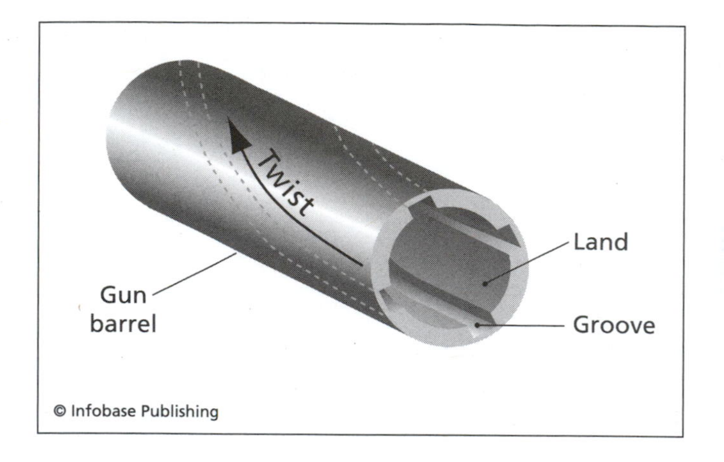

Lands and grooves in a gun barrel

the figure, the land is the portion of the barrel that remains after cutting, with the grooves being cut between them. As the bullet engages the grooves, corresponding marks are etched into the bullet that can be seen in detail using a forensic COMPARISON MICROSCOPE. The number of lands and grooves as well as their dimensions and the direction and degree of twist are CLASS CHARACTERISTICS that can be used to identify the make and manufacturer of a weapon used to fire a bullet. The FEDERAL BUREAU OF INVESTIGATION (FBI) maintains a database that describes the rifling characteristics of a large number of weapons.

These class characteristics alone are not sufficient to identify a specific weapon. To accomplish this, the firearms examiner needs the gun in question as well as bullets known to have been fired from it. It is the microscopic detail and unique STRIATION patterns that allow a bullet to be matched to a gun.

Landsteiner, Karl (1868–1943) Austrian *Physician* Landsteiner was a prolific and gifted researcher in the area of immunology who discovered the ABO BLOOD GROUP SYSTEM in 1900–01. Typing of this group would be the mainstay of forensic SEROLOGY until DNA TYPING finally supplanted it in the early 1990s. In the late 1800s the first attempts at blood transfusions had led to many deaths when the blood of the donor caused the red blood cells of the recipient to clump together (agglutinate). Landsteiner noted that this reaction was not universal in that the blood of some individuals was compatible. These observations coupled with his research led to the identification of the ABO system, which was the first blood group system identified. Landsteiner also recognized that a person's blood group was inherited and thus would be useful in paternity cases. Because the ABO system is inherited, it also served as a GENETIC MARKER system in forensic analysis of blood and body fluids for nearly a century. Landsteiner's discovery led to systematic typing for blood transfusions and saved untold thousands of lives. As a result, he was awarded the Nobel Prize in medicine in 1930. Landsteiner eventually moved to New York and continued to work in the field of immunology, participating in the discovery of several more blood group systems, including the Rh system.

Further Reading

Nobel e-Museum. "The Nobel Prize in Physiology or Medicine 1930." Available online. URL: http://www.nobel.se/medicine/laureates/1930/. Downloaded May 5, 2007.

Karl Landsteiner, winner of the Nobel Prize for his discovery of the ABO blood group system ***(National Library of Medicine, National Institutes of Health)***

larvae *See* ENTOMOLOGY, FORENSIC.

lasers Light sources that emit very intense radiation at a specific wavelength. Lasers are being used in many forensic science arenas including FINGERPRINTS and PHOTOGRAPHY and as components of instrumentation used in forensic CHEMISTRY. Lasers can be used to induce fluorescence, which in turn can be exploited in several ways. Fluorescent detectors are among the most sensitive available for analytical instruments such as HIGH PRESSURE LIQUID CHROMATOGRAPHY (HPLC) and can be exploited as part of an IMMUNOASSAY used in forensic toxicology. Lasers can also be components of ALTERNATIVE LIGHT SOURCES (ALS) used in fingerprint visualization. Finally, lasers can be used to vaporize a surface for introduction into instruments such as mass spectrometers. This technique, called laser ablation, is suited for the examination of paint layers and any other samples in which a characterization of the surface is needed.

latent prints/latent fingerprints FINGERPRINTS (or other prints such as palm prints) left at a scene on an object that are not visible, barely visible, and/or can be made visible. This definition is somewhat loose, and the term *latent prints* is often used to refer to any type of fingerprint regardless of how visible it is.

Lattes, Leone (1887–1954) Italian *Forensic Scientist* Lattes, who was a professor at the Institute of Forensic Medicine in Turin, Italy, was instrumental in applying KARL LANDSTEINER'S discovery of the ABO BLOOD GROUP SYSTEM to forensic casework. In 1915, he developed a test that came to be known as the "Lattes procedure" or the "Lattes crust test" in which red blood cells were added to dried bloodstains to determine the ABO blood type of the stain. For example, if the stain came from a person with Type A blood, the stain will contain anti-B antibodies. When Type A cells are added to the stain, nothing happens, but when Type B cells are added, they clump together (agglutinate). Although novel at the time, the Lattes procedure did not work well with old stains, and interpretation of results was difficult. However, it laid the groundwork for later more sensitive and robust testing methods such as ABSORPTION-ELUTION that were widely used until DNA TYPING supplanted ABO typing in forensic serology.

lead This heavy metal, symbol Pb, is of forensic interest primarily as a poison and as the metal used to make bullets. As a poison, lead has a long and infamous history and has frequently been suggested as a contributing factor in the fall of the Roman Empire. Because it is relatively soft, it is easily smelted and extracted and thus became one of the first metals to be exploited by humans. Most lead poisoning is inadvertent or accidental versus homicidal or suicidal, probably because there are many other materials such as ARSENIC and CYANIDE that are more deadly and thus better suited to those purposes. Modern exposures to lead come primarily from old paints (leaded paints were banned in 1978), lead from gasoline (although use of tetraethyl lead as an additive ceased in the 1990s), and mining and pollution. Lead has also been used in pipes and is an ingredient in solder. Chronic lead poisoning is a danger for children, whose health, especially of the central nervous system, is at particular risk. Toxicologists have several tools available to detect lead in the blood including ATOMIC ABSORPTION (AA), INDUCTIVELY COUPLED ATOMIC EMISSION (ICP-AES), and INDUCTIVELY COUPLED MASS SPECTROMETRY (ICP-MS).

Lead is also used in BULLETS, although many modern bullets are partially or completely jacketed with other metals such as copper. Lead is the metal of choice because it is easy to melt and cast and because it is so dense and heavy. With a projectile shot from a firearm, its power is defined by the energy that the projectile can transfer to its target, which in turn depends on the mass of the projectile. This energy, the kinetic energy, is expressed as $KE = \frac{1}{2} mv^2$. Thus, a dense projectile made of lead will be more effective than a projectile of the same size made of a less dense metal such as aluminum. Ammunition that uses depleted uranium, an even denser metal, is used by the military for this same reason. Lead is also an ingredient in many primers and as a result is a component of GUNSHOT RESIDUE (GSR).

lethal dose The amount of a substance such as a poison or toxin that is necessary to cause death. This idea is not as simple as it may first appear since the exact lethal dose for any one individual will be unique. Factors that influence the size of a lethal dose include health of the person, age, family history and genetics, and weight. For example, two extra-strength aspirin tablets are a normal dose for an adult, but that same dose could be fatal to an infant. To account for this size dependence, dosages of drugs are normally defined in units of milligrams of the substance per kilogram body weight. Toxicologists often refer to a value such as the LD_{50}, or "lethal dose 50," a value most often derived from animal studies. In such studies, a large population of experimental animals such as mice or rats is given increasing doses of the substance being tested. The dosage that is fatal for half the population is the LD_{50}. Although a useful estimate, an LD_{50} is specific only to the animal of interest and is not always directly transferable to human beings. Some LD_{50} values exist for humans, but this data is normally derived from studies of accidental or unintentional exposures. Even in those cases, the factors mentioned above can greatly influence the lethal dose for an individual. Thus, the LD_{50} value is an average. For a highly poisonous material such as potassium CYANIDE (KCN), the LD_{50} is on the order of about 1 mg/kg of body weight, which translates to a lethal dose of about 68 mg for a person who weighs 150 pounds. By way of comparison, a typical chewable "baby aspirin" tablet contains about 80 mg of material. Similarly, a fatal dose of ARSENIC trioxide (AsO_3) for the same size person would be about 123 mg.

Alternatively, a fatal dose can be expressed as a concentration in blood or other body fluids, as is the case with CARBON MONOXIDE (CO). Carbon monoxide poisoning accounts for a large number of accidental deaths each year. Its toxic effects are caused by the compound's ability to bind with hemoglobin much more strongly than does oxygen. Thus, a person exposed to carbon monoxide will have hemoglobin that is bound to the poison rather than to oxygen. Toxic effects, including a lethal level, are related to the %COHb, or percentage of the hemoglobin that is bound to CO instead of oxygen. Fatalities can occur in the range of 50 percent or above, but the exact value of %COHb that is fatal will vary among individuals.

leucomalachite green A PRESUMPTIVE TEST for blood that is not as frequently used as are the KASTLE-MEYER (K-M) and LUMINOL tests. It is prepared by combining leucomalachite green and sodium perborate ($NaBO_3$) in water and acetic acid.

lidocaine Also known as Xylocaine, this white powder is sometimes used as a CUTTING AGENT for COCAINE because, like cocaine, it is an anesthetic that produces localized numbness. It also reacts with COBALT THIOCYANATE, a reagent commonly used as a presumptive test for cocaine. However, the color change is different from that observed with cocaine and would not produce a FALSE POSITIVE if interpreted by an experienced forensic chemist.

Lieberman test (Liebermann reagent) A PRESUMPTIVE TEST (a "spot test") occasionally used in DRUG ANALYSIS. The reagent is prepared by dissolving potassium nitrite (KNO_2) in concentrated sulfuric acid, and it is used to test for COCAINE (a yellowish color indicates the possible presence of the drug) and MORPHINE (a black test is positive).

lie detector *See* POLYGRAPH.

ligature The object used to cause STRANGULATION in a murder, suicide, or accidental death. Ligatures are often ropes or other cords but can be anything that will encompass the neck and to which pressure can be applied. Towels, scarves, belts, sheets, and phone cord have all been used as ligatures, and the KNOTS used to tie or secure them can become critical evidence. Ligatures can also produce distinctive impressions on the neck that can be physically matched to the ligature. For example, if a belt with a woven pattern is used to strangle someone, the weave pattern may be visible in the skin in much the same way a bite mark is. A ligature is also any cording that is used to bind a victim.

likelihood ratio A quantity useful in the statistical analysis, presentation, and interpretation of forensic analyses. What is unique and relatively new about this approach is that it requires at least two different theories or hypotheses concerning given scenarios, which are usually referred to as "competing hypotheses." Assume, for example, that a single hair is found on the clothing of a homicide victim and that a suspect has been identified. Investigators, and ultimately a court, will want to know if the hair is the suspect's or not. However, since hairs cannot be linked to a unique COMMON SOURCE in the way fingerprints can, the answer to that question becomes more complex than a simple "yes" or "no." Instead, the results of the analysis can be reported based on probabilities and a likelihood ratio. One hypothesis is that the hair is that of the suspect, whereas one competing hypothesis might be that the hair is very similar to the suspect's but came to be on the victim as a result of some random process. The relative likelihood of these two hypotheses can be expressed numerically using a likelihood ratio. The likelihood ratio could take into account several separate probabilities such as the probability that the victim and the suspect came into contact, the probability that a transfer occurred, the probability that the hair remained in place long enough to be detected, and so on. Similar probabilities could be estimated for competing hypotheses and the summation of these probabilities used to create the likelihood ratio that would be expressed as: LR = probability that the suspect is the source of the hair / probability that a random person is. The type of evidence involved and the amount of hard data and databases available are among the many factors that determine the reliability and usefulness of a likelihood ratio. For example, fingerprints or DNA TYPING results are supported by much larger databases and knowledge of frequencies than are hairs, fibers, glass, and other kinds of physical evidence. Despite these limitations, likelihood ratios can be useful for examining alternative explanations and encouraging the generation of alternative explanations for forensic findings.

See also BAYESIAN STATISTICS.

limnology, forensic Limnology is a subdiscipline of biology devoted to the study of freshwater bodies and ecosystems. Thus, forensic limnology is an extension of this study into civil and criminal investigations and cases. Freshwater systems includes lakes, streams, ponds, and marshes, and a type of case that might involve limnology would be something such as a body being dumped in a lake and using the principles of limnology to determine what time of the year the dumping occurred. Typical types of organisms that are used are freshwater DIATOMS, other types of algae, larger aquatic plants, and microscopic and small animals.

Further Reading

Siver, P. A., W. D. Lord, and D. J. McCarthy. "Forensic Limnology: The Use of Freshwater Algal Community Ecology to Link Suspects to an Aquatic Crime Scene in Southern New England." *Journal of Forensic Sciences* 39, no. 3 (1994): 847.

Lindbergh kidnapping A 1932 crime that, until the O. J. SIMPSON trial, was considered to be the "crime of the century." Like the Simpson trial, scientific evidence played a critical role, and the atmosphere and media frenzy surrounding both trials were similarly described as circuslike.

Charles A. Lindbergh became a national and international hero in 1927 after flying the Atlantic alone in the *Spirit of St. Louis*. He later married Anne Morrow,

Arthur Koehler testifying at the Lindbergh trial *(Bettman/CORBIS)*

and their first child, a son named Charles, Jr., was 20 months old when he was kidnapped around 9:30 P.M. on March 1, 1932. He was taken from his nursery located on the second floor of the Lindberghs' home in Hopewell, New Jersey. The kidnapper left a homemade ladder at the scene that would later provide critical evidence, but footprints also left were not properly documented or preserved. A ransom note, one of 14 that would be sent by the kidnapper, demanded $50,000 in ransom for the boy's safe return. Mailed notes all had postmarks from the New York City area. The investigation by the New Jersey State Police was headed by Colonel Norman Schwartzkopf, father of the general who would later lead coalition forces to victory in the 1991 Gulf War. The case took many bizarre turns and became the subject of a worldwide media frenzy. An elderly retired teacher named John Condon became the intermediary between Lindbergh and a man who called himself "John," and eventually a ransom of $50,000 was paid, the majority of it in gold certificates, which were used as currency at the time. Serial numbers of the bills were recorded before the money was delivered. Sadly, the body of the child was found a month later in the woods close to the Hopewell home. The cause of death was listed as a skull fracture, most likely the result of an accidental fall from the ladder during the kidnapping. However, given that a forensic pathologist or medical examiner did not conduct the AUTOPSY, that conclusion has been questioned. It is possible that the kidnapper killed the boy shortly after the abduction and dumped the body near the home. The kidnapper was never seen or heard from after the ransom was paid.

In 1933, President Franklin D. Roosevelt ordered that all gold certificates be exchanged for standard currency. Law enforcement agents still working the case hoped for a break, anticipating that the kidnapper would have to turn in a good portion of the ransom money, most of which had not yet surfaced. The amount that had been spent was in the New York City area, allowing law enforcement to concentrate efforts there. The break came in September 1934, when an exchange bank received a $20 gold certificate from a man who matched the description of the kidnapper. More important, a license plate number had been written on the bill, which was quickly traced to a truck belonging to Bruno Richard Hauptmann, who lived in the Bronx. He was arrested, and a subsequent search of his garage uncovered a gun and several thousand dollars of the ransom money gold certificates. Hauptmann at first denied all knowledge of the money but soon changed his story, claiming that a friend had given him the money. The friend had previously returned to Germany and died there, making it impossible to investigate that claim.

During the time between the kidnapping and murder of the child and Hauptmann's arrest, forensic investigations were undertaken on the physical evidence, including the ransom notes (QUESTIONED DOCUMENTS), trace evidence, psychological and psychiatric studies, and perhaps most damning for Hauptmann, analysis of the ladder. ALBERT S. OSBORN, a pioneer in the field of questioned documents, performed analysis of the handwriting found in the ransom notes. Arthur Koehler, a wood expert employed by the Forest Service, undertook a meticulous evaluation of the ladder, including the wood, construction techniques, and toolmarks found on the wood. Eventually, he was able to trace the lumber used to a lumberyard and mill located in the Bronx. Marks made by planers on the wood in the ladder matched a planer at the yard. A search of the attic above Hauptmann's apartment revealed a missing floorboard. Nail holes and tree ring patterns from a rail of the ladder lined up perfectly where the floorboard had been. Furthermore, Koehler was able to demonstrate this at the trial as well as to show how the planer marks from the ladder matched Hauptmann's planer. Hauptmann was convicted and, after a series of appeals and reviews, including one by the Supreme Court, was executed on April 3, 1936. As with any case that captures the public's imagination, conspiracy theories have been and continue to be put forward. However, the consensus among forensic scientists who have reviewed the case is that the analyses were sound and that Hauptmann was guilty as charged.

Further Reading

Baden, M. M. Plenary Session: "The Lindbergh Kidnapping Revisited: Forensic Sciences Then and Now." *Journal of Forensic Sciences* 28, no. 4 (1983): 35.* *This issue contains several articles concerning the crime and the forensic analysis it entailed.

Graham, S. A. "The Anatomy of the Lindbergh Kidnapping." *Journal of Forensic Sciences* 42, no. 3 (1997): 368.

Horan, J. J. "The Investigation of the Lindbergh Kidnapping Case." *Journal of Forensic Sciences* 28 no. 4 (1983): 1,040.

linguistics, forensic The evaluation of the use of language in forensic applications. Techniques of forensic linguistics are used in such areas as attempting to identify a region where a person is from, determining the author of a document or determining if two documents were written by the same person, and attempting to clarify the meaning of statements made in court or to law enforcement officials. Linguists can also work with QUESTIONED DOCUMENT examiners and VOICE RECOGNITION experts. Threatening letters or phone calls, ransom notes, e-mail communications, and disputed wills are just a few examples of the types of evidence forensic linguists analyze. One of their most common tasks is in the area of speaker or author identification, comparison, authentication, and analysis. Some authors consider forensic stylistics to be a separate specialty focusing on the style of speech (oral or written) characteristic of a group or individual.

Analysis of documents or speech involves study of the types of words used, word choice (for example, "pop," "soda," or "coke" to describe a carbonated beverage), grammar, accent or dialect, spelling, error patterns, and sentence structure. Statistical analysis and comparison to general population usage patterns are sometimes employed in an attempt to determine where a speaker might have come from or where he or she might live. Similarly, if a will is suddenly changed on the eve of a person's death, linguists can compare known writings of the person with the will to see if the questioned writing follows the same pattern as that of the old will. Many of these types of analyses are used as investigative tools more than in courtroom testimony directly. For example, linguistic analysis alone would not be able to prove that the author of a threatening letter lived in a certain region, but knowing that it is likely could prove a great help to investigators.

Another area considered to be a part of forensic linguistics is "discourse analysis," which evaluates courtroom transcriptions or other legal statements such as confessions for accuracy, perception, intent, and meaning. For example, the exclamation "drop it!" might refer to a weapon or be a request for someone to drop a topic of conversation. How such a statement was meant and how it was interpreted could be critical. Discourse analysis can also be important when translations are involved since translators must often use their judgment to select words or phrases in one language to express the meaning of another. Thus, two translators can take exactly the same statement and derive two different translations with differences that might seem subtle in one language but substantial in the other. Thus, word selection, sentence construction, and other linguistic elements become critical in conveying meaning.

Further Reading

Pickett, P. "Linguistics in the Courtroom." *FBI Law Enforcement Bulletin*, October 1993.

Varney, M. "Forensic Linguistics." *English Today* 13, no. 4 (1997): 42.

lip prints/lipstick A form of impression evidence that can be left on glasses, cigarettes, and other surfaces. There are two aspects of lip prints that can be of forensic interest. First is the pattern itself and second is the composition of the material, such as lipstick or lip balm, that might have been the medium by which the impression was made. The study of lip patterns, also called celioscopy, might seem similar to dactyloscopy, the study of FINGERPRINTS. However, research to date indicates that, unlike fingerprints, lip prints are not unchanging and that there appears to be some inherited characteristics such that relatives can have similar lip prints. In contrast, even identical twins have different fingerprints. Thus, the patterns of lip prints may have some investigative use, but this type of forensic evidence has not proven to be of much value.

The chemical composition of materials with which the print is made may be much more valuable. Using techniques such as microscopic analysis and instrumental analysis, including GAS CHROMATOGRAPHY/MASS SPECTROMETRY (GC/MS), the chemical composition of a lipstick can be identified in part or in whole. If a comparison standard is available, comparisons can be made in an effort to link them to each other. Furthermore, manufacturers of cosmetics change their ingredients frequently, so knowledge of chemical composition might help establish a broad time frame in which the product was available.

Further Reading

Moenssens, A. A., J. E. Starrs, C. E. Henderson, and F. E. Inbau. *Scientific Evidence in Civil and Criminal Cases.* Westbury, N.Y.: The Foundation Press, 1995.

liquid chromatography *See* HIGH-PERFORMANCE LIQUID CHROMATOGRAPHY (HPLC).

lividity Also referred to as livor mortis, this is the settling of blood that occurs in a body after the heart stops beating. At the time of death, circulation ceases, and blood is no longer being pumped under pressure to all areas of the body. In addition, since blood is not reaching the lungs and is not being oxygenated, it takes on a bluish-purple tint. Exceptions can occur if the victim has been poisoned with substances that alter the color of the blood such as CARBON MONOXIDE (CO). CO imparts a distinctive cherry red color that will also alter the appearance of the lividity stain. Lividity occurs in the areas of the body in which gravity naturally causes settling, and lividity stains can look similar to bruises. The only places blood will not settle is in locations where pressure being applied. For example, if a person dies while seated, pressure being applied to the buttocks will prevent blood from pooling there.

The first signs of lividity begin to appear about an hour after death and reach a maximum after three to four hours. After about 12 hours, no additional lividity will occur. Thus, lividity stains are useful in determining the time since death, or the POSTMORTEM INTERVAL (PMI). The stain pattern can also be used to determine if a body was moved during the period in which lividity was developing. If a stain has already formed and the position of the body is altered within the 12-hour window after death, a new pattern can form while the old stains partially fade. Similarly, if the body is moved after 12 hours, the lividity pattern may not match the position in which the body is finally discovered. Such inconsistencies are usually uncovered at AUTOPSY, when the clothing of the deceased is completely removed.

livor mortis *See* LIVIDITY.

Locard, Edmond (1877–1966) French *Forensic Scientist* Locard was instrumental in taking new theoretical ideas of what was then called police science and applying them to casework. Locard was trained in both law and medicine and was influenced by the writings of HANS GROSS as well as the fiction of SIR ARTHUR CONAN DOYLE. In 1910, Locard established a forensic laboratory in Lyon, France. The lab was primitively equipped, but even so, Locard was able to establish a reputation and to increase the visibility of forensic science in Europe. Locard was interested in microscopic evidence, particularly DUST, and he believed that such trace evidence was crucial in linking people to places. Although he apparently never used the exact phrase himself, Locard is most famous for Locard's exchange principle that evolved from his studies and writings. The principle is stated simply as "every contact leaves a trace" and reflects his belief that every contact between a person and another person or a person and a place

Edmund Locard *(© Collection Roger-Viollet/The Image Works)*

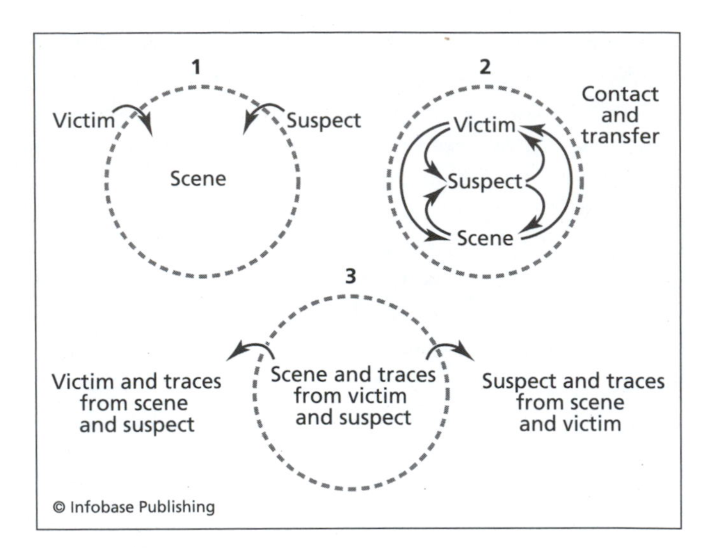

Locard's exchange principle

results in the transfer of materials between the entities involved. Most of this TRANSFER EVIDENCE, such as dust, is microscopic, and it may not last long, but the transfer does occur, and it is the task of the forensic scientist to find those traces and use them to establish the link. An example is shown in the figure, in which a victim and a suspect come into contact at a crime scene, resulting in transfers of trace evidence among all elements present. When the contact is finished, the scene will contain evidence of both individuals, perhaps in the form of blood, hairs, fibers, or fingerprints. Likewise, each person will carry away traces of the scene (dust, carpet fibers, and so on) and of the other person. Not all of these may be detectable, but they do occur, according to Locard's exchange principle.

The success of his laboratory and methods encouraged other European nations to form forensic science laboratories after the conclusion of World War I. In Lyon, he founded and directed the Institute of Criminalistics located at the University of Lyon, and he remained a dominant presence in forensic science into the 1940s.

See also LACASSAGNE, ALEXANDER.

Locard's exchange principle *See* LOCARD, EDMOND.

logP A quantity used to measure the relative solubility of compounds such as drugs in water as compared to fats and lipids. In TOXICOLOGY, this coefficient is important because it reveals in part how easily a drug can cross fat-based membranes (lipid layers) found in cell walls. It is also called the octanol/water partition coefficient (K_{ow}) to describe how the value was originally obtained in the laboratory. To determine K_{ow} of a drug using this method, a known amount of it is placed in a special glass container called a separatory funnel. Water is added along with octanol, a long-chain alcohol that is not soluble in water. In the separatory funnel, the layers separate with the water on the top and octanol on the bottom, just as oil and vinegar separate. When the funnel is vigorously shaken, the liquids mix with the drug and the drug dissolves in one or both depending on its chemical structure. Molecules that are nonpolar such as oils favor the octanol layer, while polar drugs or drugs that ionize will dissolve in the water to a much larger degree than in the octanol. When the layers separate, each is analyzed to determine the concentration of the drug in both. Drugs that prefer the octanol are termed lipophilic or hydrophobic, while drugs that prefer the water layer are lipophobic or hydrophilic. In the body, hydrophilic drugs are often excreted quickly in the urine, while lipophilic drugs are retained longer and may concentrate in fatty tissues. A familiar example are the vitamins C and E. Vitamin C is hydrophilic, and any excess amounts taken (within reason) are quickly excreted from the body. In contrast, vitamin E (an oil) lingers in the body and can build up over time.

London subway bombings *See* "Identification of the Dead" essay.

lorazepam *See* ATIVAN.

Love Canal A milestone in the history of the environmental movement and subsequent environmental legislation and one of the earliest implementations of the principles of ENVIRONMENTAL FORENSICS. The canal in Niagara Falls, New York, had been started in the 1890s but was never completed and never used as a transportation waterway. Instead, the large ditch was used by Hooker Chemical to bury waste materials during the 1940s and early 1950s, before laws regulated these practices. Wastes were placed in barrels or poured into the soil, and the ditch was eventually covered. In 1953, the property was given to the local school district, which built a school on it. Following wet years in the 1970s, waste began to seep into the school and into residences built alongside the converted canal. After national publicity and a prolonged dispute, residents were eventually bought out and resettled. The incident was crucial in the passage of subsequent legislation, such as the "Superfund."

luminescence The emission of ELECTROMAGNETIC RADIATION (EMR) by an atom or molecule as the result of some form of excitation. The emitted electromagnetic energy can be of any kind, but the most familiar is that in the visible range of the electromagnetic spectrum. There are three forms of luminescence based on how the excitation occurs: chemical (chemoluminescence or chemiluminescence); heat (thermal luminescence); and photoluminescence (caused by absorption of light). An example of thermal luminescence can be seen (literally) when table salt (sodium chloride, NaCl) is placed in a flame. The heat is absorbed by the sodium atoms and causes them to emit a distinctive yellow light. This light is the same as that produced in sodium vapor lamps commonly used as streetlights. "Glow in the dark" watches are based on photoluminescence—the material on the watch face absorbs visible light, which is later emitted. In these cases, the emitted light is of a longer wavelength (lower energy) than the light that was originally absorbed. Thus, if a material absorbs energy in the ultraviolet region of the electromagnetic spectrum, it may emit in the visible region, which is lower energy and has longer wavelengths. In forensic science, chemoluminescence is exploited to detect traces of blood using the LUMINOL reagent (see color insert). Luminol reacts chemically in the presence of heme (in hemoglobin), and in the process of the reaction, some of the chemical energy is converted to light that can be seen as long as the surroundings are dark. FLUORESCENCE, another form of luminescence, is exploited in many areas of forensic chemistry, toxicology, biology, and microscopy.

luminol A PRESUMPTIVE TEST for blood that is favored because of its sensitivity, particularly to old blood or traces that have been left after an attempted cleanup of a crime scene. Luminol, unlike other tests for blood, does not work by producing a color change in the presence of hemoglobin. Rather, the reaction produces light in a process called chemoluminescence—a "glow in the dark" effect. Since the glow is faint, this test is not appropriate for areas that cannot be made dark such as large outdoor scenes. However, indoors luminol is one of the most useful tests for blood. In addition, the chemicals used in the luminol reagent (3-aminophthalhydrazide, sodium carbonate [Na_2CO_3], and sodium perborate [$NaBO_3$]) do not interfere with any subsequent DNA TYPING that is performed on bloodstains that are discovered. Finally, since no color change need be observed, the color of the material on which the blood has been deposited is not a limitation as it can be for other reagents. Luminol is sometimes referred to as an "enhancement chemical," and it can be sprayed over large areas. An example of luminol application is found in the color insert.

lysergic acid diethylamide (LSD) A potent hallucinogen in the same family as MESCALINE (PEYOTE) and PCP. Doses as low as 25µg (one microgram [µg] is equal to one one-millionth of a gram) are sufficient to induce the drug's effects, which include alteration of sensory perceptions, hallucinations, a feeling of floating or being "out of body," and extreme mood swings. Physical symptoms such as dilation of the pupils, sweating, and increased heart rate also occur. LSD is an ergot ALKALOID that is produced from lysergic acid. Lysergic acid is a compound produced by a fungus that attacks grasses, a family that includes grains such as wheat and rye. Another PRECURSOR chemical, lysergic acid amide, is found in the seeds of the morning glory flower. There has been speculation that historical incidents of mass hysteria or otherwise unexplained behavior, such as the Salem witch trials, might have been partially attributed to grains that were infected with the fungus and subsequently used in food such as bread in which the preparation and baking process produced LSD.

LSD was first made synthetically in 1938, and its hallucinogenic properties were unearthed by accident when the chemist who made it accidentally ingested a small amount, resulting in the first confirmed LSD "trip." For a short period, it was used in conjunction with psychotherapy, but that use was soon abandoned. LSD is now listed on Schedule I of the CONTROLLED SUBSTANCES ACT (CSA). Although not as popular now as in the 1960s and 1970s, the drug is still encountered, sold most often on tiny blotter papers with patterns on them or soaked into sugar cubes or tiny tablets that come in many different colors. These tablets are often named "haze" based on the color of the tablet; the term "purple haze" made famous by rock musician Jimmi Hendrix in the 1960s may well refer to a purple LSD tablet and the effects it produced. LSD does not produce physiological dependence, but there does appear to be psychological dependence. A unique aspect of LSD use are "flashbacks" that can occur for years after the drug was used.

Forensic chemical analysis and toxicological analysis of LSD is complicated by the very small quantities that make up the typical street dose. An analysis of suspected LSD samples usually begins with a PRESUMPTIVE TEST (chemical spot test) using the EHRLICH'S REAGENT. Because LSD fluoresces strongly under ultraviolet light (UV, or black, light), it is possible to exploit that property in further extraction and analyses. THIN LAYER CHROMATOGRAPHY (TLC) combined with UV visualization or other developers is a common tool used, as is HIGH PRESSURE LIQUID CHROMATOGRAPHY (HPLC) combined with a UV or fluorescent detection system. To obtain positive identification, it is often necessary to isolate the LSD from other materials present in the sample, and this can be achieved using HPLC or column chromatography. In both cases, the extract is placed on the chromatographic column and allowed to move through it. This process separates the LSD from other compounds, and by the time the LSD emerges from the sample, it is relatively pure. In the case of column chromatography, it is possible to monitor the movement of the LSD under UV light. By collecting this fraction of the column effluent, the chemist can obtain a pure enough sample to perform INFRARED SPECTROSCOPY (IR), which provides definitive identification. GAS CHROMATOGRAPHY (GC), so valuable for the analysis of most every other illicit drug, is not applicable to LSD since it is not stable at the elevated temperatures used in the instrument.

Toxicological analysis of LSD present in the body is difficult due to the very low concentrations typically present in blood, urine, or other fluids and tissues. IMMUNOASSAY techniques such as RIA and ELISA are the preferred analytical methods.

Further Reading

Drug Enforcement Agency (DEA) Web site. "Lysergic Acid Diethylamide (LSD)." Available online. URL: http://www.usdoj.dea.gov/concern/lsd.htm. Downloaded May 4, 2007.

M

MacDonald, Jeffrey A U.S. Army Green Beret captain and physician who was convicted in 1979 of the murder of his pregnant wife, Colette, and two young daughters. The case was the subject of a popular book and miniseries entitled *Fatal Vision* (written by Joe McGinniss), and discussions and legal maneuvering continue more than 20 years after the fact. From the forensic perspective, the case was notable because of extensive physical evidence such as hairs, fibers, blood spatter patterns, and murder weapons (an ice pick and wooden club). At the trial, fiber evidence, blood spatters, and a bloody footprint were crucial in reconstructing the crime scene and suggesting the course of events during the killings. MacDonald, now serving a life sentence, continues to maintain his innocence and recently won the right to DNA TYPING (unavailable at the time of the murders) of some of the evidence. The case continues to draw attention, due in no small part to the horrible brutality of the crime.

The murders took place on the night of February 17, 1970, at Fort Bragg, North Carolina. MacDonald claimed to have been sleeping in the living room of the family home when the screaming of his wife and daughters awakened him. Three men stood over him, he maintained, and a struggle followed in which MacDonald was slightly injured. He also claimed to have seen a woman in a floppy hat in the house. When the military police arrived, Colette's body was found on the floor of the master bedroom with a blue pajama top covering her chest. The two girls were found dead in their rooms. On the headboard of the bed in the master bedroom, the word "PIG" was written in the wife's blood. A copy of a magazine that described the recent murders of Sharon Tate by Charles Manson and his accomplices was found in the living room.

The pajama top found covering Colette's chest had numerous puncture holes that MacDonald claimed were inflicted as he struggled with the intruders. After the struggle, he claimed to have removed the shirt and draped it across her. However, he did not have any ice pick wounds, and he explained this by telling investigators that he had partially removed the shirt, pulling it off his chest and on to his arms so that the ice pick did not hit flesh. However, many of the holes correlated closely with ice pick wounds found on his wife's chest, and the lack of tearing of the holes meant the pajama top was likely stationary when the wounds were inflicted. This would be inconsistent with holes inflicted during a struggle, and the most likely explanation was that MacDonald had placed the pajama top across Colette's chest while she was on the floor. She was not moving when the ice pick was thrust through the pajama top and into her chest. Sixteen additional stabbing wounds were inflicted in the chest by a knife, and it was the combination of these wounds that caused her death.

Additional evidence in the form of blood spatter patterns, a crucial bloody footprint leading out of one of the children's rooms, and the location of fibers associated with the pajama top led to MacDonald's conviction. The scenario forwarded by the prosecution and accepted by the jury was that an argument began between MacDonald and his wife in the master bedroom that centered on the eldest daughter, also in the room with them. A 2″ × 4″ club was used to strike the girl, fracturing her skull. Colette came to her child's defense and in the resulting struggle was struck and injured. MacDonald then went to the room of the younger daughter and was followed by his injured wife. The final struggle took place there, where Colette was rendered unconscious and the younger girl was killed. The physical evidence further supported the hypothesis that MacDonald then moved his wife's body back to the master bedroom, where he removed the pajama top, placed it over her chest, and inflicted the fatal ice pick and knife wounds. The key elements in the crime scene reconstruction of MacDonald's movements were the dispersal of fibers from the pajama top and the blood stain and spatter evidence, the interpretation of which was greatly simplified since all four family members had different ABO BLOOD GROUP SYSTEM types.

MacDonald was convicted of two counts of second-degree murder and one count of first-degree murder for the killing of the youngest daughter. He was sentenced to life imprisonment. In the 30 years since the murders, MacDonald has never wavered from his claim that the four intruders were the murderers. Over the years, legal challenges have been raised based on many different aspects, including a challenge to the analysis of the hair and fiber evidence performed by the FEDERAL BUREAU OF INVESTIGATION (FBI). The analyst involved took part in the 1997 investigation of the FBI LABORATORY that led to widespread reform in the system, and the defense argued that the shadow cast on the analyst was sufficient to bring into question the results of his analysis of the critical fiber evidence. However, the final chapter may be finished when DNA analysis is completed.

Further Reading

Murtaugh, B. M., and M. P. Malone. "'Fatal Vision' Revisited: The MacDonald Murder Case." *The Police Chief,* June 1993.

MacDonnell, Herbert *See* BLOODSTAIN PATTERNS.

Madrid bombings Bracketed by the attacks of September 11, 2001, and Hurricane KATRINA in the United States, Europe endured two bombing attacks, one in Madrid, Spain, on March 11, 2004, and one in London on July 7, 2005. The attack in Madrid targeted four commuter trains, killing 191 people and injuring 2,000. The bombs used stolen commercial EXPLOSIVES detonated by cell phones, and the perpetrators blew themselves up to avoid capture. The London attack targeted the Underground subway system and involved three bombs that went off deep in the tunnels within less than a minute. A fourth bomb went off nearly an hour later, perhaps by mistake or by mishandling. In contrast to Madrid, the London bombings were suicide attacks. Some of the recovery workers who endured the heat of the Underground tunnels were veterans of tsunami recovery efforts. The Madrid attacks spawned an international controversy over a mistaken FINGERPRINT match and the wrongful arrest of a Muslim lawyer in Portland, Oregon.

See also "Identification of the Dead" essay; MAYFIELD CASE.

maggots *See* ENTOMOLOGY, FORENSIC.

magnetic flake powders *See* FINGERPRINTS.

Magnus, Albertus (1193–1280) German *Alchemist and theologian* One of the first milestones in chemistry that had direct forensic implications occurred in 1250. In that year or close to it, a true chemical separation of elemental (metallic) ARSENIC from minerals containing arsenic occurred. Albertus Magnus, a German scholar, alchemist, and theologian managed the feat, which disproved assumptions held since the time of ancient Greece. The Greeks had thought that arsenic minerals such as white arsenic (As_2O_3) and yellow arsenic (As_2S_3) were actually single elements and not combinations. By isolating the metal arsenic from an arsenic mineral, Magnus proved that these minerals were combinations of elements and not pure elements themselves.

Magnus noted that if sufficient heat is applied to minerals containing arsenic, the arsenic metal sublimes. This process is called pyrolysis or pyrolytic chemistry. Magnus trapped this vapor and allowed the metal to condense. Five hundred years later, the first forensic tests for arsenic would be based on this same principle. The same trick had been used for gold, but Magnus was the first to realize that there was such a thing as metallic arsenic and that a fire assay could be used to isolate it just as fire could be used to isolate and purify gold. For the time, this was a significant insight. Gold exists as metallic gold in nature—it is not usually locked up in ores or rocks that do not look like gold. Arsenic metal is rare in nature, and so it is not suprising that the Greeks assumed that what they did find in nature, the arsenic minerals, were also elements.

Magnus's breakthrough had enormous, if delayed, consequences in FORENSIC SCIENCE. That poisoning was rampant was an open secret, but there was no way to prove that a death was caused by a poison. The proof would lie in a definitive chemical test for arsenic found in the body. To find arsenic in the body requires separation of arsenic from tissues, stomach contents, HAIR, and BLOOD. This had to be done even if the arsenic had been chemically altered by biological processes and was no longer in the same form as when ingested. If arsenic powder just sat in the stomach, it would not be toxic; it has to be processed by the body in some way before the toxic effects are seen. Magnus's work was the first small step in that direction. A baby step is probably a better description since it took another six centuries for a specific test to evolve.

Fresco of Saint Albertus Magnus from seminary, Treviso, Italy *(Scala/Art Resource, NY)*

Magnus, also known as Albert the Great, is notable for other contributions. He became a professor at Paris University and counted among his students Thomas Aquinas. Magnus also rose to the rank of bishop in a time when early science and religion were uneasy bedfellows at best. He is quoted as saying, "The aim of natural science is not simply to accept the statements of others, but to investigate the causes that are at work in nature," a sentiment remarkable for its time. As an alchemist, he was adept at using nitric acid as a means of separating gold from silver. A similar technique would be central in separating arsenic from forensic samples a few centuries later.

See also ALCHEMY.

Further Reading

Aromaticico, A. *Alchemy: The Great Secret.* New York: Harry N. Abrams, 2000.

Moran, B. T. "Distilling Knowledge: Alchemy, Chemistry, and the Scientific Revolution." In *New Histories of Science, Technology, and Medicine,* edited by M.C. Jacob, S. R. Weart and H. J. Cook. Cambridge, Mass.: Harvard University Press, 2005.

Morris, R. *The Last Sorcerers: The Path from Alchemy to the Periodic Table.* Washington, D.C.: Joseph Henry Press, 2003.

Malpighi, Marcello (1628–1694) Italian *Botanist* Malpighi studied the structure of plants (plant morphology) as a professor at the University of Bologna. In forensic science, he is remembered as the first person to use magnification techniques to study the ridge detail and pore structure of human skin. As a result, he is considered one of the early pioneers of FINGERPRINTS, even though he neither focused on their potential to identify individuals nor their value in criminal investigations. In honor of his work, a portion of skin anatomy was named after him. The Malpighian layer (sometimes called the stratum malpighii) is found in the epidermis (the outer layer of skin) and refers to the combined basal and pickle cell layers (stratum germinativum and stratum spinosum).

Mandelin test A versatile PRESUMPTIVE TEST used in drug analysis. The Mandelin reagent is prepared by dissolving ammonia meta-vanadate in cold concentrated sulfuric acid (H_2SO_4). Addition of this reagent to cocaine results in an orange color, while codeine yields an olive green, heroin a brown, and amphetamine a bluish green. As with all presumptive tests, the results of a color test are not sufficient to identify a drug, but it is useful for screening purposes and for directing further analysis.

manner of death The manner of death refers to the classification of a death as accidental, homicidal, indeterminate, natural, or suicidal. The determination of the manner of death when it is suspicious, questioned, or unattended is determined by the MEDICAL EXAMINER (ME) or CORONER.

marijuana (cannabis) A general term for drugs derived from the plant *Cannabis sativa L,* and specifically for the leaves and flowering tops of those plants. Marijuana is the most widely abused illegal substance in the United States, with recent data showing that nearly half of all high school seniors had tried it at least once. It is classified as a hallucinogen and is listed on Schedule I of the CONTROLLED SUBSTANCES ACT (CSA). The marijuana plant can grow to more than five feet in height and is also used as a source for hemp, the fibers of which can be made into rope or clothing products. The active ingredient in marijuana and its derivatives is Δ^9-tetrahydrocannabinol ("delta-9"), usually abbreviated simply as THC. Through careful cultivation, the THC content of the plant has steadily increased and now is in the range of 3 percent to 5 percent for leaves and flowers. HASHISH or hash oil is the oily resin excreted by the flowering tops and has a higher concentration of THC, in the range of 10 percent to 20 percent. Sinsemilla, a variety of marijuana plant developed in the late 1970s, has THC content in the range of 10 percent. Within any marijuana plant, the THC content is highest in the resins and lowest in seeds.

Marijuana has been used for thousands of years both medicinally and for its hallucinogenic and intoxication effects. It is thought that Napoleon's soldiers brought it to Europe when they returned from the Middle East and that it made its way into the United States around 1910 by way of Mexican immigrant labor. The drug is almost always taken by smoking, although recently oral preparations have become available for the treatment of discomfort and nausea associated with chemotherapy. Since the drug is smoked in the form of cigarettes ("joints"), the process carries the same hazards as any smoking, including increased risk of cancer proportionate to the degree of use. Marijuana is somewhat unique in that the type

Marijuana plants with leaves and stems. ***(Courtesy of Aaron Brudenelle, Idaho State Police)***

of effect produced is proportionate to the amount ingested. Depending on dose, depressant, stimulant, and/or hallucinogenic effects may be produced, along with increases in appetite.

Forensic analysis of marijuana begins with an evaluation of botanical structures of the plant, including the presence of features called CYSTOLITHIC HAIRS. Informally, these are called "bear claws" because of their distinctive shape. The presence of these structures is strong evidence of the presence of marijuana, but not definitive evidence. The modified Duquenois-Levine test is a presumptive test for THC, which can be performed on the plant matter, resins, or seeds. Confirmatory tests for the final identification are usually THIN LAYER CHROMATOGRAPHY (TLC), GAS CHROMATOGRAPHY/MASS SPECTROMETRY (GC/MS), or HIGH-PRESSURE LIQUID CHROMATOGRAPHY (HPLC). In forensic TOXICOLOGY, the challenge is to detect THC and its metabolites in tissues and body fluids, primarily urine. This can be accomplished by IMMUNOASSAY, which is coupled to confirmatory techniques. Similar techniques can be used to detect THC in saliva as well.

Further Reading

Drug Enforcement Agency (DEA) Web site. "Marijuana." Available online. URL: http://www.usdoj.dea.gov/concern/marijuana.htm. Downloaded May 5, 2007.

Marquis test A PRESUMPTIVE TEST used in drug analysis. The Marquis reagent is prepared by adding a 40 percent solution of formaldehyde to concentrated sulfuric acid (H_2SO_4). When added to HEROIN or related opiate ALKALOIDS such as MORPHINE, the solution turns a purplish color. LSD, when present in sufficient quantities, creates a blackish color, AMPHETAMINES and methamphetamine an orange, METHADONE a yellow-tinted pink, and MESCALINE a reddish orange. As with all presumptive tests, the results of a color test are not sufficient to identify a drug, but it is useful for screening purposes and for directing further analysis.

Marsh, James (**Marsh test**) (1794–1846) English *Chemist* James Marsh was a distinguished chemist best known for the development of a reliable test for the presence of ARSENIC in the body. The test variants were used by forensic toxicologists well into the 20th century, when methods of instrumental analysis such as ATOMIC ABSORPTION (AA) eventually replaced it. Marsh was also the first to present the results of an analytical toxicology analysis in court, in the year 1836, and his test was used by M. B. ORFILA, who is considered to be the father of forensic toxicology. The Marsh test became a powerful and reliable test for arsenic at a time when arsenic poisoning (accidental, suicidal, and especially homicidal) was rampant.

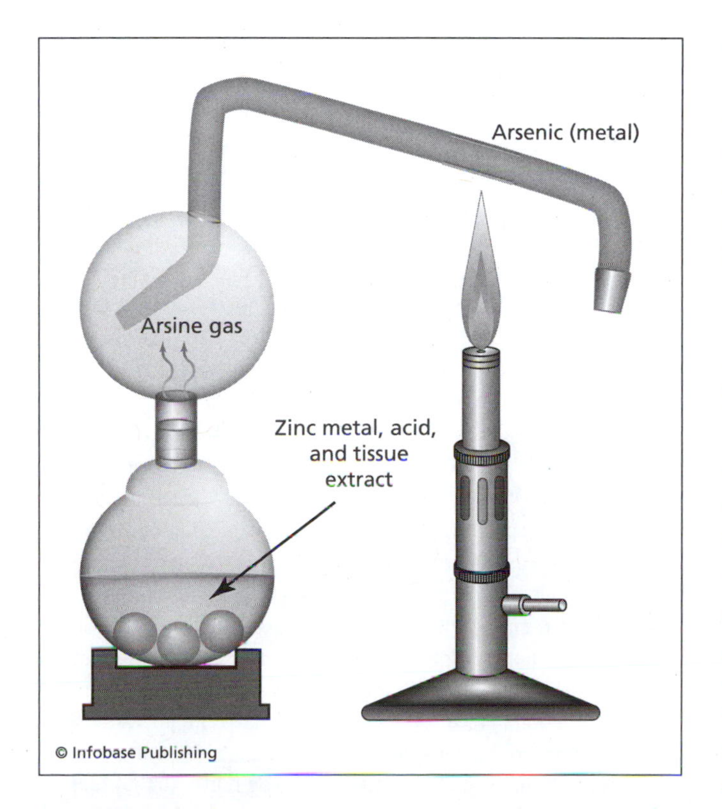

Process used for the Marsh test to detect the presence of arsenic

The test works on the basis of chemical reactions called oxidation/reduction, or redox. First, solid zinc metal is added to a glass vessel containing a powerful acid such as hydrochloric (HCl) or sulfuric (H_2SO_4). The tissue or body fluid in question is prepared and added to the vessel, where the reaction of the zinc and the acid has created hydrogen gas (H_2). The hydrogen reacts with the arsenic compound (usually in the form of arsenic trioxide, AsO_3) to produce arsine gas (AsH_3). Heating of the gas results in arsenic metal plating out on the glass or ceramic container into which the gas has flowed. The test, although reliable, required that the chemist performing it be skilled and practiced in the procedure. Tests based on similar chemical reactions are still used for screening (PRESUMPTIVE TESTS) for the presence of other heavy metals such as MERCURY (Hg).

Further Reading

Goldsmith, R. H. "The Search for Arsenic." In *More Chemistry in Crime, from Marsh Arsenic Test to DNA Profile.* Edited by S. M. Gerber and R. Saferstein. Washington, D.C.: American Chemical Society, 1997.

mass disasters One of the most daunting forensic challenges and, sadly, one that is becoming more common. These are also referred to as mass fatality incidents and are characterized by a large death toll that exceeds the capacity of local resources. Mass disasters used to be primarily associated with transportation disasters such as airplane or railroad accidents, but attacks such as those that occurred on September 11, 2001, have added terrorist acts to the list of potential causes. From the forensic perspective, mass disasters have two critical aspects: first, recovery and identification of remains and second, documentation, recovery, and preservation of evidence necessary for the subsequent investigation. On September 11, 2001, the World Trade Center in New York City became the largest crime scene ever processed. Even after the initial recovery of bodies, every load of rubble was carefully examined for additional remains as well as physical evidence. In the case of accidents such as airline crashes (for example, TWA Flight 800, described in a separate essay), evidence is essential to determine if a criminal or terrorist attack was involved, and if not, then evidence is still essential for determining possible causes.

TWA Flight 800: Mass Disaster and the Interface of Criminal Investigation, Forensic Science, and Forensic Engineering

Forensic scientists and forensic engineers often work side by side to investigate mass disasters. Many of these disasters have been the work of terrorists such as in Oklahoma City in 1995 and in New York City and at the Pentagon on September 11, 2001. On September 11, the attacks were documented and witnessed by thousands, leaving no doubt as to the cause. On other occasions, painstaking investigation coupled to forensic investigations of EXPLOSIVES and other evidence is needed. In some instances, it is not clear whether a disaster is the result of a deliberate criminal act or an accident. Such was the case of TWA Flight 800, which departed from John F. Kennedy International Airport in New York at eight o'clock on the evening of July 17, 1996, bound for Paris. At 8:31 P.M., the plane exploded as it climbed through 13,000 feet and was reduced to flaming debris and wreckage that littered a 12-square-mile area off the coast of Long Island. Two hundred and three people died. The spectacular demise of the huge Boeing 747 lit up the sky before hundreds of witnesses on the ground, on the water, and in the air.

Fearing the worst, the FBI geared up for what would become its largest and most complex investigation to date. Numerous eyewitness accounts of fiery streaks heading upward in the sky before the explosion lent credence to the theory that a shoulder-launched missile had brought down the huge plane. Later, discovery of traces of high explosives on parts of the recovered wreckage supported the theory of a missile or a bomb. Despite the preponderance of early evidence pointing toward a criminal or terrorist act, investigators from the NATIONAL TRANSPORTATION SAFETY BOARD (NTSB) converged on the site. If the explosion was somehow found to be accidental, it was the job of the NTSB to determine what caused it and how future instances could be prevented. It would take months of painstaking examination by the FBI, NTSB, FAA, and many other state and federal agencies to determine that the crash was most likely due to an accidental explosion of the center fuel tank. However, exactly what kind of malfunction or problem sparked the explosion remains unknown to this day.

There were plenty of reasons to suspect terrorists in the downing of Flight 800. The summer Olympics were being held in Atlanta, and on July 28, a bomb blast marred the festivities. Although the bombing was quickly attributed to a solitary American, law enforcement officials initially feared that the Olympics bombing and TWA 800 might be part of a sustained terrorist campaign against the United States. Thirteen years earlier, in 1983, a suicidal terrorist truck bomb had taken the lives of 241 Marines stationed in Lebanon. On December 21, 1988, a bomb exploded on Pan Am Flight 103 over Lockerbie, Scotland, killing 259 people. Seven years later, a bomb brought down an Air-India flight off the coast of Ireland, killing 329 people. Both of these incidents involved Boeing 747 aircraft on international flights. In what would prove to be a tragic rehearsal for the devastating events of September 11, 2001, terrorists parked a van loaded with explosives in the basement of the World Trade Center in 1993. They hoped that the blast would topple one of the huge towers into the other. The bomb rocked the tower but failed to collapse it, a horror that would have to wait eight years. Six people died in the attack, and, during the investigation, the FBI uncovered a wide-ranging conspiracy, the same type of conspiracy that could have brought down TWA 800.

The year 1996 had already seen a terrorist strike against U.S. personnel. On June 25, less than a month before the crash, a truck bomb was used to attack the Khobar towers in Saudi Arabia, where U.S. Air Force personnel were living. Nineteen were killed. Around the same time, U.S. law enforcement and intelligence agencies were learning of a new terrorist mastermind, Osama bin Laden, who appeared linked to the 1993 World Trade Center attacks and possibly to the Khobar Towers attack. Terrorists in Iran were also threatening a campaign against the United States, so the atmosphere was ripe to assume that Flight 800 was no accident. Still, the FBI and the NTSB kept their minds and their options open even as most involved assumed they were working on a criminal investigation that could be considered an act of war.

As the investigation proceeded, it quickly became evident that the cause of the crash was an explosion in the center fuel tank. However, the cause of that explosion remained a mystery. Investigators knew that an accidental explosion of a fuel tank was possible. The 747 aircraft that was Flight 800 was built in 1971. The 747 that had come off the assembly line immediately before it was destroyed in 1976, when a fuel tank exploded after a lightning strike. Everyone aboard was killed. A 737 aircraft, also built by Boeing, suffered an explosion of the center fuel tank in 1990 while still on the ground. Eight people were killed aboard this Air Philippines flight, and, had the explosion occurred midair, the results would have been catastrophic. On the night of July 17th, the center fuel tank of Flight 800, a massive hollow space the size of a garage, contained only 50 gallons of fuel. Complicating this was the heat of an East Coast summer day, which meant that air conditioning units below the tank were running nonstop while the jet was on the ground. These units generated considerable heat, some of which transferred to the fuel tank. The hotter the jet fuel becomes, the more of it vaporizes, and the more vapor, the more explosive the atmosphere. The flash point of jet fuel is close to 100°F, although an ignition source such as a spark or flame is still required to cause an explosion. However, under those conditions, even a tiny spark had the potential to ignite the vapor, causing a horrific explosion that would tear the plane apart. Unfortunately for investigators, such an explosion could also be created by a missile impact or bomb in or near the center fuel tank.

Bombs and missile impacts leave distinctive wounds on aircraft such as characteristic structural damage and explosive residues. Structural damage would still be detectable even after the wreckage was submerged for some time, but explosive residues were another matter. A study conducted as part of the investigation revealed that explosives, which are somewhat water soluble, are quickly dissipated by immersion in salt water. Initial recovery efforts in Long Island Sound concentrated on retrieving bodies for the grieving families. Although this clearly was the correct priority, investigators were concerned that the longer the wreckage remained submerged, the more evidence would be lost. As debris was recovered, it was hauled to a nearby hangar where painstakingly detailed examinations took place, followed by a reconstruction.

The accident scattered debris over an area approximately 4 miles by 3 1/2 miles at a depth of 120 feet. The area was divided into three "debris fields," and wreckage recovered was assigned a color code based on the field from which it was recovered. The green zone, the one farthest up the coast and thus the farthest away from JFK airport, held the wreckage that included the wings and rear fuselage. The red zone, also the largest, was the closest to JFK and contained debris from the fuselage section immediately forward of the wing. The yellow zone, smallest of the three, held the remainder of the fuselage and cockpit and was concentrated in a corner of the red zone and near its intersection with the green zone. The locations of the different portions of the fuselage as well as patterns of soot and burn damage (found on wreckage in the green zone) were essential for analysis. Computer models were used to reconstruct the last few seconds of Flight 800 and to account for the pattern of wreckage dispersal. The section of the plane where the center fuel tank was located was the first to separate from the fuselage, further evidence that the initial catastrophic event took place there. Seconds later, the front portion of the plane, including the cockpit, was ripped away, falling to the sea farther east. The rear of the plane, including the wings and engines, considerably lighter with the loss of the other two sections, apparently shot upward in flame until the engines exhausted their fuel. This, investigators felt, could explain many of the eyewitness accounts of flaming streaks ascending in the sky. The wreckage continued to break apart and plummeted to the sea farthest away from the airport.

At the wreckage site, visibility in the water was poor, and the water temperature at depth was cold and grew colder as recovery work moved into the fall. As the underwater recovery effort was getting under way, the FBI organized hundreds of agents to interview witnesses and chase down leads. Early speculation that a "friendly fire" accident occurred involving American armed forces was discounted, as was the possibility of large missiles. A smaller, shoulder-launched missile remained a possibility, but an increasingly remote one as the FBI learned more about the limitations of such a weapon. At the altitude the plane was when the explosion occurred, nearly 14,000 feet, only a missile fired from a boat directly under the plane would have had a chance to hit the target, and even then, the odds were slim. In addition, anyone below on the water would have been showered with flaming debris and fuel, an unlikely scenario, but not one that could be dismissed. Suicidal terrorists were nothing new. The FBI checked hospitals for people admitted with burns, examined boats for burns and scorches, and had divers look for the remains of boats mixed among the wreckage. No promising supporting evidence was uncovered. Thus, despite the convincing and numerous eyewitness accounts of flaming streaks in the sky, the investigation turned more toward the possibility of a bomb.

Bombs that had been used to bring down aircraft before, such as the Pan Am 103 bomb, were composed of chemical compounds such as RDX and PETN, residues of which investigators hoped would remain on materials that were near the detonation. However, the long immersion times concerned everyone and made the task of detecting, confirming, and interpreting findings exceedingly difficult. As wreckage was recovered and delivered to the hangar, forensic chemists and explosives experts from the FBI and ATF combed the pieces for any traces of explosives using dogs and portable equipment. Promising locations were swabbed or transported whole to the FBI laboratory in Washington, D.C., for further testing. Dogs and portable instruments, while excellent screening tools, occasionally identify false positives. This was the case with all the "hits" made by the dogs and hangar instruments until early August, when traces of explosives were confirmed on several samples from widely separated areas of the plane. The excitement quickly subsided when it was learned that recently an explosives sniffing dog had performed a training exercise in the same plane using relatively large amounts of explosives. Thus, finding the traces of explosives was not significant. In addition, a detailed analysis of the injuries of the passengers did not indicate any had been exposed to a concentrated explosive event such as a bomb or missile impact. As the investigation continued and more pieces of wreckage were recovered, still no structural

(continues)

TWA Flight 800: Mass Disaster and the Interface of Criminal Investigation, Forensic Science, and Forensic Engineering *(continued)*

or metallurgical evidence of a bomb or missile had been found. Similarly, the FBI probe had not uncovered definitive evidence of criminal activity. The NTSB began leaning toward an accidental cause of the explosion in the center fuel tank and pursued forensic engineering tests to examine possible accident scenarios.

The NTSB conducted numerous engineering and flight tests during the course of their investigation. For example, the cockpit voice recorder (CVR) went dead at 8:31:12 P.M., when the explosion cut power. Less than a second before this, a sound was recorded on the tape that was likely related to the explosion itself. The NTSB analyzed the CVR recordings of other accidents including the Lockerbie and Air India bombings and the accident on the ground involving the Air Philippines plane that experienced an explosion in the center fuel tank. The last recorded sound from flight 800 was found to be more similar to the accidental explosion than to those recorded from the bombed flights. Extensive tests were also conducted on wiring, fuel, fuel vapors, fuel probes, potential electromagnetic interferences from personal electronic devices, and other electrical components. Many of these tests were performed on components recovered from the sea. The NTSB also conducted test flights using heavily instrumented 747 aircraft performing under similar conditions to those on the night of the accident. To the engineers' surprise, they found that the temperature of the fuel exceeded the 100°F flash point due to the heat transferred from the air conditioning units. Although the vapors could not self-ignite until they reached much higher temperatures, the vapors were definitely flammable, making an accidental ignition more plausible.

Diving operations ended in November 1996, but since a full reconstruction of the plane had been approved, trawling operations were conducted until April 1997. When that ended, an amazing 98 percent of the aircraft had been retrieved from the bottom, wreckage that was transported to the hangar, examined, tested, and eventually reconstructed. No convincing evidence of a bomb or missile was found, either in the form of chemical residues or characteristic structural damage. This fact, coupled with other evidence and the flammability of the vapors in the center tank, led to the investigation's closing within a few months. The NTSB listed the probable cause as ignition of these vapors, but even after months of work, the agency was not able to pinpoint the source of the spark or flame that ignited those vapors. Specifically, the final NTSB report stated:

> The National Transportation Safety Board determines that the probable cause of the TWA flight 800 accident was an explosion of the center wing fuel tank (CWT), resulting from ignition of the flammable fuel/air mixture in the tank. The source of ignition energy for the explosion could not be determined with certainty, but, of the sources evaluated by the investigation, the most likely was a short circuit outside of the CWT that allowed excessive voltage to enter it through electrical wiring associated with the fuel quantity indication system.
>
> Contributing factors to the accident were the design and certification concept that fuel tank explosions could be prevented solely by precluding all ignition sources and the design and certification of the Boeing 747 with heat sources located beneath the CWT with no means to reduce the heat transferred into the CWT or to render the fuel vapor in the tank nonflammable.

The NTSB made several recommendations to the FAA concerning ways to lessen the risk in the future, but many of these recommendations, such as filling the tanks on every flight or filling nearly empty tanks with nitrogen to prevent combustion, were rejected as not feasible and/or too expensive.

See also FBI; METALLURGY, FORENSIC.

Further Reading

Milton, P. *In the Blink of an Eye, the FBI Investigation of TWA Flight 800.* New York: Random House, 1999.

National Transportation Safety Board. *In-flight Breakup over the Atlantic Ocean, Trans World Airlines Flight 800, Boeing 747-131, N93119, Near East Moriches, New York, July 17, 1996.* Aircraft Accident Report NTSB/AAR-00/03. Washington, D.C. Notation 6788G, August 23, 2000. Available online: URL: www.ntsb.org.

At the scene of mass fatalities, the first and most critical task is the recovery of remains and their identification. Since many such incidents result in partial or complete destruction of intact bodies, identification becomes much more difficult and requires the analysis of disarticulated (separated) body parts. In the worst cases such as the World Trade Center, only very small fragments may be recovered. The tools used in the identification process include analysis of dental remains (ODONTOLOGY), fingerprinting, analysis of bone (ANTHROPOLOGY), personal effects, distinguishing physical features such as scars and tattoos, and DNA TYPING. Since remains are disarticulated and may be widely scattered, several different modes of identification may be used and computer database programs employed to link together the remains of individuals. When DNA analysis is required, the remains must be properly preserved to minimize microbial degradation and decomposition processes that might complicate laboratory testing.

Problems presented by mass disasters have long been recognized. In the 1980s, the National Association of Funeral Directors proposed a program called DMORT, which stands for Disaster Mortuary Operational Teams. In the aftermath of the TWA 800 crash, Congress passed the Family Assistance Act in 1996 to provide a framework for federal assistance in mass disasters, notably airline crashes. As a result, DMORT teams have become an integral part of managing mass disasters. A given DMORT team consists of volunteers from forensic disciplines, funeral directors, and support personnel such as database and computer specialists, dental hygienists, and X-ray technicians. When their assistance is requested, DMORT teams can be quickly dispatched and can set up portable mortuary facilities on-site. Not surprisingly, the largest deployment to date of DMORT teams was in response to the September 11 attacks.

mass casualties *See* "Identification of the Dead" (essay).

mass spectrometry (MS) An instrumental technique and detector widely used in forensic science. Mass spectrometry can be applied to organic analysis (drugs and arson, for example) or to inorganic analysis (bullets, paint, and so on) depending on the type of sample introduction system it is coupled to. Mass spectrometry as used in forensic science is not a "stand alone" technique, but rather a mass spectrometer is coupled to different sample introduction instruments or devices, yielding what is called a hyphenated technique. Examples include GAS CHROMATOGRAPHY/MASS SPECTROMETRY (GC/MS), HIGH-PRESSURE LIQUID CHROMATOGRAPHY/MASS SPECTROMETRY (HPLC/MS), INDUCTIVELY COUPLED PLASMA/MASS SPECTROMETRY (ICP-MS), and PYROLYSIS mass spectrometry. Note that a slash ("/") is often used in place of a hyphen, but

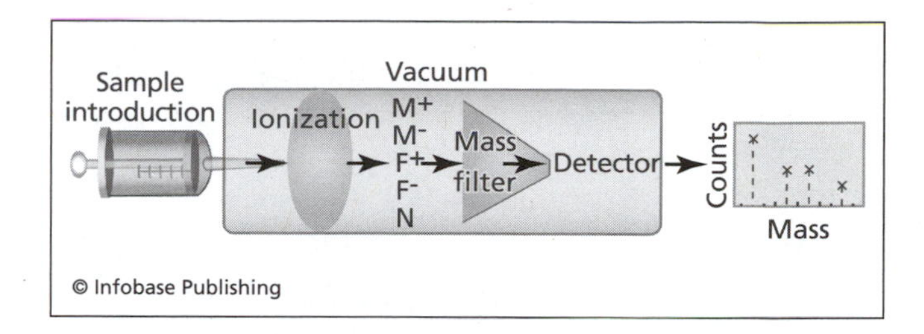

Schematic of a generic mass spectrometer showing the sample inlet, ionization, mass filter, and output. "N" indicates a neutral species.

the meaning is the same, and both notations are used interchangeably. Mass spectrometry is used in forensic science in DRUG ANALYSIS, ARSON, analysis of PLASTICS and PAINTS, TOXICOLOGY, GUNSHOT RESIDUE (GSR), and EXPLOSIVES to name a few applications. The greatest advantage of mass spectrometry is that in almost all cases, it can provide definitive identification of the compounds or elements that make up a given sample. This is called QUALITATIVE ANALYSIS, and in addition, mass spectrometry is a powerful quantitative tool in that it can be used to determine how much of each component is present, down to trace level concentrations.

Mass Spectrometer Designs

The first figure illustrates the basic design of a generic mass spectrometer. The sample is introduced through an inlet into a chamber that is kept at a very low pressure, in the range of 1×10^{-5} torr (about 1 one-billionth of normal atmospheric pressure). This low pressure is essential to prevent collisions between ions (charged particles) created by the instrument and atmospheric components. The sample molecule (if it is an ORGANIC compound) is then ionized to form charged fragments (F_1^+, and so on) of the original molecule. The M^+ ion is called the "molecular ion" in that it is created by stripping a single electron away from the original compound. In organic mass spectrometry, the type used for the analysis of drugs, for example, this molecular ion is important because it will have the same molecular weight as the parent molecule. Smaller fragments also form that can be as small as individual atoms. The process of ionization is different for inorganic materials such as MERCURY and ARSENIC, when the inductively coupled plasma creates the ions before they are introduced into the mass spectrometer. Thus, in a sample containing gold (Au), the plasma will create gold ions (Au^+) that will have the characteristic atom weight of

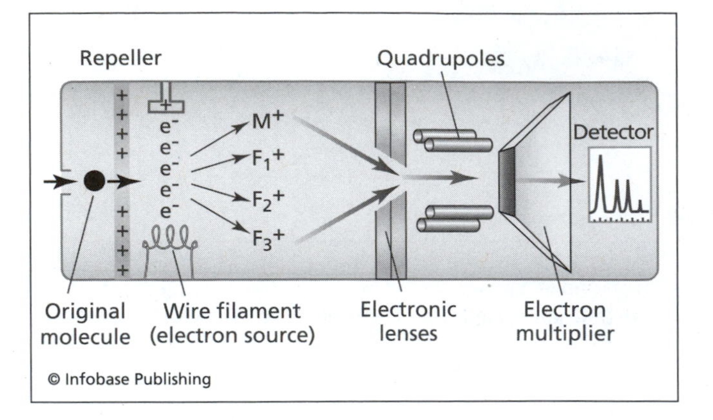

Operation of a quadrupole mass spectrometer

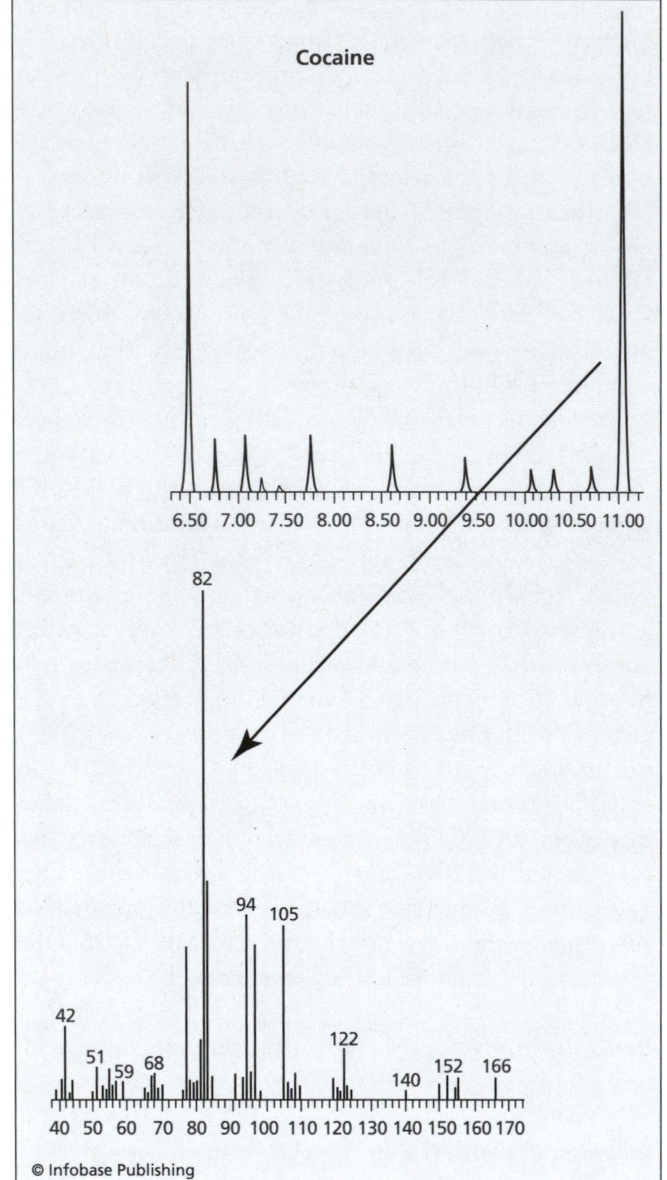

Output of a GC instrument. The sample introduced was cocaine. The top is the gas chromatograph showing separation of the sample into individual components. The bottom is a mass spectrum of the component that eluted from the gas chromatographic column at 11.026 minutes, the largest peak in the chromatograph.

gold (197 atomic mass units). However, these atoms do *not* fragment and are detected as is. Ions may be positively or negatively charged. In forensic work, most applications focus on the positive ions such as M^+ and fragments.

A number of different types of ionization schemes are available for mass spectrometers used for organic molecules. The most common is electron impact (EI), described below. Other types include chemical ionization (CI), particularly useful when a strong M^+ signal is needed; electrospray, common with HPLC systems; fast atom bombardment (FAB) and matrix assisted laser desorption/ionization (MALDI), used for large molecules like proteins; and laser ablation, for surface analysis of organic and inorganic materials.

In organic mass spectrometry, if instrument conditions are standardized and the same type of ionization is used, the pattern of fragmentation for a given molecule is reproducible from instrument to instrument. This is what provides the qualitative analysis of mass spectrometry. For example, if a cocaine sample is introduced into the mass spectrometer, the overall pattern of fragmentation will always be the same as long as the same instrumental settings are used. This standardization is the basis of spectral libraries that forensic scientists use to identify individual compounds. With inorganic mass spectrometry, the task is easier since the detected mass is the same as the mass of an element found on the periodic table of the elements. For example, if a signal is recorded at a mass of 74.9, the element is arsenic.

As shown in the figure, once ions are produced, they are directed into a region called the mass filter, where they are separated out into individual masses and the abundances recorded. The separation is based on size of the ion as well as the charge. The results are plotted in a "mass spectrum" that shows the relative abundance of each mass. The actual mass spectrum shown is for cocaine, and this spectrum will always look the same under standard instrument conditions. Thus, spectra obtained in any one forensic lab are directly comparable to those obtained by any other lab using the same standard conditions. There are a number of different types of mass filters used in mass spectrometers including TOF (time-of-flight), quadrupoles, ion traps, and magnetic sector. The most common type found in forensic labs for both organic and inorganic applications is the quadrupole, illustrated in the figure.

Quadrupole Mass Spectrometers

The original molecule is introduced into the mass spectrometer by an inlet such as a gas chromatograph (GC) or high pressure liquid chromatograph (HPLC). It passes by a plate that has a positive charge, and ionization occurs. These ions are created by collision with a stream of electrons created by a tiny filament much in the way the filament in a light bulb produces light. The electrons are drawn toward the target that is positively charged since unlike charges attract. Conversely, since like charges repel, the positive ions are driven into a stack of focusing lenses by the positive charge on the repeller. The ions are focused into a tight beam that enters the quadrupole area, which consists of four metallic rods. Manipulation of the electrical fields allows the ions to be filters so that only one mass is detected at any given time. The emerging ions are directed into an electron multiplier that amplifies the signal sent to the detector. The quadrupole is nearly standard for GC; ion traps and TOF designs are

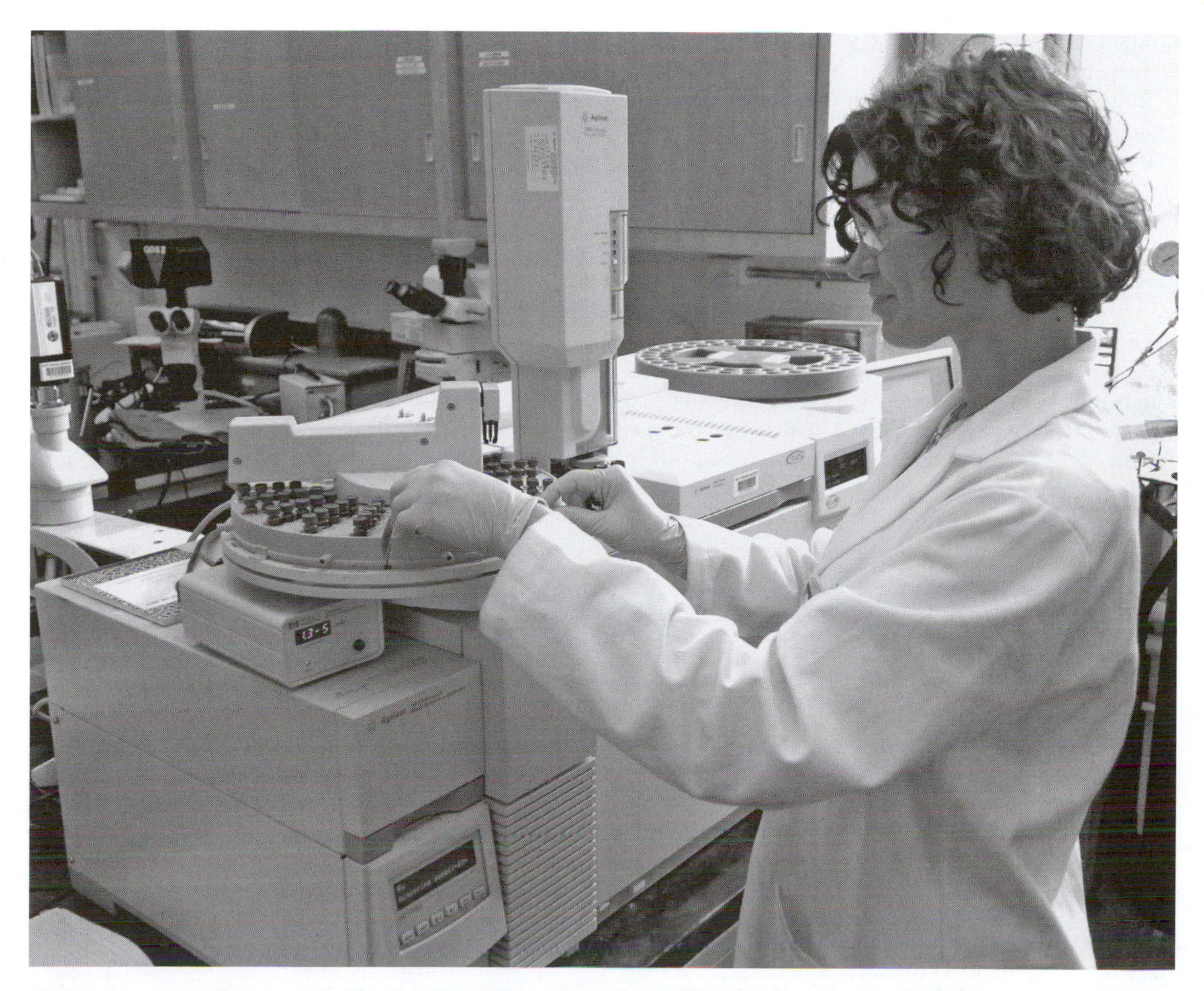

GCMS system in use ***(Courtesy of Dr. Giorgia De Paoli, Bennett Department of Chemistry, West Virginia University)***

found in many HPLC systems. There are even MS/MS systems in which fragment ions that emerge from the mass filter are directed into another mass spectrometer that repeats the process and produces another series of smaller fragments. This type of information is extremely useful when attempting to identify a material that was not found in the mass spectral library. MS/MS is also useful for very large molecules such as proteins, but this application is not widespread yet in forensic science.

Further Reading

Yinon, J. ed. *Forensic Applications of Mass Spectrometry.* Boca Raton, Fla.: CRC Press, 1995.

Mayfield case A recent case in which an incorrect FINGERPRINT match led to a wrongful arrest of an Oregon lawyer named Brandon Mayfield. It occurred in 2004 and was associated with the terrorist bombing of commuter trains in Madrid on March 11, 2004. In that case, Mayfield spent 14 days in prison as a material witness to the attack based on a fingerprint. Spanish investigators provided the FEDERAL BUREAU OF INVESTIGATION (FBI) with an image of a partial print found on a plastic bag recovered from the scene. Investigators suspected that the attackers had handled it. Search of a fingerprint database provided a list of potential matches, and evaluation by experts led to an identification of Mayfield, whose prints were on record because he had served in the army. The fact that Mayfield had converted to Islam and had acted as an attorney for a person Portland authorities had accused of operating a terrorist training camp was additionally cited, even though the matter he acted on was an unrelated child custody case.

Spanish investigators expressed misgivings about the identification and eventually linked the questioned fingerprint to another suspect. Mayfield was released, and the FBI issued a press statement and apology in May 2004, promising to review procedures and practices. The problems were ultimately blamed on poor quality of the digital image, even though several examiners looked at the print and believed it to be a match to Mayfield. The case will likely have far-reaching consequences for fingerprints and pattern identification, although it will take years for it to play out.

See also MADRID BOMBINGS.

McCrone, Walter (1916–2002) American *Chemist, Microscopist* Walter McCrone was one of the preeminent microscopists in the world and founded the McCrone Research Institute (URL: www.mcri.com) in 1960 in Chicago, Illinois. McCrone obtained his Ph.D. from Cornell University in New York in 1942 and worked for a time at the Armour Research Institute. His interests in microchemical analysis led to the formation of the McCrone Associates in 1956, where the emphasis was on this topic as well as on the analysis of crystals and general microscopy. The McCrone Research Institute followed, and it remains a premiere training facility for industrial and forensic microscopists worldwide. Of his many publications, one of the best known is *The Particle Atlas,* used by forensic microscopists to identify unknown materials. He became familiar outside the scientific community during the analysis of the Shroud of Turin in 1980. The shroud was purported to be the burial cloth of Jesus. McCrone, along with a group of other scientists, conducted numerous tests and concluded that the shroud was a clever forgery dating not to the first century, but to the 14th. He was awarded the American Chemical Society's Award in Analytical Chemistry in 2000 for his numerous contributions to MICROSCOPY and microchemical analysis.

MDMA *See* ECSTASY.

Mecke's test A PRESUMPTIVE TEST used in drug analysis, often in conjunction with other tests such as the MARQUIS test. The Mecke reagent is prepared by dissolving selenious acid in concentrated sulfuric acid (H_2SO_4). When added to HEROIN, the reagent turns yellow, which fades into a greenish color. When present in sufficient quantities, LSD produces an olive green color that turns blackish, and PSILOCYBIN produces a yellow-green that turns brownish. As with all presumptive tests, the results of a color test are not sufficient to identify a drug, but it is useful for screening purposes and for directing further analysis.

medical examiner (ME) A forensic pathologist appointed by a jurisdiction such as a state, county, or city to oversee death investigations. Medical examiners did not appear in the United States until 1877, when Massachusetts became the first state to abolish the CORONER system in favor of an ME system. New York City followed in 1915, and many jurisdictions have since converted to the newer system. However, both systems still coexist in the United States. An ME is a physician with specialized training in forensic pathology who is qualified to conduct an AUTOPSY and tissue analysis. Like a coroner, an ME is charged with determining the cause, manner, and circumstances of DEATH. The cause of death is the injury or condition responsible, such as a gunshot wound or drowning, while the manner of death is classified as natural, accidental, suicide, or homicide. At a crime scene involving a death, normally it is the ME's office that has jurisdiction over the body, while law enforcement agencies are responsible for the rest of the scene. Similarly, the ME is responsible for collecting any physical evidence discovered during the autopsy and for delivering it to the appropriate forensic laboratory.

Further Reading

Wright, R. K. "The Role of the Forensic Pathologist." In *Forensic Science: An Introduction to Scientific and Investigative Techniques*. 2nd edition. Edited by S. H. James and J. J. Nordby. Boca Raton, Fla.: CRC Press, 2005.

medicine, forensic *See* PATHOLOGY, FORENSIC.

melting point The temperature at which a substance changes from a solid to a liquid state. In forensic science, melting points of chemical compounds are occasionally used to assist in the identification of unknowns. However, the most common use is the determination of the melting points of synthetic materials such as fibers as an aid to their identification.

meperidine *See* DEMEROL.

Merck Index A handbook of chemical compounds and drugs used as a reference by forensic chemists, toxicologists, and pathologists. The index, currently in its 13th edition, is published by Merck Sharp and Dohme Research Laboratories and contains data on more than 10,000 compounds as well as tables, lists, and extensive cross referencing information.

mercury (Hg) A toxic heavy metal unique in that it is the only metal that is a liquid at room temperature. Like any metal, it conducts electric current and can be used as a component in explosive devices that use mercury switches. However, mercury is best known as an environmental pollutant and poison. Despite its toxicity, mercury is rarely encountered in homicide cases, in which ARSENIC and CYANIDE are more commonly employed. Mercury exists in many different forms besides the familiar elemental form (also called quicksilver), which is actually one of the least toxic forms unless the vapor is inhaled. Mercury can exist as solid salts such as mercuric chloride ($HgCl_2$), a corrosive salt that can be fatal in doses of one to two grams, and as mercuric sulfide (HgS), a coloring agent. One of the organic forms, dimethyl mercury ($(CH_3)_2Hg$), is also extremely toxic. People are most often exposed to mercury by consuming contaminated food such as fish, in which mercury from an aquatic environment can concentrate. Such chronic poisoning produces neurological symptoms such as tremors, memory loss, and eventually hallucinations. It is thought that the term "mad as a hatter" can be traced to hatters who worked with felt impregnated with mercury compounds.

Because mercury volatilizes at a relatively low temperature, forensic examination of tissues and body fluids is more complicated than is that of other related heavy metals such as zinc and cadmium. A simple screening test called the Reinsch test involves placing a strip of copper in an extract of the sample and looking for the formation of a blackish deposit of mercury on the surface. The required confirmation of mercury is usually accomplished by a "cold vapor" technique coupled to a specialized sensor or to an ATOMIC ABSORPTION (AA) spectrophotometer.

mescaline A hallucinogenic compound that is contained in peyote, which is the "button" found on top of a cactus found in Mexico and the southwestern United States. This button is also called a mescal button, and it consists of the flowering head of the *Lophora williamsii* cactus. The compound was first isolated in 1896 and made synthetically about 20 years later. Chemically, mescaline is related to AMPHETAMINE. Although mescaline is listed on Schedule I of the CONTROLLED SUBSTANCES ACT (CSA), peyote is legal for use in rituals in the Native American Church found primarily in New Mexico and Arizona. Mescaline can also be sold illicitly in tablets or gel capsules. The hallucinogenic effect can be pronounced and lasts about 12 hours. Given the potential severity of the effects, violent or dangerous behavior can be seen in users. Forensic analysis of the buttons involves a grinding or blending step followed by extraction into ethanol or other solvent. The extract can be tested using presumptive chemical color tests such as the MARQUIS reagent, which will turn orange in the presence of mescaline, and the MECKE test, which will turn a brownish orange. Confirmation can be obtained using CRYSTAL TESTS, THIN LAYER CHROMATOGRAPHY (TLC), and various gas chromatography analyses including GAS CHROMATOGRAPHY/MASS SPECTROMETRY (GC/MS).

metabolism and metabolites A breakdown product of a substance such as a drug or toxin produced by a metabolic process in the body. The substance ingested is called the parent compound and is classified as a "xenobiotic," meaning it is foreign to the body. For example, if someone were to ingest COCAINE, that compound is not normally found in the body and is thus considered to be foreign. In other cases, the substance may be found naturally in the body, but in trace concentrations. In forensic TOXICOLOGY, it is often the metabolites that are the target of analysis in fluids such as blood and urine.

As illustrated in the figure, the process of metabolism involves stepwise changes in the molecule using different kinds of chemical reactions. Each stage of the process is catalyzed (speeded up) by the actions of enzymes, which are biological catalysts. The first stage metabolites may undergo additional processes, leading to a large number of by-products. Final metabolic products have three principal fates. Volatile products such as carbon dioxide (CO_2) can be exhaled, as in the case of the metabolism of ALCOHOL (ethanol). Water soluble products (called hydrophilic, "water loving") are excreted in the urine or in other body fluids. Products that are fat soluble (hydrophobic, or "water hating") build up in fatty tissues. Heavy metal poisons such as MERCURY can produce fat soluble products. Most metabolic processes are concentrated in the liver, but this is not exclusive. Some metabolism can take place in the stomach and intestines (gastrointestinal, or GI, tract) and in other organs such as the kidneys. Because metabolism takes place in stages, the relative ratio of the parent substance to a metabolite (the P/M ratio) can be used to estimate the time that has passed since the substance was ingested.

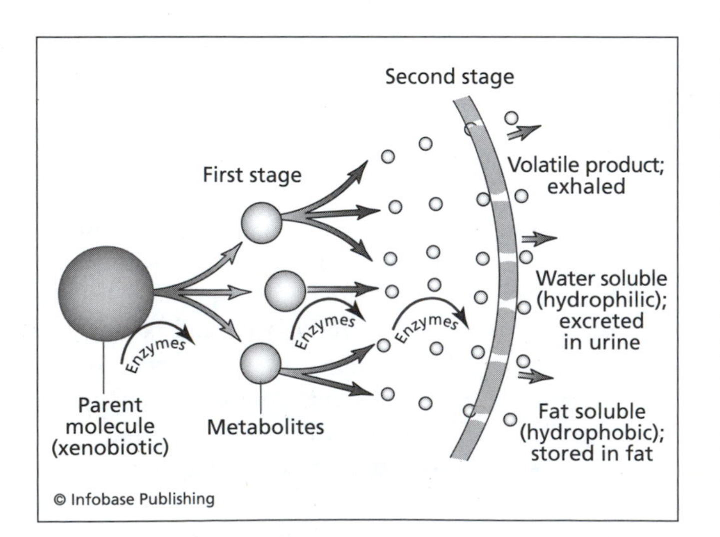

Generalized view of metabolism

See also ADME.

metallurgy, forensic This area of expertise is often part of or closely associated with forensic ENGINEERING and revolves around the analysis of metal components of objects. Steel is often the subject of forensic metallurgical analysis since steel is so commonly used as a structural component in everything from planes to buildings. Forensic metallurgy was used in the investigation of the crash of TWA FLIGHT 800 in 1996 when there was considerable doubt early in the investigation as to what caused the center fuel tank to explode. Potential causes included a bomb or missile, both of which would have left characteristic evidence. For example, if a bomb exploded from within the plane, the metal skin of the aircraft would have been peeled backward and outward, whereas a missile striking the fuselage and exploding would have left a much different signature on the fuselage. In this case, forensic metallurgy pointed to an explosion from within. Other examples in which forensic metallurgy is used include the analysis of bombs such as pipe bombs and more traditional forensic engineering analysis. Questions raised by the collapse of the World Trade Center towers on September 11, 2001, centered on what effect the burning jet fuel had on the steel superstructure of the towers and what related factors dictated when and how the collapses occurred. Tools of forensic metallurgy include nondestructive imaging techniques such as X-ray RADIOLOGY, SCANNING ELECTRON MICROSCOPY (SEM), INDUCTIVELY COUPLED PLASMA MASS SPECTROMETRY, and ATOMIC EMISSION INSTRUMENTATION (ICP-MS and ICP-AES). Metal fatigue and joint failures (welded, bolted, or other) are often a central concern in many incidents.

Further Reading

Eagar, T. W., and C. Musso. "Why Did the World Trade Center Collapse? Science, Engineering, and Speculation." *JOM: The Journal of the Minerals, Metals, and Materials Society* 53, no. 12 (2001): 8.

methadone (dolophine) A synthetic opiate that is used to treat addiction to HEROIN and other narcotics. Methadone helps to relieve withdrawal symptoms, eases cravings, and is often used as part of a detoxification ("detox") treatment. The forensic detection of methadone is accomplished by THIN LAYER CHROMATOGRAPHY (TLC) supplemented by confirmatory techniques including GAS CHROMATOGRAPHY/MASS SPECTROMETRY (GC/MS), gas chromatography with other detectors such as FLAME IONIZATION (FID) or NITROGEN-PHOSPHORUS (NPD), and HIGH-PRESSURE LIQUID CHROMATOGRAPHY (HPLC). In toxicological analysis of urine and other body fluids, IMMUNOASSAY techniques are employed as well.

methamphetamine *See* AMPHETAMINES.

methaqualone (Quaaludes) A drug classified as a sedative-hypnotic that was developed in the 1950s and the subject of abuse in the 1960s and 1970s. The drug was removed from the domestic market in the 1980s and is no longer commonly encountered in forensic drug analysis. Methaqualone in urine can be detected using IMMUNOASSAY, while GAS CHROMATOGRAPHY/MASS SPECTROMETRY (GC/MS) is used when pills or powders are recovered.

method validation A procedure or procedures used to ensure that an analytical method is reliable and that any results obtained are reproducible. Method validation is an important part of QUALITY ASSURANCE/QUALITY CONTROL (QA/QC) and is applicable in many forensic areas such as DRUG ANALYSIS and DNA TYPING. Typical method validation requires that an analyst or analysts perform the method on a sample of known composition or type to prove that the results are acceptably accurate and reproducible (precise). Once a method has been validated, it is reasonable to expect that any trained analyst using that method properly will obtain the same results as any other analyst working in any other laboratory.

methylphenidate (Ritalin) *See* DRUG CLASSIFICATION.

Michelson interferometer A device used in modern INFRARED SPECTROPHOTOMETERS (IR) that allows many wavelengths of ELECTROMAGNETIC RADIATION (EMR) to be directed at a sample simultaneously rather than sequentially. The advantage of this approach is that a sample can be scanned much faster and can be scanned multiple times to improve the quality and sensitivity of the resulting spectrum.

An interferometer works based on the principle of interference of coincident light waves, as illustrated in the figure. If two beams are in phase, the interference is constructive, and the resulting combination has the same maxima and minima at twice the amplitude (height). On the other hand, if the beams are completely out of phase, the resulting interference is destructive, and the net is a canceling out of both. Between these two extremes, there will be different combinations of constructive and destructive interference occurring. The Michelson interferometer takes advantage of this, coupled to a mathematical operation called a Fourier transform that enables a computer to break a composite pattern down into individual components, in effect undoing the interference and reconstructing the original components. The Michelson interferometer creates an output called an interferogram that is composed of a complex signal generated by multiple wavelengths and multiple cycles moving through the extremes of constructive and destructive interference. The Fourier transform is then applied to the signal to recover the intensity of the original radiation at each wavelength used.

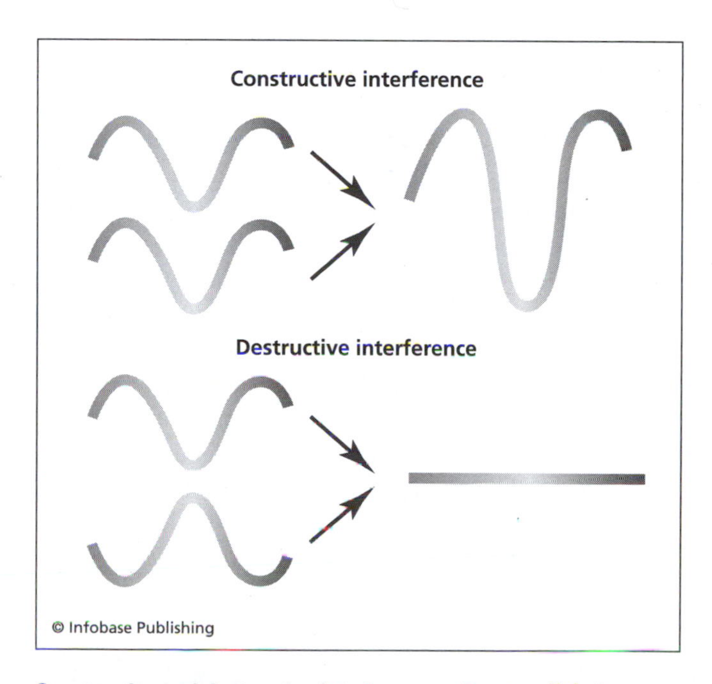

Constructive and destructive interference when two light beams are combined

microbial degradation Processes of decomposition and breakdown of materials by the actions of microorganisms, principally bacteria and fungi. This is a serious concern in forensic science given that any degradation can influence the results of laboratory analysis. Although the public usually associates this process with the decomposition of human bodies, microbial degradation is an issue for any kind of biological evidence including blood and body fluid stains. Any organic material that a microbe can use as food, from remains to ropes and cloth, is subject to degradation. The rate of degradation depends on the suitability of the environment for the microbes, which generally favor warm, moist conditions. Some bacteria are aerobic (require oxygen, or O_2), whereas others are anaerobic, and both types can play a role in degradation of biological evidence. Once degradation begins, evidence can be contaminated or altered, in some cases to a point at which laboratory analysis is difficult or impossible. However, even though bacteria do possess DNA, normally the presence of this DNA does not itself interfere with DNA TYPING, since the tests are specific to human DNA.

Microbial degradation is perhaps most obvious in the process of bodily decomposition and in this context is often referred to as putrefaction. Bacteria are present in abundance in the gut, and during life they are a critical component of the digestive process. After death, the processes that normally control this bacteria no longer function, and the bacteria multiply freely, feeding on the rich source of nutrients in blood and tissue. During this process, noxious gases are produced, leading to bloating of the body and the characteristic odors associated with decay. Soil bacteria will also attack bodies that are buried or dumped outdoors. The rate of decomposition is proportional to temperature, so the colder it is, the slower the process. Conversely, in very dry conditions, bacterial degradation may stop due to lack of water, and the remains will be desiccated or mummified and relatively well preserved. Blood and other body fluids are also subject to putrefaction, and for this reason, blood-stained materials should be dried and stored refrigerated or frozen to slow the process.

microchemical tests *See* PRESUMPTIVE TESTS.

microcrystalline tests *See* CRYSTAL TESTS.

microscopy, microchemistry, and microspectrophotometry One of the original tools of the trade and still one of the most versatile and critical components of any forensic science laboratory. When EDMOND LOCARD founded the first forensic lab in 1910, his primary pieces of equipment were microscopes. Locard was an advocate of the value of TRACE EVIDENCE and TRANSFER EVIDENCE such as DUST and FIBERS that can be adequately characterized only by examining their microscopic characteristics and components. Since then, microscopes have been used in all of the major forensic specialties including DRUG ANALYSIS, SEROLOGY/BIOLOGY, QUESTIONED DOCUMENTS, FINGERPRINTS, FIREARMS and TOOLMARKS, and ENTOMOLOGY to name only a few. Microchemistry has also been important and involves observing reactions such as color changes and crystal formation under a microscope. The latest additions to the microscopists' tools are microspectrophotometers that allow for spectrometry to be performed on tiny areas of samples. INFRARED (IR), visible, ultraviolet, and fluorescent spectrophotometry can now be performed at the microscopic level, allowing for analyses such as the characterization of the dyes used in fibers and inks used in currency. Finally, the SCANNING ELECTRON MICROSCOPE (SEM) coupled to X-ray instrumentation has become a staple in many larger labs, allowing for magnification factors of hundreds of thousands.

Optical Microscopes

All optical microscopes are based on magnification as a result of light passing through a lens of some type. The simplest approach to magnification is a magnifying glass as shown in the figure on the next page. A lens can produce two different kinds of images. A real image is one that can be projected onto a screen (like a movie) or onto the retina of the person looking at it. The second type of image is a virtual image, and that is what a magnifying glass produces. The virtual image is not physically real and can be seen only when looking through the lens. In the figure, the view is looking at a small fiber, about 25 millimeters (mm) in length through a magnifying glass that is 4X, meaning that the image seen is four times larger than the object itself. Thus, the virtual image that the viewer sees through the magnifying glass appears to be about 100 mm long. Magnifying glasses, also called hand lenses, are limited to a maximum of about 20X. This reason is practical: as the magnifying power of a lens increases, so does the curvature of the lens. As this curve becomes sharper and narrower, the area that can

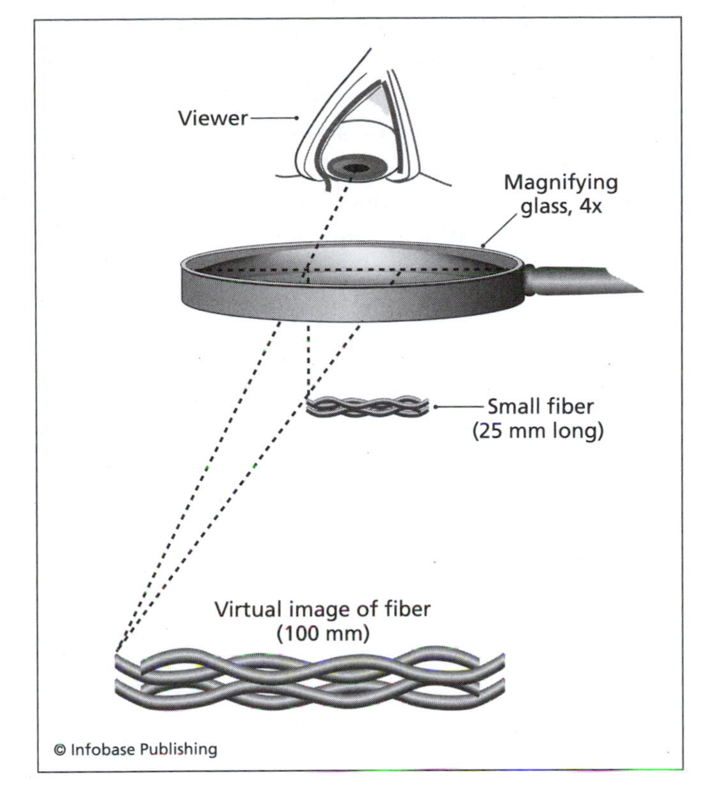

Process of magnification using a simple magnifying glass

be viewed becomes progressively smaller to the point that it is useless. However, for many forensic examinations, 10–20X is sufficient, and variations of the simple magnifying glass are used in fingerprints, questioned documents, and many other applications.

A COMPOUND MICROSCOPE, often referred to as a biological microscope, builds on this design by using two lenses in series as shown in the figure. The sample is illuminated by either transmitted (through the sample) or reflected light. The method of ILLUMINATION depends on the type of sample and is particularly important when photos are taken or comparisons are required. In either case, the light passes through the objective lens that is closest to the sample. This lens creates a real image that is projected to the ocular lens found in the eyepiece. Like a magnifying glass, the ocular lens creates a virtual image, but the virtual image is not of the object itself but rather of the real image created by the objective lens. The resulting virtual image seen by the viewer is thus magnified twice, once by each lens, and the total magnification is the sum of the magnification power of each individual lens. In this example, the sample is a tiny 1 mm spec that is first magnified and projected as an image that is about 20 mm by the 20X objective lens. This 20-mm image is magnified again by the 5X ocular to create a virtual image that appears to be 100 mm, a total magnification of 20×5 or 100X. Biological microscopes can be monocular (one eyepiece) or binocular (two). As with a magnifying glass, there are constraints to a compound microscope. Although high magnification is possible (800X), increasing magnification comes at a cost. The higher the magnification, the smaller the field of view, as was the case with the magnifying glass. Second, as magnification increases, the depth of focus (how "deep" into a sample the focus remains sharp) decreases. Typically, the resolving power of a microscope is defined by the symbol NA (numerical

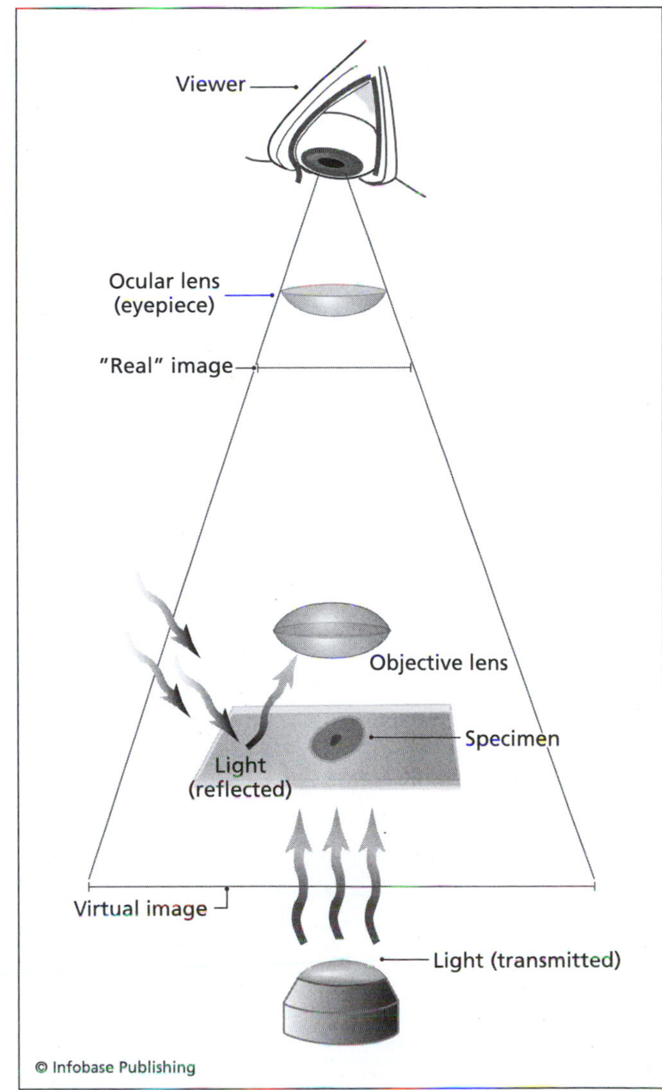

Magnification by a compound microscope

aperture), the value of which is printed on the objective. These trade-offs are illustrated in the figure below. Representative forensic uses of the compound microscope include observation of microchemical tests in drug analysis, preliminary examinations of hairs and fibers, and searching for sperm cells.

Another common microscope in the forensic lab is the STEREOMICROSCOPE, also called the stereoscopic binocular microscope or the stereobinocular microscope. Unlike the compound microscope, in which the eyepiece can be monocular or binocular, the stereoscope requires two eyepieces to create a three-dimensional image of the object. A stereomicroscope is not a high-magnification device and is used for processes such as preliminary examinations and sorting of particulates. In drug analysis, it is used to identify the CYSTOLITHIC HAIRS ("bear claws") that are characteristic of marijuana. This examination is often coupled with a microchemical reaction, the addition of hydrochloric acid to the sample. Since the cystoliths are made of calcium carbonate, they dissolve in the acid, and the release of bubbles of carbon dioxide can be seen under the microscope.

The POLARIZING LIGHT MICROSCOPE (PLM) and variants are perhaps the most versatile microscopes used in forensic science. Originally designed for use in geology and mineral identification (a petrographic microscope), PLMs are used in all manner of trace evidence analysis including fibers, soils, and paint. The key components of a polarizing light microscope are two polarizing elements, or filters, one situated below and one above the sample; a compensator slot; and a round rotatable stage equipped with a Vernier scale used to measure extinction angles. The microscope works by illuminating a sample placed on the stage with light that is propagating parallel or perpendicular to that sample. By rotating the stage or the filters, different patterns of light and color will be observed. A number of optical properties of the sample such as BIREFRINGENCE can be studied this way. Compounds and particles are easily identified using a PLM. An example is starch, which displays a distinctive internal cross pattern when viewed under crossed polarized light. Since starches are often used as CUTTING AGENTS for drugs, this is a useful capability.

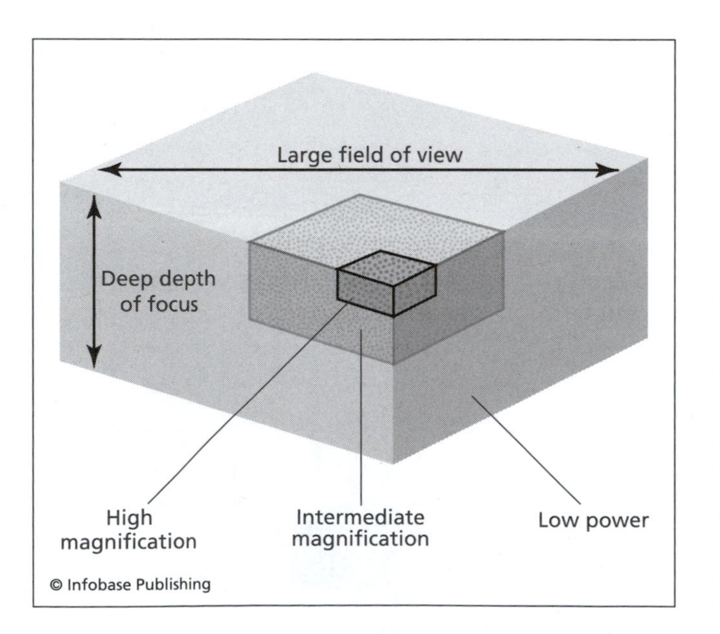

The limits to magnification in terms of image area and depth of focus

The final type of optical microscope commonly found in forensic laboratories is the COMPARISON MICROSCOPE, which allows for side-by-side comparisons of two different specimens. Comparison scopes are nothing other than two compound microscopes linked together by an optical bridge. They can be used for trace evidence analysis and in FIREARMS and TOOLMARK identification.

Other Tools

The SCANNING ELECTRON MICROSCOPE (SEM), although not an optical microscope, has become an important tool for the analysis of trace evidence. Because it images with electrons rather than with light (photons), SEM images do not have color unless it is artificially added. However, what SEM does offer is very high magnification (100,000X or more) coupled with the ability to determine the elemental composition of the surface being studied. This is because when a surface is bombarded with electrons, X-rays are emitted that are characteristic of the element or crystal being studied. With the proper accessories, it is possible to create an elemental map of an object, to study tiny differences across a surface, and to identify species with distinctive chemical compositions such as GUNSHOT RESIDUE (GSR). Instruments with these capabilities are generically referred to as SEM with an electron microprobe or SEM-EDS (energy dispersive spectroscopy). X-ray diffraction (SEM-XRD) can be used to identify the crystalline form of materials. Other types of microscopy and related techniques found in forensic labs include PHASE CONTRAST MICROSCOPY, dark field microscopy (in which otherwise obscured features may be seen), OIL IMMERSION, and hot stage determinations of melting points and refractive indexes. Photomicrographs are pictures or images obtained from microscopic examination and may be presented in court.

Microspectrophotometry
This technique is finding an expanded role in trace evidence analysis and comes about from the natural marriage of microscopic imaging and SPECTROPHOTOMETRY. Spectrometry is based on the study of how matter interacts with ELECTROMAGNETIC RADIATION (EMR), or light. Since optical microscopes work with the visible portion of the electromagnetic spectrum, it has proven relatively simple to link the two. Visible range microspectrophotometry is useful in characterization of the color of fibers, for example. In forensic work, the most widespread type of microspectrophotometry uses the NEAR INFRARED (NIR) and INFRARED region (IR). IR microspectrophotometers are used for the analysis of drugs, paints, dyes and coatings on fibers, and inks. Subtle differences in surface composition can be studied on the microscopic level and can be used to create surface composition maps similar to the elemental surface maps possible with SEM-EDS. Variations of micro-IR include ATTENUATED TOTAL REFLECTANCE (ATR), DIFFUSE REFLECTANCE (DRIFTS), and IR-polarizing microscopy. Microspectrophotometers are also available for the visible region, for ultraviolet (UV) fluorescence, and for IR scattering, a technique called RAMAN microspectrophotometry.

See also X-RAY TECHNIQUES.

Further Reading
DeForest, P. R. "Foundations of Forensic Microscopy." In *Forensic Science Handbook*. Vol. 1. 2nd edition. Edited by R. Saferstein. Upper Saddle River, N.J.: Prentice Hall, 2002.

Eyring, M. B. "Visible Microscopical Spectrophotometry in the Forensic Sciences." In *Forensic Science Handbook*. Vol. 1. 2nd edition. Edited by R. Saferstein. Upper Saddle River, N.J.: Prentice Hall, 2002.

Kubic, T. A., and N. Petraco. "Microanalysis and the Examination of Trace Evidence." In *Forensic Science: An Introduction to Scientific and Investigative Techniques*. 2nd edition. Edited by S. H. James and J. J. Nordby. Boca Raton, Fla.: CRC Press, 2005.

Nomenclature Staff, Royal Microscopy Society. *RMS Dictionary of Light Microscopy*. Royal Microscopy Society—Microscopy Handbook # 15. Oxford: Oxford University Press, 1989.

Stoney, D. A., and P. M. Dougherty. "The Microscope in Forensic Science." In *More Chemistry in Crime, from Marsh Arsenic Test to DNA Profile*. Edited by S. M. Gerber and R. Saferstein. Washington, D.C.: American Chemical Society, 1997.

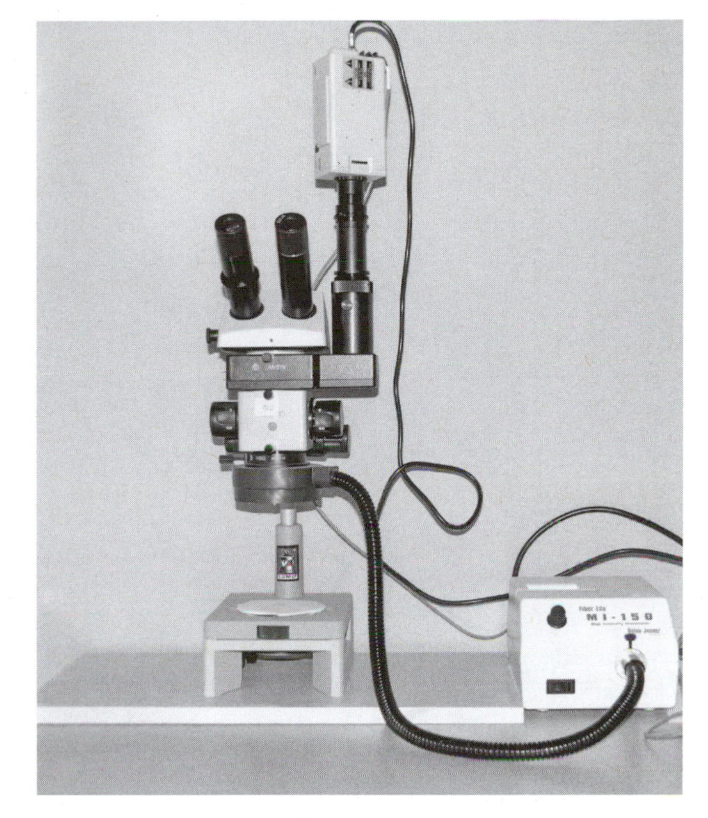

A stereomicroscope ***(Courtesy of Michael Bell, West Virginia University, Forensic and Investigative Sciences program)***

military explosives *See* EXPLOSIVE.

mitochondrial DNA (mtDNA or mDNA) *See* DNA TYPING.

MNSs blood group system *See* BLOOD GROUP SYSTEMS.

modes of ingestion In forensic TOXICOLOGY, the way in which a drug, poison, or other foreign substance enters the body. Drugs and drugs of abuse can be swallowed, injected, or snorted (absorbed through the nasal membranes), but there are other routes of exposure, and some of these may be incidental or unintentional. Poisonous gases such as CO and HCN enter

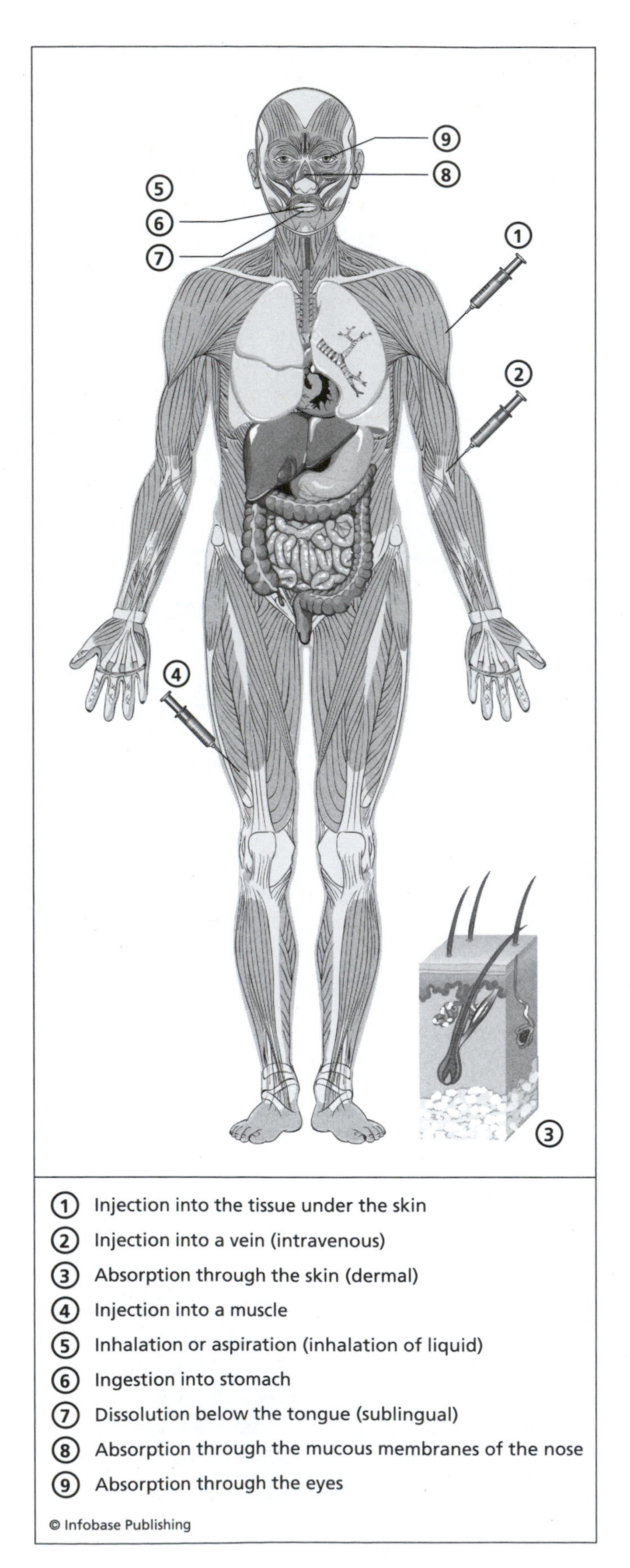

the bloodstream by way of inhalation, and many substances can be absorbed through the skin. Drugs delivered by suppository are absorbed in the lower intestine. Injections can be introduced directly into a blood vessel (intravenous), just below the skin's surface (subcutaneous), or into a muscle (intramuscular). The mode of ingestion affects how and when a drug appears in the bloodstream, which must occur before the substance can be distributed to various tissues in the body. Any mode of ingestion is possible in forensic casework.

See also ADME.

modus operandi (MO) *See* PSYCHIATRY, FORENSIC, PSYCHOLOGY, FORENSIC, AND PROFILING.

morphine A derivative of OPIUM that constitutes from approximately 5 percent to 20 percent of the extract. Morphine is also a metabolite (*see* METABOLISM AND METABOLITES) of some of the opiate alkaloids such as heroin. The name "morphine" comes from the name of the Greek god of dreams, Morpheus. Morphine was first extracted from opium in the early 1800s and by the time of the Civil War was widely used for its potent pain relieving ability desperately needed by the wounded in that conflict. Morphine is classified as a narcotic analgesic (pain reliever) that in large doses can cause respiratory depression and death. It is still used for the relief of severe pain such as that of terminal cancer. Morphine comes in many forms, including salts such as morphine hydrochloride and morphine sulfate. Illicit use is not as extensive as is that of heroin, which is easily synthesized from morphine. In forensic TOXICOLOGY, IMMUNOASSAY and THIN LAYER CHROMATOGRAPHY (TLC) are used to screen fluids (principally urine) for morphine and its metabolites. Forensic chemists also use TLC and presumptive chemical tests to screen for the presence of morphine in physical evidence. The identification of morphine is usually confirmed by GAS CHROMATOGRAPHY/MASS SPECTROMETRY (GC/MS).

morphology A term meaning structure. In forensic science, morphology could refer to any number of things including microscopic evidence (the morphology

There are numerous ways for a foreign substance to enter the human body. The mode of ingestion can have a significant effect on the toxicity and other effects of the substance.

or structure of a hair, for example) or biological structures and features.

Munsell color system A system designed to standardize colors and now used in FORENSIC SCIENCE to describe the color of some kinds of evidence such as soil. The system was conceived by an American painter, Albert H. Munsell, in 1905 with later revisions and variations. The three variables used to describe colors in the Munsell system are hue, brightness, and saturation (similar to chroma; also called value). Munsell charts and collections are used in the forensic analysis of PAINTS and SOILS. Because books and samples of color are used for color comparison, the Munsell color space is sometimes referred to as a catalog system.

See also CIE COLOR SYSTEM; COLORS AND COLORANTS.

murder A crime that involves the killing of another human being. In American law, there are six categories in which a killing can be classified, and often, forensic evidence is critical in deciding which a person will be charged with. In general, first degree murder is a willful, deliberate, and premeditated killing undertaken with "malice aforethought," that is, an intent to cause harm present before the crime is committed. Second-degree murder has all of the same elements of first degree except premeditation. Thus, in murder cases it is often important to use physical evidence to demonstrate premeditation. One example would be if a murderer purchased a gun a few days before killing someone; another would be if a person were to place a sheet in the trunk of his or her car the day before it was used to transport a body to a remote dumping ground. In both instances, the killer has demonstrated planning and thus premeditation. Manslaughter is the second broad category of killing, also divided into two degrees. Manslaughter in the first degree (more commonly called voluntary manslaughter) is an intentional killing but without malice aforethought. Many "crimes of passion" fall into this category, when one person kills another while in the grip of a sudden rage. Involuntary manslaughter is generally a killing that results from negligence, such as a vehicular homicide in which one person's negligent driving results in the death of another. The other two categories of murder are justifiable homicides (such as legitimate self-defense) and accidental death. An example of this would be one hunter accidentally killing another during a hunting trip, although evidence might be developed to elevate the charge to involuntary manslaughter if the killer was drinking.

muzzle flash and muzzle blast Muzzle blast is a shock wave that is produced when a firearm is discharged. In a gun, when the trigger is pulled, a firing pin strikes the PRIMER, causing it to explode. This tiny explosion ignites the PROPELLANT, producing a large volume of hot gas that rapidly expands, pushing the BULLET down the barrel of the weapon. When this wave of hot expanding gas contacts the atmosphere, this is the muzzle blast. The muzzle flash is also produced by the ignition of the propellant and appears as a flash of fiery tendrils exiting any openings in the gun. Any gun that accelerates the bullet faster than the speed of sound (approximately 330 meters per second, or 1,080 feet per second) will produce a cracking sound, which is a small sonic boom similar to those produced by supersonic aircraft. SILENCERS are designed to reduce or eliminate the muzzle blast and the cracking sound.

N

NAFE **(National Academy of Forensic Engineers)** *See* ENGINEERING, FORENSIC.

Napoleon *See* BONAPARTE, NAPOLEON.

narcoanalysis The use of drugs, informally referred to as "truth serums," to illicit information or a confession from a suspect. BARBITURATES such as sodium pentothal and scopolamine are the most commonly used. Courts have refused to admit the results of narcoanalysis, and the technique is rarely, if ever, used in criminal investigations.

narcotics *See* DRUG CLASSIFICATION.

National Crime Information Center (NCIC) Now a part of the FEDERAL BUREAU OF INVESTIGATION (FBI), the NCIC was established in 1967 as a national repository for data needed and used by law enforcement agencies throughout the United States and Canada. The center includes fingerprint searching, files of mug shots, and a convicted sex offender registry. NCIC is housed under the FBI's Criminal Justice Information Section (CJIS).

National Integrated Ballistic Information Network (NIBIN) *See* IBIS SYSTEM.

National Transportation Safety Board *See* MASS DISASTERS; NTSB; TRANSPORTATION DISASTERS.

Nazi method A method of making METHAMPHETAMINE in clandestine drug laboratories (*see* CLANDESTINE LABS). This method is also called the Birch method, Birch reaction, or Birch reduction after one of the chemical steps involved. There are conflicting reports as to why the "Nazi" label has been adopted. The synthesis begins when the precursors, usually obtained from over-the-counter cold medicines, are dissolved in a solvent. Next, a reactive metal such as lithium (from batteries) is added, along with anhydrous ammonia, a chemical used as fertilizer. Once ammonia is present, the solution turns a deep blue that fades to gray as the reaction proceeds. Because many of the materials can be obtained from legitimate sources or easily stolen, the method is relatively simple to use; however the ammonia gas is dangerous and fatalities have occurred.

neuropsychology, forensic *See* PSYCHIATRY, FORENSIC.

neutron activation analysis (NAA) Also called gamma ray spectroscopy, this technique has been used in cases in which ELEMENTAL ANALYSIS is needed. The most notable application was to the analysis of bullets and fragments recovered during the investigation of the KENNEDY ASSASSINATION. In NAA, a sample is placed into a nuclear reactor (research type, not those used in nuclear power plants), where it is bombarded by neutrons. These neutrons are called "thermal neutrons" since they are moving relatively slowly and thus can be absorbed by atoms in the sample. This step is called activation and produces a radioactive isotope of the original atom. Isotopes are atoms that have the same number of protons but a different number of neutrons in their nucleus. For example, the hydrogen atom consists of one proton and no neutrons. Deuterium is an isotope of hydrogen, and its nucleus contains one proton and one neutron. Tritium, another hydrogen isotope, contains one proton and two neutrons. Since it is the number of protons that determines the chemical identity of an atom, all three are hydrogen, and all three are chemically identical. However, they differ in mass. Contrary to popular belief, not all isotopes are radioactive. Deuterium is not, but tritium is. NAA works based on absorption of neutrons to produce a radioactive isotope.

Radioactive isotopes are unstable and will begin to decay, giving off gamma (γ) ray radiation. Other types of radiation (alpha and beta) may also be given off but are not the radiation measured in NAA. The rate of decay of the activated species depends on the isotope and is expressed as the HALF-LIFE, which is the time required for half of the radioactive atoms to decay. For example, if an isotope has a half-life of 10 minutes and 120 atoms are initially activated, after 10 minutes, 60 will still be radioactive. After 20 minutes (2 half-lives), 30 will remain, and so on. In NAA, a short half-life is desired; otherwise the sample will remain radioactive for too long a period of time. Because the wavelength of gamma rays emitted is characteristic of the element, the identity of the elements can be determined using gamma ray spectroscopy. In some NAA applications, radiochemical separation of the activated samples is required.

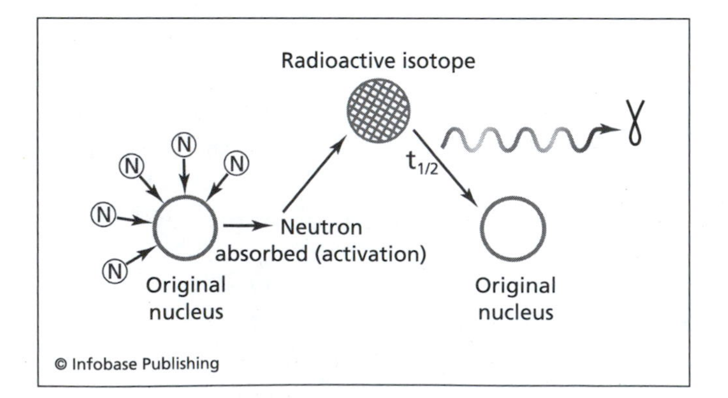

Neutron activation analysis

Although NAA is a NONDESTRUCTIVE technique that is very sensitive, it is rarely used in forensic analysis. The primary drawback is that access to a research nuclear reactor is required, along with specialized detectors. Expertise in analysis and interpretation of the data is also needed. Finally, other instruments designed for elemental analysis are now available to most forensic laboratories including ATOMIC ABSORPTION (AA), INDUCTIVELY COUPLED PLASMA ATOMIC EMISSION SPECTROSCOPY (ICP-AES), INDUCTIVELY COUPLED PLASMA MASS SPECTROMETRY (ICP-MS), and X-ray techniques associated with SCANNING ELECTRON MICROSCOPY (SEM).

See also X-RAY TECHNIQUES.

Newton, Sir Isaac (1627–1727) English *Physicist and alchemist* Although not a forensic scientist, Sir Isaac Newton's work was critical in developing the scientific foundations of FORENSIC SCIENCE. Of particular forensic importance was Newton's work with visible light, which eventually led to the development of instrumentation still routinely used in forensic laboratories. Newton proved that white light consists of a combination of light of different colors. He did so by building the first prism, which took daylight and broke it into its components. Newton also worked extensively in the field of optics, essential for everything from microscopes to modern analytical instruments indispensable to forensic science. He gave a series of lectures at Trinity College in Cambridge entitled *Of Colours* during the years 1670–72, and he published a book series, *Opticks* (1704). Newton believed that light, like air, was composed of corpuscles. This concept fits with the modern concept in several ways. Light shares the nature of both energy and particles called photons, which are discrete packets of electromagnetic energy loosely analogous to corpuscles.

Among Newton's discoveries was that the ability to break white light into components was based on refractive index, a principle fundamental to forensic MICROSCOPY. Newton also described concepts such as chromatic aberration, familiar to microscopists and photographers. Because the refractive index of a material such as GLASS is a function of the wavelength, a glass lens will bend blue light more than it will bend red light. When working on the small scale of a microscope, this difference can be significant, but it can also be exploited to aid in identification. Newton's work was fundamental in developing microscopy as well as a host of other analytical tools routinely used in forensic laboratories.

next-of-kin software Software used in DNA TYPING analysis to link samples and types to family relationships. The terrorist attacks of September 11, 2001, emphasized the need for this capability and drove much of its development. Kinship software is useful when reference samples from the missing person are not available. It also increases the confidence of identification of a family member, invaluable when dealing with mass casualities and grieving families. Given that DNA typing methods and databases exist and are proven, the initial challenges came in developing software and statistical methods to link several different DNA types to a probability of kinship. Undoubtedly, kinship techniques will be invaluable in future mass casuality incidents.

See also DNA; "Identification of the Dead" essay.

Further Reading

Biesecker, L. G., J. E. Bailey-Wilson, J. Ballantyne, et al. "DNA Identification After the 9/11 World Trade Center Attack." *Science* 310 (5751) (2005): 1122–1123.

NIBIN *See* IBIS SYSTEM.

ninhydrin A versatile compound used to visualize latent FINGERPRINTS on porous surfaces such as paper and cardboard. It can also serve as a developing agent for use in THIN LAYER CHROMATOGRAPHY (TLC) in DRUG ANALYSIS and TOXICOLOGY. Ninhydrin is also sometimes referred to as triketohydrinhene hydrate. Because of its versatility and sensitivity, it has become one of the primary tools in latent fingerprint visualization and has displaced iodine fuming as the method of choice for porous materials. The first synthesis of ninhydrin is attributed to Ruhemann in 1910, and the characteristic purple produced by ninhydrin reacting with latent prints is called Ruhemann purple. However, ninhydrin was not recognized as a developer for latent prints until the 1950s. Since then, use of and research into ninhydrin has expanded, leading to several breakthroughs.

Ninhydrin reacts with the amino acids and their degradation products that are part of any latent print. Amino acids are an especially attractive target for development since they adhere to cellulose, the primary ingredient in wood-derived products such as paper and cardboard. They are also fairly stable, allowing prints to be developed long after they were originally deposited. Ninhydrin can be swabbed or sprayed onto a surface, or the entire article can be dipped into a solution. NFN is a designation for a nonflammable spray containing ninhydrin and a propellant that will not burn such as a hydrochlorofluorocarbon. Once ninhydrin is applied to an article, development of the prints can take time (hours or even days), but increasing heat and moisture in the development environment can accelerate it. Ninhydrin can also be used in a sequence of several reagents, the most common being DFO (a ninhydrin derivative), followed by ninhydrin, and a mixture known as physical developer (PD). Ninhydrin and related compounds such as DFO can also be coupled with laser luminescence and metallic salts such as zinc chloride ($ZnCl_2$) used with alternate light sources and lasers. As a result, very faint fingerprints can often be developed. Research into ninhydrin and related compounds is continuing at a rapid pace, resulting in ever more sensitive techniques, reagents, and sequences.

Further Reading

Almog, J. "Fingerprint Development by Ninhydrin and Its Analogs." In *Advances in Fingerprint Technology.* 2d ed. Edited by H. C. Lee and R. E. Gaensslen. Boca Raton, Fla.: CRC Press, 2001.

nitrocellulose (NC) A low explosive used in SMOKELESS POWDER and the PROPELLANT used in modern AMMUNITION. If nitrocellulose is the only ingredient, the powder is called single base, and if it is mixed with NITROGLYCERINE, the powder is called double base. Nitrocellulose is also an ingredient in varnishes and lacquers and was at one time used in automotive paints. There are also formulations of dynamite that use nitrocellulose as an ingredient. When mixed with nitroglycerine in the proper proportions, the resulting mix is a gel that is water resistant and can be used in wet environments. The plastic Celluloid is derived from nitrocellulose and camphor. The forensic analysis of nitrocellulose is accomplished principally by THIN LAYER CHROMATOGRAPHY (TLC).

nitrogen phosphorus detector (NPD) A specialized detector used with GAS CHROMATOGRAPHY (GC) that is used in TOXICOLOGY and DRUG ANALYSIS. This detector, also called a thermionic or alkaline flame

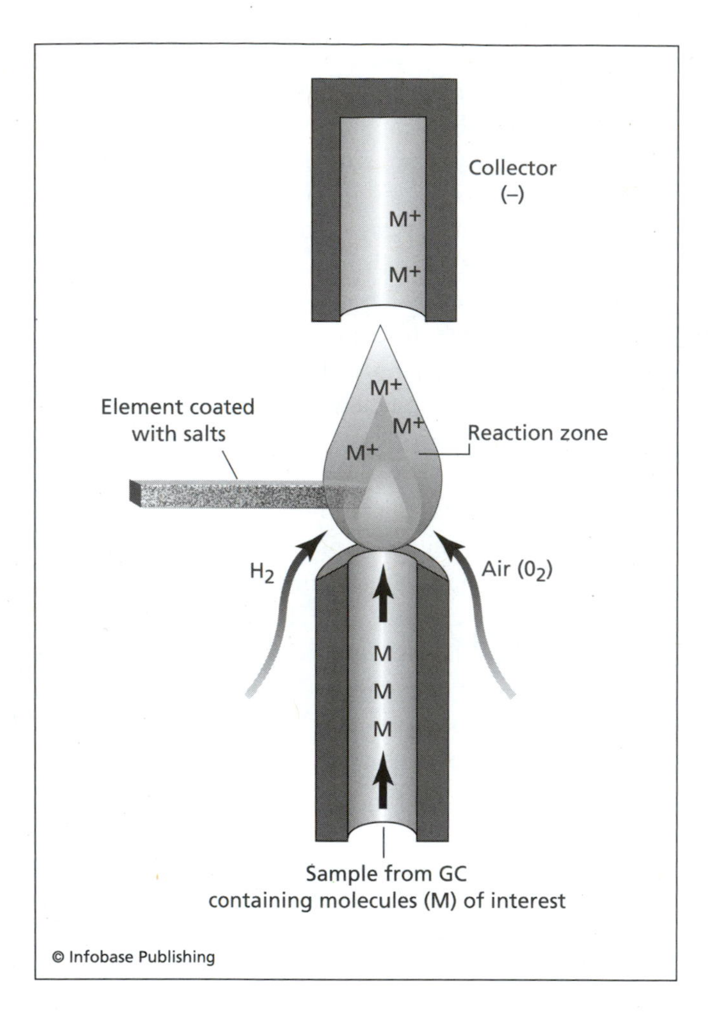

Simplified schematic of a nitrogen phosphorus detector (NPD) used in gas chromatography (GC)

detector, has a similar design to a FLAME IONIZATION DETECTOR (FID). As shown in the figure, the sample exits the GC column and enters into the detector, where hydrogen gas (H_2) and air (containing oxygen, O_2) are mixed into the stream and into a flame that burns while the detector is operating. An element coated with an alkaline salt such as rubidium sulfate (Rb_2SO_4) is inserted into the reaction area and allows for efficient ionization of molecules containing nitrogen or phosphorus in the flame. Once the molecules are ionized, they are attracted to a collector that registers an electrical current that is based on the number of ions that strike it. Since so many pharmaceuticals, illegal drugs, and their metabolites contain nitrogen, the NPD is useful in forensic applications. In addition, since the detector only responds to a limited number of compounds, the need for extensive and complex sample preparation and clean up is reduced.

nitroglycerin (NG) A shock sensitive material used as an explosive and also as a drug to combat a heart condition called angina pectoris. Nitroglycerin, informally called "nitro," was synthesized in 1847, and its potential for mining and other operations was immediately realized. However, the instability and sensitivity of the compound caused many accidents and prevented its widespread use until 1867, when Alfred Nobel found that it could be stabilized by mixing it with a diatomaceous earth formulation. Nobel went on to become a wealthy man, and a portion of his estate was used to establish the Nobel Prize administered from his native Sweden. Different formulations of dynamite have evolved and appeared over the years, but, in general, the term *straight dynamite* refers to a composition that ranges from about 15 percent to 60 percent nitroglycerine. Gelatin dynamite is a combination of nitroglycerin and NITROCELLULOSE (NC) and is useful in wet environments.

As shown in the figure, nitroglycerine is a relatively small molecule with nitrogen groups substituted for the -OH groups on a glycerin molecule. Glycerin is a common ingredient in consumer products and is used as a moisturizing agent in lotions, as an ingredient in soaps, and as a sweetener. It is a thick, oily liquid, as is nitroglycerin, and old dynamite is often found with an oily seepage on the outer coating. Some dynamite formulations now have some of the NG replaced with EGDN (ethylene glycol dinitrate), a related compound. Nitroglycerin can also penetrate the skin in much the same way that glycerin in cosmetics and lotions does, and for treatment of angina pectoris, NG can be administered as an ointment that penetrates the skin and acts as a vasodilator, relieving the chest pain associated with angina.

The forensic analysis of nitroglycerin (primarily from EXPLOSIVES and GUNSHOT RESIDUE [GSR]

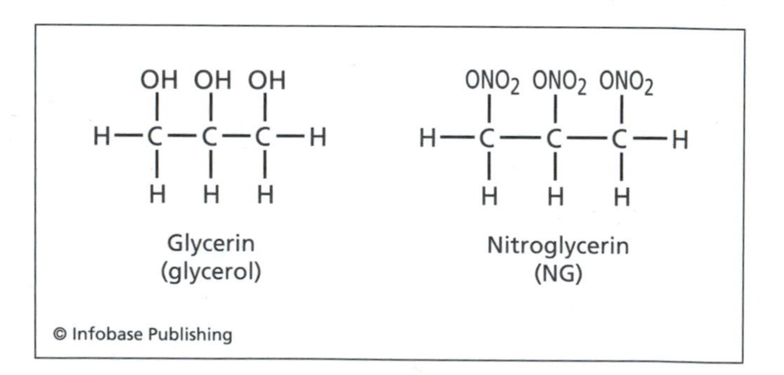

Comparison of the structures of glycerin and nitroglycerin

evidence) relies on color tests (PRESUMPTIVE TESTS), THIN LAYER CHROMATOGRAPHY (TLC), GAS CHROMATOGRAPHY/MASS SPECTROMETRY (GC/MS), and HIGH-PRESSURE LIQUID CHROMATOGRAPHY (HPLC).

nitroguanidine An ingredient used in propellants for high-powered firearms such as artillery. Propellants that contain nitroglycerin, nitrocellulose, and nitroguanidine are referred to as triple-base powders, while what is commonly referred to as gunpowder or smokeless powders contain two propellants, nitrocellulose and nitroglycerine. Triple-base powder is used in very large caliber weapons not typically encountered in forensic contexts.

nondestructive testing A test that does not alter or damage the sample. Nondestructive tests are valued in forensic science since they preserve the evidence in the same condition in which it was recovered. Truly nondestructive tests are rare, but many types of forensic testing come close to this goal. For example, consider the analysis of FIBER evidence. Visual and simple microscopic examination of the fiber will not damage it, nor will simple analysis such as MICROSPECTROPHOTOMETRY. However, if the analyst wishes to determine the melting point of the fiber, the sample will be altered and potentially destroyed during heating. Similarly, if the solubility of a fiber is studied, if it dissolves in a solvent, it is destroyed, although valuable data might be obtained. DNA TYPING is an example of a destructive test in that some small portion of a stained material must be destroyed to obtain results. However, using PCR techniques, only very small samples are required. Thus, although the analysis is destructive, enough sample usually remains to allow for further testing.

Norris, Charles (1868–1935) American *Physician and Medical Examiner* The second chief MEDICAL EXAMINER of New York City and the one most remembered for establishing the office as one of best in the United States. The first man selected for the job was an ironic choice, Dr. Patrick J. Riordan. He had been the CORONER during the time when the system had come under the scrutiny that ultimately led to its demise. Outcry ensued, and Riordan's tenure lasted a month, during which he interviewed other candidates including Dr. Charles Norris, who had trained extensively in Europe (including at Vienna and Edinburgh) and was considered an expert in forensic MEDICINE. He was also a New York native and grasped the problems faced by the enormous city and the surrounding areas. As it turned out, Norris was the right man in the right place at the right time, and he served with distinction until his death in 1935. Colleagues described him as firm yet dignified, favoring frock coats and firm decisions.

His tenure spanned a horrific explosion on Wall Street in 1920 and the implosion of the stock market in 1927, resulting in many suicides. Gang warfare was rampant, and Prohibition came and went, but the seemly routine deaths dominated the workload. Gone were the days when insurance companies and families could negotiate and bribe the coroner into a ruling favorable to them, and many cases that would have never been investigated got the attention they desperately needed. In one example, a policeman saw a longshoreman throw a bag into the water; further investigation revealed body parts strewn along the river. The man was arrested, and he confessed to killing his girlfriend at her apartment, where the head and torso of the woman were found. The man told police that he had awoken after a night of intoxication and found his girlfriend dead on the floor next to him. He assumed that he had killed her while drunk and to hide the crime had dismembered the body and attempted to dump it. Investigation by the ME told a different tale, one of death by carbon monoxide poisoning, and the man was released.

Norris's tenure was not without problems and conflicts. He resigned twice and twice returned after the mayor pleaded with him to do so. Both incidents were precipitated by budgetary issues. In the first, the city refused to cover the cost of an office car and driver, an expensive item that Norris paid for personally. The second incident began when city officials had the clocks removed from the morgue, no doubt to encourage longer workdays. Norris was quick to make his displeasure known in both cases. He was succeeded by Dr. THOMAS GONZALES, another veteran of the 1918 birth of the office.

Further Reading

Eckert, W. G. "Charles Norris (1868–1935) and Thomas A. Gonzales (1878–1956), New York's forensic pioneers." *The American Journal of Forensic Medicine and Pathology: Official Publication of the National Association of Medical Examiners* 8 (4) (1987): 350–353.

———. The forensic or medicolegal autopsy. Friend or foe? *The American Journal of Forensic Medicine and Pathology:*

Official Publication of the National Association of Medical Examiners 9 (3) (1988): 185–187.

Eckert, William G. "Forensic Sciences and Medicine." *The American Journal of Forensic Medicine and Pathology* 11 (4) (1990): 336.

NTSB (National Transportation Safety Board) A federal agency formed in 1967 to investigate aviation, railroad, pipeline, marine, and other types of transportation accidents. The NTSB absorbed the functions of the CIVIL AERONAUTICS BOARD (CAB) when it was formed and remained under the funding umbrella of the Department of Transportation (DOT) until 1975, when it became completely independent. The board maintains a "go team" that can rapidly deploy to the sites of accidents and call in any needed expertise, including that of manufacturers and other companies whose equipment or personnel were involved in the incident. The five board members that oversee the NTSB are all presidential appointees and serve for five years. The NTSB maintains a comprehensive Web site at www.ntsb.gov.

The NTSB investigates accidents, announces findings, and makes recommendations to avoid future incidents, but the board has no regulatory power. NTSB investigations usually fall under the category of forensic engineering, and, increasingly, the NTSB works alongside law enforcement agencies when it is possible that a criminal or terrorist act was responsible for an accident. If a criminal act is confirmed, the FBI assumes responsibility for the investigation. Two recent examples include the crashes of TWA FLIGHT 800 and Egypt Air Flight 990. In the second case, the plane crashed nose-first into the Atlantic off the coast of Nantucket Island, Massachusetts, on October 31, 1999. After extensive review by the NTSB (among other agencies) and the elimination of mechanical failures, the most likely cause was assigned to the copilot, who apparently purposely crashed the airliner, killing 217 people.

nuclear magnetic resonance (NMR) An instrumental technique used in organic chemistry and biochemistry to study molecular structure. NMR has also been adapted for medical use as magnetic resonance imaging (MRI). NMR has been used for the analysis of drugs and in a few other areas, but the expense and complexity of the instrumentation has limited its forensic applications. NMR is primarily a qualitative technique and is most frequently used to reveal the structure of the carbon-hydrogen "skeleton" of organic molecules. Like INFRARED SPECTROSCOPY (IR), NMR requires a pure sample of the compound of interest. However, occasionally NMR is performed on complex samples to generate a chemical "fingerprint" that can be used for comparison purposes.

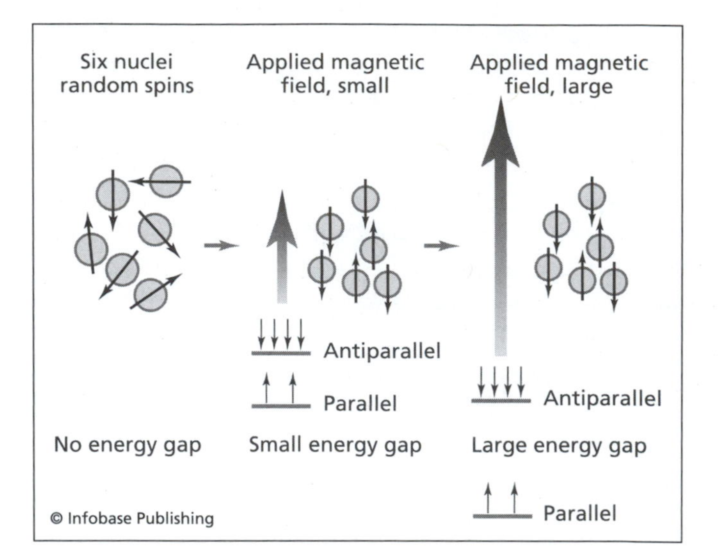

Alignment of spinning atomic nuclei in a magnetic field

NMR is based on the absorption of ELECTROMAGNETIC ENERGY (EMR) in the radio frequency range by spinning atomic nuclei. This contrasts with other types of SPECTROSCOPY such as IR in which the electrons (outside the nucleus) absorb energy. Any atom that has an odd mass such as hydrogen (mass of 1, indicated as ^{1}H) is amenable to NMR, as is any atom with an even mass but an odd number of protons. Nitrogen is an example of this type of atom, with seven protons in the nucleus and a mass of 14. NMR is most often applied to hydrogen and a naturally occurring isotope of carbon, ^{13}C. The bulk (~99 percent) of naturally occurring carbon is ^{12}C, with six protons and six neutrons in the nucleus. About 1 percent is ^{13}C, with six protons (an even mass number) and seven neutrons, leading to an odd mass overall. When NMR targets the hydrogen, it is often referred to as "proton NMR" since a hydrogen atom consists of one proton. By coupling proton NMR with ^{13}C NMR, the backbone of carbon and hydrogen of an organic compound can be deduced. Other elements such as ^{19}F and ^{31}P can be studied using NMR.

As nuclei spin, they generate a tiny magnetic field. As shown in the figure, the orientation of these fields, indicated by the arrow, are random. When a magnetic field is applied, the nuclei can align parallel to the applied field or opposite it (antiparallel). Those that are parallel are at a lower energy state than are those that are antiparallel, creating an energy gap between them. The larger the applied field, the larger the gap. When the nuclei are exposed to radio frequency energy (about the same range as FM radio stations), the nuclei in the parallel configuration absorb the radiation and are promoted to the higher antiparallel state. The degree of absorbance and the frequency at which it occurs is then plotted and interpreted.

Further Reading

Lee, G. S., et al. "A Methodology Based on NMR Spectroscopy for the Forensic Analysis of Condoms." *Journal of Forensic Sciences* 46, no. 6 (2001): 808.

null hypothesis A fundamental component of statistical testing and of the scientific method. The null hypothesis provides a starting point for the analysis of data and for testing a given idea that arise from a forensic analysis. Simply put, the null hypothesis assumes that two quantities or entities are the same or come from the same source (a COMMON SOURCE). Any differences between them can be attributed to small random errors and not to any statistically significant problems with the hypothesis itself. In drug analysis, for example, a chemist might be asked to determine the percentage of COCAINE in two samples of a white powder. If one sample is analyzed and found to contain 51.0 percent while the second contains 51.2 percent, the null hypothesis would be that the percent cocaine in each is the same and the difference is due only to small random errors and not to any significant differences in the cocaine concentrations. In the case of the analysis of a bloodstain found at a scene and the blood from a suspect, the null hypothesis would state that the source of the blood is the same, namely, from the suspect. A similar case would be encountered in firearms from which a bullet fired from a seized weapon is compared to a bullet recovered during an autopsy. The null hypothesis would be that the two bullets were fired from the same weapon.

One of the key aspects of a null hypothesis is that it must be stated in such a way that it is possible to disprove it. In the case of the gun and bullets, an examiner would carefully study both bullets, and if significant discrepancies are found, then the null hypothesis has been falsified. The two bullets could not have come from the same gun. When a null hypothesis is falsified, then a new hypothesis must be developed and similarly tested until all attempts to falsify it fail. When this situation occurs, the null hypothesis is accepted as the correct explanation. Consider the bloodstain example. If DNA TYPING were performed and the results showed that the suspect's blood was of a different type than that of the stain, the null hypothesis would be incorrect. Further investigation might uncover another suspect, leading to a new null hypothesis. Comparative testing would continue until the null hypothesis could no longer be falsified, meaning that the type found in the stain matched that of the suspect, be it the second one identified or the 20th.

In some types of analysis, it is possible to assign a level of certainty or probability to an evaluation of the null hypothesis. This is accomplished by calculating confidence intervals. In such cases, a probability level is selected such as 5 percent. This means that 95 times out of 100, the rejection of the null hypothesis is correct, and in 5 percent of the cases it would be incorrect. In the above example, a chemist obtained two different concentrations for cocaine, 51.0 percent and 51.2 percent, which are close but not identical. If these values were obtained by analyzing several small subsamples of each cocaine sample, then these percentages represent the average values for all of these tests. Statistical tests can be used to determine if the difference between these two average values is significant and what level of certainty applies to that determination. For example, the chemist may be able to say that at the 99 percent confidence level, the two means are really representing the same value. The chemist might state it this way: "I am 99 percent certain that the two means are the same and that the small difference in percentage is not statistically significant." This also means that there is only a 1 percent chance that this statement is false. Note that it is never possible to have 100 percent confidence using these types of tests. Also, use of confidence levels requires numerical data, something that is not always available in forensic analysis.

See also FALSIFIABILITY.

nursing, forensic The application of nursing skills to legal matters and law enforcement. Nurses are often the first to treat or see people who have been the victims of crime and violence, but they also have contact with those who are suspects in criminal activity. Relatives and friends of both victims and suspects may also be involved in such scenarios and thus fall under the purview of forensic nursing. Nurses in these situations need to be familiar with evidence collection, not only physical but also behavioral and in the form of communication, written or otherwise. Clinical forensic nursing (CFN) focuses on nursing in settings such as emergency rooms and other treatment facilities such as prison clinics and forensic (psychiatric) hospitals. Forensic nurses may also be associated with MEDICAL EXAMINER's (ME) or CORONER's offices and act as part of death investigations. The types of investigations that forensic nurses can become involved in include civil and criminal cases and span the gamut from crimes of violence to car accidents, workplace accidents, substance abuse, product tampering, neglect, and medical malpractice.

A sexual assault nurse examiner (SANE) is a nurse trained to collect evidence and to counsel victims of sex crimes. Nurses involved in sexual assault cases were a crucial element leading to the recognition of forensic nursing as a specialty within the larger profession. The International Association of Forensic Nurses (IAFN) was formed in 1992, and in 1995, forensic nursing was officially recognized as a subspecialty of nursing by the American Nurses Association. By 2002, the IAFN had nearly 2,000 members and offered the first certification examinations for SANE nurses. The association maintains a Web site at www.IAFN.org.

Further Reading

Goll-McGee, B. "The Role of the Clinical Forensic Nurse in Critical Care." *Critical Care Nursing Quarterly,* May 1997.

nylon The first widely successful synthetic FIBER and one that is still encountered as trace and TRANSFER EVIDENCE. Nylon was developed by Wallace Carothers and first synthesized by his team of researchers at the Du Pont Company in 1938. Carothers committed suicide in 1937. Two years later, nylon was introduced to the public and became famous as an ingredient in ladies stockings, which are still called "nylons." As a polymer, nylon consists of long chains of small chemical units called monomers that are chemically bonded together and arranged in such a way as to give the polymer its physical and chemical characteristics. There are several types of nylon fibers that are named based on the type of monomer units from which they are made. Nylon 66 (also called nylon 6-6), as an example, is made of monomers that each contain six carbon atoms.

Nylon is also classified as a condensation polymer, meaning that when the monomers are chemically linked together, small molecules such as water (hence "condensation") are given off. The type of bond that forms to link the monomers together is called an amide bond, which is the same type of bond that links AMINO ACID monomers together to form PROTEINS. Synthesis of nylon also requires heating, and the nylon is drawn out into fibers while still gooey. As a result, the long polymer molecules align side by side and can chemically interact, creating a fiber that is strong and has the characteristic sheen of nylon. Because it is synthetic, nylon is durable and resistant to attack by fungus, microbes, and insects. Brands of nylon available include Anso and hyrdofil, which are used for insulation.

The durability and widespread use of nylon also means that it is a common form of TRANSFER EVIDENCE. The analysis of nylon fiber is accomplished by microscopic examination, including examination of the morphology (structure) of the fiber, its cross section, its BIREFRINGENCE, and its spectral properties. The latter can be accomplished using MICROSPECTROPHOTOMETRY. Nylon and other fiber evidence was crucial in the WAYNE WILLIAMS case.

O

oaths and ordeals Methods of obtaining and validating oral testimony in an era when documented and scientific evidence was unused or unavailable. Ancient Egypt and Mesopotamia law shared a reliance on oral testimony supported by oath. When oral testimony was not sufficient to make a decision and supporting evidence was lacking, the other option was the ordeal, in which the court deferred to God. Trial by ordeal in Mesopotamia involved such procedures as binding the accused and tossing him or her into the river. The ancients assumed that the guilty would sink and that God would save the innocent. This was the opposite of the later ordeals used in Europe, where the guilty floated and the innocent sank. The Egyptians summarized reliance on otherworldly judgment through the story of Osiris, the god of the underworld and judge of the dead.

The type and severity of the ordeal varied among cultures, but ordeals were used in Mesopotamian, Roman, Greek, Jewish, Chinese, Indian, and Islamic law. The ordeal was often a form of combat or a physical test that invoked God to act as referee. The verdict and sentence depended on the result of the ordeal.

Despite the superstitious aspects, ancient laws did recognize the importance of high standards of proof and evidence in criminal proceedings, particularly in capital cases. This precedent survives in modern law, where criminal cases typically require proof of guilt beyond a reasonable doubt. To increase the reliability of evidence, ancient practices required more witnesses, more oaths, and additional ordeals. In some cases, the witnesses called were physicians or scientists of a sort, but their participation rarely had anything to do with their professional skills. As legal systems evolved across the ancient world, supporting and physical evidence came to play a larger role. This led to the formalization of procedures governing what the trier-of-fact would be able to consider and what he or she would not.

The reign of the Catholic Church under the various popes was important in the history of FORENSIC SCIENCE and evidence in another way. Much of the church's effort was devoted to rooting out heresy, a crime difficult to prove and impossible to prove with any objectivity. Scientific and physical evidence rarely played a part in such proceedings, but sadly torture did. Equally sad, this was nothing new. The Greeks considered torture to be the best method of validating oral testimony. The Inquisition made extensive use of torture for the same reason. Often torture came after conviction to elicit information for later investigations.

Ironically, the church, an early and fervent adopter of torture, eventually grew uncomfortable with the concept of the ordeal. By the 13th century, ordeals were uncommon, and the courts returned to a reliance on sworn oral testimony as the best evidence. Ironically, this was a period when humans were considered innately sinful and untrustworthy. Partially in recognition of this tension, the concept of a HALF-TRUTH appeared in English law during the Middle Ages. Calling something a half-truth was not derogatory, but rather indicated what value should be placed on that evidence. Alone, a half-truth was not enough to prove or disprove an issue, but it did have value in supporting one version of it over another. A half-truth in medieval law did not imply a purposeful omission of key information; rather it weighted the evidence as at somewhere between definitive and useless.

obliterations This term has two meanings in forensic science. First are obliterations encountered during the analysis of QUESTIONED DOCUMENTS, in which someone has erased or otherwise removed or hidden some kind of writing or printing. The other type of obliteration is the purposeful removal of serial numbers on evidence such as stolen property or guns. This type of obliteration is addressed in SERIAL NUMBER RESTORATION.

Any type of writing, be it handwriting or printing by commercial presses, computer printers, typewriters, or copiers, can be obliterated or altered. The obliteration can be achieved by erasing and overwriting, crossing out or scribbling over, cutting or scraping the paper surface away, using erasing substances such as a "whiteout" compound or correction ribbons on typewriters, and finally using chemical means such as a bleaching agent to chemically remove the writing from the paper. Document examiners have several tools at their disposal to find out what writing was obliterated. First is the use of a microscope, by which damage to the paper fibers from erasure or scraping and cutting can be seen. Different types of lighting and illumination are also used. As with INDENTED WRITING, oblique lighting (lighting from a side angle) can sometimes reveal details of the obliterated writing.

Perhaps the most valuable tool for visualizing obliterated writing is LUMINESCENCE and the use of illumination and photography in the ultraviolet and infrared regions of the electromagnetic spectrum. For example, if writing is obliterated by marking over it with a different ink, infrared radiation may be able to penetrate the marking over and make the original writing visible. Photography in the infrared region is often valuable, as is ultraviolet (black light) illumination followed by fluorescence. Different filters and lighting can also be exploited to enhance contrasts and better show what the original obliterated writing was.

odontology, forensic The application of dentistry to legal matters in the areas of personal identification, age determination, BITE MARKS, evaluation of wounds and trauma to the jaws and teeth (particularly in potential child abuse cases), and evaluation of alleged dental malpractice or negligence. A mature adult possesses a set of 32 teeth collectively referred to as dentition. Baby teeth, those replaced by the permanent teeth, are called deciduous teeth or primary dentition. Both types of dentition can be and are utilized in forensic applications. Because teeth are physically and chemically resilient, they endure and are likely to survive the severe trauma associated with MASS DISASTERS. Since each tooth has five different surfaces (four sides and a top) and since teeth have different spacings, orientations, distinctive gaps, and widths, dentition offers an ideal method of identification. Given the fact that each person will develop or acquire unique wear patterns and dental work such as fillings over their lifetime, the overall pattern is generally considered to be unique to each person. The American Board of Forensic Odontologists (ABFO) together with the odontology section of the AMERICAN ACADEMY OF FORENSIC SCIENCES (AAFS) are the principal professional organizations in the field. The AFBO publishes guidelines for use in identification and bite mark cases.

Dental identification tasks can fall into one of two categories: identification by comparison and identification when no comparison is available. The former is the type of identification required in mass disasters such as an airline crashes. In such cases, the usual identification methods such as appearance and personal effects often cannot be applied. Who was aboard the plane is usually known, and the identification task becomes one of separating the commingled remains and assigning them to the proper person. Forensic odontologists work with dentition and rely principally on comparisons of antemortem dental records (obtained while the deceased was alive) with postmortem data. DNA analysts are often part of the identification team, while forensic ANTHROPOLOGISTS can be called in to work with any skeletal remains. Often, governments organize teams of such professionals ahead of time to deal with mass disasters and fatalities.

The other type of identification the odontologists participate in is that of unknown remains in which there is no comparative data available, at least initially. This would occur if a badly decomposed body were found in a shallow grave in a remote location with no clues as to the identity of the victim. Odontologists can contribute by estimating age (AGE-AT-DEATH ESTIMATION) based on the condition of the teeth. For infants (even fetuses) through young adults, dentition is a fairly reliable method of determining age and is based on the well-known succession and development patterns of teeth. However, once a full set of permanent teeth are in place, dentition becomes less reliable since genetic factors, diseases, oral hygiene, diet, and many other factors contribute to the general condition of teeth. In some cases, an alternative is to use a technique called AMINO ACID RACEMIZATION (AAR) using portions of teeth. To further help identify a victim, odontologists can study the victim's dental work and restorations and offer suggestions and insights as to where and when such techniques were or are used.

One of the more recent uses of odontology has been in the area of bite mark comparison and analysis. The same characteristics that lend uniqueness to teeth for identification can be exploited for the analysis of bite marks. Courts have accepted bite mark evidence since the 1950s, but it has not been without controversy. The primary challenge in bite mark evidence is preservation and accurate documentation, since the material in which a bite mark is found is rarely stable. For example, a bite mark found on the victim of a homicide will change as decomposition proceeds. If the victim survives, the appearance of the wound will change as it heals. Thus, photography and occasionally casting (as is done with other forms of IMPRESSION EVIDENCE) must be rigorous and timely. The ABFO has developed specialized measuring devices that are recommended for use in photographing bite marks. Different angles and illumination (such as with ultraviolet light) can also be useful.

Finally, odontologists may be called upon to evaluate wounds to the jaw and mouth in cases such as assaults and child abuse. In a related role, they may get involved in cases of alleged malpractice or negligence by other dentists.

See also "Identification of the Dead" essay.

Further Reading

Clark, D. H., ed. *Practical Forensic Odontology.* Oxford: Butterworth-Heinemann, 1992.

Glass, R. T. "Forensic Odontology." In *Forensic Science: An Introduction to Scientific and Investigative Techniques.* 2nd edition. Edited by S. H. James and J. J. Nordby. Boca Raton, Fla.: CRC Press, 2005.

Moenssens, A. A., J. E. Starrs, C. E. Henderson, and F. E. Inbau. *Scientific Evidence in Civil and Criminal Cases.* Westbury, N.Y.: The Foundation Press, 1995.

Stimson, Paul G., and C. A. Mertz, eds. *Forensic Dentistry.* Boca Raton, Fla.: CRC Press, 1997.

oil immersion *See* MICROSCOPY.

Oklahoma City Bombing *See* ENGINEERING, FORENSIC.

opiate alkaloids *See* ALKALOIDS.

opium A mixture of compounds obtained from the unripened seed pods of the plant *Paperver somniferum.* This poppy plant grows in large areas of Asia, especially southwest Asia, and Mexico. To obtain opium, the seedpod is cut, and a brownish milky substance is extracted. This extract is then dried and can be crushed into a powder that gradually lightens in color. Opium consists of a mixture of ALKALOIDS, including MORPHINE and CODEINE, and can be consumed directly by ingestion or smoking. Opium was first seen in large quantities in the United States in the 1800s, when immigrants brought it from China and the Orient. "Opium dens" were widespread, and opium was an ingredient in patent medicines that were commonly available. Abuse of raw opium is no longer common, and the substance is now listed on Schedule II of the CONTROLLED SUBSTANCES ACT (CSA).

Opium is now used as a source of morphine, codeine, and other alkaloids that can be chemically altered to make related narcotic drugs. Raw opium can contain morphine in concentrations of about 5 percent to 20 percent, and this compound can be extracted and used directly or easily converted to HEROIN. Codeine can also be extracted, but this compound is not typically abused on the scale that heroin is. Thebaine, another alkaloid found in opium, can be converted to OXYCODONE, a prescription drug used to relieve moderate to severe pain.

Although opium is rarely encountered as a street drug, it is used as a PRECURSOR to many other illegal and abused prescription drugs. The forensic analysis of opium using techniques such as GAS CHROMATOGRAPHY/MASS SPECTROMETRY (GC/MS) and HIGH PRESSURE LIQUID CHROMATOGRAPHY (HPLC) can be useful in linking a particular batch of opium to a region in the world in which it might have been grown.

ordeals *See* OATHS AND ORDEALS.

Orfila, Mathieu (M. J. B or Mathieu Joseph Bonaventure) (1787–1853) Spanish, French *Forensic Toxicologist* Orfila, considered the founder of

M. J. B. Orfila, the father of forensic toxicology ***(National Library of Medicine, National Institutes of Health)***

forensic TOXICOLOGY, was born in Spain but moved to France, where he worked and became professor of forensic chemistry and dean of the medical faculty at the University of Paris. He began publishing early, with his first paper on poisons in 1814, when he was 26 years old. He spent a good deal of time studying poisons, particularly ARSENIC, which was frequently used for murder in that era. As a toxicologist, he concentrated on methods of analysis of poisons in blood and other body fluids and tissues. He became involved in a famous arsenic poisoning case in 1839, when a young woman, Marie Lafarge, was accused of using arsenic to murder her much older husband. Initial results of the analysis of the husband's remains were negative, but Orfila was able to detect arsenic in the exhumed remains. Marie was eventually convicted of the crime. His testimony in the case was one of the earliest examples of sound scientific testimony by a recognized scientific expert in a court of law.

Further Reading

De Forest, P. R., R. E. Gaensslen, and H. C. Lee. "About Forensic Science." In *Forensic Science: An Introduction to Criminalistics*. New York: McGraw Hill, 1983.

organic compounds and organic analysis Organic compounds are defined as those that contain primarily carbon (C) and hydrogen (H), a class of compounds centrally important in forensic science. Examples of organic compounds are sugars, petroleum products, drugs, plastics, and proteins. Carbonate (the ion CO_3^{2-}) and carbon dioxide (CO_2) are not considered organic compounds since they do not contain hydrogen. Organic chemistry is the study of organic compounds and is the foundation of biochemistry, which revolves around the organic chemistry that occurs in living organisms. Likewise, organic analysis involves detection, quantitation, and study of organic compounds. Organic compounds may be natural or synthetic. For example, MORPHINE and CODEINE occur naturally as part of OPIUM, while HEROIN is synthesized from morphine. DESIGNER DRUGS are new synthetic organic compounds that are made in CLANDESTINE LABS.

Organic compounds are the focus of forensic DRUG ANALYSIS and TOXICOLOGY. The majority of solvents used as ACCELERANTS in ARSON are organic compounds. Gasoline, for example, is a complex mixture of organic compounds derived from crude oil and contains organics such as octane, benzene, and toluene. PLASTICS are POLYMERS, most of which are also derived from petroleum products, as are many synthetic fibers. Paints are blends of organic and inorganic materials suspended in water or an organic solvent, as are inks. Even GUNPOWDER has organic ingredients.

Organic analysis can be conducted using a myriad of techniques. In forensic science, the most common include simple chemical tests such as PRESUMPTIVE TESTS for different drug classes, instrumental techniques such as INFRARED SPECTROSCOPY (IR), and many different types of chromatography. Examples include THIN LAYER CHROMATOGRAPHY (TLC), GAS CHROMATOGRAPHY/MASS SPECTROMETRY (GC/MS), and HIGH-PRESSURE LIQUID CHROMATOGRAPHY (HPLC).

orthotolidine test (**tolidine test**) A PRESUMPTIVE TEST for blood that is also referred to by the shorthand notation *o*-tolidine. This test works similarly to most other tests for blood in that the *o*-tolidine reacts with hemoglobin in blood in the presence of hydrogen peroxide (H_2O_2) to cause a bluish-green color to form. Like other presumptive tests, the tolidine test is not specific for blood and can produce FALSE POSITIVES with substances such as horseradish. Tolidine is also a carcinogenic material, and thus this test is not as widely used as the KASTLE-MEYER TEST.

Osborn, Albert S. (1858–1946) American *Questioned Document Examiner* Osborn was a pioneer of early forensic document examination (QUESTIONED DOCUMENTS) and wrote a text in 1910 that is still considered a foundational work and reference in the field. He also was a founding member of the American Society of Questioned Document Examiners (ASQDE) and served as its first president from 1942 until 1946, the year of his death. Osborn's sons continued in the field, and both Albert and a son were involved in the LINDBERGH KIDNAPPING CASE.

O'Shaughnessy, Sir William Brooke (1809–1889) English *Forensic Chemist* Sir William Brooke O'Shaughnessy was one of the first forensic chemists, and typical of the time, a generalist. His career began at age 21, and he was well known for questioning and correcting the status quo. His first notable achievement was to debunk a common test using indigo. Chemists exploited indigo for analytical purposes in the 1800s, using a reagent consisting of indigo dissolved in sulfuric acid. If the sample tested contained nitric acid, the indigo was bleached from the characteristic blue to a clear solution. This was important because some poisoners used nitric acid. O'Shaughnessy published his first paper in the medical journal *Lancet* in 1830, describing the shortcomings of the test. O'Shaughnessy pointed out that other compounds besides nitric acid reacted with indigo. In noting the test's problems, he was contradicting some of the early English forensic scientists of note, including Robert Christison, who had a medical degree from the University of Edinburgh. It was a courageous action for such a young chemist.

Not satisfied with just debunking the existing method, O'Shaughnessy went on to describe a method to take its place. This method consisted of three tests that were specific for nitric acid. First, nitric acid would turn orange in the presence of MORPHINE (forming another DYE); second, it would form a solid when added to urea nitrate; and third, it would facilitate formation of silver fulminate. Detection of the latter was simple, obvious, and hazardous. In publishing the paper, O'Shaughnessy was following Christison's adage to forensic chemists (then principally toxicologists) that the job of the analyst was more than to provide some evidence; it was to provide the best evidence possible given the limits of scientific knowledge of the time. For a short period, indigo dye and other tests represented those limits.

The emergence of a reliable set of classification and identification tests for drugs typified forensic CHEMISTRY in the decades to come. In a classic example of the interplay of public health and legal matters in the early history of FORENSIC SCIENCE, O'Shaughnessy took on another project in 1830 at the request of the *Lancet* editor. He developed a systematic method of testing candy for the presence of organic and inorganic poisons, put there by accident or design. Mindful of the tendency of children to suck on the candy wrappers, he considered those as a possible mode of ingestion. He eventually identified a number of contaminants and adulterants, including oxides of lead and antimony, mercuric sulfide, copper carbonate, calcium sulfate, and Prussian blue. To confirm the presence of the metals, he isolated the elemental form, presaging the work of JAMES MARSH. The work spurred public concern and further investigation of adulterated products, with offenders having their names published in the *Lancet*.

O'Shaughnessy had an abundance of self-confidence and the will to use it. He would take on any authority, and in retrospect he was right more often than he was wrong. He once wrote of the French chemist MATHIEU ORFILA's work related to lead detection in poison, "Nothing can be more practically absurd than Orfila's directions in this instance." In contrast, he embraced the Marsh test, a test developed by James Marsh used to detect ARSENIC, and was among the first to recognize its value in forensic TOXICOLOGY, writing in 1842, "The moment I read Mr. Marsh's notice I saw at once its extra-ordinary value . . . That week I applied it in the investigation of two cases of arsenic poisoning received into Police Hospital, and I was subsequently the first to publish the practical results of its application to legal analysis." He was not in fact the first to do so; Orfila had beaten him by two years with the Lafarge case (a case in which Marie Fortunée Lafarge was convicted of poisoning her husband with arsenic). O'Shaughnessy had a good excuse for the lapse. Frustrated by not being named to a post in medical jurisprudence at the University of London, he joined the East India Company as an assistant surgeon and went to Calcutta in 1833. He had wanted to return to his practice of medicine, but his forensic chemical skills kept intruding, and he was given the post of chemical examiner at Calcutta Medical College in addition to his medical duties. It was there that he demonstrated the Marsh test and began to use it. True to form, he was the first to point out one of its shortcomings.

When the Marsh test was first introduced, chemists assumed that any form of arsenic would respond. In Europe and England, white As_2O_3 was the most common form of arsenic used as a poison. But in India was yellow arsenic (As_2S_3, Kings Yellow) that was easily available in street bazaars and thus accessible to would-be murderers. It was this yellow form of arsenic used in the John Bodle case in 1832 that motivated Marsh to devise the better test.

In this case, Marsh was called to test for the arsenic Bodle was accused of using to poison his grandfather. Marsh was able to detect the arsenic, but because the arsenic used was yellow, as opposed to the white arsenic Marsh commonly tested, the results deteriorated before they could be presented to the jury. Bodle was acquitted, and Marsh created a better test. (Bodle later confessed that he had in fact killed his grandfather by poisoning his coffee.) O'Shaughnessy enters the story in 1838, two years after publication of the Marsh test. Using it, O'Shaughnessy examined the stomach contents and stomach lining of a victim of suspected arsenic poisoning. He could clearly see the yellow solid clinging to the stomach lining, but the Marsh test yielded negative results. To prove his suspicions, he dissolved the yellow solid in nitric acid and then performed the Marsh test, obtaining the positive result. He later pointed out, as did others, that the element antimony would also yield a positive result with the Marsh test. More important, he pointed out the forensic implications of this finding. At the time, one treatment used in the case of suspected poisoning was to administer an emetic to induce vomiting; one such emetic was potassium antimonyl tartrate.

Despite O'Shaughnessy's contributions to forensic chemistry, there was much more to the man. His role as physician led him to pioneer the use of cannabis (MARIJUANA) to English medical practice as a treatment for tetanus, cholera, and convulsive problems. Having worked with many cholera victims early in his career, he became adept at looking at stomach contents and judging if a death was due to the disease or to arsenic poisoning. In India, he became an assayer at the Calcutta Mint and was instrumental in bringing telegraphic service to the country. He wrote a book on forensic chemistry in the form of a medical manual for students in Calcutta, and was knighted in 1856.

osteology *See* ANTHROPOLOGY, FORENSIC.

oxycodone (OxyContin®) A narcotic pain reliever made from thebaine, an alkaloid derived from OPIUM. Oxycodone acts in the same manner as morphine and codeine, and, as with these substances, abuse can lead to physical and psychological addiction. OxyContin® was introduced in 1995 and contains a time-released form of oxycodone. Since it is time released, the amount of oxycodone in each tablet is much higher, and this has led to increasing abuse and diversion of OxyContin® for illegal use. Abusers crush the tablets and by doing so destroy the time releasing properties of the drug. This allows them to get a large dose immediately by ingestion, smoking, or injection, with effects that mimic heroin. If whole tablets are available, forensic analysis of oxycodone and OxyContin® consists of visual inspection and comparison of the tablets with examples found in the *PHYSICIAN'S DESK REFERENCE (PDR)*. Subsequent analysis relies on THIN LAYER CHROMATOGRAPHY (TLC) and instrumentation such as GAS CHROMATOGRAPHY/MASS SPECTROMETRY (GC/MS) and INFRARED SPECTROPHOTOMETRY (IR). For TOXICOLOGY analysis in samples such as BLOOD and URINE, similar approaches are used and can be combined with IMMUNOASSAY techniques.

P

p30 A protein that is specific to seminal fluid and detectable for about eight hours after sexual intercourse has occurred. Also known as prostate specific antigen (PSA), detection of this material in a stain shows that it is or contains semen. IMMUNOLOGICAL TESTS such as IMMUNOASSAY and IMMUNOELECTROPHORESIS are used in these tests.

paint Because of its widespread use, one of the most common types of TRANSFER EVIDENCE. Most often seen are automotive paints and paints used in homes and building, also referred to as architectural paint. Thousands of formulations have been developed and used, but in simple terms, paint can be thought of as a coating of coloring agents suspended or dissolved in a solvent containing other additives such as binders and drying agents. Pigments were at one time primarily INORGANIC compounds such as titanium dioxide (TiO_2, a white), Fe_2O_3 (rust), and other compounds made up of cadmium, lead, and other metals. Because of their toxicity, many of these compounds have been replaced with ORGANIC pigments. However, lead paints still present a POISON hazard in many older and inner-city buildings, where children eat it. Solvents employed in paint formulations can be organic, such as toluene, or water-based, such as those used in latex paints. Water-based paints can be cleaned up with soap and water and are popular for interiors of homes. The solvent, be it water or an organic, is referred to as the "vehicle." Finally, polymers or materials capable of polymerization (called binders) are included and, when dry, form a protective coating over the pigments. The process of applying paint and forming the coating is summarized in the figure.

From a forensic perspective, paint is a potentially rich source of information. All components, from pigments to the polymer coating, can be subjected to visual and chemical analysis using techniques such as POLARIZING LIGHT MICROSCOPY (PLM), INFRARED SPECTROPHOTOMETRY (IR), X-ray fluorescence (XRF), and SCANNING ELECTRON MICROSCOPY (SEM). Other noninstrumental procedures include testing the solubility of a paint sample using compounds such as hydrochloric acid (HCl) and acetone and simple microchemical tests. However, these examinations are destructive, which presents a problem with very small flakes of paint.

Even a complete battery of chemical and instrumental tests cannot prove that two paint samples are from the same COMMON SOURCE. At best, these tests by themselves can show that two samples are similar or dissimilar. However, since many paints are applied in layers, the analyst has another feature that can be used in comparison. Two paint chips that have seven matching layers are much more likely to have come from one source than chips with only one or two layers. A PHYSICAL MATCH can prove conclusively that two separate paint samples were once part of the same coating. This can occur in a hit-and-run accident when paint from the vehicle is transferred to the victim. If the chips are carefully recovered, it may be possible to fit them back into the damaged portion of the vehicle like pieces of a puzzle.

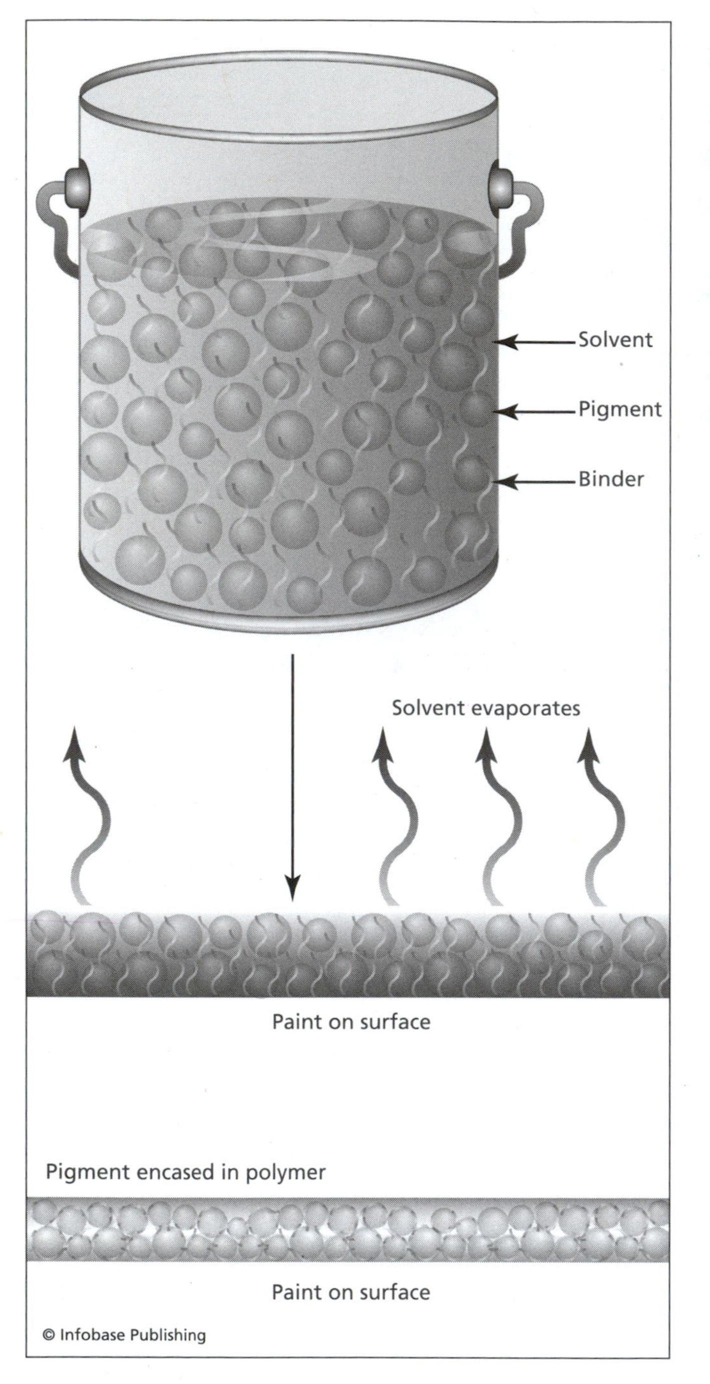

Paint, application and drying

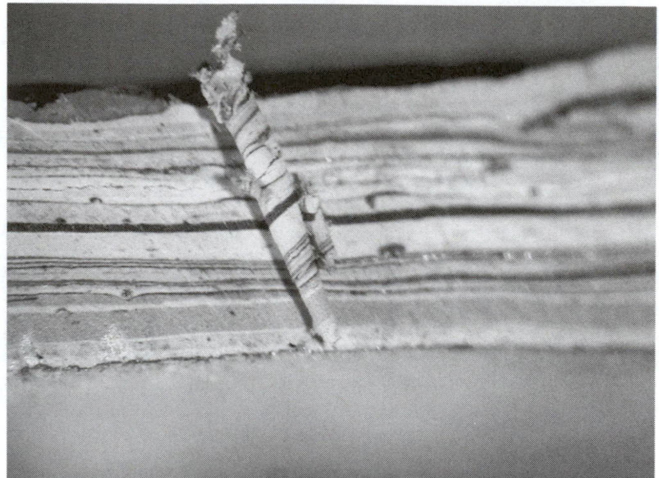

Paint chip and a curled section showing the layer structure ***(Courtesy of William M. Schneck, Microvision Northwest—Forensic Consulting, Inc.)***

Collection of paint chips can be performed using tweezers, a knife, or a TAPE LIFT. Once collected, the chips must be carefully protected to preserve features and edges that might be essential for a physical match. Plastic pill bottles such as those used for prescription medicine are often used to hold paint chips, as are envelopes and carefully folded paper. In the lab, paint examination begins with a STEREOMICROSCOPE or COMPOUND MICROSCOPE and progresses from there depending on the case, amount collected, and other factors. Since color is one of the key properties of paint, particularly for automobiles, identification of a specific color is essential to linking two samples. The "PDQ" system (Paint Data Query) is a database of automobile paint colors used since 1975 for that purpose. The system was established by the Royal Canadian Mounted Police (RCMP) and is now accessible to other nations including the United States. Thus, with a chip of paint, it is possible to significantly narrow down potential sources by manufacturer, year, and model.

See also X-RAY TECHNIQUES.

Further Reading

Geradts, Z. "Use of Computers in Forensic Science." In *Forensic Science: An Introduction to Scientific and Investigative Techniques*. 2nd edition. Edited by S. H. James and J. J. Nordby. Boca Raton, Fla.: CRC Press, 2005.

Thornton, J. I. "Forensic Paint Examination." In *Forensic Science Handbook*. Vol. 1. 2nd edition. Edited by R. Saferstein. Upper Saddle River, N.J.: Prentice Hall, 2002.

painting with light A photographic technique used when ILLUMINATION is difficult. At large outdoor CRIME SCENES, it may be necessary to photograph large areas at night for example. Taking a photo of the inside of a gun barrel is also difficult. Placing an intense light source pointing down the barrel is not a

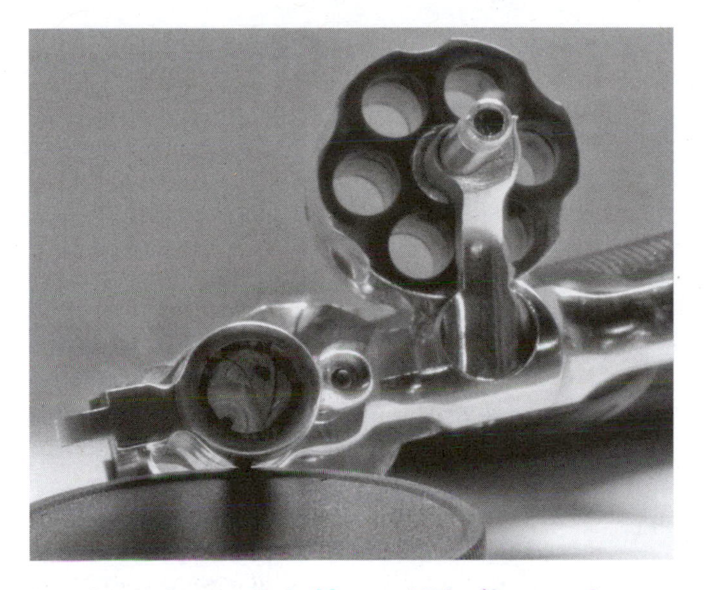

A gun barrel photographed with normal light ***(Courtesy of Michael Bell, West Virginia University, Forensic and Investigative Sciences program)***

good choice because the lighting will be uneven and can produce glare. To paint with light in this case, the camera is mounted on a tripod and aimed down the gun barrel. The room lights are dimmed and the shutter opened. To illuminate the barrel, a small flexible light is used to direct light in all areas of the barrel. Since the shutter is open, the camera will collect all of the light. The result is a smoothly illuminated photo, as seen here. Another example is found in the color insert.

The same barrel images using painting with light techniques ***(Courtesy of Michael Bell, West Virginia University, Forensic and Investigative Sciences program)***

palynology, forensic *See* BOTANY, FORENSIC.

paper A mass-produced consumer item often of interest in forensic investigations. Like other types of such evidence, physical matching may be all that is needed to link a piece of paper to a source. For example, if a note is written on a piece of paper that is torn out of a tablet, the rip and tear markings will be distinctive and in many cases, the torn page can be lined up directly with the ragged edges left on the tablet. In other cases, other forms of analyses can be attempted to learn more about the paper. Paper is made from wood pulp that consists of FIBERS that are processed ("beaten") to incorporate water into the matrix. Dewatering removes much of the moisture and allows hydrogen bonds to form between fibers, the interaction that gives paper mechanical strength. Chemical additives are introduced at several points in the process. Paper manufacturing is an example of mass production, and as a result the batch-to-batch variation is small by design. From a forensic perspective, this affords few opportunities to individualize paper, but assigning class characteristics is feasible.

The paper production cycle starts with wood chips or other ground starting material. Once reduced in size, the chips are moved to the pulping stage, where lignins are broken down and the fibrils are separated. Pulping is typically accomplished by treating raw pulp under alkaline conditions. This process is referred to as Kraft process or Kraft pulping, but there are many alternatives. Once the lignin is broken down, the pulp remaining is a brownish color that can be bleached by oxidizing agents to a whiter color.

The beating stage serves several purposes. Large fiber structures are broken down mechanically, resulting in a roughened surface and internal changes in the fibers. These changes increase the potential for hydrogen bonding and create a matrix of fibers that can hold water and other materials. The beating stage is important for facilitating these associations and encouraging fibers to "swell." Fillers can be added at this stage to fill voids in the fiber matrix and to give paper desired physical characteristics and appearance. Additives include starches, binding agents, fibrous components, finishing agents, whiteners, cotton fibers, and colorants (DYES). Fillers range from clay materials such as kaolin (hydrated aluminum silicate) to $CaCO_3$ and TiO_2. Fillers have many functions including altering the "feel" of paper, its weight, and optical properties such as reflectivity (gloss) and absorbance. Cotton is often an ingredient in paper because it is high in cellulose but low in lignin. In general, the higher quality the paper, the higher the cotton or "rag" content. Fine writing papers are high in cotton content and have a different thickness, strength, texture, and feel than does paper used for copiers, printers, and other mechanical printing applications. Some papers are treated with buffers to prevent gradual acidification of the media. Wet strength of paper, particularly products such as paper bags and cardboard, can be increased by addition of binders and agents that increase the strength of the attractive forces between fibers.

Once the pulp has been treated and lignin broken down, the preparation process involves combining remaining ingredients, mixing, and dilution into thin slurry. The slurry is poured over screens that capture the fibers as they are shaken. The fibers form a network held together by hydrogen bonds. Since excess water will disrupt hydrogen bonding between fibers, a heating and pressing stage follows, but not all water is removed. Coatings can also be added as required. A watermark may be placed in the paper at this stage. A watermark is a thinner area of the paper that is emblazoned with a characteristic manufacturer's mark that is visible when the paper is held to the light. Cutting is part of the process as well. Thus, in addition to chemical characteristics, striations and toolmarks may be useful in classifying papers. Dyes and PIGMENTS can be incorporated into paper in many ways, including impregnation and surface treatments. Solubility is an important concern, as is the mechanism of binding. Other materials that are part of the paper-making process include foam and slime control agents. These are required to prevent problems with the machinery used to manufacture the paper.

Forensic Analysis of Paper

In questioned document cases, the analysis of paper is usually associated with the analysis of INK or other media used to create writing or printing. Chemical analysis of paper by itself utilizes the traditional three types of forensic tools: visual and microscopic examination, organic analysis, and inorganic analysis. Even with these tools available, it is extraordinarily difficult to match one paper sample to another. This is one of the conundrums of FORENSIC SCIENCE that is applicable to all types of mass-produced materials. Forensic scientists use difference to individualize; mass-produced items are purposely designed to minimize difference. Paper manufacturers strive to achieve consistency among batches so that the paper that leaves the plant on Monday is the same as paper leaving on Tuesday. When differences are small, the only way to compensate is by the depth and detail of the analysis, but this must be done with minimal sample destruction. This concern is particularly acute in questioned document cases where DESTRUCTIVE ANALYSIS may be out of the question.

Since paper contains fibers, microscopic examinations can be useful. Cotton fibers have a distinctive appearance under polarized light and are easily recognized. Counting of fibers can be used to estimate the cotton rag content of the paper. If a coating has been applied to increase strength, slickness, or add glossiness, this can be observed as well. Pulp fibers found in the paper can be identified by specific source wood (pine, birch, etc.), but this requires specialized skills and training. Paper can also be exposed to UV light to see if it fluoresces, not unusual with papers treated to be of high whiteness. The same treatment is used to make white clothing appear brighter and whiter.

Organic characterization of paper has focused on infrared techniques and some use of thin layer chromatography in the case of dyed papers. Inorganic analysis using ICP-MS has also been used to discriminate between groups of papers although using destructive techniques. As with inks, paper ages in chemically predictable ways, although the timeframe of that decay depends on many factors. Paper ages principally as a result of the breakdown of cellulose that eventually produces yellowing and brittleness in the paper. Cellulose also degrades via oxidation reactions with the atmosphere, creating more acidic products and perpetuating the cycle. However, while dating papers and treating them to prevent deterioration has been of interest to historians, archaeologists, and conservators, there has not been a great deal of forensic work in dating relatively recent paper items.

Further Reading

Brunell, R. L. "Questioned Document Examination." In *Forensic Science Handbook*. Vol. 1. 2d ed. Edited by R. Saferstein. Upper Saddle River, N.J.: Prentice Hall, 2002.

Paracelsus (1493–1541) Swiss *Alchemist and Medicinal Chemist* A key figure whose work would directly affect FORENSIC SCIENCE, an alchemist, philosopher, and writer with unconventional ideas and experimental skills. Paracelsus sought to understand why treatments

Portrait of Paracelsus in the Louvre, Paris ***(Ablestock/ Jupiterimages)***

worked so that existing medicines could be improved and new ones developed. Paracelsus was a colorful and controversial figure, born into a Swiss family and named Philipus Aureolus Theophrastus Bombastus von Hohenheim. The name Bombastus was appropriate. He modestly assumed the name Paracelsus, which means "greater than Celsus," a physician in Rome during the first century C.E. From his early teens, Paracelsus moved frequently between universities, gathering knowledge, and moving on.

Over the years, Paracelsus learned and practiced medicine, gaining respect despite his personality. Most of his ideas about healing and CHEMISTRY were wrong, but he managed to upset the rusted foundations of medicine and chemistry, which was still being taught based on Roman texts. The modern equivalent would be going to medical school and learning from a book published in 1000, before the plagues swept Europe. Science needed a good shaking up, and Paracelsus was the man to do it.

Paracelsus's work in medicinal chemistry led him to state, "What is there that is not poison? All things are poison and nothing [is] without poison. Solely the dose determines the thing that is not a poison." Without a fundamental chemical understanding of medicines, there was no way to know what an appropriate dose was. Above a threshold, any therapeutic agent can become toxic and a poison. For its time, this was a revolutionary idea and one that started a chain of events that led to effective tests for ARSENIC in the body. Paracelsus was one of the pioneers of experimental science when science was more philosophical than experimental.

Paracelsus laid the groundwork for the broader field of TOXICOLOGY both experimentally and philosophically. He wrote widely, and his works were popular and thus widely disseminated and studied. As the first appearance of modern science, this existing understanding, even if faulty, formed the basis for the next round, leading (ideally) to a continually self-correcting and improving knowledge of the world.

Paracelsus described a method for isolating arsenic metal from arsenic sulfide compounds, building on the work of Arab alchemists as well as that of ALBERTUS MAGNUS. Paracelsus also wrote a treatise on the effect of gaseous metals as poisons in the mining industry, focusing on arsenic and MERCURY. Medicines he prepared and prescribed included low levels of bismuth, arsenic, antimony, and mercury salts, and he is credited with formulating the first viable treatment for syphilis. For this, he concocted an ointment based on mercury. ROBERT BOYLE was also in the hunt for an arsenic test suitable for death investigations. He tried bubbling gases through solutions containing dissolved arsenic but never achieved a reliable and reproducible method.

See also ALCHEMY; ORFILA, MATHIEU.

Further Reading

Aromaticico, A. *Alchemy: The Great Secret.* New York: Harry N. Abrams, Inc., 2000.

Moran, B. T. *Distilling Knowledge: Alchemy, Chemistry, and the Scientific Revolution.* In *New Histories of Science, Technology, and Medicine,* edited by M. C. Jacob, S. R. Weart and H. J. Cook. Cambridge, Mass.: Harvard University Press, 2005.

Morris, R. *The Last Sorcerers: The Path from Alchemy to the Periodic Table.* Washington, D.C.: Joseph Henry Press, 2003.

paraffin test *See* DERMAL NITRATE TEST.

partitioning and affinity Chemical and physical processes that are used extensively in FORENSIC SCIENCE as part of sample preparation. For partitioning to occur, a phase boundary must exist:

$$\text{Analyte}_{p1} \leftarrow\rightarrow \text{Analyte}_{p2}$$

where p1 and p2 refer to different phases. The two phases may be insoluble liquids (i.e., water and hexane) or a boundary between solid and liquid or liquid and gas. Partitioning relies on the analyte having a greater affinity for one phase over the other due to charges, polarity, and other chemical properties. Partitioning is also at the heart of CHROMATOGRAPHY, which is used in forensic CHEMISTRY for sample preparation, cleanup, screening tests, and as a first step in hyphenated instruments such as GCMS.

See also LOGP.

paternity testing Determination of parentage of a child that often involves or revolves around a legal issue such as life insurance, inheritance, or child support. Although the term "paternity" in a strict sense applies to determination of the father, "paternity testing" generally refers to either parent. The same techniques that are used in forensic SEROLOGY and forensic BIOLOGY are utilized in paternity testing, principally DNA TYPING. Prior to the advent of these techniques, traditional blood typing such as the ABO BLOOD GROUP SYSTEM and red cell ISOENZYMES were standard, along with HUMAN LEUCOCYTE ANTIGEN (HLA) typing. One of the more famous early paternity cases involved Charlie Chaplin, who in 1946 was excluded from being the father of a child based on his ABO blood type.

Determination of parentage comes down to exclusion, inclusion, probabilities based on known gene frequencies (POPULATION GENETICS), and the laws of heredity. For example, consider the isoenzyme system PGM. Ignoring subtypes for the sake of simplicity, three types can be identified: 1-1, 2-1, and 2-2. One of the two is contributed by the mother, one by the father, and so the child of a 1-1 mother and a 1-1 father could not be of type 2-1 or 2-2. Similarly, if a child is 2-1 and the mother is 2-2, then the father must be either a 1-1 or a 2-1, since the 1 had to have been contributed by him. If a putative father is identified and typed as 2-2, that is an example of exclusion—he could not possibly have fathered the child. If, however, he is typed as 2-1, that is inclusive, meaning he could have. Inclusive is not the same as conclusive; only DNA typing of multiple loci can provide what is considered to be conclusive evidence of paternity.

The use of mitochondrial DNA (mtDNA) has proven useful in paternity cases when good samples are not available from one or more people involved. This could occur if one parent is dead or missing, for example. Mitochondrial DNA is unique in that it is passed on directly from the mother to the child with no contribution from the father. Thus, mitochondrial DNA from anyone in the mother's line can be used in place of that of a given individual.

See also ANASTASIA AND THE ROMANOVS.

Further Reading

Melvin, J. R., et al. "Paternity Testing." In *Forensic Science Handbook*. Vol. 2. Edited by R. Saferstein. Englewood Cliffs, N.J.: Regents/Prentice Hall, 1993.

Moenssens, A. A., J. E. Starrs, C. E. Henderson, and F. E. Inbau. *Scientific Evidence in Civil and Criminal Cases*. Westbury, N.Y.: The Foundation Press, 1995.

pathology, forensic The study of the cause, manner, and mechanism of death undertaken by a medical doctor. Once a doctor has obtained an M.D., he or she attends a residency in pathology followed by additional training in forensic pathology and death investigation. Once this training is completed, he or she can take the examination given by the American Board of Pathology. Forensic pathologists may work as a MEDICAL EXAMINER (ME), CORONER, or in the offices of one of these officials. The primary job of the forensic pathologist is performing autopsies of questioned or suspicious deaths and determining the cause, manner, and circumstances of death. This is done based on the AUTOPSY results and also on information gathered as part of the death investigation. The pathologist is also tasked with identification of the body and with estimating the time of death (POSTMORTEM INTERVAL, PMI).

Any death can be classified by manner into one of four categories: natural, accidental, suicidal, or homicidal. The cause of death is that action or injury that started a chain of events (short or long) that ultimately resulted in the death. A proximate cause is the first event in the chain, whereas the immediate is the final event. For example, if a person is stabbed, goes to the hospital, has surgery, but later dies as a result of infection, the cause of death is the stab wound. The proximate cause is the injury itself, the immediate cause is infection, but the infection would not have happened had not the stab wound been inflicted. The mechanism of death is the actual medical, physiological, or biochemical process that resulted in death. If a person receives a blow to the head with a club, brain swelling (edema) is the actual process that results in death. The cause remains the blow to the head. Finally, the manner of death is the circumstance. For example, if a man's body is found at the base of a large building, the cause of death is the fall. The circumstances would be whether he jumped, fell, or was pushed. Thus, the forensic pathologist must combine information from the autopsy with information gathered during the investigation to draw conclusions about the death and to determine if criminal activity was involved.

See also GONZALES, THOMAS J.; HELPREN, MILTON; SPILSBURY, SIR BERNARD.

Further Reading

Dix, J., and R. Calaluce. *Guide to Forensic Pathology.* Boca Raton, Fla.: CRC Press, 1999.

Knight, B. *Simpson's Forensic Medicine.* 10th edition. London: Edward Arnold, 1991.

Spitz, W. U., ed. *Spitz and Fisher's Medicolegal Investigation of Death.* 3rd edition. Springfield, Ill.: Charles C. Thomas, 1993.

Wright, R. K. "The Role of the Forensic Pathologist." In *Forensic Science: An Introduction to Scientific and Investigative Techniques.* 2nd edition. Edited by S. H. James and J. J. Nordby. Boca Raton, Fla.: CRC Press, 2005.

pattern matching and pattern recognition A technique used in forensic comparisons in areas such as FINGERPRINTS, FIREARMS, BITE MARKS, and TOOLMARKS. Some of these procedures can be automated, such as those that exist for fingerprints and firearms evidence. However, many pattern recognition techniques cannot be easily encoded. For example, when a fingerprint is submitted to an automated system and a list of potential matches is returned, it is still up to the latent print examiner to make the final judgment as to whether the prints match. Similarly, when an impression of a shoeprint is made at a scene and a possible source shoe is located, an analyst compares them and uses pattern matching techniques to determine if that shoe created that impression. In most pattern recognition applications, analysts employ some type of "point comparison" technique in which they locate interesting, significant, or unique features on one of the pieces of evidence and then attempt to locate that same feature on the other piece. However, even if many points are compared, there is some element of subjectivity in any pattern matching analysis. In some cases, this subjectivity has been challenged in court. In particular, fingerprint identification has recently come under increasing scrutiny, as has the examination of bite marks. Although computers can be programmed to automate many of these comparisons, in the end a human being still makes the final judgment as to what matches and what does not. As such, the courts are expected to scrutinize pattern matching evidence more closely in the future, with potentially serious consequences in some forensic analyses.

PCP *See* PHENCYCLIDINE.

peer review A process of evaluation of research results used in science. Peer review of journal articles, laboratory reports, and proposals is commonly used in forensic science and, in the case of review of laboratory work, represents a form of QUALITY ASSURANCE/QUALITY CONTROL (QA/QC). Peer review of research is one criterion that can be used by trial judges to determine if new scientific techniques and procedures should be accepted. In the case of an article submitted to a professional journal, peer review would mean that the editor of the journal would send the article to two or three independent reviewers who would read the manuscript, comment on it, and recommend publication, further work, or no publication. The editor considers the opinions of all the reviewers and makes a decision. The author receives the feedback anonymously and uses it (ideally) to refine or rework the article until it is acceptable. The goal of all peer review is to ensure the reliability, validity, and quality of scientific work.

peptidase A *See* SEMEN.

periodic table A table developed by the Russian Dmitri Mendeleev (1834–1907) by which all known elements are organized in an orderly row and column arrangement. The current version of the periodic table, still the foundational tool of chemistry, is found in Appendix III. Without knowing about nuclear structure or electron arrangements, Mendeleev was able to assign elements to families (the columns of the table) of elements that shared the same type of chemical behavior. Across the table (rows), elements are organized by increasing atomic number. Within a given family, chemical behavior is similar, such as in Group 8, the noble gases, all of which are nearly inert. Similarly, in Group 1, each element tends to react violently with water and to form salts with Group 7 elements (halogens) such as NaCl, LiBr, and so on. Modern versions of the periodic table can include much more information such as crystal form, ionization potential, and density.

PETN A high EXPLOSIVE used commercially, by the military, and in mixtures employed by terrorists. PETN (pentaerythritol tetranitrate) was synthesized in the 1940s and is used by the military as a sheet explosive (a common form is Detasheet), in grenades, and as a propellant in small caliber weapons. Detonation cord, often called "det cord" or Primacord, contains a core of PETN or RDX covered in cotton and enclosed in a weatherproof casing. Although it is less sensitive than TNT, PETN must be detonated by a shock wave from some type of initiator such as a blasting cap before it will explode. Small amounts of it in air will burn rather than explode. Terrorists have used Semtex, a commercial mixture of PETN and RDX. Evidence suspected to contain PETN is usually first extracted with acetone and the residual tested using chemical PRESUMPTIVE TESTS such as the GRIESS TEST. THIN LAYER CHROMATOGRAPHY (TLC), HIGH-PRESSURE LIQUID CHROMATOGRAPHY (HPLC), and INFRARED SPECTROSCOPY (IR) are used to confirm any identification.

peyote *See* MESCALINE.

pH scale The scale used to describe the acidity or alkalinity (basicity) of a water-based solution. Water can dissociate into two ions, H^+ and OH^- (hydroxide), and the pH of a solution refers to the balance of those two species. The pH is defined mathematically as $pH = -\log[H^+]$, where the $[H^+]$ denotes the concentration of hydrogen ions (also called protons) measured in units of molarity (number of moles H^+ per liter, M). The pH scale centers on 7, which is a neutral solution in which the concentration of hydrogen ions equals the concentration of hydroxide at 1×10^{-7}M. A pH of less than 7 is acidic (H^+ in excess), with the lower the number the more acidic the solution. A pH of greater than 7 is basic (alkaline, excess OH^-), with higher numbers corresponding to increased alkalinity. Lemon juice, which is acidic, has a pH of about 4, skim milk about 6.6, blood about 7.4, milk of magnesia about 10.5, and lye (NaOH, sodium hydroxide) about 13.

pharmacodynamics *See* PHARMACOLOGY, FORENSIC.

pharmacokinetics *See* PHARMACOLOGY, FORENSIC.

pharmacology, forensic The study of drugs and their effects on organisms (primarily humans) in legal matters. Forensic pharmacology addresses such questions as when a drug was taken, how much was taken, and what effects it produced in an individual. It is intimately related to forensic TOXICOLOGY, PATHOLOGY, and DRUG ANALYSIS. Forensic pharmacology involves two broad topic areas: pharmacokinetics, the movement of drugs through the body, and pharmacodynamics, the effects of drugs and their metabolites on the body.

Pharmacokinetics can be divided into four stages, as is illustrated in the figure. After the drug is introduced, it is absorbed and distributed to tissues in the body by the bloodstream. Portions of the drug are metabolized and in effect act as new drugs that enter into the same cycle. As an example, heroin is metabolized to monoacetylmorphine, which then reenters the cycle as a "new" drug that may be further metabolized and/or eliminated. Elimination of the drug and metabolites follows, although some portions of some compounds may accumulate in fatty tissues. Introduction can occur in many ways, including inhalation of gaseous substances, injection, ingestion, sublingual (dissolving under the tongue), and absorption directly through the skin. If a drug is swallowed, absorption occurs in the gastrointestinal tract. Once the drug is introduced into the bloodstream in a soluble form, it spreads to any tissues that are in contact with the blood, although this can be somewhat selective. For example, there are substances that cannot cross the blood-brain barrier. Elimination can be by several routes, principally in urine, but also

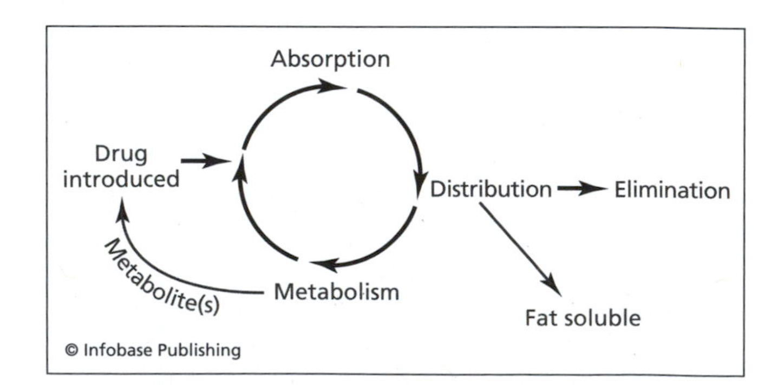

Generalized metabolic steps of interest in pharmacokinetics

in sweat, feces, or breath. For example, a portion of any alcohol that is ingested is eventually exhaled as gaseous carbon dioxide. Pharmacologists utilize mathematical models to describe these processes.

Many factors determine how a dose of a drug will affect an individual. Different drugs have different rates of absorption, different metabolites, and different speeds of elimination. One measure of how long a drug remains in the system is the half-life, symbolized by $t_{1/2}$. This information, combined with toxicological findings can be useful in determining how much of a drug was taken and roughly when. Heroin is metabolized to morphine and has a half-life of 1 to 1.5 hours. Evidence of heroin ingestion can be detected in urine for two to four days after ingestion. For cocaine, $t_{1/2}$ is between 45 minutes and 1.5 hours and can be detected for one to two days. PCP ("angel dust" or PHENCYCLIDINE) has a $t_{1/2}$ of seven to more than 50 days and is detectable for about a week.

See also ADME; MODES OF INGESTION.

Further Reading

Benjamin, D. M. "Forensic Pharmacology." In *Forensic Science Handbook*. Vol. 2. Edited by R. Saferstein. Englewood Cliffs, N.J.: Regents/Prentice Hall, 1993.

Spiehler, V., and B. Levine. "Pharmacokinetics and Pharmacodynamics." In *Principles of Forensic Toxicology*. 2nd edition. Edited by B. Levine. Washington, D.C.: American Association of Clinical Chemistry, 2003.

phase contrast microscopy A specialized microscopic technique that increases the contrast between items that would otherwise be very difficult to see, such as transparent or nearly transparent materials. Phase contrast techniques are particularly useful in forensic science to visualize sperm cells and fine structure in hair and soil. A phase contrast microscope is similar to a COMPOUND MICROSCOPE in design but includes specialized objectives, lighting, plates, and light condensers. When a specimen and oil (or other mounting medium) are both similar in refractive index, it is difficult to see surface relief. The purpose of phase contrast techniques is to enhance the edges and increase the contrast between the object and the mounting medium, making it much easier to see. However, resolution will decrease since this enhancement is achieved by inserting a "phase plate" into the optical path. Occasionally, a technique called dark field illumination is coupled to phase contrast to further improve the image quality.

Further Reading

Nomenclature Staff, Royal Microscopy Society. *RMS Dictionary of Light Microscopy.* Royal Microscopy Society—Microscopy Handbook # 15. Oxford: Oxford University Press, 1989.

phencyclidine (PCP) An abused drug that is easily synthesized and was at one time used as a veterinary anesthetic. The compound was synthesized in 1926 and used as a cat anesthetic in the 1950s. In the early 1960s, it was introduced as a human anesthetic but soon withdrawn from the market after bizarre behavior and other side effects were noted. Production of the compound ended in 1979. As an illicit HALLUCINOGEN, PCP first appeared in the late 1960s, and it was placed on Schedule II of the CONTROLLED SUBSTANCES ACT (CSA) in 1970. Because it is relatively easy to make, its PRECURSORS are also controlled. PCP is sold as a powder usually called "angel dust" that varies in color depending on the conditions under which it was made. It is also encountered dissolved in ether or other solvents. The liquid form is typically absorbed into materials such as oregano, parsley, or even MARIJUANA that are rolled into cigarettes and smoked. The effects of PCP on an individual are often unpredictable, with stimulant, anesthetic, and hallucinogenic symptoms possible. Most deaths that occur from PCP are not due to the drug itself but rather to actions that a person takes while under its influence, such as jumping from a height. PCP can be analyzed using THIN LAYER CHROMATOGRAPHY (TLC), HIGH-PRESSURE LIQUID CHROMATOGRAPHY (HPLC), IMMUNOASSAY, GAS CHROMATOGRAPHY (GC) using a MASS SPECTROMETER (MS), NITROGEN PHOSPHORUS DETECTOR (NPD), or INFRARED SPECTROPHOTOMETRY (IR).

Further Reading

Jenkins, A. "Hallucinogens." In *Principles of Forensic Toxicology*. 2nd edition. Edited by Barry Levine. Washington, D.C.: American Association of Clinical Chemistry, 2003.

Drug Enforcement Agency (DEA) Web site. "PCP." Available online. URL: http://www.usdoj.dea.gov/concern/pcp.html. Downloaded August 21, 2002.

phenolphthalein *See* KASTLE-MEYER COLOR TEST.

phenotype In genetics, the outward appearance of a trait as opposed to the genotype, which is the actual combination of genes. It is the expression of a given genotype in a particular environment. For example, in the ABO BLOOD GROUP SYSTEM, a person inherits one gene from the mother and one from the father. If both genes are A, the the person's genotype is AA, and their phenotype is A, meaning that their blood will be Type A. If a person inherits an A gene from one parent and an O gene from another, their genotype is AO, but their phenotype is A and they will have Type A blood.

phenyl-2-propanone (P2P) A PRECURSOR in a CLANDESTINE LAB synthesis of METHAMPHETAMINE. This material was added to the CONTROLLED SUBSTANCES ACT (CSA) (Schedule II) and now is rarely encountered.

phonetics, forensic The study of the sound components of speech, usually with the intent of identifying the speaker or to learn things about the speaker. In general, the type of evidence of interest in forensic phonetics is called a "disputed utterance" or "questioned utterance," in the same way "disputed documents" are the subject of QUESTIONED DOCUMENT analysis. For example, the speech of a native of the southern United States is usually quite different from that of someone from Boston or New Jersey, even though all speak English. These are extreme examples but illustrate the idea that the sound of speech can provide useful information. Other aspects of interest relating to speech include pitch, speed, intonation, emphasis, and accent on syllables. Aside from helping to uncover information, phonetics can be helpful in interpreting meaning from the sound of speech. Again, a simple example would be the difference in meaning between the following statements: "*I* like it" versus "I like *it*" versus "I *like* it." All three are written the same but spoken quite differently and because of different emphasis carry a different meaning.

Forensic phonetics is often applied to recorded speech such as taped messages or threatening phone calls. Specialized tools exist to analyze recorded speech in an area often referred to as "voice recognition." Voice recognition consists of listening to the speech with the ear (auditory analysis) as well as an evaluation of the frequency structure of the speech. As shown in the figure, the ear detects pressure waves of air created by the vocal structure of another person. The strength, or amplitude, of the wave reflects the loudness, or volume, while the frequency reflects pitch. High-frequency noise, such as a dog whistle, has a wave pattern

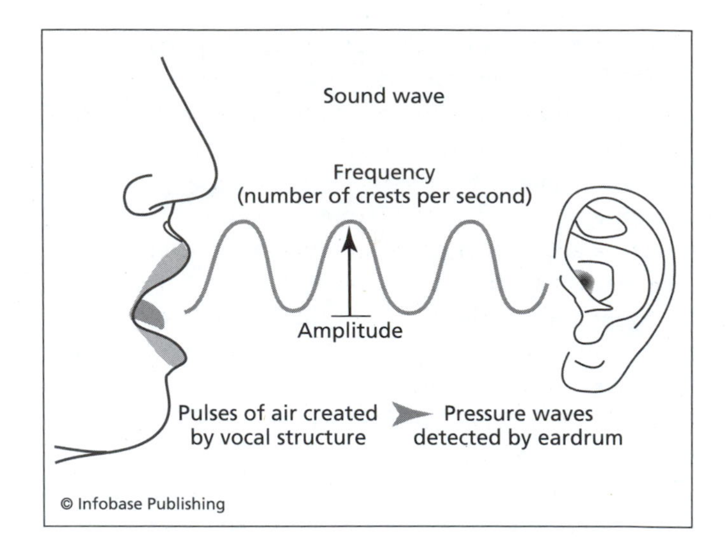

Human speech and hearing, which is based on the production and detection of pressure waves in air

with more "crests" per second than does a low-pitch sound. The unit of frequency is the Hertz (Hz), with 1 Hz equal to one cycle (one crest) per second.

A graph of the frequency structure of a voice produced by a sound spectrograph is called a spectrogram, sound spectrogram, or voiceprint. The voiceprint was developed in 1941 at Bell Labs by Lawrence Kersta. Since each person has a different vocal structure, it was thought that each voice pattern would be unique. However, acceptance of voiceprints has never obtained the acceptance that fingerprints have, and voiceprint analysis remains a controversial subject. Nonetheless, the sound spectrographs can produce useful evidence including matching of two samples of recorded speech. The sound spectrograph has also been used to study stress patterns, although this, too, remains highly controversial.

Further Reading

Baldwin, J., and P. French. *Forensic Phonetics*. London: Pinter Publishers, 1990.

phosphine gas *See* RED COOK METHOD.

photocopiers Copy machines that use a photographic process to produce a copy of a document. These are of interest in some QUESTIONED DOCUMENT cases. The term *copier* is generic and usually refers to copiers that work much as laser computer printers (xerographic process). Informally, the three terms (*copy, photocopy,* *and Xerox*) are used interchangeably, although there is a distinct difference between how a photocopier works and how a xerographic copier works. In a photocopy process, an electrostatic charge is placed on a surface and is selectively discharged by exposure to light, creating a negative image. Projecting light through the negative creates a print, much as "printing" a negative creates a photograph.

photography An indispensable tool for the documentation of crime scenes and physical evidence that has been used since the late 1800s in forensic science. Recent developments in digital and microscopic imaging (photomicrography) as well as in videotaping (VIDEOGRAPHY) have expanded the scope of forensic photography.

The word *photography* means "light writing" and is based on exposure of light-sensitive compounds such as silver chloride (AgCl) to light. The light causes the salt to darken, and when developed, the pattern forms a negative of the original image. Color photography employs layers of these light sensitive compounds, each of which responds to different colors. Projection of light through the negative is used to create the prints. One of the earliest types of photography was called the daguerreotype, named after its French inventor, L. J. M. Daguerre. These were also known as "tintypes" because of the backing, and by the 1870s daguerreotype portraits were common. One of the most famous was a tintype made of Billy the Kid. Around mid-century, photos were being used for identification of cadavers and of criminals. Application of photography to forensic work became commonplace in the latter part of the century, and ALPHONSE BERTILLON was one of its well-known advocates, both for crime scene documentation and for identification of individuals. His early work with identification cards (Portrait Parle) can be considered one of the precursors of today's "mug shots."

Photographs can also be created using ELECTROMAGNETIC RADIATION (EMR) in regions of the spectrum besides the visible range. Infrared radiation can produce "thermographs," or images based on the amount of heat emitted by the subject. Black light illumination using ultraviolet radiation (UV) can induce fluorescence in the visible range, while illumination of other materials with visible light can produce fluorescence in the infrared region. Radioactive decay can also be used to create images on X-ray film. In all these examples, a permanent visible image results.

Fundamentally, the role of forensic photography is to faithfully record a scene or person such that the image is a true and reliable representation of the subject. In a sense, a photograph preserves reality so that others can see exactly what the person taking the photograph saw. Photography is used to document crime scenes and evidence found, to show fingerprints, to document laboratory analyses, and for surveillance. For example, photos can be taken of gels used in DNA TYPING and of developed plates used in THIN LAYER CHROMATOGRAPHY (TLC), since both are not permanent. Also, cameras can be attached to microscopes and used to record images of hairs, fibers, and any other material on the microscope stage. These images may be based on film, but increasingly they are created using digital imaging techniques.

Photography is also used at AUTOPSY, to document injuries, and for identification purposes. Since the photos are meant to faithfully represent reality, several issues must be considered when the photo, or more likely a series of photos, is taken. First is the lighting and angle of illumination. Changes in lighting can dramatically alter the appearance of the subject, as can the angle of illumination. This is why series of photographs are used: to ensure that all possibilities are covered and all variations in appearance are documented. Scale is also important, and wherever possible, a measuring or other scaling device is included in the picture. Use of filters such as polarizing filters to reduce glare or colored filters to increase contrast may

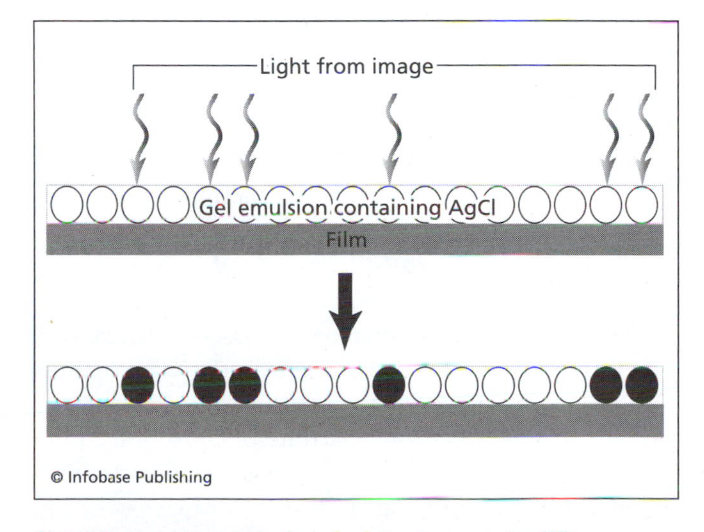

Simplified process of black and white photography. When exposed to light, the compound AgCl darkens.

be needed. However, a key consideration is always the avoidance of distortions or illusions that are not representative of the subject at the time the photograph is taken. The photographer must also decide what type of film to use (black and white versus color, speed, quality, and so on). Digital imaging, increasingly the standard in forensic photography, has its own requirements for lenses, settings, and exposure. It is common to have replicate sets of photos taken by different people and/or different cameras as insurance against any one failing. Increasingly, videotaping of scenes is being used, but it does not replace traditional photography.

Finally, digital imaging has introduced other caveats. Since images are recorded electronically, a host of issues are raised that were not a concern with traditional film-based photography. Digital images can be modified using software, so steps must be taken to ensure that the CHAIN OF CUSTODY of the electronic files is secure and that any alteration of an image is electronically traceable. Conversely, digital images of evidence such as fingerprints and firearms evidence can be processed electronically and stored in large databases for retrieval. The same considerations apply to digital images as to film-based photographs, namely, that illumination, angles, and so on be correct and complete based on the type of evidence being documented.

See also PAINTING WITH LIGHT.

Further Reading

Bodziak, W. J. "Photography of Footwear Impressions." In *Footwear Impression Evidence*. New York: Elsevier, 1990.

Redsicker, D. *The Practical Methodology of Forensic Photography*. Boca Raton, Fla.: CRC Press, 1994.

photomicrography *See* PHOTOGRAPHY.

physical evidence Broadly speaking, any type of tangible evidence as opposed to something such as the testimony of an eyewitness. Physical evidence can be anything, from a microscopic trace of dust to a car, but there are some generalizations that can be made. Physical evidence must be documented, collected, marked, transported, and stored in a manner consistent with its type. It is subject to a CHAIN OF CUSTODY to ensure its safety and integrity from the time of collection to use in court. Analysis of physical evidence involves IDENTIFICATION, comparison, determination of CLASS CHARACTERISTICS, and INDIVIDUALIZATION when possible, and occasionally it is used in reconstructions.

physical matching Linking pieces of evidence together that exist as separate pieces but once were part of the same item. Reconstructing a broken window is an example of physical matching, as is linking a specific match back to the matchbook from which it was torn based on the unique tearing pattern.

See also COMMON SOURCE; INDIVIDUALIZATION.

physical properties Properties of matter that can be measured by physical means rather than chemical means. Measuring a physical property can be accomplished without changing the chemical composition of the material being studied. On the other hand, chemical properties can only be determined by processes that alter the chemical composition of the sample. For example, the freezing point of water is a physical property since it can be determined by freezing the water and measuring the temperature at which that occurs. The chemical formula of liquid water (H_2O) is the same as the chemical formula of ice, so no chemical change has occurred. The flammability of butane, a chemical property, can only be determined by an experiment that involves the COMBUSTION of butane, resulting in the conversion of butane to carbon dioxide and water. Determination of physical properties is used in forensic science to characterize various kinds of evidence, and examples of physical properties include color, density, and melting points. The analysis of a FIBER might involve a determination of the color and the melting point, which will assist the analyst in classifying the fiber.

Physician's Desk Reference (PDR) A reference book used by medical doctors that lists PRESCRIPTION DRUGS and their composition, actions, indications, and other important information. In forensic science, the *PDR* is an indispensable reference manual in forensic CHEMISTRY (DRUG ANALYSIS), PATHOLOGY, and TOXICOLOGY. The first edition of the *PDR* was published in 1946 and annually thereafter by the Medical Economics Company. For identification of prescription drugs seized as physical evidence, the large collection of actual size photographs is invaluable, as is the extensive cross listing and referencing of the drugs and active ingredients.

picric acid *See* EXPLOSIVE POWER.

pigments *See* PAINT.

pKa A quantity used in forensic TOXICOLOGY to describe the acidity or basicity of drugs and metabolites. Similar to the LOGP value, pKa can be alternatively listed as -logK_a where Ka is an equilibrium constant describing the dissociate of a weak acid. Acid/base CHEMISTRY plays a central role in drug chemistry and thus in DRUG ANALYSIS, toxicology, and sample preparation. Indeed, drugs are often classified as acidic, basic, or neutral. The functionalities in drugs that define their classes are amino groups (bases), phenolic groups (acids), and carboxyl groups (acids), which are all weak acids or bases. It is not unusual for a drug molecule to have more than one acid or base group, each of which is called an ionizable center. Accordingly, acid/base character is intimately related to solubility in water and organic phases. When ionized, a drug molecule is water-soluble and when un-ionized, it is soluble in solvents such as octanol and chloroform.

plane polarized light *See* POLARIZED LIGHT.

plastics A form of POLYMER that can be encountered as a form of physical, trace, or transfer evidence. Plastics and related compounds are found in storage bags, food containers, bottles, paints, fibers, and automobile parts to name just a few. The term *plastic* generally is used to refer to synthetic polymers derived from petroleum products. Polyethylene and polypropylene are two examples.

platinic bromide and chloride *See* GOLD BROMIDE AND GOLD CHLORIDE.

poaching *See* FISH AND WILDLIFE SERVICE.

poison Any substance capable of causing a toxic (harmful) response in an organism. In forensic science, detecting a poison is the responsibility of the MEDICAL EXAMINER (ME) or pathologist, the toxicologist, and the forensic chemist. An oft-repeated phrase in toxicology is "the dose makes the poison," which means that every substance can cause harm depending on how much is ingested. Even such familiar substances as water and table salt can cause death if sufficient quantities are ingested over a short period of time. The toxicity of a poison is measured by a quantity called the LD_{50}, "lethal dose 50." The LD_{50} of a substance is that dose (based on body weight) that results in death for half of an experimental population such as laboratory rats or mice. For any given individual, the fatal dose of a poison will depend on general health, age, weight, and a number of other variables.

Poisons can be subdivided a number of ways. Biological poisons, broadly defined, are those obtained from living organisms such as bacteria or plants. Hemlock, strychnine, venom from snakes, and botulism are examples. Inorganic poisons include hydrochloric acid (HCl), CYANIDE (both salts and gaseous HCN), lye (sodium hydroxide, NaOH), ammonia, MERCURY, lead, thallium, and ARSENIC, which is perhaps the most notorious poison of all. Indeed, it was the prevalence of arsenic poisoning in the 18th and 19th centuries that was critical in driving the development of forensic toxicology. The term *organic poisons* usually refers to ORGANIC COMPOUNDS including organic insecticides and pesticides such as DDT, toluene, and other petroleum distillates, and many drugs. Methanol (wood alcohol) poisoning can occur with homemade or adulterated liquors in which the methanol is substituted for ethanol. Of the gaseous poisons, CARBON MONOXIDE (CO), HCN, and forms of arsenic are the most familiar.

Poisoning can be accidental, suicidal, or homicidal. At a scene, the fact that the death has been caused by a poison may not be obvious, and so a thorough search of the area is required with attention focusing on empty containers, glasses, residues, and medication bottles. In homicides, food is a common means of administration. In Chicago in 1982, cyanide was placed into Tylenol capsules, resulting in the death of seven people. To date, no one has been convicted of the crime. In the wake of recent terrorist activity, there is increasing concern about mass poisonings of food, drugs, or water supplies, presenting a significant forensic and law enforcement challenge.

See also ORFILA, MATHIEU; PARACELSUS; PRODUCT TAMPERING.

polarized light (plane polarized light) Light that has been filtered such that the waves of ELECTROMAGNETIC RADIATION (EMR) are oscillating in only one direction. For this reason, the term "plane polarized light" is often used. One familiar application is in sunglasses; if the lenses have been polarized (have a Polaroid™ coating), the glare from reflections is eliminated, because light rays that do not accord with the filter's direction are blocked. Polarization of light is exploited in forensic science in POLARIZED LIGHT MICROSCOPY

(PLM) and occassionally as part of TOXICOLOGY and DRUG ANALYSIS.

polarized light microscopy (PLM) One of the most versatile microscopic techniques used in forensic science, principally for the analysis of TRACE EVIDENCE. The technique was originally used in geology for the analysis and identification of minerals, and polarizing light microscopes are sometimes still referred to as "petrographic" microscopes. The design of polarizing light microscopes is similar to that of a COMPOUND MICROSCOPE with the addition of a few key components. The light that comes from below the sample first passes through a filter called the polarizer, which converts the beam to plane POLARIZED LIGHT. This passes through the sample and then through a second polarizing filter called the analyzer. Initially, these two filters are set such that the light passing through the analyzer is polarized in a plane that is perpendicular to the light that can pass through the analyzer. As a result, with no sample in the light path, the view appears completely dark, since the analyzer blocks all the polarized light from the polarizer, that is, the filter orientations are perpendicular to each other. However, when a sample is placed on the stage, the oscillations of light that emerge from it are no longer polarized in one direction. Thus, some of this emerging light will be able to pass through the analyzer, where the beams can interfere with each other. The result is a viewable image (magnified) with distinctive contrast that may be highly colored.

PLM is valuable for studying the optical properties of materials that are ANISOTROPIC. A material that is *isotropic* for a given optical characteristic will have the same value of that characteristic regardless of the direction from which the light is coming. Solid materials that are made up of molecules that are randomly placed or molecules that are not symmetric will be isotropic. Thus, a material that is isotropic will have the same refractive index regardless of the direction from which the light is coming. In contrast, *antistropic* materials will display different properties depending on the direction of propagation of the light. The difference in the refractive index of anisotropic materials is the BIREFRINGENCE, which can be determined quantitatively and used to identify materials. Using related techniques, PLM can determine a number of optical properties of a substance that can aid in comparison and identification. In forensic science, PLM is used for evidence such as synthetic fibers, soil, glass, paint, building materials, and analysis of CUTTING AGENTS such as starch in drugs.

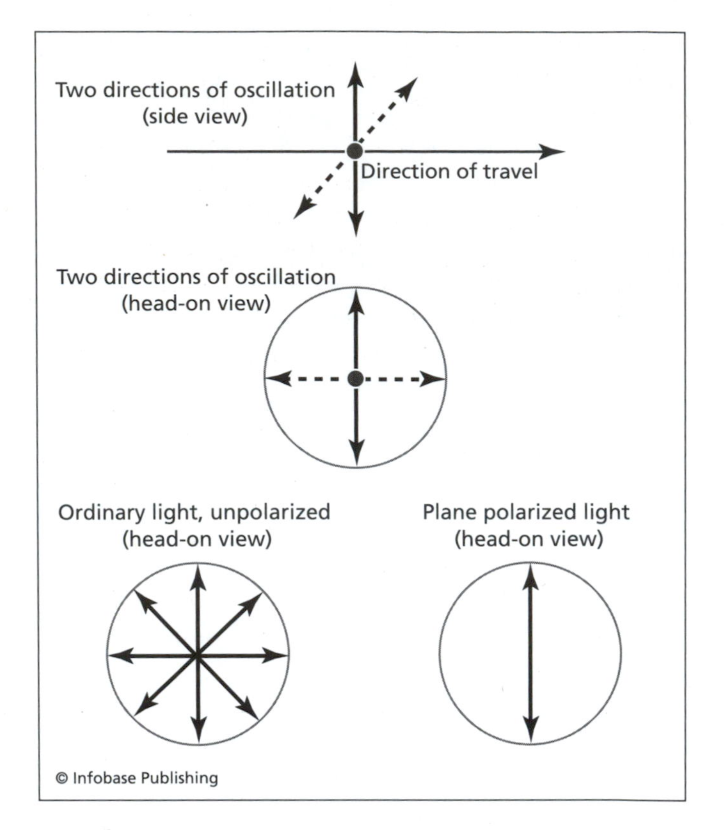

Plane polarized light

Further Reading

De Forest, P. R. "Foundations of Forensic Microscopy." In *Forensic Science Handbook*. Vol. 1, 2nd edition. Edited by R. Saferstein. Upper Saddle River, N.J.: Prentice Hall, 2002.

McCrone, W. C., L. B. McCrone, and J. G. Delly. *Polarized Light Microscopy*. Ann Arbor, Mich.: Ann Arbor Science Publishers, Inc., 1978.

Nomenclature Staff, Royal Microscopy Society. *RMS Dictionary of Light Microscopy*. Royal Microscopy Society—Microscopy Handbook # 15. Oxford: Oxford University Press, 1989.

pollen *See* BOTANY, FORENSIC.

polygraph Informally called a "lie detector," the polygraph is an instrument that consists of several sensors that respond to selected physiological functions. A polygraph does not detect truthfulness or lying per se; the examiner studies the chart of data and makes a determination of deceptiveness. The functions monitored are the heart rate, blood pressure, respiration

The Future of Lie Detection

The mainstay of deception analysis, the polygraph, or lie detector, is more than 80 years old and remains controversial. Although routinely used as an investigative tool, the results of polygraph tests are rarely admissible in courts, and there are a small number of individuals that can "beat" the polygraph. The polygraph is designed to detect bodily responses that in turn reflect an emotional response. There is no question that the acquisition of these measurements is reliable, since heart rate, blood pressure, and similar responses are not hard to measure. Rather, the issue is how well changes (or lack of changes) in these responses correlate with deception. If someone has rationalized a lie or does not associate any emotion with deception, the results of a polygraph test are of little value.

The use of polygraphs for other than law enforcement investigations is spreading, and not without protest. Employers such as the Department of Defense (DOD) and the Department of Energy (DOE), as well as others involved in intelligence and law enforcement, are turning to the polygraph as a preventive and general screening tool. Job applicants and even current employees are screened with the test as part of a background check or to root out potential security violations. In part as a response to these controversial policies, a review of the polygraph by the National Academies (formally the National Academy of Sciences) was published in November 2002. The study found that the polygraph was not reliable enough to be used for screening people for security purposes. Along with the possibility of the FALSE NEGATIVE results mentioned above, FALSE POSITIVES can occur as well. Since the polygraph measures emotional response rather than knowledge, guilt, or innocence, a person may show a positive (deceptive) response as a result of nervousness or other factors. Although trained examiners are taught how to minimize these problems, they still occur. These shortcomings have prodded researchers to examine newer, and hopefully more reliable, methods for detecting deception.

One alternative is a technique known as voice stress analysis (voiceprint or voice print analysis). The method relies on the detection of "microtremors" in the voice that are involuntarily produced under stress, such as when a lie is being told. Little research supports the reliability of voice stress analysis, and courts have yet to accept it. Narcoanalysis ("truth serums") has also failed to gain widespread acceptance. However, high-tech approaches on the horizon may obtain general acceptance if research supports initial claims of success.

Trained interviewers and interrogators learn to study body language in an attempt to discern whether a person is telling the truth. Many of the signals of deception are found in facial expressions such as fleeting smiles and eye movements. One new technique attempts to harness this ability into an expert system that combines recordings of interviews with computer analysis of facial responses, some of which are nearly undetectable. The idea is to train the computer, using recordings and expert input, to detect signs of deception. The work is still in the preliminary stages. Another similar technique under study is thermal imaging analysis that would detect increased blood flow and thus warmth (i.e., flushing) that may accompany a lie.

Like the polygraph, thermal imaging, voice stress, and facial analysis target the "symptoms" of lying and could be considered "emotion detectors" rather than lie detectors. The information of interest, the raw material of truth or lie, is stored in the brain. Two new techniques are attempting to go to the source to find the brain signature of deception. The first exploits NUCLEAR MAGNETIC RESONANCE (NMR) imaging of the brain, the same technology used in medical MRI scanning. Functional MRI (fMRI) attempts to image brain activity and to correlate it with the stimulus that produced it. In this application, portions of the brain that are activated while telling a lie are mapped and studied. Unlike emotional responses, this activity should not be subject to false negative or false positive responses, but research is still in the preliminary stages. The other technique receiving publicity and attention is the use of MERMER and so-called brain fingerprinting to detect deception. This work has been championed by Dr. Lawrence Falwell and is not without its critics.

Falwell believes that MERMERs (multifaceted electroencephalographic responses) are "brain memories" that are created by witnessing or participating in an event such as a crime. Thus, if a person is present at a crime scene, a corresponding MERMER is produced; if he or she is not at the crime scene, none is. Falwell's technique is based on detecting MERMERs related to the event in question. The subject is fitted with a headband that contains electrical sensors as would be used to obtain a medical EEG (electroencephalograph). These sensors detect small electrical currents and electrical activity within the brain. One response in particular, a spike that occurs within a short time after a familiar object or image is seen, is targeted. This response, the p300, is named because it is observed about 300 to 800 milliseconds after a familiar image is seen. Thus, detection of the p300 spike should be indicative of familiarity. If the image shown were a crime scene, a p300 spike would indicate that the person is familiar with the location, although that does not mean the person is guilty of the crime. As before, much more research and validation is needed before MERMER science is likely to become an accepted part of forensic deception analysis.

As new deception detection devices are proven and added to the arsenal, there is little chance that any one will prove to be 100 percent reliable. Lie detectors, no matter

(continues)

The Future of Lie Detection *(continued)*

what type, are best used with a clear understanding of their strengths and limitations and as just one part of a thorough investigation. A lie detector is no substitute for good detectives; rather, it is just one more tool at their disposal.

Further Reading

Committee to Review the Scientific Evidence on the Polygraph. *The Polygraph and Lie Detection.* Washington, D.C.: The National Academies Press, 2002.

Falwell, L. A., and S. S. Smith. "Using Brain MERMER Testing to Detect Knowledge Despite Efforts to Conceal." *Journal of Forensic Sciences* 46, no. 1 (2001): 135.

Felder, B. J. "Truth and Justice by the Blip of a Brainwave." *New York Times,* October 9, 2001.

Hansen, M. "Truth Sleuth or Faulty Detector?" *ABA Journal,* May 1999, 16.

Holden, C. "Panel Seeks Truth in Lie Detector Debate." *Science,* February 9, 2001, 967.

Perina, K. "Truth Serum: Brain Scans May Be Foolproof Lie Detectors." *Psychology Today,* (January/February 2002): 15.

Moenssens, A. A., J. E. Starrs, C. E. Henderson, and F. E. Inbau. *Scientific Evidence in Civil and Criminal Cases.* Westbury, N.Y.: The Foundation Press, 1995.

Wen, P. "'Brain Fingerprints' May Offer Better Way to Detect Lying." *Boston Globe,* July 5, 2001.

rate, respiration volume, and galvanic skin response. The last is a measure of how well the skin conducts electric current. The controversy that always has swirled around the device is not the reliability of the sensors—there is no question that these functions can be accurately measured and recorded by the instrument. Rather, the issue is whether the pattern of the readings can be directly and unequivocally related to truth or deception. In turn, this relies upon the training and skill of the person administering the test. For this reason, polygraph evidence is rarely accepted in court, although one state, New Mexico, routinely does admit polygraph findings. The polygraph test has evolved from what was once considered part of forensic science to the category of investigative tool.

A polygraph examination can be divided into three phases. The first is a pretest interview in which the process is explained and the issue in question is discussed. The purpose is to help the subject relax, relieve any nervousness, and ensure that the subject is not surprised by any of the questions. The next phase is the examination itself, which can be conducted using one of three techniques. The relevant/irrelevant (R/IR) question approach is based on the assumption that a truthful person will show little difference in the amount of emotion and attention given to both types of questions, while someone who is being deceptive should invest more emotion in deceptive answers to relevant questions than to responses to irrelevant questions. Most polygraph examinations today rely not on this technique but on the CQ (control question) technique, which retains some elements of the R/IR approach. Here, the subject is given irrelevant questions as well, but they are designed to produce a deceptive answer. Finally, the detectable lie control (DLC) can be used. Some examiners combine methods as well. After the questions are delivered comes the final part of the examination, scoring of results by the examiner. The subject may be judged to be truthful or deceptive, or the results may be inconclusive.

Several issues surround polygraphs that have yet to be resolved, but perhaps the most critical, particularly in the eyes of courts, is the ability to objectively determine the accuracy (or error rate) of the test. Laboratory tests can be performed, but they do not closely relate to what occurs during a real examination. For example, volunteers in a study can be told to lie, but since they know it is a study and there are no consequences for lying, their emotional and physiological responses are not likely to mimic a deceptive criminal's. Also, some people are able to rationalize lies to the point that they do not believe they are being deceptive and respond accordingly. Purposely altering breathing rate or applying pain to oneself can also create patterns that can prove inconclusive. The lack of uniform standards of examination and scoring further add to the subjectivity of the procedure. Although it is estimated that a skilled examiner can obtain reliability rates in the range of 90 percent, the uncertainties of the procedure make it unlikely that it will be uniformly accepted by courts in the way that forensic evidence such as DNA TYPING is.

See also DECEPTION ANALYSIS, FORENSIC; "The Future of Lie Detection" essay.

Further Reading

Gallai, D. "Polygraph Evidence in Federal Courts: Should It Be Admissible?" *The American Criminal Law Review* 36, no. 1 (1999): 87–116.

Moenssens, A. A., J. E. Starrs, C. E. Henderson, and F. E. Inbau. *Scientific Evidence in Civil and Criminal Cases.* Westbury, N.Y.: The Foundation Press, 1995.

polymerase chain reaction (PCR) *See* DNA TYPING.

polymers Large molecules (macromolecules) that are created by linking together tens, hundreds, or thousands of subunits called monomers by strong chemical bonds. Polymers are common in nature and include PROTEINS (polymers of AMINO ACIDS). Cellulose is a glucose polymer that makes up about 90 percent of the weight of cotton, for example. Synthetic polymers were first produced in the mid-1800s and included cellulose nitrate and its derivative, celluloid, both of which are readily combustible and thus of limited commercial use. Bakelite, a rigid plastic polymer, was synthesized in 1907 by mixing together urea (the pungent compound that gives urine its characteristic odor) and formaldehyde. NYLON was introduced in 1939, followed by an explosion in the development and utilization of synthetic polymers. In forensic science, polymers are encountered as FIBER evidence (NYLON, rayon, polyesters, and so on) and in BUILDING MATERIALS (polyvinyl chloride, PVC pipe), paints, plastics, ropes, as well as many other forms.

polymorphic Literally "many forms." In forensic serology/biology, this term is used to describe variants in blood that exist in different forms that are determined genetically. These variants include the ABO BLOOD GROUP SYSTEM, red cell ISOENZYME systems such as PGM, and the loci used in DNA TYPING. For example, a person can have a blood type of A, B, or O, so the ABO blood group system is polymorphic. Since the variant is determined by heredity and thus by the genes, such systems are called GENETIC MARKER systems. Typing of genetic marker systems, particularly the ABO and isoenzymes, was the tool used by forensic serologists to individualize blood (to the extent possible) prior to the advent of DNA typing.

population genetics and databases The information and techniques that are used to estimate the frequency of types (typically blood or DNA types) in a population. For example, the B blood type of the ABO BLOOD GROUP SYSTEM occurs in approximately 3 percent of the population. Obviously not every person in the world has been typed, so this value is an estimate based on a database of people already typed and statistical and genetic calculations. Selection of the appropriate population database is critical since frequencies can vary among Caucasians, African Americans, Asians, and Hispanics, to name just a few. In addition, there are subpopulations as well, often based on origin, such as Japanese versus Chinese and Mexican Hispanics versus Puerto Ricans. When frequency estimates are combined to give an overall frequency, it is important that the traits be independent of each other as well. This is an issue in DNA TYPING, by which currently 13 different types can be determined and combined to estimate an overall frequency of a given combination. Thus, using proper statistical and genetic principles and calculations, it is possible to estimate overall frequencies based on a relatively small database of typed individuals.

portrait parle *See* BERTILLON, ALPHONSE; PHOTOGRAPHY.

postal inspection service *See* UNITED STATES POSTAL INSPECTION SERVICE.

postmortem examination *See* AUTOPSY.

postmortem interval (PMI) The time that has elapsed since death occurred. Contrary to popular belief, it is rarely possible to assign an exact time of death, and reliable estimates of the PMI require the use of multiple tools as opposed to one single method. Typically, the MEDICAL EXAMINER (ME) is tasked with determining the PMI and is often aided by other forensic professionals, particularly in cases in which the death occurred weeks or months prior to the body being discovered. For forensic purposes, the death event is divided into two processes, brain death and cell death (autolysis), but the time of death generally refers to brain death.

Techniques that can be used to estimate the PMI include the following.

Algor Mortis

This is the cooling of the body that starts after death. At room temperature, a body will cool about 1 to 1.5°F per hour until it reaches ambient temperature, but this rate is variable and depends on factors such as the environment, amount of body fat on the person, the person's height-to-weight ratio (a measure of surface area), and the person's body temperature at death, which may well be above or below normal. Use of cooling to estimate PMI is best accomplished with multiple measurement points (including in the liver) and measurements over time, neither of which is always possible.

Rigor Mortis

This is the stiffening of the joints (actually muscles) that is often referred to simply as rigor. Rigor begins to appear within a few hours of death, peaking at about 12 hours. The speed of the process is the same for all muscles but will be noticeable first in smaller muscles such as fingers. The process reverses and is usually gone within 36 hours of death. The degree of rigor is tested using the jaw, elbow, and knee joints. Variables that influence the rate of formation of rigor include temperature (warmer temperatures speed the process) and the amount of exercise the person had just prior to death. The more vigorous or strenuous it was, the quicker rigor will set in.

Livor Mortis

Once the heart stops, blood is no longer circulating and no longer picking up oxygen in the lungs. As a result, the blood becomes bluish and will settle in the body under the influence of gravity, with the process starting about an hour after death. Livor mortis appears as a bluish-purple discoloration at the lowest points and looks similar to a bruise. Settling will not occur in areas in which pressure seals off blood vessels, so, for example, if someone dies sitting on a hard wooden chair, blood will not collect where the buttocks are in contact with the chair. The pattern of lividity normally does not change after about 12 hours. Lividity patterns can be seen until DECOMPOSITION processes disguise them. Like rigor, lividity can aid investigators in determining if a body has been moved at some point after death but before they arrive at the scene.

Decomposition

The stage of decomposition can provide clues to the PMI, but it varies based primarily on the environment around the body since death. Once autolysis occurs, body chemistry begins to change rapidly, and enzymes become active that promote degradation of molecules such as proteins and carbohydrates. In these early stages of decomposition, microorganisms already present in the gut multiply and spread. This results in gas production that produces a bloating in the abdomen and then generalized bloating that can swell the body to twice or three times its original size before collapsing. The bloating can cause the eyes and tongue to swell and protrude as well. The appearance of the skin begins to change, passing through stages of green before reaching black, a stage appropriately named black putrefaction. Bacteria flourish in blood, which is an excellent medium for supporting them, accelerating the breakdown process. During these changes, the characteristic odors of decomposition are given off. Skin and hair can slough off during these processes, and the body eventually becomes unrecognizable. Stages of advanced decomposition can be reached in 12 to 18 hours in warm and humid environments, but in cold or freezing areas the process slows dramatically, stretching into weeks, months, or even years depending on how cold it is.

Putrefaction leads next to a stage of dry decay or mummification, again depending on environmental conditions. As skin dries, it shrinks, and this shrinkage has led to the popular misconception that hair and nails continue to grow after death. ADIPOCERE may form in this stage as well. Eventually, most or all tissue will be dried and will decay away, leaving only bones. In addition to the environmental factors, other considerations for the evaluation of decomposition include insect activity (ENTOMOLOGY) and animal scavenging. Insects such as blowflies are attracted to blood and will lay eggs in wounds and orifices such as the nose and ears. Maggot feeding can distort features and accelerate decay to a remarkable degree, and in areas of considerable insect activity, a fresh corpse can be skeletonized within a week or two. Submergence of a body will also have an impact on decay processes. A determination of the postmortem submergence interval, like the PMI, requires the use of several methods considered in light of the environmental conditions.

Miscellaneous

Often the best clues for determining time of death can be gleaned from careful examination of the death scene. Type of clothing on the victim can also be helpful. Contents of the stomach are sometimes helpful in determining the composition of the final meal, which can provide clues to the PMI. Finally, the level of potassium in the VITREOUS HUMOR (fluid of the eye) has been used in some cases.

Further Reading

Dix, J., and M. Graham. *Time of Death, Decomposition, and Identification, An Atlas.* Boca Raton, Fla.: CRC Press, 2000.

Haglund, W. D. "Forensic Taphonomy." In *Forensic Science: An Introduction to Scientific and Investigative Techniques.* 2nd edition. Edited by S. H. James and J. J. Nordby. Boca Raton, Fla.: CRC Press, 2005.

postmortem submergence interval (PMSI) *See* POST MORTEM INTERVAL.

power of discrimination (P_d) *See* DISCRIMINATION INDEX.

precipitin tests *See* IMMUNODIFFUSION.

precursors In forensic science, a term most often used to describe the ingredients necessary to synthesize illegal or illicit drugs. For example, PHENYL-2-PROPANONE (P2P) was once a common ingredient in the clandestine manufacture of METHAMPHETAMINE. However, once access to this substance was limited, it was replaced with other more readily obtainable precursors. Similarly, MORPHINE is a precursor to HEROIN.

predator drugs *See* DATE RAPE DRUGS; DRUG FACILITATED SEXUAL ASSAULT.

prescription drugs A common type of evidence encountered by forensic chemists. Prescription drugs can be obtained fraudulently or otherwise diverted for illegal purposes. For the analysis of prescription medication, often the first task is a visual examination of the medication using the *PHYSICIAN'S DESK REFERENCE* (PDR). This is followed by a confirmatory test such as GAS CHROMATOGRAPHY/MASS SPECTROMETRY (GC/MS) or INFRARED SPECTROSCOPY (IR). Prescription drugs may also be encountered in forensic TOXICOLOGY, and in these cases the original medication (tablet or otherwise) may not be available. Screening tests are used to identify common medications, followed by confirmatory techniques.

Further Reading

Geradts, Z. "Use of Computers in Forensic Science." In *Forensic Science: An Introduction to Scientific and Investigative Techniques.* 2nd edition. Edited by S. H. James and J. J. Nordby. Boca Raton, Fla.: CRC Press, 2005.

preservation of evidence A multifaceted concern in forensic science that starts with collection and includes proper packaging, storage, and documentation. If evidence is collected at a scene, the first step in preservation is photography and documentation. Once this is complete, evidence is collected using the method most suitable to the specific type, and a CHAIN OF CUSTODY document is started. For example, if bloodstained clothing is found at a scene, it should be air dried and stored in paper, which will promote air circulation and will retard MICROBIAL DEGRADATION. However, for something like a lifted fingerprint or cartridge case, storage in a plastic bag would be acceptable. From the moment of collection, the chain of custody ensures and preserves the integrity of evidence, while proper packaging and storage preserve the evidence in the same state as when it was collected (or as much as this is possible). Blood and body fluid evidence is stored in a refrigerator or freezer in a secured area, while drugs might be stored in a room temperature locker. During an analysis, all efforts should be made to preserve at least half of the sample for additional analysis by another forensic scientist.

presumptive tests Preliminary chemical tests widely used in the analysis of blood, drugs, GUNSHOT RESIDUE (GSR), and EXPLOSIVES. A presumptive test does not provide definitive identification; rather, it provides information useful for directing further analysis. For example, if the question is "Is this red stain blood?" a positive result with a presumptive test would mean "more likely than not." Since these tests involve addition of a chemical reagent and looking for a color change, these tests are sometimes referred to as "color tests." Other terms used are "screening tests" and "spot tests." Presumptive tests are subject to both FALSE POSITIVES and FALSE NEGATIVES, but these tend to be limited and recognized.

Presumptive Tests for Blood

Tests for blood are based on the ability of the heme group in hemoglobin to act as a catalyst in chemical reactions that involve a color change. This ability is referred to as "peroxidase like" activity in reference to biological enzymes that react with peroxide groups such as those found in hydrogen peroxide (H_2O_2). One of the oldest such test for blood is the BENZIDINE test, developed around the turn of the last century. However, this substance has been shown to be carcinogenic and is no longer used. Related tests using TETRAMETHYLBENZIDINE and *o*-TOLIDINE also use carcinogenic reagents and are rarely used.

Safer reagents are found in the KASTLE-MEYER, LEUCOMALACHITE GREEN, and LUMINOL tests, which are currently the most widely used reagents. Also used are commercial test strips called Hemastix that detect blood in urine. The luminol test is unique in that the reaction does not cause a color change but rather results in the emission of light. Sensitivities of these tests vary, but it is usually possible to detect blood that has been diluted hundreds- or thousands-fold. The following table summarizes the common presumptive color tests for blood.

	Color for positive result
Benzidine	Blue
o-Tolidine	Blue
Kastel Meyer (phenolphthalein)	Pink
Leucomalachite green	Blue-green
Hemastix	Blue-green

Presumptive Tests for Drugs

Perhaps in no other forensic area are preliminary color tests used more than in DRUG ANALYSIS. These tests are designed to narrow down the possibilities when an unknown substance is delivered to the laboratory for identification. The following table lists some of the more common color tests for drugs and what they are used for. All are listed as separate entries in this volume.

Reagent	Uses
Duquenois	Marijuana
Marquis	Opiates and amphetamines
Cobalt thiocyanate (Scott or Ruybal test)	Cocaine
Dilli-Koppanyi	Barbiturates
Ehrlich (Van Urk test)	LSD

Presumptive Tests for Gunshot Residue and Explosives

The other large group of screening tests targets the components of GUNSHOT RESIDUE (GSR) and EXPLOSIVES, which share many common elements. The GRIESS and DIPHENYLAMINE tests react with nitrate and nitrite ions (NO_{3-} and NO_{2-}) found in both. Residues of lead (Pb) can be detected using sodium rhodizonate.

Further Reading

Cox, M. "A Study of the Sensitivity and Specificity of Four Presumptive Tests for Blood." *Journal of Forensic Sciences* 36, no. 5 (1991): 1,503.

Siegel, J. A. "Forensic Identification of Controlled Substances." In *Forensic Science Handbook*. Vol. 2. Edited by R. Saferstein. Englewood Cliffs, N.J.: Regents/Prentice Hall, 1993.

Spalding, P. R. "Identification and Characterization of Blood and Bloodstains." In *Forensic Science: An Introduction to Scientific and Investigative Techniques*. 2nd edition. Edited by S. H. James and J. J. Nordby. Boca Raton, Fla.: CRC Press, 2005.

Sutton, P. T. "Presumptive Testing for Blood." In *Scientific and Legal Applications of Bloodstain Pattern Interpretation*. Boca Raton, Fla.: CRC Press, 1999.

primary transfer (direct transfer) A classification of a type of transfer or trace evidence. For example, if a person owns a pet cat, a direct transfer of cat hair could occur between the cat and the person's sweater. If this person then brushed against someone else and cat hair from their sweater was transferred to the other person, this is a secondary, or indirect, transfer. Because indirect transfers are common, one of the challenges of interpretation of evidence such as hairs or fibers is sorting out which of them are significant and which are not. One tool that is used to assist in these interpretations is exemplars, or CONTROL SAMPLES. If the person with the cat became the victim of a crime, sample hairs would be needed from the cat to help in determining if the transfer of cat hair to the victim were significant or the result of some other indirect transfer. Occasionally, the secondary transfer may be the one of interest. In practice, it can be difficult if not impossible to distinguish direct from indirect transfers.

primers In forensic applications, a term that can refer to a coating or to AMMUNITION. When a new surface such as bare metal in a car body or new sheetrock in a wall is created, it must be primed before the paint is applied. As such, the primer is a form of PAINT evidence subject to the same types of analyses. However, the more common meaning of the word *primer* refers to a component of modern ammunition for use in FIREARMS.

In ammunition, a primer is a primary EXPLOSIVE that is used to ignite the PROPELLANT, which burns and creates the gases that force a bullet forward. Primers consist of a shock-sensitive explosive (typically lead styphnate, also called lead trinitroresorcinate), an oxidizer (barium nitrate, $BaNO_3$), and a fuel such as antimony sulfide (Sb_2S_3). This is the same combination of ingredients used in COMBUSTION, and the ignition of a primer produces an intense flame that is directed through vents into the chamber of the CARTRIDGE CASE, where the propellant is stored. This flash ignites the propellant and results in the acceleration of the bullet down the barrel of the weapon.

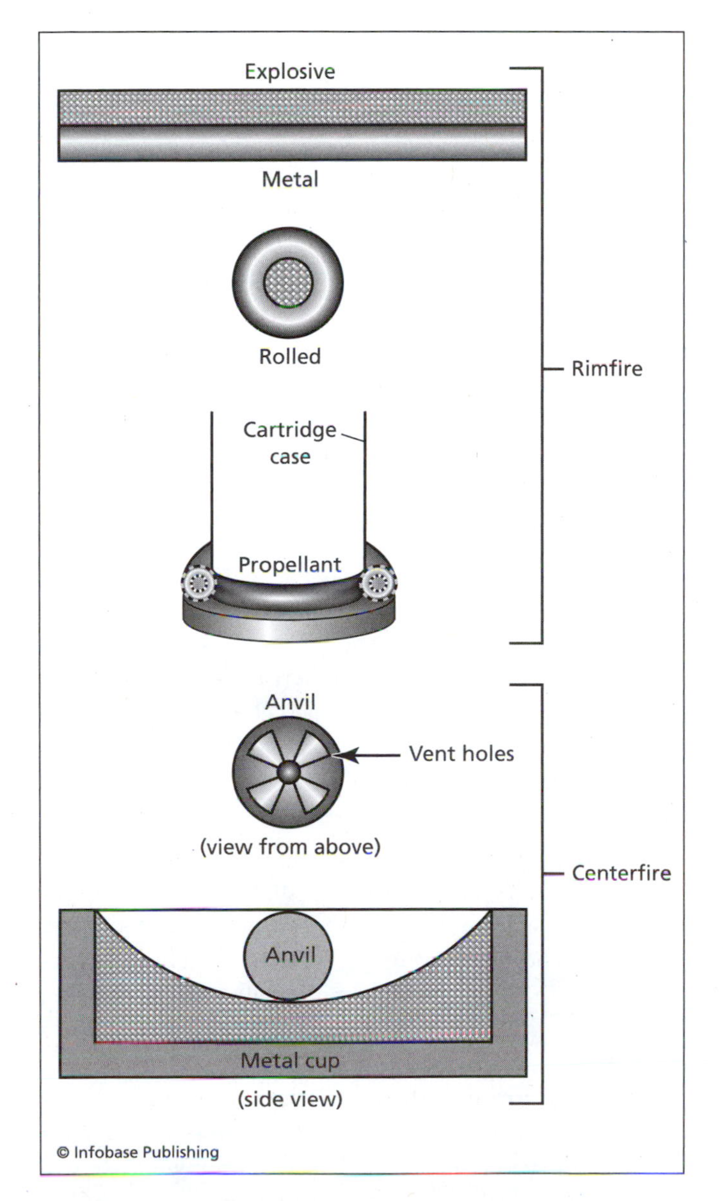

Rimfire and centerfire primers used in modern ammunition

The first primers were simple devices that burned or produced a spark. Flintlock guns and wheel guns took advantage of sparks produced by friction against a flint to ignite the powder. In the mid-1800s, the design was improved, and a primer consisting of mercury fulminate and perchlorate oxidizers was incorporated into the cartridge case rather than being part of an external assembly. These materials were eventually replaced with the modern components to overcome corrosion and fouling problems. The centerfire primer dominates modern ammunition, although small-caliber weapons use the rimfire design. Both are shown in the figure. In a rimfire primer, the initiator is contained inside a roll of metal that wraps around the base of the cartridge. When the trigger is pulled, the firing pin strikes the rim and creates the flash that is directed into the cartridge through a series of vents. With a centerfire primer, the firing pin compresses the propellant against an anvil and creates the flash. In addition to the initiator, fuel, and oxidizer, primers can contain many other ingredients including sensitizers (to make the initiator more sensitive to shock), binders, and traces of other explosive materials. The compounds used in primers are important components of GUNSHOT RESIDUE (GSR).

printers, copy machines, and typewriters *See* PHOTOCOPIERS; QUESTIONED DOCUMENTS.

product liability The legal procedures involved when a product causes, or is thought to have caused, injury. The injury may be physical ("personal injury") or financial and can involve an individual or an organization. The product that caused the injury may have been poorly designed, improperly assembled, or otherwise damaged during production. Forensic engineers often play a role in product liability procedures, although expert witnesses from many disciplines may be called. Product liability cases may involve a crime or criminal conduct, but most are civil cases. Many evolve into class-action lawsuits in which many complaints are combined into one. Perhaps the best known product liability cases of the last few years have involved the tobacco companies and payments for deaths and injuries caused by their products.

product tampering A crime that came to national attention in 1982, when seven died when cyanide powder was placed into Tylenol capsules. As a result of this and other incidents around the same time, the way products were packaged was changed dramatically to include safety seals and sealed packages. In many forensic drug labs, the day after Halloween is particularly busy, when adulterated food (actual or suspected) arrives for analysis. Razor blades and pins in candy are among the many variations of product tampering. Forensically, TOXICOLOGY and DRUG ANALYSIS are the two sections most directly involved, although fingerprints and other kinds of trace evidence may also be found. After September 11, 2001, the U.S. government became increasingly concerned about product tampering as a form of terrorism, particularly in the area of adulteration of the food supply. The federal FOOD AND DRUG ADMINISTRATION (FDA) is the principal responsible government agency, and it maintains a forensic facility in Cincinnati.

Further Reading

Logan, B. "Product Tampering Crime: A Review." *Journal of Forensic Sciences* 38, no. 4 (1993): 918.

professional associations *See* Appendix I.

profiling *See* PSYCHIATRY, FORENSIC AND PSYCHOLOGY, FORENSIC, AND PROFILING.

propellant A low EXPLOSIVE material used in AMMUNITION. Known informally as GUNPOWDER, propellant is placed in a cartridge case packed between the PRIMER and the projectile. BLACK POWDER (a mixture of charcoal, potassium nitrate, and sulfur) was used from ancient times until about the mid-1800s, when it was replaced by SMOKELESS POWDER. Modern smokeless powder propellants are either "single base," consisting of NITROCELLULOSE, or "double base," consisting of cellulose nitrate and NITROGLYCERIN. As shown in the illustration, the impact of the firing pin on the primer cases a flash, which ignites the propellant. The granules usually burn progressively and from the outside in as shown. The rapid COMBUSTION that follows creates large volumes of hot expanding gas resulting in high pressures that are confined within the barrel, forcing the bullet forward at high speed.

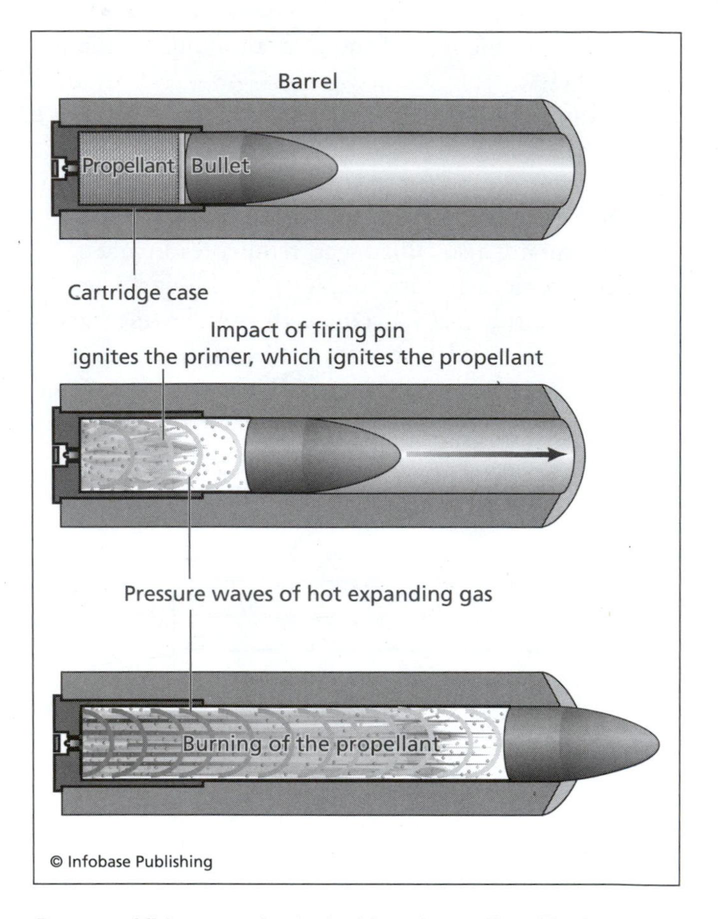

Process of firing a gun by the ignition of propellant. The burning propellant produces hot expanding gases that impart movement to the bullet.

proteins Large complex molecules ("macromolecules") formed by linking together tens to thousands of AMINO ACID subunits. Proteins are biological POLYMERS and are also referred to as polypeptides. Because of the myriad attractions and repulsions between the subunits, proteins twist and fold into complex three-dimensional structures that are critical in determining their functions. The primary structure of any protein is the sequence of amino acids that constitute it, while the secondary structure is defined by repeating subunits or sequences of amino acids along the chain. For example, the familiar helix (twist) of DNA is an example of secondary protein structure. Proteins in hair that form long fibers are another type of secondary structure. These structures can in turn interact with each other, forming clumps or globular shapes that define the tertiary structure, and when several of these combine to form a functional protein, the quaternary structure results. Examples of proteins encountered in forensic science include HEMOGLOBIN (the basis of many PRESUMPTIVE TESTS for blood), antigens such as those found in the ABO BLOOD GROUP SYSTEM, and antibodies used in IMMUNOLOGICAL tests.

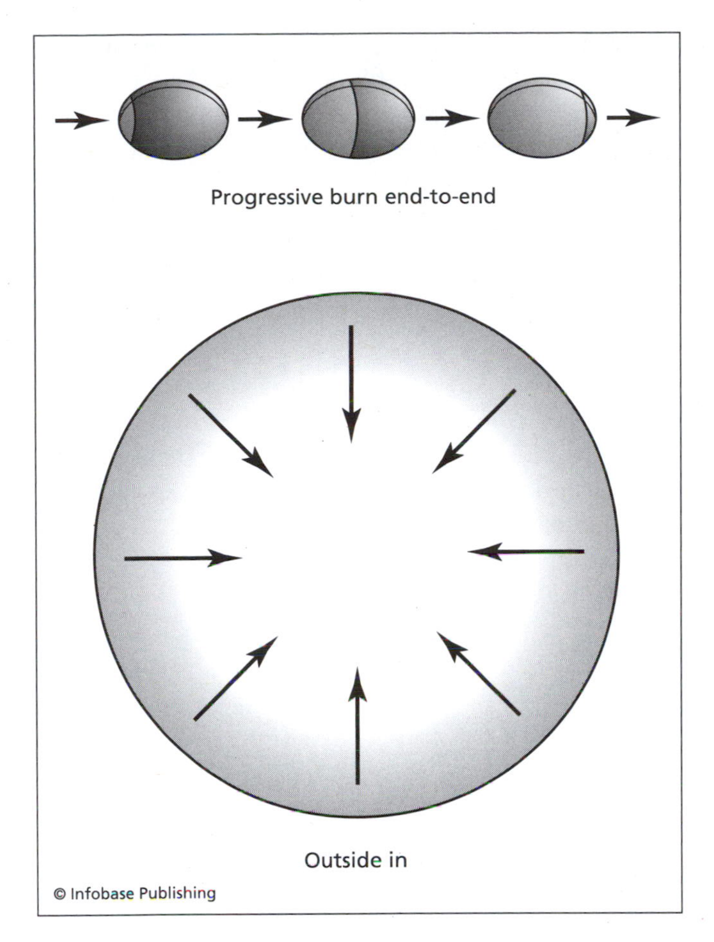

Ways a granule of propellant can burn

Modern gunpower seen under low magnification. Different shapes of granules are commonly seen in commercial mixtures. ***(photo by Suzanne Bell)***

pseudoephedrine *See* EPHEDRINE.

pseudoscience Theories, ideas, or explanations that are represented as scientific but that are not derived from science or the scientific method. Pseudoscience is a constant concern in forensic science since it is critical that evidence presented to a court as scientific really is. Scientific information is obtained using the scientific method and involves collection of data by experiment and direct observation. These experiments and observations must be designed so that others can repeat and verify them. Pseudoscience often springs from claims of folk wisdom or selective reading without independent data collection or validation. Scientific statements are specific and well defined, while pseudoscience is vague and variable. One of the key differences between the two is that a scientific statement or theory is put in such as a way as to be FALSIFIABLE. For example, if a toxicologist writes a report that states that a blood sample contained cocaine, another toxicologist could repeat the test and falsify it if indeed the results were false. The mechanism of validation, new experimentation, and refining of theories to fit the evidence means that scientific thinking has survived challenge and represents the best explanation possible. In contrast, pseudoscientific statements are usually not falsifiable using objective experimental or observational evidence. Pseudoscience provides no room for challenge and tends to dismiss contradictory evidence or to selectively decide what evidence to accept. Thus, pseudoscience is usually nothing more than a claim, belief, or opinion that is falsely presented as a valid scientific theory or fact. UFOs and astrology are pseudoscience, and in forensic science, GRAPHOLOGY, forensic hypnosis, and voice stress analysis are often placed in this category as well. The *DAUBERT* DECISION and its extension in the *KUMHO TIRE COMPANY* decision have greatly assisted the courts in defining pseudoscience and keeping it out of the courtroom.

Further Reading

Lee, J. A. *The Scientific Endeavor: A Primer on Scientific Principles and Practice.* San Francisco: Addison Wesley Longman, 2000.

psilocyn and psilocybin Hallucinogens contained in mushrooms found principally in Mexico. In this respect, these compounds are similar to MESCALINE, which is obtained from peyote cactus. Psilocybin can also be synthesized from psilocyn, but most seizures are of the naturally occurring materials. In mushrooms, there is more psilocybin than psilocyn, but the psilocyn is about twice as potent. The drugs are taken by chewing the dried mushrooms, and the effects start within an hour and can last for four hours or more. These compounds are listed on Schedule II of the CONTROLLED SUBSTANCES ACT (CSA). Analysis of suspected psilocyn and psilocybin begins with PRESUMPTIVE TESTS such as the MARQUIS TEST, resulting in a yellow color, and Froehde's reagent, which turns a greenish yellow. Further testing usually involves THIN LAYER CHROMATOGRAPHY (TLC) and INFRARED SPECTROSCOPY (IR).

psychiatry, forensic; psychology, forensic; and profiling Disciplines involved with what is broadly categorized as behavioral evidence. Unlike physical evidence, behavioral evidence is intangible and subject to different perspectives and interpretations. Examples of the use of behavioral evidence include determining criminal responsibility and the validity of an insanity plea, assessing a person's danger to society and likelihood to reoffend, treatment of both victims and offenders, studies of violent behavior, and the evaluation of eyewitness testimony. Behavioral evidence is used in civil as well as criminal cases in areas such as child custody and conflict resolution. Psychiatrists are physicians with an M.D. and additional training in behavior, neurology, and pharmacology. They can prescribe medications. Psychologists have a college degree (usually an M.S. or Ph.D.) and do not have medical training. However, both are active in the justice system and perform many of the same functions. Related to these are neuropsychiatry, which blends neurology with psychiatry, and neuropsychology, which focuses on behavior as indicators of brain and nervous system function.

Profiling has become one of the better-known aspects of behavior evidence, principally because of media portrayals such as Thomas Harris's novel *Silence of the Lambs* that was made into an Oscar-winning movie in 1991. Many people are familiar through fiction with the FBI's Behavioral Science Unit, which is part of the National Center for the Analysis of Violent Crime (NCAVC) at the FBI Academy in Quantico, Virginia. Profiling is not a distinct discipline and is practiced by many different types of people, not all reputable. The emergence of profiling in law enforcement can be traced back to JACK THE RIPPER, who captured and still holds the public fascination. From that time (the late 1800s) to well into the next century, profiling was done mostly by psychiatrists or similarly trained individuals. The FBI's unit was formed in the early 1970s. Although there are no standards for profiling, the process normally includes examination and review of physical evidence and crime scenes as well as psychological evaluation of victims (victimology). Like other forensic disciplines, a large part of profiling is seeking of patterns, such as consistent elements of crimes, and any signature behaviors shown during the commission of a crime. A signature consists of actions that are consistent across all crimes in a series, although the modus operandi (MO) can change or evolve. Identification of a signature can help investigators determine if the same person likely committed several crimes or if the crimes are unrelated.

Further Reading

Keppel, R. C. "Serial Offenders: Linking Cases by Modus Operandi and Signature." In *Forensic Science: An Introduction to Scientific and Investigative Techniques*. 2nd edition. Edited by S. H. James and J. J. Nordby. Boca Raton, Fla.: CRC Press, 2005.

Napier, M. R., and K. P. Baker. "Criminal Personality Profiling." In *Forensic Science: An Introduction to Scientific and Investigative Techniques*. Edited by S. H. James and J. J. Nordby. Boca Raton, Fla.: CRC Press, 2003.

Sadoff, R. L. "Forensic Psychiatry." In *Forensic Science: An Introduction to Scientific and Investigative Techniques*. Edited by S. H. James and J. J. Nordby. Boca Raton, Fla.: CRC Press, 2003.

Schlesinger, L. B. "Forensic Psychology." In *Forensic Science: An Introduction to Scientific and Investigative Techniques*. Edited by S. H. James and J. J. Nordby. Boca Raton, Fla.: CRC Press, 2003.

Turvey, B. E. *Criminal Profiling: An Introduction to Behavioral Evidence Analysis*. San Diego, Calif.: Academic Press, 1999.

psychology, forensic *See* PSYCHIATRY, FORENSIC, PSYCHOLOGY, FORENSIC, AND PROFILING.

pupea *See* ENTOMOLOGY, FORENSIC.

Pure Food and Drug Act In the United States, the first recreational drugs were the opiates such as MORPHINE and OPIUM, introduced by Chinese immigrants in the mid-1800s. San Francisco was the first city to pass a law regulating drugs in 1875. The first federal law regarding drugs was the Pure Food and Drug Act of 1906, which required labeling of patent medicines. The first federal agency with responsibility for drug control, the Bureau of Revenue, was formed in 1915. This organization was a precursor to the DRUG ENFORCEMENT ADMINISTRATION (DEA).

putrefaction A stage in the DECOMPOSITION process. Putrefaction begins with a greening of the skin along with a surge in MICROBIAL DEGRADATION leading to bloating and purging of gases and fluids from the body. Putrefaction is marked by the characteristic foul odors of decomposition. The stage ends when all soft tissue has disappeared.

pyrolysis gas chromatography (GC) A type of GC used in the forensic examination of FIBERS, PAINTS, and similar samples. The word *pyrolysis* means "fire cutting" and refers to the method in which a solid sample such as a fiber can be introduced into the GC instrument. Normally, samples are dissolved in a solvent and injected into the instrument. In pyrolysis, the sample is rapidly heated, resulting in decomposition and the release of gaseous materials that can be directly introduced into the carrier gas stream of the gas chromatograph. The resulting output is sometimes called a "pyrogram." Pyrolysis is not as reproducible as other methods of GAS CHROMATOGRAPHY and thus is not used often to obtain quantitative data on amounts and concentrations of individual components. However, the technique can be valuable in obtaining a qualitative "fingerprint" of a sample that can be used to compare separate fiber or paint samples to determine if they are consistent or different.

Quaaludes *See* METHAQUALONE.

qualifications A combination of education, training, and experience that allows a forensic scientist to testify as an expert witness in an area of expertise. The expert's qualifications are introduced and reviewed during the process of VOIR DIRE, during which both sides have the opportunity to ask the expert questions and to judge if he or she has the necessary background to be accepted by the court as a reliable expert in some aspect of the case at hand. What represents reasonable credentials varies with discipline. For example, a MEDICAL EXAMINER (ME) at a minimum would be expected to have an M.D. degree and specialized training in forensic PATHOLOGY. A forensic anthropologist would need a Ph.D. in ANTHROPOLOGY along with specialized forensic training. In contrast, a criminalist working in a state laboratory on HAIR and FIBER analysis would need a B.S. or similar degree coupled with on-the-job training.

Qualifications are divided into the categories of education (academic credentials), experience, on-the-job and continuing education, membership in professional associations, publication record, and applicable certifications, and licenses. Although both the prosecution and defense can ask questions during qualifications procedures, it is the judge who makes the final decision as to whether a person qualifies as an expert. Even if a person is found qualified in one area, say DNA analysis, that does not qualify him or her to testify on fingerprints or firearms, although both are forensic disciplines.

See also *DAUBERT* DECISION; *FRYE* DECISION.

qualitative analysis and qualitative evidence Qualitative analysis involves observations and testing that identify the components or constituents of a sample. However, qualitative analysis does not determine how much of each component is present. Similarly, qualitative evidence is evidence that is analyzed without producing quantitative data. Examples of qualitative evidence include HAIRS, FIBERS, BULLETS, and CARTRIDGE CASINGS, all of which are analyzed using microscopy. For example, a firearms examiner uses a microscope to study markings on bullets; the exact concentrations of lead, copper, and so on of bullets are rarely of interest. In this example, the microscopic examination is qualitative; a chemical analysis of the bullet to determine the exact concentration of lead would be quantitative.

In many forensic applications, qualitative analysis is sufficient or serves as a starting point for further comparison and attempts at INDIVIDUALIZATION and linkage to a COMMON SOURCE. If a fiber were recovered on the clothing of a suspect, a qualitative analysis would be used to determine that the fiber is a synthetic fiber impregnated with a green DYE, for example. The examiner would conduct further observation and testing to attempt to link that fiber to another person or place. Quantitative analysis is not necessary; the exact concentration of dye present is not crucial, nor would it be helpful.

In other cases, qualitative analysis is the necessary starting point for a QUANTITATIVE ANALYSIS. After all, it is necessary to know *what* is present before determining *how much* of it is there. Quantitative analysis is most often undertaken in forensic drug analysis. For example, if an unknown white powder is submitted for drug analysis, the results of a qualitative analysis might show that the powder contains COCAINE mixed with flour. A quantitative analysis might show that the powder is 10 percent cocaine and 90 percent flour. Depending on the jurisdiction, a quantitative analysis may be required by law, although only for the cocaine or other illegal substance that is present.

quality assurance/quality control (QA/QC) An inclusive term for procedures and protocols used in forensic analysis to ensure that any data or results produced are accurate, reliable, and trustworthy. The term *quality assurance* is typically used to describe the foundation in place to ensure acceptable laboratory performance. In this usage, QA refers to things such as training, documentation, laboratory policies and procedures, and METHOD VALIDATION. Quality control reflects all the actual practices used to ensure the trustworthiness of data such as CONTROL SAMPLES, instrument logs, use of "good laboratory practices" (GLP), and proper documentation. Regardless of how the terms are applied, aspects of QA/QC include (but are not limited to) facilities, maintenance procedures and documentation, training and continuing education, CERTIFICATION of analysts, ACCREDITATION, proper preparation of any reagents or standards, control samples, documentation and record keeping, safety protocols, and peer review. In this sense, anything that adds to the trustworthiness of data or laboratory results can be considered to be part of QA/QC. Guidelines for QA/QC protocols are published by entities such as the National Institutes of Standard and Technology (NIST) and by the AMERICAN SOCIETY OF CRIME LABORATORY DIRECTORS Laboratory Accreditation Board (ASCLD). Many professional societies also publish or recommend procedures and guidelines that could be considered part of QA/QC.

quantitative analysis Testing that leads to determination of how much of a given component or components are present in a sample. This is in contrast to QUALITATIVE ANALYSIS, in which only the identity of components is necessary, not amounts or concentrations. Quantitative analysis is most often an issue in TOXICOLOGY and DRUG ANALYSIS, when the amount of an illegal substance present is critically important. For example, in the analysis of blood alcohol, the first stage is the definitive identification of ethanol (qualitative analysis) followed by an accurate determination of the quantity of alcohol present. A blood alcohol level below 0.08 percent in most states is not considered illegal, while concentrations over 0.08 percent can result in charges of driving under the influence.

questioned documents The specialty in forensic science that deals with suspicious, forged, or damaged documents. Handwriting analysis is considered part of questioned document examination as well and should not be confused with GRAPHOLOGY. Questioned documents began to emerge as a forensic discipline around 1870, and an early practitioner was ALPHONSE BERTILLON. Unfortunately, Bertillon was not always as good as his reputation, and mistakes he made in the field cast a shadow over document evidence for many years.

ALBERT S. OSBORN is considered to be a pioneer of document examination in the United States and wrote a book entitled *Questioned Documents* in 1910, along with a later revision. These books are still cited in the field, and Osborn's sons continue to be active examiners. Albert and a son were involved in the LINDBERGH KIDNAPPING case (1932), a crime that was instrumental in bringing forensic evidence, including documents, to the public's attention. In the United States, the prin-

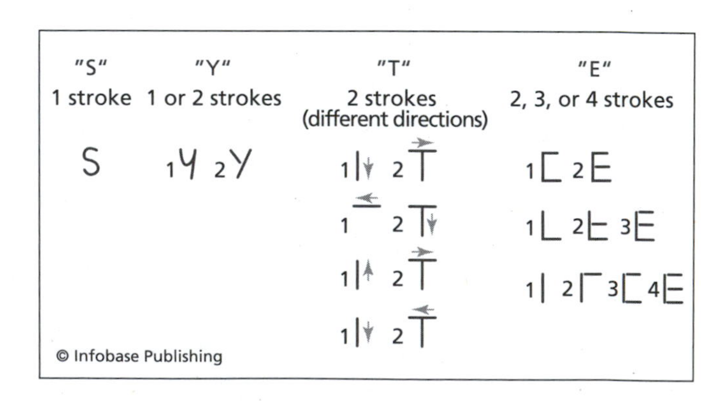

Different handwriting methods for creating printed letters. Note the different order and number of strokes that are possible.

cipal professional associations for document examiners are the American Society of Questioned Document Examiners (ASQDE) and the Questioned Document section of the AMERICAN ACADEMY OF FORENSIC SCIENCES (AAFS). Together, the ASQDE, AAFS, and the Canadian Society of Forensic Science formed the American Board of Forensic Document Examiners (ABFDE) in 1977 to provide professional certification of document examiners. There are no degree programs specifically for document examination, but certification with the ABFDE requires that the examiner have a bachelor's degree along with the requisite training, which retains a strong element of traditional apprenticeship work.

Types of Document Evidence
Examiners are faced with a wide variety of evidence such as documents produced by handwriting, typewriters, copiers (photo and xerographic), computer printers (laser, inkjet, and dot matrix), machine printers (check writers and cash register receipts), and facsimile machines (FAX). The question may be who wrote a document, who did not, when it was written, in what order things were added, and authenticity, to name a few. Alterations are a common theme, be they additions or erasures/obliterations. Document examiners can also be involved in cases of suspected or questioned balloting and lottery tickets. Charred, folded, damaged, or torn documents can also be important items of physical evidence. Stamps such as those found on certified mail or dated receipts can be disputed or questioned, and a document examiner may analyze items such as envelopes or postage stamps. Advances in technology such as color laser printers and high-resolution scanners and copiers broaden the scope of questioned documents and increase the challenges presented. However, the central principles of document examination remain the same and are comparable to methods used for most physical evidence.

Document examination, be it of handwriting or a printed document, typically begins with the evaluation of CLASS CHARACTERISTICS and proceeds to individual characteristics. Photography, microscopy, and spectroscopy are important tools at the examiner's disposal, supplemented with large databases maintained by professional organizations such as the ASQDE and law enforcement agencies such as the SECRET SERVICE. Increasingly, digital imaging and enhancement is becoming an important element of the analysis of evidence.

Handwriting Analysis
Although young students learn the same or similar systems for writing, the process quickly becomes individualized and unique. However, the uniqueness of handwriting cannot be equated to the uniqueness of fingerprints or DNA type because there is such a range of "normal" writing for any one person. Try writing your own signature 20 times and observe the differences—all are yours, but not all are identical. Document examiners often are faced with determining, based on handwriting, who wrote or who did not write a given signature or other entry on a document. As with most forensic investigations, the evaluation usually starts with a study of class characteristics that moves into a detailed search for INDIVIDUALIZATION.

There are several types of handwriting including printing or block lettering, cursive writing, disconnected cursive or script, and numerals. All can be evaluated to determine authorship, even block printing and numerals. Most signatures are done in cursive, although some individuals print theirs. On the other extreme, many signatures are illegible but are still considered to be an individualized and authentic representation of one person. Other signatures acquire

The order in which two lines are written may be determined using low magnification microscopy. ***(Suzanne Bell)***

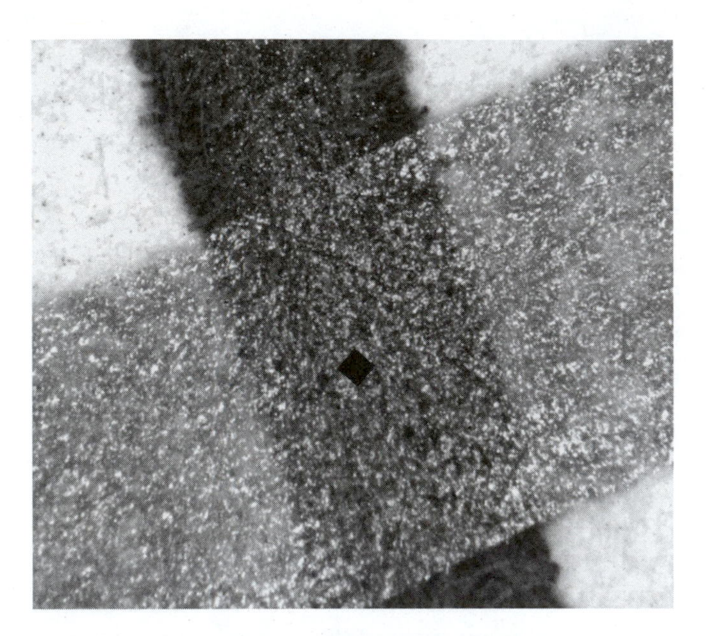

The order is reversed from the previous photo. ***(Suzanne Bell)***

flourishes and stylistic traits that are not transferred to a person's normal writing. For these reasons, the signature is considered to be a specialized case of handwriting and one of the most frequently contested elements of a questioned document.

Even when a handwritten entry is printed, many factors contribute to variations seen among different writers. For example, the way the writing instrument is held, the angle to the paper, the pressure applied, the type of backing (rough or smooth, even or uneven), and the type of writing instrument can all affect the appearance of the writing. As shown in the figure on page 300, there are many methods for constructing individual letters as well. Some, such as *s* and *c,* require a single stroke and no lifting of the pencil. A *y* can be made with a single stroke or two. With a *t,* only two strokes are needed, but the order and direction of each stroke varies. With a letter such as *E,* several methods exist. Another factor that influences the appearance of hand printing is the way in which a stroke is ended. As shown, lines tend to thin out and become lighter toward the end of a stroke. In some people's printing, the pencil may not completely leave the paper, resulting in faint linkages between adjacent letters. Thus, even simple block printing can become highly individualized, and these same factors extend to handwriting of numerals. With cursive writing, the same considerations apply, along with others. The angle of slant of cursive varies, as does the relative proportion of letters and spacing. Accordingly, cursive writing is generally more individualized than is block printing.

Other variables can arise in a handwritten document that may be of use in determining authorship. Layout of writing on a page tends to be consistent for one person, and an example is addressing an envelope. A person will usually place the destination address and the return address in the same place every time and will use the same format. Some people put the zip code on a separate line, others do not. Word choice can be distinctive, as can misspellings and grammatical errors. Studies of these characteristics can be useful and can fall under the domain of forensic LINGUISTICS and/or forensic PHONETICS. Occasionally the style of writing can point to education in a specific region or country.

Unlike fingerprints and blood types, handwriting style evolves and can be influenced by factors such as age, vision, tremors, drugs, and alcohol. Teenage girls often have florid and stylized signatures that fade over the years. Still, the variation over a lifetime for one person is normally less than the variation between people, allowing document examiners to identify authorship in many cases. However, the task becomes more difficult in cases in which handwriting and particularly signatures are forged or copied for criminal purposes. This kind of writing falls into two broad categories—disguised writing and forgery. In disguised

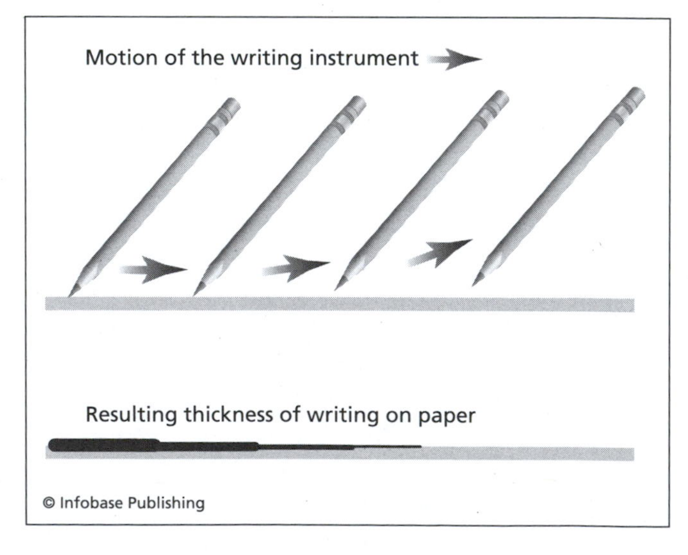

Different line thicknesses produced during lifting of a writing instrument

writing, a person is trying to hide his or her identity rather than impersonate another specific person. An innocent example familiar to most people is writing with the "wrong" hand. Ransom notes can fall into this category as well.

Forging of writing and signatures is more common and one of the largest sources of document evidence. In these cases, a person deliberately and for criminal purposes re-creates the signature of someone else. To successfully forge a signature, it is necessary to reproduce both the appearance of the signature and to do so with a smooth, continuous motion. To accomplish both is nearly impossible. If a person simply tries to mimic a signature by looking at it while copying, it is considered to be a form of drawing, and as a result the pen strokes tend to be slow, deliberate, and not necessarily in the same order that the real signature was created. Another approach is to trace a signature by various means (bright light, carbon paper, and so on). Evidence of the tracing is usually detectable, again by speed of writing or by identification of faint tracing lines or indentations. The introduction of high-resolution scanners and copiers complicates the job of the handwriting expert in that signatures can be scanned or copied and then inserted into a document. Fortunately, all these approaches create clues and evidence that can show the signature was illegally obtained and used.

In any handwriting analysis and comparison, the examiner works from a set of known writings. These can include collected writings that are known to have been written by either the suspect or the person whose writing has been forged. Examples of known writing include cancelled checks, lists, and notes. Exemplars are writings obtained after the incident in question, collected by police or the document examiner. Exemplars include copying of presented material that can be given in written form or dictated. The latter can be important, as often the police or examiner will not want the suspect to know what writing has been forged. Thus, in a case of a forged will, a suspect would not be shown the will in question but rather would be requested to produce exemplars crafted to contain similar letter combinations, words, and phrases. Usually an exemplar will include several repetitions to determine normal variability, a key component in any type of handwriting comparison. Using the questioned writing and the known writings from the victim and suspect, the examiner will study class characteristics and individual traits in an attempt to determine authorship.

Writing Instruments and Paper

In handwriting cases, the instrument used to produce the writing can provide important evidence. Pencil lead comes in many different formulations, as do INKS. Although these are produced in such large quantities that they are not unique, chemical and spectrochemical analysis of these components can be useful in dating documents or in eliminating possible sources. Pens as writing instruments have evolved from quill feathers through fountain pens, ballpoints, felt tips, and gel tips, all of which leave characteristic patterns on paper. For example, not all felt tips are the same width (wide point versus fine point), and these class characteristics can be important in a questioned document case. Occasionally, inks can be dated to provide an earliest possible date of use or a bracketed range in which the ink was available and could have been used.

The type of paper used to create a document can also be analyzed. Paper is made from wood pulp, resins and binders, bleaching agents and colorants, and cotton fiber. The better the quality of paper, the higher the cotton content. The ingredients are mixed in a slurry and then dried on frames or rollers that may leave distinctive marks, as may cutting tools. Some companies put a watermark on their paper, which is a physical indentation created by thinning of the fiber content in certain areas. Class characteristics that a document examiner can study include thickness of the paper, watermark if present, and fluorescence created by dyes, bleaches, and other chemical additives. As with ink and pencil evaluations, paper is not unique, but paper type can be useful in dating and determining if papers are similar or dissimilar. For example, if a three-page will written in 1955 is altered by the addition of a forged page added in 2002, the papers will be dissimilar, and that is sufficient evidence to prove that a forgery has occurred. In some cases, paper can be used for a PHYSICAL MATCH to prove a common source. For example, if a torn page from a document is found in one place and the torn corner recovered in another, a physical match of the two pieces proves that they were once part of the same document.

Ideally, paper analysis is performed using nondestructive techniques such as microscopy and spectroscopy. However, in some cases more aggressive destructive methods may be needed. Small portions of the paper may be removed and the fibers studied under a microscope. Chemical analysis of trace elements and organic compounds can also be undertaken. To view a faded watermark, specialized photography can be performed. One method involves placing the paper on top of a sheet impregnated with radioactive carbon-14, the same isotope used in radiochemical dating. Left overnight, the ^{14}C emits beta radiation that passes through the paper to a detector. In the areas where the watermark is present, the paper is thinner, allowing more beta particles to reach the detector and producing a brighter image compared to thicker areas of the page.

Typewriters, Copiers, Faxes, and Computer Printers

The first typewriters were introduced around 1870 and became the most important piece of office equipment until word processors became widely available more than a century later. Like handwriting, typewriters have class and individual characteristics that can be used in comparison and identification cases. Typewriters use rollers to move paper in front of an inked ribbon. The image of a letter is placed on the surface of the paper by a key striking the ribbon. All of these elements can be used in document analysis. For example, some rollers leave distinctive marks on paper, and these marks can change over time as the rollers wear. Likewise, the individual keys may have distinctive wear patterns that are transferred to the page each time the ribbon is struck. Collectively, these kinds of marks are called "machine wear" or "machine defects," and the concept carries through to copiers and computer printers. These are the characteristics that can individualize typewriting and that can be used to link a document to a specific typewriter. A single-use ribbon retains the image of typed letters and may also be of use if recovered.

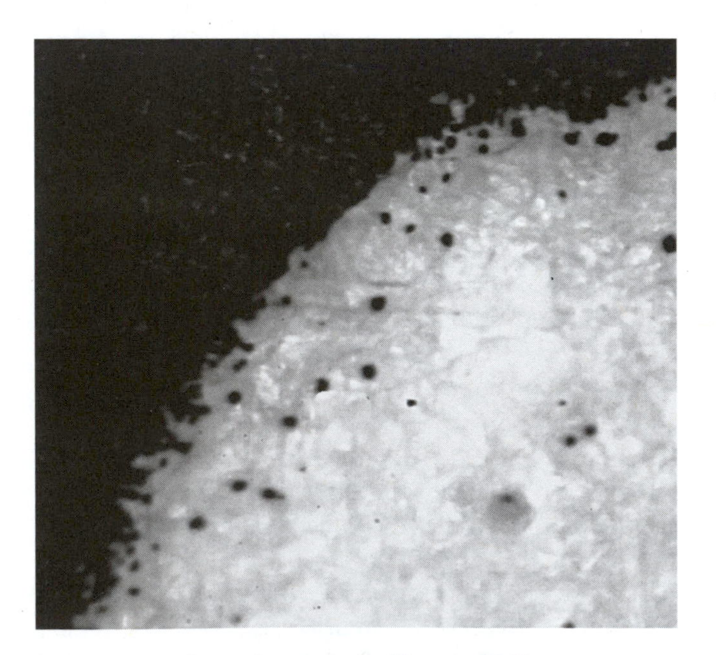

Toner particles from a laser printer ***(Suzanne Bell)***

The fonts used in a document are examples of class characteristics. Pica spacing allows for 10 characters per inch, while elite allows for 12. Different companies created slightly different typefaces over the years, and by comparison with a databases of typefaces, it is sometimes possible to identify the typeface, make, and model of a typewriter used to create a document. The construction of letters over the years has evolved as well. For example, older typewriters often created *M* and *W* with the middle reaching to the same height as the end lines. Later typefaces changed to have the middle line about half the length of the two end lines. Proportionally spaced fonts were also introduced, with variable spacing between letters as opposed to the fixed spacing of pica and elite. IBM introduced the Selectric typewriter in 1961 that used an interchangeable ball element instead of individual keys, and as the century drew to a close, the functions of newer typewriters began to merge with those of computer word processors.

For the analysis of typewritten documents, a number of tools are available including a grid method in which the examiner grids the paper and evaluates the spacing, font, and defects of characters. A comparison projector can be used to project large images of documents for side-by-side or overlapping comparisons. Inks in ribbons can be evaluated, as can carbon paper and correction inks and fluids. Use of alternative lighting and fluorescence can also be used.

One of the first computer printers available, and one that is still used, was the dot-matrix printer. This is considered, along with typewriting, to be a form of impact writing. This is in contrast to copiers, ink-jet printers, and laser printers that impregnate paper with ink. The function of a dot-matrix printer is shown in the figure on the next page. Like a typewriter, an inked ribbon is used to make the letter. However, the letter is created by a series of dots mounted on a print head. Multiple passes of the print head coupled to selective activation of the pins create the letters. Like a typewriter, a dot-matrix printer will have class characteristics as well as machine defects that can be used for

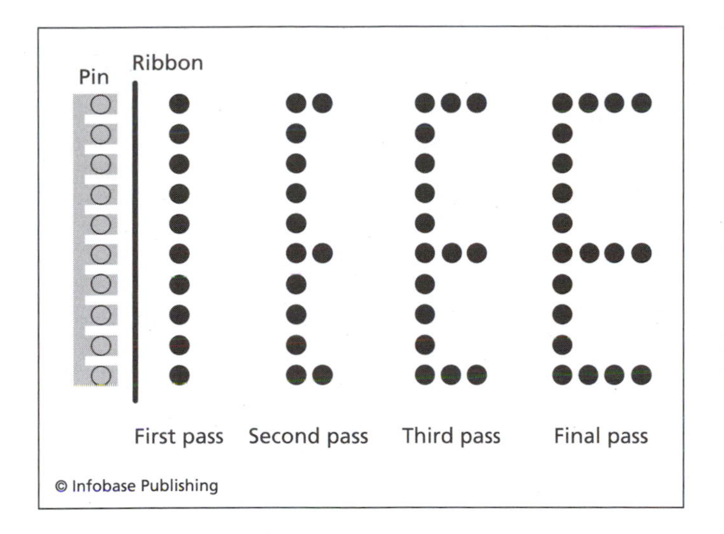

Operation of a dot-matrix printer

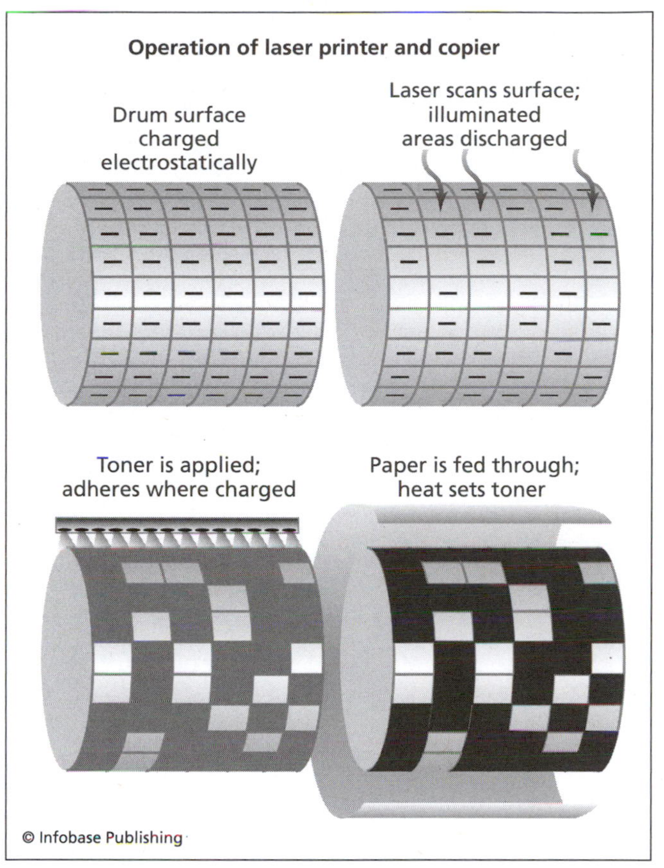

Simplified depiction of the operation of laser printers and copiers

comparison. Ink-jet printers work on much the same principle except that ink is sprayed from the print head onto the paper in selected small areas. Laser printers and copiers work on a different principle, as shown in the figure. Initially, a drum surface is charged electrostatically in a fine grid pattern. A laser selectively scans the grid, discharging any grid that it strikes. Toner is applied to the drum and adheres wherever the charge remains, and paper is then fed through using a roller or similar system. Toner adheres to the paper and is set in place using heat. Photocopiers work on a similar principle but project an image rather than scan it with a laser. Toner particles can often be seen on paper; an example is found on the previous page.

Analysis of documents created using printers relies on the same principles of class and individual characteristics. Rollers and drums may contain scratches or imperfections that can individualize them. Ribbons, inks, and toners can also be analyzed and compared in much the same way inks from pens and other writing instruments are. Thus, while computer technology has complicated the work of the document examiner, it has not made it impossible.

Further Reading

Ellen, D. *The Scientific Examination of Documents, Methods and Techniques.* 2nd edition. London: Taylor and Francis, 1997.

Hilton, O. *Scientific Examination of Questioned Documents,* revised ed. Boca Raton, Fla.: CRC Press, 1993.

Koppenhaver, K. M. *Attorney's Guide to Document Examination.* Westport, Conn.: Quorum Books, 2002.

Norwitch, F. H., and H. Seiden. "Questioned Documents." In *Forensic Science: An Introduction to Scientific and Investigative Techniques.* 2nd edition. Edited by S. H. James and J. J. Nordby. Boca Raton, Fla.: CRC Press, 2005.

R

radioimmunoassay (**RIA**) *See* IMMUNOASSAY.

radiology, forensic (**radiography, forensic**) The application of medical and dental radiology (X-ray techniques) to forensic work. Forensic radiology is considered a branch of forensic medicine that is related to but separate from forensic PATHOLOGY. Although the most frequently used tool in the field is the familiar X-ray film, all radiological techniques are available and are seeing increasing use in forensic cases. These techniques include ultrasound; magnetic resonance imaging (MRI, based on NUCLEAR MAGNETIC RESONANCE [NMR]); fluorescent imaging; CT scans (computated axial tomography, or CAT scans); and nuclear imaging techniques in which radioactive materials are used. Positron emission tomography, or PET scans, are an example of this kind of technique. When many such imaging techniques are used, the procedure is sometimes called "virtual autopsy."

The first X-rays of a human being were taken in 1895, and word quickly spread in the medical and scientific community. The first forensic applications of X-rays occurred within that same time frame. Forensic radiology is used as part of death investigations, in autopsies, and for the identification of unknown remains. Radiological identification requires that antemortem (before death) images be available for comparison. Aside from the frequent use of dental X-rays, other features such as healed fractures, shape of the frontal sinuses, and other bony structures can be used comparatively. Radiology can also help locate projectiles such as bullets, particularly in bodies that are decomposed or burned. In such cases, there may be no remaining evidence of soft tissue injury or trauma that would be characteristic of a gunshot wound. Forensic radiology may also be applied in other criminal and civil cases such as child abuse and malpractice.

See also ANTHROPOLOGY, FORENSIC; DENTISTRY, FORENSIC.

Further Reading

Brogdon, B. D. *Forensic Radiology.* Boca Raton, Fla.: CRC Press, 1998.

Raman spectroscopy A technique related to INFRARED SPECTROMETRY (IR) that is being used in forensic science principally associated with MICROSCOPY and applied to DRUG ANALYSIS and TRACE EVIDENCE. The Raman effect was discovered by Dr. Sir C. V. Raman (1888–1970), who was awarded the Nobel Prize in 1930 for that work. Unlike traditional infrared spectroscopy, which relies on measuring how much infrared radiation is absorbed by a given sample or molecule, Raman spectroscopy is based on how much light is scattered. Since this scattering behavior is specific to a given molecule, Raman spectroscopy (like IR spectroscopy) produces a unique spectrum for each compound. Thus, Raman spectroscopy can be used to identify specific molecules or for the qualitative analysis of complex mixtures.

The first figure on the next page illustrates how Raman activity arises. Bonds between atoms can be thought of as springs connecting two atoms. The

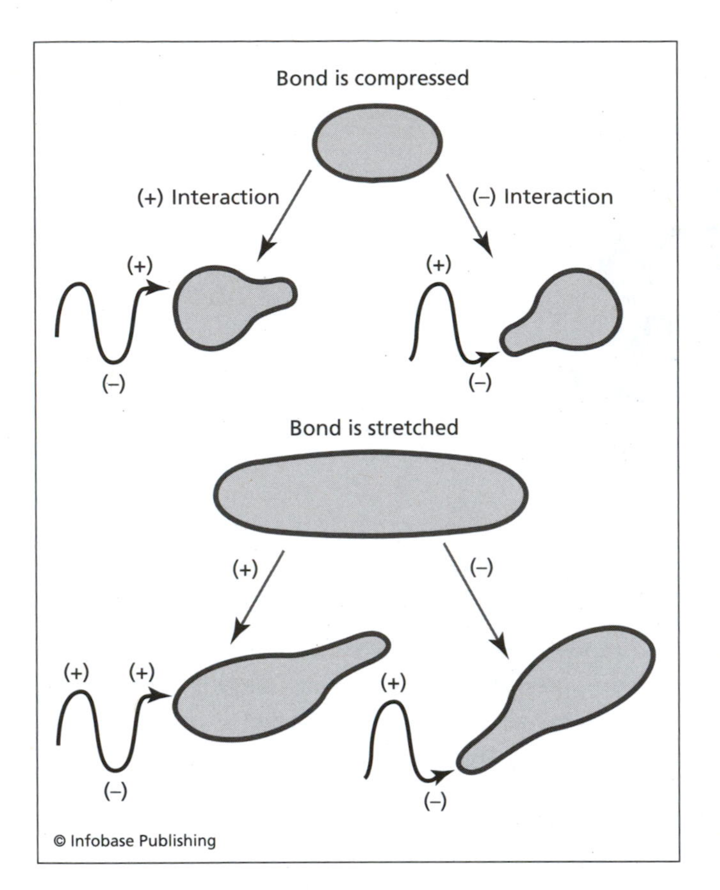

Changing polarizability of chemical bonds and interaction with electromagnetic energy

spring moves, sometimes bringing the two nuclei closer together (compression) and sometimes spreading them farther apart (stretching). Infrared spectroscopy is based on absorbance of infrared radiation, the energy of which is used to cause this bond stretching or other similar movements. For a compound to be infrared active, it must have a dipole moment, which in simple terms means an imbalance or unevenness in the electron cloud that surrounds the atoms of a molecule. However, for Raman activity, these bonds must be polarizable by incoming electromagnetic energy. Furthermore, the degree of polarizability must change as the bonds stretch and contract. Interestingly, this leads to a "mutual exclusion" property—a bond that is IR active will be Raman inactive, and vice versa.

A simplified pictorial representation of polarizability is shown in the second figure. Carbon dioxide (CO_2) is IR inactive since the bonds stretch symmetrically, resulting in no net stretch in the molecule. Around the three nuclei that constitute the molecule is a cloud of electrons that make up the actual chemical bonds holding it together. In the figure, the cloud is shown as a collection of negative (-) charges. In this case, that cloud can be polarized by incoming ELECTROMAGNETIC RADIATION (EMR), which consists of oscillating electrical and magnetic fields. Since like charges repel each other and unlike charges attract, the electron cloud responds to the passage of the (+) and (-) portions of waves. When the wave is (+), electrons move toward it; when it oscillates to the (-) portion, the electrons move away. Additionally, the degree of polarization changes as the electron cloud compresses and expands. It is this change in polarizability that makes the bonds, and in this example the molecule, Raman active.

In modern instruments, a laser light source is used to induce Raman activity as illustrated in the third figure. For Raman spectroscopy, the detector is placed out of line with the laser so that it picks up only scattered light. Three types are encountered. The first and most intense is called Rayleigh scattering, which will be seen at the same wavelength as the original incident radiation. Changes in polarization of the bonds (here using carbon tetrachloride molecule [CCl_4] as an example) create different energy levels resulting in "Stokes lines," which occur at longer

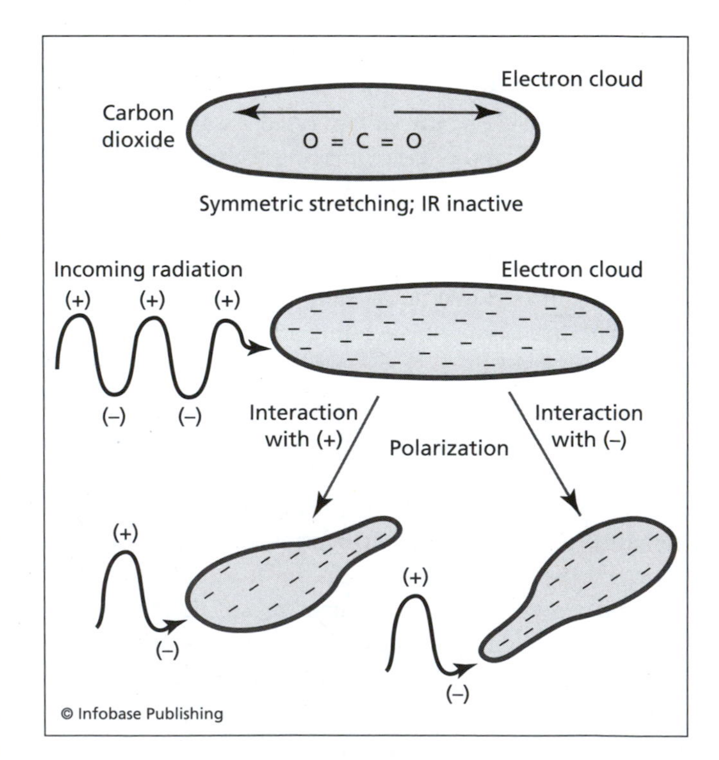

Polarization of a molecule

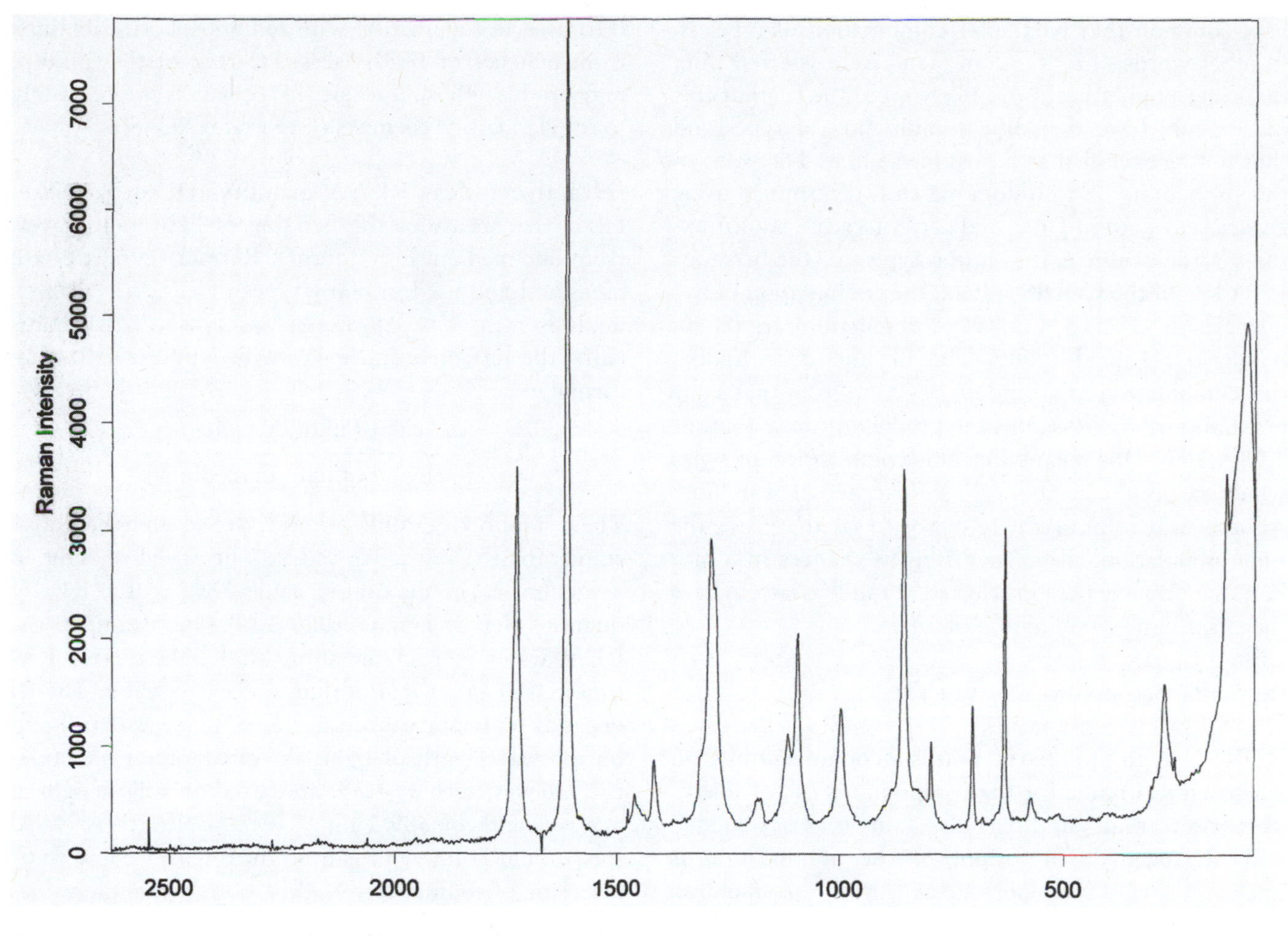

A Raman spectrum of polyester fiber ***(Suzanne Bell)***

wavelengths (lower energy) than the incident light. The anti-Stokes lines represent scattering at shorter wavelengths (higher energy) than incident. The Stokes lines are always more intense than the anti-Stokes.

One of the first forensic applications of Raman spectroscopy was for explosives and drugs, both pure samples and mixtures. The Raman spectrum of a mixture, although rarely useful for identifying a single compound, can provide what is commonly (and generically) referred to as a kind of "chemical fingerprint." Thus, if two different drug samples are seized, their Raman spectra would be useful in determining if they have a similar chemical composition and thus may have been part of the same batch. Recent advances in instrumentation have produced Raman microscopes that have been used for FIBER analysis as well. Given that Raman is usually nondestructive and can be applied to very small quantities of material, its use in forensic science is likely to increase.

Further Reading

Imelda, P. K., G. W. White, and P. M. Fredricks. "Characterization of Fibers by Raman Microprobe Spectroscopy." *Journal of Forensic Sciences* 43, no. 1 (1998): 82.

Ingle, J. D., and S. R. Crouch. "Molecular Scattering Methods." In *Spectrochemical Analysis*. Englewood Cliffs, N.J.: Prentice Hall, 1988.

Nobel e-Museum. "The Nobel Prize in Physics 1930." Available online. URL: http://www.el.sics/reates/0/. Downloaded May 5, 2007.

Sands, H. S., et al. "UV-Excited Resonance Raman Spectroscopy of Narcotics and Explosives." *Journal of Forensic Sciences* 43, no. 3 (1998): 509.

random match (random man not excluded, RMNE) A measurement most often used in DNA TYPING (but applicable to other types of evidence) that states how likely a given combination of frequencies is to occur in a population. For example, if a bloodstain is typed for several different DNA loci, it is important to know how common that particular combination of types is. In other words, if a "random man" were selected from the same population, how likely would it be that this man would have the same combination of types and thus not be excluded as a possible source? The value of this probability of a random match is determined using POPULATION GENETICS and POPULATION FREQUENCIES. As an example, if a stain is typed for the first four DNA loci of the CODIS system, the combination of frequencies (in the U.S. Caucasian population) are 0.080, 0.068, 0.041, and 0.080. The likelihood of finding this combination in a random selection from that same population is expressed as the product of these four, or 1.78×10^{-5}. This means that this combination of types would be found in about 18 people out of a million. Another way to phrase it would be to say that using the same population, there would be 18 chances in a million that any one person selected at random would have the exact same combination of types.

rape kit *See* SEXUAL ASSAULT KIT.

RDX A high EXPLOSIVE that is a component of C-4. The chemical name for the compound is cyclotrimethylenetrinitramine, and it is also known as "cyclonite" and "hexogen." The origin of the abbreviation is unclear, variously reported as "*R*oyal *D*emolition *E*xplosive" or "*R*esearch *D*epartment *E*xplosive." RDX is a secondary high explosive, meaning that it is shock insensitive and must be detonated by a primary high explosive. It is the principal component (about 90 percent) of C-4, a military explosive that is moldable and is sometimes referred to generically as "plastic explosive." The forensic analysis of RDX and other explosive residues includes the use of presumptive tests such at the GRIESS test and DIPHENYLAMINE, THIN LAYER CHROMATOGRAPHY (TLC), microscopic examination using POLARIZING LIGHT MICROSCOPY (PLM), INFRARED SPECTROSCOPY (IR), and HIGH-PRESSURE LIQUID CHROMATOGRAPHY (HPLC).

red cook method One of two currently favored chemical routes used to make METHAMPHETAMINE in clandestine drug laboratories (*see* CLANDESTINE LABS). The other method is called the NAZI METHOD. The red cook procedure is reasonably quick producing high yields and can be done via a "one-pot" method. Heating is sometimes used, but is not absolutely essential. The red cook method involves generation of hydroiodic acid (HI) used in conjunction with red phosphorus obtained from matches or road flares. Heating of the mixture may produce phosphine gas (PH_3), which can be deadly to the clandestine chemists or to first responders.

refractive index (RI) A quantity (PHYSICAL PROPERTY) that measures the bending of light as it travels from one medium into another. Refractive indexes are measured and used in many types of TRACE EVIDENCE analysis such as GLASS, FIBERS, and soils. Mathematically, the refractive index is defined by the following formula:

$$RI = \frac{\text{velocity of light in a vacuum}}{\text{velocity of light in medium}} = \gamma$$

The symbol γ (gamma in the Greek alphabet) also stands for the refractive index. The speed of light in a vacuum is the maximum achievable, so the RI is a quantity that is greater than 1.00. For example, the RI of water is 1.33, meaning that light travels 1.33 times faster in a vacuum than it does in water. The RI depends on temperature (20°C (68°F) is standard) and on the wavelength of light; the cited values of refractive indexes, such as 1.33, are based on yellow light at a wavelength of 589.3 nm. This wavelength is called the sodium D line and is a bright intense yellow characteristic of sodium streetlights. Other example values of refractive indexes include air at 1.00029 and quartz (fused) at 1.46.

As shown in the following figure, the change in velocity of light going from one medium to another (here air to water) results in a bending of the light rays at an angle symbolized by θ (theta in the Greek alphabet). This bending in water causes objects to appear in a different location from where they really are. In a familiar experiment, inserting a pencil into a glass of water makes it appear to bend as a result of this refraction of light. Because the RI varies with wavelength, visible light (white light) can be broken up into its constituent colors of red (R), orange (O), yellow (Y), green (G), blue (B), indigo (I), and violet (V) using a prism. The light is refracted for the first time at initial contact with the glass and again when emerging from the glass back into air. The slight differences in RI of the separate wavelengths are sufficient to break the light into a rainbow. As a result, prisms were used in older SPECTROPHOTOMETERS as a means to isolate wavelengths.

The dependence of RI on temperature is exploited in forensic MICROSCOPY using hot stage techniques. To

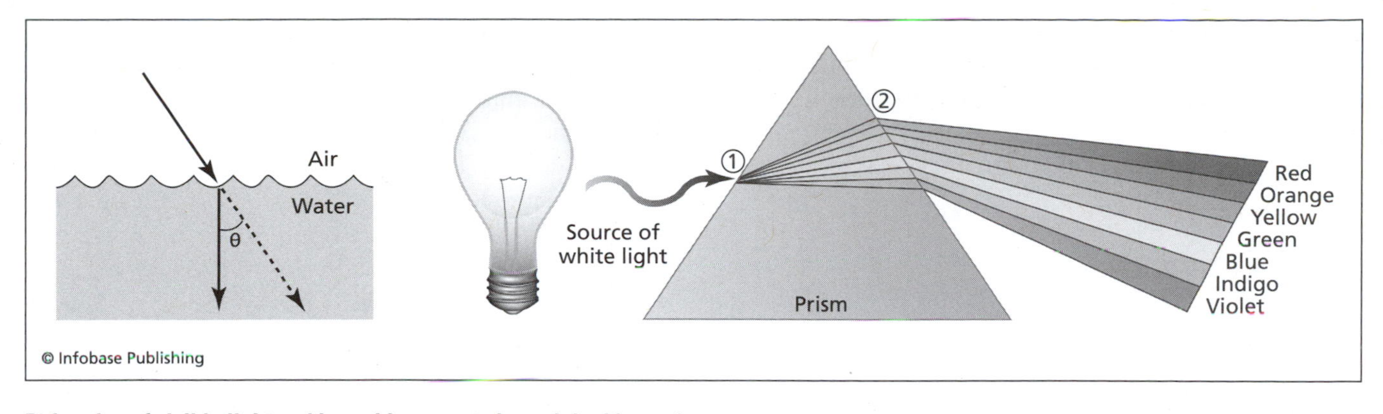

Refraction of visible light and how this property is exploited in a prism

determine the RI of a glass sample, it can be mounted in oil on a HOT STAGE and viewed under the microscope. As the oil is heated, the refractive index changes, and when it matches that of the glass sample the piece of glass will appear to vanish. Refractive index is also used as a detector for HIGH-PRESSURE LIQUID CHROMATOGRAPHY (HPLC). Finally, differences in refractive indexes within fibers (BIREFRINGENCE) are invaluable in forensic fiber analysis.

See also ANISOTROPY; GEOLOGY, FORENSIC.

Reinsch, Egar Hugo (1809–1884) German *Chemist* Egar Hugo Reinsch introduced a simple procedure in 1841 used to detect heavy metals used as poisons. The procedure requires only simple reagents and a copper wire and was much easier to perform than the MARSH TEST. Like many of the early tests, the Reinsch test was fairly sensitive but not foolproof. Despite the concerns, variations of the test were widely used in forensic TOXICOLOGY well into the 20th century and still are in some cases. Other variations of these tests, more improvements than innovations, evolved throughout the 1800s. One of the most famous post–Marsh test ARSENIC poisonings was that attributed to MARY ANN COTTON (1832–73). Her victims, at least 20, included most of her close relatives, including her mother, several children and stepchildren, three spouses, and one lover. Her motive was insurance money and her downfall excess. She killed her children (whose deaths were initially attributed to gastric distress) when their existence interfered with impending relationships and marriages necessary for her to obtain the money. The last child she killed was her son Charles. The attending doctor was suspicious and obtained samples from the boy's body. Using the Reinsch test, he found arsenic, leading to Mary's arrest. The trial itself had to await the birth of yet another child who owed his life to fortunate timing, a timely conviction, and the gallows. The jury rejected Mary's suggestion that arsenic vapors given off from wallpaper killed her son.

representative sampling The process of obtaining samples for testing that will accurately reflect the composition of the bulk material in question. This is an issue any time evidence is collected and subjected to forensic analysis. For example, if a drug arrest leads to the seizure of 20 bricks of a green plantlike material, the forensic chemist must obtain representative samples of the seizure, perhaps by selecting five samples from each brick with each subsample obtained from a different area of the brick. If DNA TYPING is to be undertaken on a bloodstained garment, the analyst must decide what part or parts of the stain to test. The inevitable tradeoff is that more subsamples allow for more complete (not necessarily better) characterization of a sample. However, each subsample requires an additional laboratory analysis. Thus, the need for representative sampling must be balanced with the need to preserve evidence for any later analysis and the need to keep analysis time to a reasonable level. Statistical techniques can be used to guide in selection of representative samples.

Further Reading

Frank, R. S., G. W. Hinkley, and C. G. Hoffman. "Representative Sampling of Drug Seizures in Multiple Containers." *Journal of Forensic Sciences* 36, no. 2 (1991): 350.

revolver A type of handgun or pistol in which the cartridges are housed in a rotating cylinder. Revolvers can hold from five to seven cartridges and include calibers from .22 to .45. The forensic examination of revolvers follows the same general protocols as those for any rifled weapon.

See also AMMUNITION; BULLETS; CALIBER; CARTRIDGES; FIREARMS.

RFLP *See* DNA TYPING.

Rh factors (Rh blood group) A blood group system that, like the ABO BLOOD GROUP SYSTEM, is based on antigens located on the surface of the red blood cell. Rh type is best known for potential effects on babies. If a mother who is Rh^+ gives birth to a child that is Rh^-, her body will produce antibodies to the Rh^- factors. If she conceives another child that is also Rh^-, severe complications are possible. The Rh system actually consists of seven main types plus other rarer types, which would seem to be ideal for individualizing. However, Rh was rarely used in forensic applications primarily due to difficulties in typing it in stains. The system was useful in PATERNITY testing, but as in forensic cases, advances in DNA TYPING have eliminated the infrequent use of Rh typing.

Further Reading

Melvin, J. R., et al. "Paternity Testing." In *Forensic Science Handbook*. Vol. 2 Edited by R. Saferstein. Englewood Cliffs, N.J.: Regents/Prentice Hall, 1988.

Rhohypnol *See* DATE RAPE DRUGS.

ridge characteristics (ridge detail) Detailed features of the ridges found on the hands and feet, and most particularly for forensic purposes, on the fingers. There are several different characteristics that can be used to individualize a FINGERPRINT. Collectively, these details are referred to as MINUTIAE, and they are the key to fingerprint analysis. Once the fingerprint analyst has been provided with prints to compare, he or she selects a set of ridge characteristics to examine on both prints. Their characteristics, number, direction, location, and location relative to each other are all part of this comparison. This process is referred to as a point comparison, after which the analyst will determine that the compared prints match or do not match. Currently, there is no universal standard on the minimum number of comparison points required for a match to be pronounced.

rifle A high-powered FIREARM designed to be fired from the shoulder with two hands. Rifles have long barrels and impart high muzzle velocity to the bullets fired. Types of rifles include single-shot, lever-action (seen in western movies), bolt action, semiautomatic, and automatic. Representative calibers of rifles include 30-06, 45-70, and .223. Assault rifles are a type of automatic rifle that includes the famous M-16, AK-47, and Uzi weapons developed by the U.S., Soviet, and Israeli militaries, respectively. These rifles can be fired in fully automatic or semiautomatic modes. The forensic examination of rifle evidence follows the same general protocols as those for pistols and other rifled weapons.

See also AMMUNITION; BULLETS; CALIBER; CARTRIDGES.

rifling *See* LANDS AND GROOVES.

rigor mortis The stiffening of a body that occurs shortly after death and that can be used to estimate the POSTMORTEM INTERVAL (PMI). Rigor begins to set in two to six hours after death, starting in the small muscles of the jaw and progressing through the trunk and out to the arms and legs. The stiffness remains for two to three days and then releases in the reverse order. The rate of stiffening is temperature dependent, with colder temperatures accelerating it and warmer temperatures slowing it. The rate can also be influenced by physical activity before death. If a person ran or engaged in strenuous activity right before death, the rate of stiffening increases. Rigor mortis is not the same thing as cadaveric spasm, which can lock a joint in the position it was in at the moment of death. If a person commits suicide by a gunshot wound to the head, a tight grip on the weapon can remain after death, but this is not the same thing as rigor mortis.

Rita, Hurricane *See* IDENTIFICATION OF THE DEAD.

rope and cordage This type of evidence may be found in cases in which a victim was restrained or in cases such as hanging and strangulation when a cord is used as the LIGATURE. If a KNOT is involved, such as in a hanging, proper crime scene collection procedure is to not untie the knot but rather to cut the rope somewhere else to preserve it. Class characteristics such as type of material, number of strands, twisting, and colors can be determined, while the constituent FIBERS can be analyzed using the normal protocols for fiber analysis. Any evidence clinging to or embedded in the cordage such as hairs or blood can also be collected and studied.

See also ASPHYXIA.

Ruhemann's purple *See* FINGERPRINTS; NINHYDRIN.

rules of evidence (federal) *See* FEDERAL RULES OF EVIDENCE.

Ruybal test *See* COBALT THIOCYANATE.

saccharides *See* SUGARS.

Sacco and Vanzetti case *See* FIREARMS.

saliva A common form of BODY FLUID evidence that has become much more useful as DNA TYPING techniques have improved. Saliva can also be used in forensic TOXICOLOGY, as it is considered to be a filtrate of the blood, similar to URINE. Three pairs of salivary glands produce saliva.

As physical evidence, saliva is usually encountered as a stain that may be difficult to see. There are obvious areas where saliva can be expected, such as on moisture-activated adhesives on stamps and envelopes, around BITE MARKS, on cups and glasses used for drinking, and on or in gum and food. However, dried saliva, unlike stains of BLOOD and seminal fluid, is usually colorless and not easily visualized. The most common PRESUMPTIVE TEST for saliva involves the detection of the enzyme amylase, which catalyzes the breakdown (specifically the hydrolysis) of STARCH in food to glucose. Amylase is found in other body fluids, but the concentration in saliva is generally the highest. However, this means that the intensity of the response must be considered.

One form of amylase test, radial diffusion, is shown in the figure on the next page. The test is conducted in a shallow circular dish, such as a Petri dish, that has been filled with agarose gel. Agarose is a starch product and as such will be degraded by amylase. To evaluate a stain extract, a hole is punched in the gel and the extract added. The sample is incubated and kept moist while the sample diffuses into the gel. If amylase is present, it will degrade the starch and turn the gel from a cloudy appearance to clear. The more amylase present, the larger the area of degraded starch. To better visualize the extent of the amylase activity, an iodine stain is placed on the gel. Where starch remains, a deep blue color will be seen; areas of starch degradation (amylase activity) will remain clear. For the tentative identification of stains on materials such as clothing, the test can be modified. For example, to detect a stain on a shirt, a wetted filter paper can be pressed against it and then the paper tested for amylase activity. The pattern on the paper will show the outlines of the saliva stain, if present.

Prior to DNA typing using amplification techniques, comprehensive typing of saliva stains was not possible. If the donor of the stain was a SECRETOR, then the saliva contained ABO BLOOD GROUP SYSTEM factors that could be typed. However, this alone cannot individualize, although it could be useful for exclusion purposes. With current DNA procedures, saliva, which is rich in epithelial cells (buccal cells), can be typed and requires less sample than if the material were blood. This is because DNA typing requires white blood cells, which are found in lower concentrations in blood than are buccal cells in saliva.

Finally, as a medium for toxicological analysis, saliva can be valuable for the detection of drugs ranging from caffeine to cocaine to ethanol (ALCOHOL). The collection of saliva is a simpler procedure than is

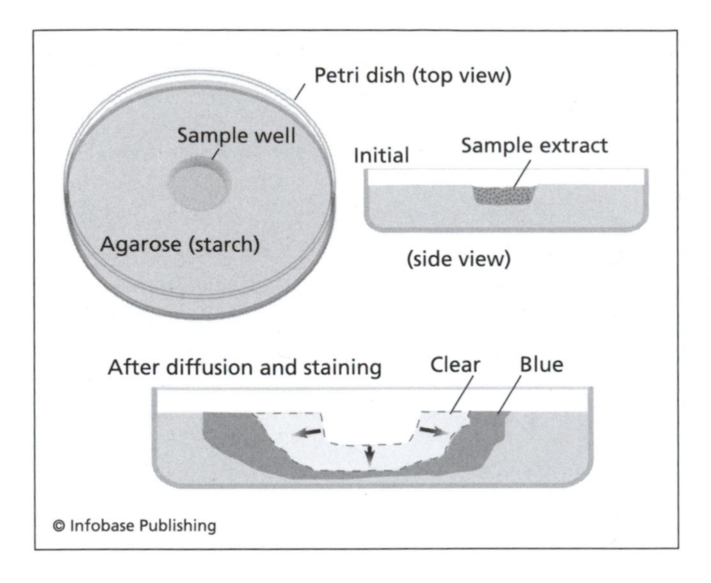

Method of using an amylase presumptive test for saliva. A photograph of an actual test is shown in the color insert.

that for blood, which involves an invasive technique. Saliva also has advantages over urine for drug testing since the collection of saliva can be observed directly and does not involve the privacy issues of observing the collection of urine.

Further Reading

Greenfield, A., and M. M. Sloan. "Identification and Biological Fluids and Stains." In *Forensic Science: An Introduction to Scientific and Investigative Techniques*. 2nd edition. Edited by S. H. James and J. J. Nordby. Boca Raton, Fla.: CRC Press, 2005.

saltpeter *See* GUNPOWDER.

sampling *See* REPRESENTATIVE SAMPLING.

scanning electron microscope (SEM, SEM-EDX, SEM-EDS, SEM-XRD) A form of MICROSCOPY that uses beams of electrons instead of beams of light and is able to achieve very high magnification, up to 1,000,000X (magnification by a factor of a million). SEM also has high depth of field, meaning that a large portion or depth of the image remains in focus no matter how high the magnification. This contrasts with traditional light microscopy, with which the higher the magnification, the shallower the depth of focus. SEM and the associated X-RAY TECHNIQUES (electron dispersive spectroscopy [also called energy dispersive] or X-ray diffraction [XRD]) are becoming increasingly important in forensic microscopy and have been used for GUNSHOT RESIDUE (GSR), BUILDING MATERIALS, PAINT, DUST, and other types of TRACE EVIDENCE.

A simple schematic of an SEM system is shown in the figure. Inside a vacuum chamber, an electron gun supplies a tightly focused beam of incident electrons that interact with the sample. Although many types of interactions result, it is the emission of back-scattered and secondary electrons that is used to create an image. On the sample's surface, elements with higher atomic numbers (refer to the periodic table, Appendix III) will scatter more incident electrons and appear brighter than elements of lower atomic numbers. This difference will be discernible in the final image. Secondary electrons, which are actually emitted from the sample rather than scattered, are used to obtain information about surface features (topography). To generate an image, the electron beam is moved over the surface, scanning it as the back-scattered and secondary electrons are collected. The image is a display that shows the relative intensity of the electrons collected at a given location. Older systems used cathode ray tubes (CRTs, similar to older televisions and to CRT computer monitors) to display the image, while newer systems typically incorporate some type of digital imaging. Since the signal is only related to electron

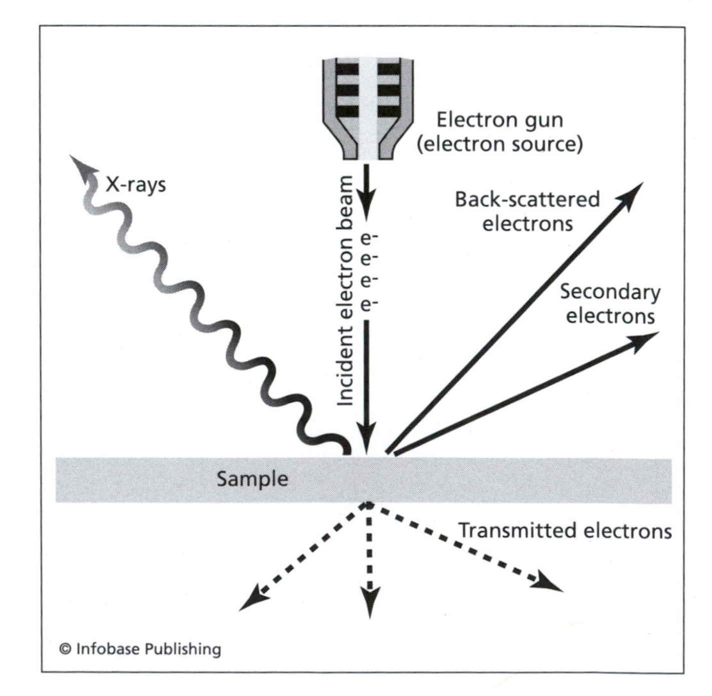

Simplified process of electron microscopy

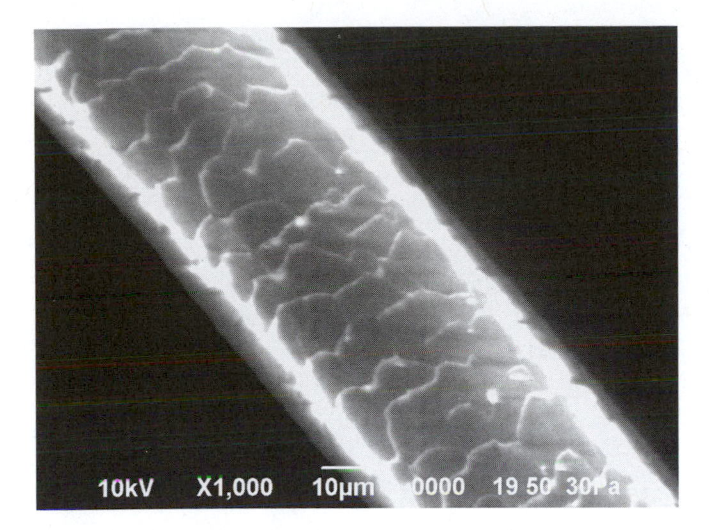

An SEM image of human hair ***(Courtesy of Heidi Barron, Bennett Department of Chemistry, West Virginia University)***

detection and not to the detection of light, as in traditional microscopy, the image is not colored. However, coloring (called false coloring) can be added to improve the visualization.

Another advantage of SEM is its ability to determine the elemental composition (ELEMENTAL ANALYSIS) of the sample. When the incident electron beam interacts with atoms on the surface, inner shell electrons are ejected, and other electrons "fall" into those shells to fill the gaps. As a result, electromagnetic energy in the X-ray range of the electromagnetic spectrum is released. The wavelength and energy of the emitted electrons are characteristic of the elements from which they came, and this relationship allows for the identification of the elements. Energy dispersive spectroscopy (EDS) determines the element by measuring the energy of the emitted X-ray, while wavelength dispersive spectroscopy (WDS) does so by measuring its wavelength. Most SEM instruments use the EDS system, while electron microprobe instruments often incorporate the WDS detection system. These devices use electrons for elemental analysis but not for imaging, as in SEM. An SEM system can also be used for X-ray diffraction (XRD) to identify crystalline compounds.

Further Reading

Schwoeble, A. J., and D. L. Exline. "Methods for the Examination of Gunshot Residue." In *Current Methods in Forensic Gunshot Residue Analysis.* Boca Raton, Fla.: CRC Press, 2000.

science In general, science refers to a method of study of the natural world and the universe using observation, experimentation, and experience. Science attacks problems and seeks understanding using experimentation and observation based on measurable criteria. Temperature, pressure, and volume are measurable quantities, whereas criteria such as "worthiness" are subjective and not measurable. Other qualities that distinguish science from other areas as well as from PSEUDOSCIENCE include the use of HYPOTHESIS AND THE SCIENTIFIC METHOD and the principle of FALSIFIABILITY, which requires that any scientific finding or theory be stated specifically enough that it can be tested, replicated, and, if possible, proven false. For example, the statement "gas expands as pressure decreases" is specific enough that experiments can be designed to disprove (falsify) it. The results must be consistent and reproducible as well. If one experiment finds that gases do not expand when pressure decreases, the theory would have to be changed to reflect the new knowledge. Furthermore, a statement such as "the alignment of the stars when you were born determines what kind of day you will have today" cannot be falsified, tested, or measured. In forensic science, the results of analyses are stated such that they can be tested and verified. For example, the statement "the stain was found to be human blood of type AB" can be tested by other experts using another portion of the stain.

Further Reading

Lee, J. A. *The Scientific Endeavor: A Primer on Scientific Practices and Principles.* San Francisco: Addison Wesley Longman, 2000.

scientific method *See* HYPOTHESIS AND THE SCIENTIFIC METHOD.

scientific working groups (SWG) Groups sponsored by the FBI LABORATORY or related federal agencies that consist of FBI scientists as well as those from academia, forensic laboratories, national laboratories, and other affiliations. Members can be from the United States or other nations. Example working groups include Scientific Working Group for the Analysis of Seized Drugs, under the DRUG ENFORCEMENT AGENCY (SWGDRUG), Scientific Working Group for Firearms and Toolmarks (SWGGUN), Scientific Working Group for Forensic Document Examination (SWGDOC), Scientific Working Group for Materials Analysis (SWGMAT), Scientific Working Group on Bloodstain Pattern Analysis

(SWGSTAIN), Scientific Working Group on DNA Analysis Methods (SWGDAM), and Scientific Working Group on Friction Ridge Analysis, Study and Technology (SWGFAST).

Typically, an SWG meets yearly to discuss issues related to their focus area, and they develop and publish standards and guidelines to assist forensic practioners working in that area. The publications make recommendations and assist laboratories and forensic scientists to develop and implement the best current practices in their forensic discipline.

Further Reading

Federal Bureau of Investigation Laboratory Division. "FBI Laboratory 2005 Report" (FBI Publication # 0357). Available online. URL: http://www.fbi.gov/hq/lab/labannual05.pdf. Downloaded April 14, 2007.

scopolamine A BARBITURATE that can be used as a sedative or as a poison. Scopolamine was originally obtained from extraction from plants in the Solanaceae family, with a chemical synthesis described in 1956. The synthetic version is sometimes referred to as "Atroscine." It has the same molecular formula ($C_{17}H_{21}NO_4$) as COCAINE, but shares none of its physiological properties. Scopolamine is listed on Schedule II of the CONTROLLED SUBSTANCES ACT (CSA). It has also been used during the course of "narcoanalysis," known informally as the use of "truth serums," an aspect of forensic DECEPTION ANALYSIS (FDA). Courts do not generally accept testimony that arises from the use of truth serums, including scopolamine.

The forensic analysis of scopolamine begins with the use of PRESUMPTIVE TESTS, principally the DILLIE-KOPPANYI test and occasionally CRYSTAL TESTS. This is followed by confirmatory analyses including THIN LAYER CHROMATOGRAPHY (TLC), GAS CHROMATOGRAPHY/MASS SPECTROMETRY (GC/MS), gas chromatography with a NITROGEN PHOSPHORUS DETECTOR (NPD), UV/VIS SPECTROPHOTOMETRY, HIGH PERFORMANCE LIQUID CHROMATOGRAPHY (HPLC), and INFRARED SPECTROSCOPY (IR). In toxicology, IMMUNOASSAYS are also employed.

Scotland Yard A term that is used to describe the Metropolitan Police, the city police department of London. The main office is located in a complex called New Scotland Yard.

Scott test *See* COBALT THIOCYANATE.

scrapings Samples taken from under the fingernails of victims in murders, assaults, and sexual assaults, among other crimes. Fingernail scrapings are part of many SEXUAL ASSAULT KITS and are routinely collected at AUTOPSY when foul play might have occurred. Scrapings can be collected from suspects as well. During a struggle, skin, hair, and other kinds of TRANSFER EVIDENCE can accumulate under the nails of the victim or the suspect. This evidence can be analyzed and/or subjected to DNA TYPING (if appropriate). If a rape victim's skin is found under the fingernails of a suspect, or the suspect's skin is found under the fingernails of the victim (as confirmed by DNA ANALYSIS), there is no question that the two had contact with each other.

screening tests *See* PRESUMPTIVE TESTS.

sculpture, forensic *See* ART, FORENSIC; FACIAL RECONSTRUCTION.

search patterns *See* CRIME SCENES.

secondary crime scene *See* CRIME SCENES.

secondary transfer When TRACE EVIDENCE is transferred from one place (the source) to another (the recipient), that evidence may be transferred again. This can complicate the interpretation of TRANSFER EVIDENCE such as HAIRS, FIBERS, DUST, and SOIL, which by their nature move easily. For example, if a burglar breaks into a home in which a cat lives, he or she will likely get cat hair on his or her clothing, since cat hair clings to fabric. The burglar may then leave the residence and sit in a car, depositing a portion of the cat hair in the car. This is a secondary transfer, which can lead to subsequent transfer. If, at some later time, a different person uses the car and gets the hair on him, a tertiary transfer has occurred. Consequently, transfer and trace evidence found on a person or a body reflect his or her most recent environment. The more time that passes after contact, the more trace evidence is likely to be lost.

See also PRIMARY TRANSFER (DIRECT TRANSFER); WAYNE WILLIAMS CASE.

Secret Service (United States Secret Service, USSS) A branch of the DEPARTMENT OF HOMELAND SECURITY that is responsible for protection of federal officials and for the investigation of counterfeiting crimes. The Secret

Service was created in 1865 to battle the counterfeiting rampant after the Civil War. After the assassination of President William McKinley in 1901, the duties expanded to include fraud and presidential protection. In 1908, a group of eight Secret Service agents were transferred to the Justice Department, the first step leading to formation of the FBI. The protective functions of the Secret Service expanded to include the White House, vice president, visiting dignitaries, and diplomats, as well as government facilities such as the White House.

The Secret Service has a significant forensic capability emphasizing, but not limited to, support of counterfeiting investigations. The Forensic Services Division includes QUESTIONED DOCUMENTS, FINGERPRINTS, photographic and related services, POLYGRAPH, and a general chemical instrumentation section. The division maintains a reference collection of INKS numbering more than 7,000 as well as related standards for toners, copiers, and other devices relevant to questioned document investigations. The Secret Service is home to the FISH system, a large and growing reference collection of handwriting samples similar to reference collections maintained by the FBI and other federal agencies.

Further Reading

United States Secret Service. "History." Available online. URL: http://www.secretservice.gov/history.shtml. Downloaded on May 5, 2007.

United States Secret Service. "Forensic Services Division." Available online. URL: http://www.secretservice.gov/history.shtml. Downloaded on May 5, 2007.

secretors The roughly 80 percent of the population that secretes antigens from the ABO BLOOD GROUP SYSTEM into their BODY FLUIDS. Prior to the ascendance of DNA TYPING, secretor status was an important consideration in forensic SEROLOGY, particularly in sexual assault cases. As was discovered in the 1930s, a person's secretor status is under genetic control and is an inherited characteristic. Depending on the population studied, approximately 75 percent to 85 percent are secretors. In a secretor, body fluids that have the ABO substances are SEMEN, SALIVA, vaginal fluid, and, to a lesser extent, URINE, SWEAT, and gastric juices. Using tests such as ABSORPTION-INHIBITION and ABSORPTION-ELUTION, it is possible to type these fluids just as blood is typed. However, the difficulty arises when potential contamination and mixing is considered. For example, in sexual assault cases, stains or other samples will likely contain fluids from both the suspect and the victim. If both people are secretors (which is the most likely scenario), then the ABO type obtained will be attributable to both. Interpretation can be further complicated when the secretor status of either one or both contributors is unknown, a common occurrence if a suspect was unidentified. However, once DNA typing replaced ABO in routine casework, the issue of secretor status was no longer relevant.

Further Reading

De Forest, P. R., R. E. Gaensslen, and H. C. Lee. "Body Fluids." In *Forensic Science: An Introduction to Criminalistics.* New York: McGraw Hill, 1983.

Greenfield, A., and M. M. Sloan. "Identification and Biological Fluids and Stains." In *Forensic Science: An Introduction to Scientific and Investigative Techniques.* 2nd edition. Edited by S. H. James and J. J. Nordby. Boca Raton, Fla.: CRC Press, 2005.

selected ion monitoring (SIM) A technique used in MASS SPECTROMETRY (MS) to increase the sensitivity of the analysis for a selected compound or small set of compounds. In the most common form of mass spectrometer operation, the mass detector scans through a large range of atomic masses, collecting fragment ions at a given mass for only a fraction of a second before moving to the next mass. The wider the mass range scanned, the less time spent collecting ions of each individual mass. When SIM is employed, the detector collects data at only a few masses or a single mass. As a result, the detector collects ion fragments at the targeted masses for longer times, increasing the sensitivity and allowing for detection of smaller quantities. The inevitable trade-off is that since only a few fragments are collected, identification of unknown compounds, which require a large range of mass fragments, is more difficult. In forensic DRUG ANALYSIS, SIM methods have been used to detect the drug LSD, normally found in very small quantities on tiny blotter papers. It has also been used to detect THC, the active ingredient of MARIJUANA, in urine. Other compounds targeted include MORPHINE and PHENCYCLIDINE (PCP).

semen (**seminal fluid**) A body fluid central to SEXUAL ASSAULT cases that until recently was difficult or impossible to INDIVIDUALIZE since it is often found diluted and/or mixed with vaginal secretions. Semen is most often found on items collected as part of a SEXUAL ASSAULT KIT such as swabs of the vagina, mouth, and anus or as a stain found on clothing or sheets. A single ejaculation typically consists of a few milliliters (2 to 6) of fluid that, like BLOOD, has a serum component (seminal plasma) and a cellular component (sperm cells, or spermatozoa). The number of cells found in a milliliter of the ejaculate is variable but is usually in the range of 100,000,000. The cellular component composes about 10 percent of the volume of the ejaculate. Semen is a thick milky liquid that dries as a crusty, somewhat shiny material that acquires a slight yellowish tinge as it ages. Because it is white, it can be difficult to detect visually, although careful tactile examination (gloved to prevent contamination) may reveal crusty patches.

Examination of stains progresses from location of a stain to PRESUMPTIVE TESTS, followed by confirmatory testing and, if comparison standards are available, specialized extraction techniques followed by DNA TYPING. Techniques used to locate a stain include use of a UV lamp, since components of semen fluoresce. However, many other materials, including many fabrics, also fluoresce, limiting the utility of this technique. A widely used procedure to both locate stains and provide tentative identification as seminal fluid is a presumptive color test for seminal acid phosphatase (SAP). Along with choline, SAP is a major component of seminal fluid, and its presence is indicative of seminal fluid, but not conclusive. Like all presumptive tests, the test for SAP is subject to FALSE POSITIVES and FALSE NEGATIVES. In addition, other secretions such as vaginal fluid contain acid phosphatase, although VAP can be distinguished from SAP under certain analytical conditions. Regardless, the SAP test is useful for screening large items and for detecting stains in areas where they are difficult to see.

To perform the SAP test, a substrate containing phosphate such as alpha (α) napthyl phosphate is combined with a dye such as Fast Blue B. When a stain containing SAP is treated with these reagents, a purplish color results. Tests can be performed directly on portions of stains and swabs, on extracts, or by rubbing a wetted filter paper on a suspected stained area. The reagents are then applied to the filter paper, not the stain itself. This method is particularly valuable for screening large items such as bedding. A few CRYSTAL TESTS have been and still are used occasionally to identify components of seminal fluid. The Florence test utilizes an iodine solution to produce crystals with choline, while the Barberio tests reacts with spermine.

To confirm the identification of semen, the simplest test is microscopic identification of intact sperm cells using PHASE CONTRAST or dark field MICROSCOPY. Staining and standard light microscopy may also be used. A common staining protocol is called the Christmas Tree stain for the red and green color combination it creates. Sperm is often difficult to extract from materials, and thus agitation is required. However, there are many situations in which sperm cells will be absent or present in such low concentrations as to be undetectable. Some men, either due to diseases or a vasectomy, will not have sperm in their ejaculate, a condition called aspermia. Oligospermia is a low sperm count that can also result in nondetection of sperm cells. In addition, sperm do not remain indefinitely where deposited; lifetime in the mouth is only a few hours, as is that of motile (active) sperm in the vagina. Nonmotile sperm can be detected longer after intercourse, but if the victim is dead, the decomposition process will complicate detection.

The P30 or PSA test avoids the problems of low or no sperm in the ejaculate by confirming the presence of seminal fluid. Prostate specific antigen is found only in seminal fluid, although it will be absent if the man's prostate has been removed. PSA is detected using IMMUNOLOGICAL methods, most commonly crossed-over electrophoresis, or IMMUNOASSAY. Finding of PSA is considered to be conclusive for seminal fluid even in the absence of sperm cells.

Prior to the development of DNA techniques, typing of seminal fluid was dependent on SECRETOR status of the parties involved and targeted, ABO BLOOD GROUP SYSTEM types, and a few ISOENZYMES such as PGM. However, DNA techniques, coupled to selective extraction procedures that rely on the hardiness of sperm cells relative to other cells, are generally able to obtain DNA types from seminal fluid as well as mixed samples. Although with mixed samples the separation and isolation is rarely 100 percent, interfering materials are generally found at a low concentration, simplifying interpretation of results.

Further Reading

Greenfield, A., and M. S. Sloan. "Identification of Biological Fluids and Stains." In *Forensic Science: An Introduction to Scientific and Investigative Techniques*. 2nd edition. Edited by S. H. James and J. J. Nordby. Boca Raton, Fla.: CRC Press, 2005.

Shaler, R. C. "Modern Forensic Biology." In *Forensic Science Handbook*. Vol. 1. 2d ed. Edited by R. Saferstein. Upper Saddle River, N.J.: Prentice Hall, 2002.

semiautomatic firearms *See* FIREARMS.

seminal acid phosphatase (SAP) *See* SEMEN.

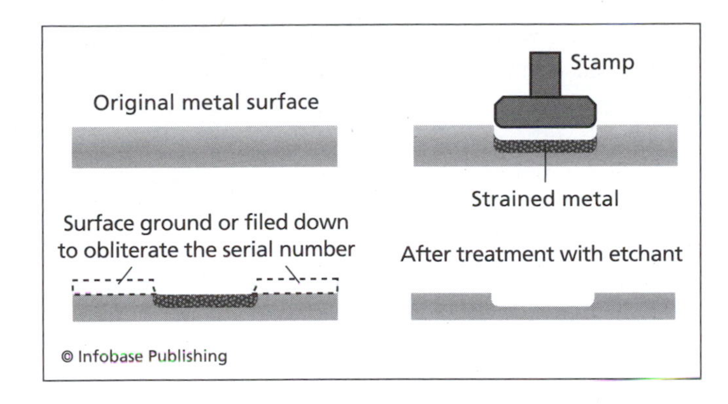

Process of recovering an obliterated serial number that was stamped into a metal surface

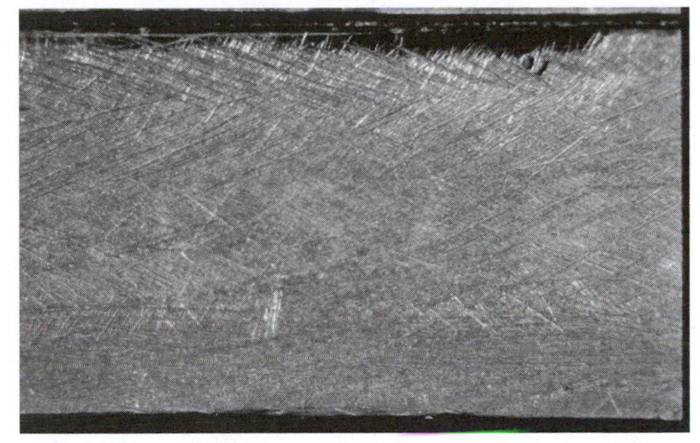

A metal surface from which a serial number has been removed *(Courtesy of William M. Schneck, Microvision—Forensic Consulting, Inc.)*

serial number restoration The process of revealing serial numbers that have been filed, scraped, polished, or otherwise obliterated. Although most common in FIREARMS cases when serial numbers on weapons have been removed, any metal object with a stamped serial number such as engine blocks, tools, and equipment can be treated in a similar manner. As shown in the figure, when a serial number is stamped into a metal object, the force of compression is transferred to the metal below the indentation, causing imperceptible damage and strain. The depth of this strain can be several times the depth of the original impression. This damage to the metal makes it more susceptible to attack by oxidizing agents such as strong acids.

When someone removes a serial number by filing or grinding, typically he or she stops when the number is no longer visible, leaving a portion of the surface in which the weakened metal still exists. When the surface is treated with an etching material (also called a chemical etchant), the strained metal will dissolve faster than the undamaged metal. If successful, such treatment will allow visualization of the original number. Similar techniques can be used when a serial number is altered by "overstamping" or other methods of chemical and physical obliteration.

The type of reagent used to restore the number will depend on the type of metal substrate. For example, aluminum may be treated with sodium hydroxide (NaOH), while steel calls for a mixture of hydrochloric acid (HCl) and copper (II) chloride ($CuCl_2$). Other metals may require reagents and acids such as chromic, sulfuric, or nitric. Regardless of the reagent used, the first step is always photographic documentation of the surface followed by cleaning and polishing. The reagents may be applied with a cotton swab, cotton ball, or by building a "dam" of clay around the area of interest and pooling the reagent on it. Careful monitoring is essential since the strained metal region may be very thin and may dissolve away quickly. A variation of the manual application of reagents is an electrolytic method in which a battery is attached to the metal (positive terminal), while the negative terminal is attached to the swab or cloth soaked in the reagent. The result is faster, requiring even closer scrutiny to avoid overprocessing.

For ferrous metals (those containing iron), another approach can be used that has the advantage of being nondestructive. The strained metal beneath the stamp will contain tiny cracks and imperfections that will

The same surface after treatment showing the restored serial number *(Courtesy of William M. Schneck, Microvision—Forensic Consulting, Inc.)*

lead to unevenness in the magnetic field produced when the metal is placed between the poles of a magnet. Tiny flakes of iron (iron filings) can be lightly dusted on the surface or applied using a thin oily suspension. If successful, the filings will congregate over the stamped area and reveal the original number. X-rays, heating, cooling, and ultrasonic methods have also been used.

Further Reading

Rowe, W. F. "Firearm and Toolmark Examinations." In *Forensic Science: An Introduction to Scientific and Investigative Techniques*. 2nd edition. Edited by S. H. James and J. J. Nordby. Boca Raton, Fla.: CRC Press, 2005.

Thorton, J. I., and P. J. Cashman. "The Mechanism of the Restoration of Obliterated Serial Numbers by Acid Etching." *Journal of the Forensic Science Society* 16 (1976): 69.

serology, forensic A term that is being replaced by *forensic* BIOLOGY. Serology is a subdivision of IMMUNOLOGY, the science that focuses on the reactions between antigens and antibodies such as are found in the ABO BLOOD GROUP SYSTEM. The word *serology* arises from the serum portion of the BLOOD, in which the antibodies are found. As DNA TYPING has all but replaced traditional forensic serology, the more inclusive term *forensic biology* is being used. In addition to DNA work, forensic biology includes areas such as forensic ENTOMOLOGY and ECOLOGY.

serum proteins GENETIC MARKER SYSTEMS in the serum portion of blood that were occasionally used in efforts to individualize bloodstains and some types of body fluid evidence. Haptoglobin (Hp) is one example of a serum protein that has more than one form (POLYMORPHIC) and that can be typed using simple gel ELECTROPHORESIS techniques. Others that have been used in forensic serology include transferring (Tf) and group-specific component (Gc). With the advent of DNA TYPING techniques, serum protein typing is no longer used.

sex determination (gender determination, sexing) The analysis of BLOOD, BLOODSTAINS, BODY FLUIDS, or skeletal remains to determine the sex of the person the evidence originated from. Although current DNA TYPING methods make this procedure relatively easy for blood and body fluids, this was not always the case. Sex determinations were possible under some circumstances using Barr bodies, but the results were not always reliable, and few labs did the test routinely. Other methods were developed that targeted the hormones unique to males and females. However, it was not until the late 1980s and early 1990s that genetic approaches were developed and tested. Currently, the sex of a donor is determined as part of DNA typing using STR systems. The amelogenin gene that is located on the X chromosome codes for tooth pulp and is different in males (XY) and females (XX). The female amelogenin gene is shorter than the male version and thus can be separated based on this size difference. Analyzed using automated sequencers (CAPILLARY ELECTROPHORESIS, CE), a male sample will display two peaks on the output, while a female sample will show one.

The sexing of skeletal remains is the responsibility of forensic anthropologists and relies on the morphology and size of bones. Of most use in determining sex are structures of the pelvic bones and the cranium (skull). Often, remains have been scattered or scavenged, and all that remains are fragments and pieces. In such case, analysis of the remaining portions can be performed and compared to large collections to provide the best possible sexing of the remains. The program FORDISC, created by the department of anthropology at the University of Tennessee, Knoxville, is widely used for this purpose.

Further Reading

Shaler, R. C. "Modern Forensic Biology." In *Forensic Science Handbook*. Vol. 1. 2d ed. Edited by R. Saferstein. Upper Saddle River, N.J.: Prentice Hall, 2002.

Inman, K., and N. Rudin. *An Introduction to Forensic DNA Typing*. 2nd edition. Boca Raton, Fla.: CRC Press, 2005.

sexual asphyxia (autoerotic death) An accidental death that is a form of STRANGULATION. Usually the victim is a male who uses a noose or other type of ligature to decrease blood flow to the brain to increase sexual pleasure. Clues are often evident at the death scene. The victim is often nude, partially clothed, or dressed in women's clothing, and pornography may be close by. The ligature is usually soft, such as a towel. Evidence of earlier episodes, such as bruises or concealment of earlier injuries, may be found.

sexual assault A crime that typically produces a great deal of physical evidence. Often, only the victim and the perpetrator are present when the crime occurs, so this evidence is critical. Sexual assault includes rape, both heterosexual and homosexual, child rape and abuse, and elder abuse. Sexual assault can also be part of other crimes such as burglary and murder. The key points of evidence collection are at the scene and also at the hospital or other medical facility where the victim is examined. This collection can also occur during AUTOPSY. In either case, sexual assault evidence collected by medical personnel centers on the SEXUAL ASSAULT KIT, which consists of blood, body fluids (principally semen and saliva), and many other items. TRACE EVIDENCE can be significant as well since an assault of any kind lends itself to rich transfer of fibers, broken fingernails, and forcibly removed hair, for example. With the increasing use of DATE RAPE DRUGS, TOXICOLOGY is playing an increasingly important role in sexual assault cases.

sexual assault kit One of several commercially available kits used to collect evidence in sexual assault cases. In rapes, physical evidence is crucial since there are rarely any witnesses and trials often are reduced to one person's testimony against another's. Thus, the evidence collected from the victim using the sexual assault kit is crucial in determining the truth. Generally, these kits consist of a whole blood sample, swabs of any dried secretions, and swabs of the vagina, genital area, thighs, anus, and mouth. In addition, smears on slides are made from these swabs, and all must be air-dried. A vaginal rinse may also be collected. Any visible HAIRS and FIBERS are collected, and the victim may be asked to undress over a large sheet of paper so that any other TRACE EVIDENCE that falls off in the process can be collected and preserved. All clothing worn also becomes part of the physical evidence.

Hair samples are plucked from the scalp and pubic area as CONTROL SAMPLES, while pubic combings are used to collect hairs from that area. In some cases, successive collections are taken starting with removal of loose hairs and ending with a vigorous combing of the pubic area, which should contain hair from the victim only. Fingernail scrapings and clippings may be collected, as may blood for TOXICOLOGY analysis, particularly if DATE RAPE drugs may have been involved. Sexual assault kits are most frequently collected from female victims, but male victims are not unheard of; similar procedures apply, with particular attention paid to oral and anal swabs.

See also NURSING, FORENSIC.

Sheppard, Sam A physician convicted of murdering his wife in 1953. The trial in 1954, like the LINDBERGH KIDNAPPING case before and the O. J. SIMPSON case to follow, was called the "trial of the century." However, it was to become a series of three trials, one involving the U.S. Supreme Court. Because the crime was so horrific and the couple was young, prominent, and attractive, the case got a tremendous amount of media attention and became a cottage industry for books, conjecture, and gossip. Questions immediately arose about the original conviction, and the continuing controversy was the inspiration for the 1960s television series *The Fugitive* and the 1993 movie with the same name.

The case had many interesting forensic aspects, and because of its length, advances in forensic science from the late 1950s to the present day, particularly in DNA TYPING, played a continuing role in the Sheppard saga. Everything from crime scene analysis to BITE MARKS, TOOLMARKS, BLOODSTAINS, PROFILING, BLOODSTAIN PATTERNS, forensic ANTHROPOLOGY, and forensic ODONTOLOGY has been applied to the evidence at one time or another, with a variety of findings, many contested or controversial. Unfortunately, as technology improved, the evidence deteriorated as memories dimmed and principals died. Thus, it is doubtful that the entire truth will ever be known.

Sam Sheppard was an osteopathic physician living in a suburb of Cleveland with his wife, Marilyn, who was 31 and pregnant on the night of the murder, July 4, 1954. Sheppard, then 30, was involved in an extramarital affair, although that information would come to light only later. According to Sheppard, he was asleep on the couch when he awoke to what he thought were his wife's screams. He raced upstairs and found Marilyn on the bed, severely beaten, with blood soaked into the bedclothes and spattered all around. He claimed to have wrestled with a "bushy-haired man" (as opposed to the "one-armed man" of *Fugitive* fame) and was knocked briefly unconscious. He awoke, checked his wife's pulse, and then went in pursuit of the intruder. He caught the man on the beach, scuffled again, and again was knocked unconscious. He awoke later and called a neighbor, who discovered the body and notified police. While all of this transpired, the Sheppards' young son remained asleep in the room next to his parent's bedroom, where the murder occurred. The way in which the body was found suggested sexual assault, but hints of burglary and drug-related robbery were also discovered. A small green bag was found later on the beach that contained

Sheppard's watch and other items. The watch was spattered with blood. The fact that it was spattered and not smeared supported the idea that whoever was wearing the watch (presumably Sheppard) was wearing it in the room while blood was literally flying.

Sheppard's defense was hurt most by inconsistencies and unexplainable elements in his story, such as why he did not call police immediately and what occurred during the two hours between the final scuffle and his calling for help. He also could not account for a T-shirt he was wearing during the scuffles. Initially, forensic evidence was limited principally to the results of the autopsy and testimony of the coroner, all of which would become important later. Sheppard was convicted December 21, 1954, and sentenced to life imprisonment. However, in 1966 the U.S. Supreme Court ordered Sheppard released on the basis that his trial was hopelessly prejudiced by publicity and errors. A second trial began in October 1966, and in the 12 years that had passed, additional forensic analysis was conducted. PAUL KIRK, an eminent forensic scientist in the 1950s, had reviewed the evidence, principally blood spatter, and concluded that Sheppard was likely innocent. However, in the days before DNA TYPING, the amount and type of information that could be extracted from bloodstains was limited to ABO BLOOD GROUP SYSTEM type and stain patterns. The second trial was also notable for the presence of F. Lee Bailey as the defense lawyer. The case was instrumental in making Bailey famous and ironically set the stage for his appearance in the O. J. Simpson trial decades later. Sheppard was found not guilty in November 1966. His health quickly deteriorated due to drinking, and he died in 1970.

Around 1990, Sam Reese Sheppard, son of Sam Sheppard, became active in trying to have his father's name cleared completely by having him declared innocent. This grew into yet another court case in which forensic science played an even larger role. Anthropologists analyzed the exhumed remains of Marilyn Sheppard and detailed the trauma she suffered to the face. DNA analysis of some of the evidence suggested the possibility of a third party present, but the age of the materials and storage techniques not geared for DNA evidence lent question to the validity of the results. A forensic odontologist debunked a theory put forth by Kirk that Marilyn had bitten her attacker. A former FBI profiler analyzed the case in detail and determined that the crime scene had been staged and that Sheppard was the killer. The third trial, a civil case, began in January 2000 and returned a decision of "not innocent," an unusual verdict that allowed legal action to continue.

shoe prints (shoe impressions, footprints, footwear impressions) A type of impression evidence commonly encountered in indoor as well as outdoor CRIME SCENES. Footwear impressions can be two dimensional (such as a shoe print seen on a dusty floor) or three dimensional, such as a shoe print in mud or snow. The former are treated much the same as latent FINGERPRINTS, while the latter are treated similarly to TIRE IMPRESSIONS. As with all types of impression evidence, good photographic documentation is essential and requires the use of oblique (angled) lighting to show all features. Having the original impression for laboratory analysis is preferable to photographs or casts. Thus, whenever possible, the impression itself, along with the underlying substrate, is removed from the scene and transported to the lab. When that is not possible, two-dimensional impressions can be recovered using chemical development (much as with latent fingerprints) followed by lifting using adhesive materials or electrostatic lifting devices. Three-dimensional footwear patterns are preserved using CASTING methods, which can be adapted to work in mud and snow and even underwater. Some types of shoe prints, such as those found in carpet, are much harder to preserve, meaning that analysis and comparisons will rely principally on photographic and other crime scene documentation.

Shoe prints are handled much as any type of physical evidence, progressing from determination of CLASS CHARACTERISTICS (physical measurements, tread patterns, manufacturer, size, and model) to comparison with known samples and ultimately, if possible, to INDIVIDUALIZATION. Footwear is mass-produced, and the majority of it is imported. However, despite mass production, shoes can acquire unique markings from flaws during manufacturing and from individual WEAR PATTERNS. For example, the soles of shoes can be made by pouring liquid plastic material into a mold or by cutting large sheets of tread material to fit the shoe body. When liquids are poured into a mold, air bubbles can form that will be unique to that particular sole even though the mold is used hundreds or thousands of times. Similarly, when a sole is cut, the cutting can impart unique markings. Once two pair of nearly identical shoes are purchased, each person will gener-

A shoe print on paper; see also the color insert ***(Courtesy of Michael Bell, West Virginia University, Forensic and Investigative Sciences program)***

ate unique WEAR PATTERNS based on their weight, gait, and how they use and wear the shoes. Thus, it is possible to link a shoeprint to a specific shoe if the impressions, both from the scene and collected from a suspect's shoe, are of adequate quality. Comparison of footwear impressions is similar to latent print examination in that individual points are compared to determine if the impressions match. As in the case of fingerprints, often the impressions left at a crime scene are partials or are damaged, complicating comparisons. Similarly, the time that elapses between the creation of the crime scene impression and recovery of the shoe is critical since wear patterns continue to develop each time the shoe is worn. To assist analysts, the FBI maintains a database of footwear.

Further Reading

Bodziak, W. J. *Footwear Impression Evidence.* New York: Elsevier, 1990.

Geradts, Z. "Use of Computers in Forensic Science." In *Forensic Science: An Introduction to Scientific and Investigative Techniques.* 2nd edition. Edited by S. H. James and J. J. Nordby. Boca Raton, Fla.: CRC Press, 2005.

shooting distance *See* DISTANCE DETERMINATIONS.

short tandem repeats (STR) *See* DNA TYPING.

shotguns *See* FIREARMS.

signature (signature analysis) *See* PSYCHIATRY, FORENSIC, PSYCHOLOGY, FORENSIC, AND PROFILING.

signatures *See* QUESTIONED DOCUMENTS.

significant figures In reading data from an instrument, the number of digits that are certain plus one. For example, when reading a typical needle-style bathroom scale, the dial is marked with lines at each pound. If weight falls partway between the line for 130 pounds and 131 pounds, the weight would be read as 130.5 pounds. The first three digits are certain since the reading is clearly greater than 130 but less than 131, but since the dial is not calibrated for anything smaller than a pound, the 0.5 is an estimate and thus an uncertain digit. The value 130.5 has four significant digits. Significant figure considerations are crucial when reporting data obtained from instrumentation.

Rules for how to handle significant digits in calculations are important and determine the number of digits that can be reliably reported. For example, if a car travels 335.5 miles on 11.6 gallons of gas, a calculator will report a mileage of 28.922413793 miles per gallon. However, this is an absurd number of digits when one considers that the gas tank is calibrated to a tenth of a gallon (3 significant figures) and the odometer recording the mileage to tenths of a mile (4 significant digits). Under this scenario, the correct answer would be 28.9 miles per gallon, a number that is rounded to the same number of significant figures as in the value with the fewest number of significant figures. Rules for other combined operations become more complex, but fundamentally, significant figure considerations arise from limitations in instruments or other types of measuring devices. In forensic science, any time a QUANTITATIVE ANALYSIS is undertaken, the issue of significant figures arises. For example, if a state has set a blood alcohol limit of 0.08 percent, how significant figures are handled and how the results are reported are very important.

silencers FIREARM accessories used to reduce the noise associated with firing the weapon. Silencers can be commercially produced or homemade and can be applied to pistols and rifles. The use of silencers is important from a forensic perspective because they can affect the way bullets are marked and the distribution and patterns of GUNSHOT RESIDUE (GSR). Silencers can also have an effect on wound stippling and general appearance.

When ammunition is fired in a weapon, two processes can contribute to the sound produced, one applicable to all types of weapons. As PROPELLANT ignites and produces a pressure wave of hot expanding gas that propels the bullet down the barrel. When this wave of high pressure reaches the atmosphere, a loud sound, called the report, is always heard. The sound is also referred to as the muzzle blast. The second sound is a sonic boom that results when the velocity of the projectile exceeds the speed of sound, 1,100 feet per second (roughly 770 mph at room temperature). The small sonic boom actually sounds like a sharp crack.

To alleviate the cracking sound, all that needs to be done is reduce the muzzle velocity of the bullet to below 1,100 ft/sec. Most often this is accomplished by reducing the propellant load in the cartridge. However, this is not always feasible. Rifles, which rely on small projectiles traveling at high velocity, are not always amenable to this kind of modification. For pistols, this modification is easily implemented. To silence the muzzle blast, alterations to the barrel are required. The key to successful silencing of this sound is to allow the hot, pressurized gas to expand and cool before venting from the weapon. Silencers that screw onto the end of the barrel (the ones most often seen in movies and on television) consist of baffles and absorbent materials to accomplish this. If the bullet comes in contact with the material in the silencer as it travels through it, characteristic marks can be transferred to it. Similarly, holes can be drilled in the barrel, and an enclosure containing baffles and/or absorbents may be used.

Simon test/Simon reagent A color-based PRESUMPTIVE TEST used in drug analysis. The Simon test is one of the more intriguing color tests that combines elements of DYES and transition metal complexes used in PIGMENTS and other colorants (*see* COLOR AND COLORANTS). It is a variant of the sodium nitroprusside test that has been used in college-level organic qualitative analysis courses for decades; the typical formulation used in forensic testing is a nitroprusside solution containing acetaldehyde (4) with a second reagent consisting of 2 percent Na_2CO_3. This two-step test is used to differentiate METHAMPHETAMINE from AMPHETAMINE, which both give a similar orange color when treated with the Marquis reagent (*see* MARQUIS TEST).

Simpson, O. J. A former football star and celebrity acquitted of the brutal murder on the night of June 12, 1994, of his ex-wife, Nicole Brown Simpson, and Ronald Goldman. The case was noteworthy, even notorious, for many reasons including the forensic aspects, which focused around CRIME SCENE documentation, CHAIN OF CUSTODY, and DNA TYPING results. Along with the LINDBERGH KIDNAPPING trial in 1932 and the SAM SHEPPARD murder trial of 1954, the Simpson trial was hailed as the "trial of the century." As with its predecessors, the outcome has generated a cottage industry of books, commentary, and speculation. In all of these trials, scientific evidence played a crucial role.

Bloodstains and bloodstained evidence were recovered on the walkway of Nicole Brown's home (the scene of the murder), in Simpson's white Bronco vehicle, in his home, and in areas leading to his home. Three different laboratories employed DNA analysis using both RFLP and PCR techniques, and where duplicate samples were analyzed independently, results were the same. At the scene, a trail of blood drops was found on the left side of a series of bloody shoeprints, indicating that the perpetrator had been injured on his or her left side. DNA typing of this blood was consistent with O. J.'s type to 1 in 240,000 for the PCR results and 1 in 5,200 for the RFLP results. One spot was consistent to 1 in 170,000,000. Shortly after the crime, Simpson was wearing a bandage over a deep cut on a left knuckle which he attributed to two wounds, one the night of the murder and one the next day while in a hotel in Chicago. The bloodstains at the scene were collected on cotton fabric patches (called swatches) on June 13th, allowed to dry, and packaged on the 14th. However, this chronology was put in question when Dr. Henry Lee, a defense expert, reported findings that indicated that the swatches were packaged when wet, not when dry. It was seemingly a small inconsistency, but one that would prove pivotal. However, some of

Lee's findings and testimony were strongly contested and controversial.

A bloody left glove was found at the scene, with DNA blood types consistent with Simpson, Nicole Brown, and Goldman. The right hand glove was not discovered until later in the bushes outside of Simpson's home. A dark pair of socks was found in Simpson's master bedroom and was documented and collected June 13. Several weeks later, bloodstains were identified on the socks and were found to have a DNA type consistent with Nicole Brown. However, questions arose as to how the stains were deposited and why the discovery came so long after the socks were originally collected. One defense expert testified that he had detected the presence of EDTA, a preservative used in freshly drawn blood, in these stains; a prosecution expert testified that he found none. If indeed EDTA had been confirmed, this would have been critical. It suggested the possibility that the blood was not deposited during the crime but rather was planted at a later date using blood drawn from Simpson after the crime had been committed. In fact, the defense relied not on disputing the actual DNA results but rather on questioning the validity of the blood evidence from which it was obtained.

In Simpson's white Bronco, a number of bloodstains were identified and typed, including what appeared to be a bloody shoeprint on the floor. The stains typed consistently with various combinations of blood from Simpson, Brown, and Goldman. Finally, three bloodstains were belatedly collected from the rear gate of Nicole Brown's house, and, although officers present on the night of the murder testified that the stains had been there originally, the defense argued that these too had been planted. Their expert also claimed to have found EDTA in these stains, also disputed by the prosecution expert. The prosecution also presented a significant body of FIBER evidence supporting its case.

The defense never argued with the results of DNA analysis directly; rather, their case rested on the supposition that members of the Los Angeles Police Department with racial motivations framed Simpson. Their attacks centered on issues of potential contamination, deliberate alteration and planting of evidence, poor crime scene documentation, and poor evidence collection and analysis techniques. If any of these occurred, it was argued, than a reasonable doubt as to Simpson's guilt had been raised. After months of courtroom presentation and endless media attention and analysis, the jury took a short time to acquit Simpson on October 3, 1995.

Implications for forensic science were significant, and the community was divided over the conduct of scientists, particularly defense experts. There is no question that mistakes were made in the course of the investigation, but mistakes are inevitable in any complex case. The key question is the significance of such errors, or lack of it. The trial also pointed to an increasingly important issue of complexity. As the intricacy of analyses, particularly for DNA, increases, the procedures and results become harder to explain to a jury composed of citizens who likely will have little relevant background. The concern in the Simpson case is that the jury may have ignored the DNA evidence at least in part because it was too difficult to understand, or that it was not well presented. Another possibility is that its very complexity made it easy for the defense to attack and raise doubts about it. This issue will worsen as forensic science advances. Finally, the outcome of the trial was never dependent solely on the evidence but was influenced by many other factors including jury selection and socioeconomic and racial issues. Thus, whether one agrees or disagrees with the verdict, the trial raised questions that will reverberate well beyond the forensic science community.

Further Reading

Bugliosi, V. *Outrage: The Five Reasons Why O. J. Simpson Got Away with Murder.* New York: W.W. Norton and Company, 1996.

Lee, H. C., T. Palmbach, and M. T. Miller. "Case Study 3. Limited Reconstruction (Focus Point: Location of Physical Evidence)." In *Henry Lee's Crime Scene Handbook.* San Diego, Calif.: Academic Press, 2001.

Levy, H. "O. J. v. DNA, What the DNA Really Showed." In *Postmortem: The O. J. Simpson Case.* Edited by Jeffery Abramson. New York: HarperCollins, 1996.

Rudin, N., and K. Inman. "The Simpson Saga: The Blood in the Bronco." In *An Introduction to Forensic DNA Typing.* 2nd edition. Boca Raton, Fla.: CRC Press, Inc., 2005.

single nucleotide polymorphism (SNP) At a given location on a gene, variations of a single nucleotide across a population. Such loci are polymorphic (literally, many-formed), meaning that different people have different nucleotides at the same loci. The variation can arise from insertions of a single base (ATGC), deletions, or substitutions at the position of interest.

Forensic Science and Juries: How Complex Is Too Complex?

Cases such as the O. J. Simpson trial point out a critical issue facing the courts. As science and technology play ever-increasing roles in society, the nature of many cases, both criminal and civil, is becoming more complicated. This problem is not isolated to cases involving DNA TYPING, although this issue tends to get the most publicity. As cases become more technically complex, it becomes more difficult for judges to adequately perform their gatekeeper role as defined by the *DAUBERT* DECISION. Compounding this is the challenge of explaining scientific and technical information to a jury of people who, like judges, likely have little or no scientific backgrounds. Many lawyers are equally unprepared. If any of the participants become mired in scientific debates and conflicting expert opinions, the danger increases that they will ignore it. Consequently, verdicts may be rooted in emotional responses or other subjective criteria. This phenomenon is known as "jury nullification," and some observers argue this occurred in the O. J. Simpson case. Another example is the Microsoft antitrust case, in which defense lawyers have argued that the circumstances are too complex to be dealt with fairly by one judge. On the horizon are even thornier issues such as cloning, genetic causes of behaviors, genetic medicine, genetic alteration of fetuses, selection of fetuses based on genetic considerations, genetic privacy issues, and new techniques in biotechnology. If courts fail to adapt to scientific and technical complexity, the faith of the public in the system could be jeopardized.

One organization taking a proactive approach is the Advanced Science and Technology Adjudication Resource Center (www.einshac.org). Educational programs at EISHC focus on molecular biology, genetics, and biotechnology, and both scientists and judges carry on work at the institute. Their efforts, along with that of other forward thinking states such as Arizona, have led to a number of recommendations that would allow the courts to be responsive and reliable in the new world of DNA typing and biotechnology.

One suggestion is a redefinition of the court as a place where jury education is of prime importance. Robert Myers and his coauthors have listed several ideas for court reform, starting with the selection of educated professionals for inclusion on a jury. They note, "There does seem to be an unwritten rule of practice that professionals should be struck [during the jury selection process] when possible." However, as they point out, in complex cases, educated jurors benefit all sides. Another simple change would be to allow jurors to take notes and submit written questions for witnesses, practices that are rarely allowed. Generally, note taking is prohibited because of a fear that it will distract jurors from testimony, but the authors observe that this seems arbitrary and even foolish in complex cases. They point out, "it is difficult to imagine an academic setting in which taking notes and asking questions would not be permitted." If the court is to become a place where juror education is central, it has to become a type of academic environment.

Courts and judges could also be encouraged to appoint independent scientific experts to evaluate and judge conflicting expert testimony from prosecution and defense scientists. This independent review would help judges in their gatekeeper role and could also limit courtroom scenes in which opposing experts undergo intensive cross-examinations conducted by lawyers who lack the scientific background to ask reasonable and pertinent questions. Myers and his coauthors express the opinion that the scenarios of "dueling experts" does more to distract juries than to obtain the truth and that cross-examination is based primarily on an attempt to discredit the witness. When jurors are faced with such distractions heaped on top of scientific complexity, it is not surprising that they may give up and ignore vital evidence simply because it has been made so difficult to understand and interpret.

Although there is reason to be concerned about this situation, there is equal reason for confidence that it will be successfully addressed. The judicial system is designed to adapt, albeit slowly. However, deliberate decisions are always preferable when the stakes are so high. Also, if society has been able to embrace technological advancement, then the courts, which are an integral part of that society, likewise will be able to do so. As Myers points out, "jurors, rather than giving up in the face of voluminous evidence and conflicting expert opinions, take their fact-finding and decision-making responsibilities seriously. . . . Jurors . . . use their common sense, have individual and common experiences that inform their decision making, and form opinions as the trial proceeds."

Further Reading

Blakeslee, S. "Genetic Questions Are Sending Judges Back to Classroom." *New York Times,* July 9, 1996.

Brick, M. "Technology; When the Judge Cannot Really Judge." *New York Times,* September 11, 2000.

Human Genome Project Information. "Genetics in the Courtroom." Available online. URL: http://www.ornl.gov/sci/techresources/Human-Genome/courts.html. Downloaded March 3, 2008.

Myers, R. D., R. S. Reinstein, and G. M. Miller. "Complex Scientific Evidence and the Jury." *Judicature: Genes and Justice, The Growing Impact of the New Genetics on Courts.* American Judicial Society 3, no 3 (1999).

This simple and short base difference contrasts with more complex polymorphisms such as used in nuclear DNA TYPING and mitochondrial DNA typing. Current estimates are that more than 80 percent of the genetic variation between human beings arises from SNPs.

SNPs are being used forensically but are not likely to replace the current 13 loci nuclear DNA typing methods in the near term. Partly this is due to cost, availability of instrumentation, and the already established and extensive databases such as CODIS. Another drawback of SNP typing is that to individualize a sample, many more types are required than in traditional nuclear DNA typing, which currently uses 13 loci in the United States. For comparable discrimination power, more than 50 SNP loci would have to be typed. Even so, there are aspects of forensic identification where SNP methods are proving useful.

SNP methods are amenable to mitochondrial DNA, given that mtDNA methods rely on sequencing the base pairs. It is also finding use in mass fatality incidents where other forms of DNA may be highly degraded. SNP procedures targeting the Y-chromosone have been used to help delineate evolutionary relationships. An extension of this is use of Y-SNPs to link a sample to a possible population group, as long as the sample comes from a male (XY) rather than a female (XX).

See also "Identification of the Dead" essay; NEXT-OF-KIN SOFTWARE.

Further Reading

Amorim, A., and L. Pereira. "Pros and cons in the use of SNPs in forensic kinship investigation: a comparative analysis with STRs." *Forensic Science International* 150 (1) (2005):17–21.

Budowle, B. "SNP Typing Strategies." *Forensic Science International* 146S (2005):S139–S142.

Sobrino, Beatriz, Maria Brion, and Angel Carracedo. "SNPs in forensic genetics: a review on SNP typing methodologies." *Forensic Science International* 154 (2) (2005):181 (14 pages).

skeletal identification *See* AGE-AT-DEATH ESTIMATION; ANTHROPOLOGY, FORENSIC; STATURE ESTIMATION.

smokeless powder A low EXPLOSIVE that is used as the propellant in AMMUNITION and as a component of pipe bombs and other homemade explosive devices. Smokeless powder was a military innovation introduced in the late 1800s to replace BLACK POWDER, which created large amounts of smoke that obscured battlefields. Smokeless powders were found to increase the performance of firearms as well, leading to widespread acceptance and near universal use. As a propellant, black powder is now used principally in historical weapons owned by collectors and hobbyists. There are two kinds of smokeless powder, single base and double base. Single base powder consists of nitrocellulose (made by treating wood shavings or cotton with nitric and sulfuric acids, HNO_3 and H_2SO_4), diphenylamine, and other additives. The amount of nitrogen in the nitrocellulose is reported by weight; "guncotton" is 13.3 percent nitrogen by weight. To prepare the single base powder, an emulsion is made with solvents, which is dried and further treated before it is ground or otherwise powdered for use. Double base powder consists of nitrocellulose and nitroglycerin along with various additives. In addition to use as propellants, smokeless powders are encountered in the forensic context as a component in an INCENDIARY DEVICE or homemade bomb such as a pipe bomb.

Granules of smokeless powder that are produced are spherical, cylindrical, or disk-shaped and are treated with materials such as deterrents. The treatments and shaping are done to control the burn rate, which is important for controlling the performance of the ammunition. Different size weapons (calibers) and the necessary projectile velocities, along with other considerations, determine optimal burn rate. Fundamentally, the role of the propellant is to burn rapidly to produce large quantities of hot, expanding gas that propel the projectile down and out of the barrel of the weapon at optimal speed. Too fast a burn will stop producing gases too quickly, while too long a burn will continue burning even after the projectile has exited the barrel.

The forensic analysis of smokeless powder employs microscopic examination to look for residues and granules, presumptive tests such as the GRIESS and DIPHENYLAMINE, and other methods used for GUNSHOT RESIDUE (GSR).

Further Reading

Rowe, W. F. "Firearm and Toolmark Examination." In *Forensic Science: An Introduction to Scientific and Investigative Techniques*. 2nd edition. Edited by S. H. James and J. J. Nordby. Boca Raton, Fla.: CRC Press, 2005.

SNP *See* SINGLE NUCLEOTIDE POLYMORPHISM.

sodium rhodizonate test A PRESUMPTIVE TEST for lead found in GUNSHOT RESIDUE (GSR). A solution of the reagent (typically 2 percent) can be added to the extract of a swab or filter paper or be sprayed on clothing, and the development of a violet red color is indicative of lead. However, as with all presumptive tests, FALSE POSITIVES can occur, and supplemental analysis is required to confirm the presence of GSR. Spraying this reagent in combination with others onto cloth or clothing is useful for estimating distance from the shooter to the target based on the amount and dispersal of the lead present (DISTANCE DETERMINATIONS). A photo of this test is shown in the color insert.

soils *See* GEOLOGY, FORENSIC.

solid phase extraction (SPE) **and solid phase microextraction** (SPME) A group of related techniques used to extract compounds from water, other liquids, or gases. It has developed as an alternative to SOLVENT EXTRACTION for sample preparation. In forensic science, solid phase methods are used for sample preparation in ARSON cases, DRUG ANALYSIS, and TOXICOLOGY. A simple illustration of solid phase extraction is the use of charcoal filters to purify drinking water. The charcoal removes contaminants from the water by adsorbing them. This adsorption of compounds on a sorbent is similarly exploited in solid phase extraction, in which the sample, usually water, is passed over the sorbent to remove the compounds of interest. The sorbent is then rinsed to remove unwanted materials to the extent possible. The compounds of interest are then desorbed (eluted) off the sorbent using heat or a solvent. This solvent extract is then concentrated down to a small volume that is ready for analysis, most often using GAS CHROMATOGRAPHY (GC) or HIGH-PERFORMANCE LIQUID CHROMATOGRAPHY (HPLC). The process cleans the sample and preconcentrates it since a very large volume of sample can be passed over the sorbent. Even if this sample is very dilute, a large volume will still deposit a significant amount of the compound(s) of interest on the sorbent.

SPE employing charcoal has been used for many years, but more sophisticated methods appeared in the 1970s and accelerated into the 1980s. The driving force for these advancements was the analysis of water in environmental applications. A number of different sorbent resins were developed, broadening the applications significantly. The majority of these are based on a silica substrate to which long chain organic molecules have been bound. Different resins are used depending on the application. SPE can be performed using prepacked cartridges or disks through which the sample is poured or drawn through by vacuum. In forensic toxicology, this approach can be used to separate acidic drugs from basic/neutral drugs, a common task when screening blood or body fluids for a large number of drugs, poisons, and metabolites.

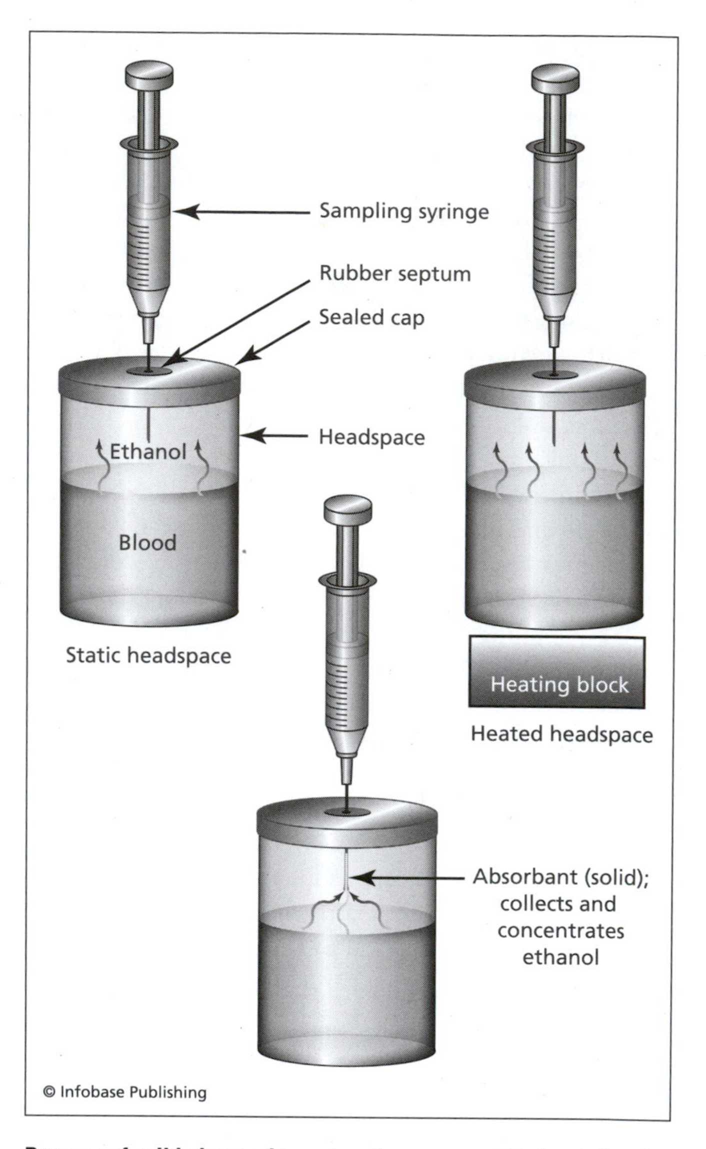

Process of solid phase microextraction compared to heated and static headspace techniques. Sampling of alcohol in a blood sample is used as the example application.

A recent innovation is miniaturization by placing sorbents into capillary tubes or coating them onto thin plates. This is called solid phase microextraction (SPME) and is now commonly used for the analysis of ARSON evidence via a HEADSPACE technique. To determine that arson has occurred, presence of an ACCE-

LERANT must be shown. The most common of these are liquids such as gasoline that can remain even after the fire has burned and been extinguished. Debris is collected and sealed in a can or other specialized container such that any residual vapors will be trapped in the headspace above it. A plate coated with charcoal or a capillary coated with resin can be lowered into the headspace and sealed such that the container remains airtight. Gentle heating drives vapors from the debris into the headspace, where the compounds of interest concentrate on the sorbent. The sorbent is then removed and in the case of the capillary design can be inserted directly into the injection port of a gas chromatograph.

Further Reading

Fritz, J. S. *Analytical Solid-Phase Extraction.* New York: Wiley-VCH, 1999.

solubility testing A technique occasionally used in the forensic analysis of PAINT, FIBERS, PLASTICS, EXPLOSIVES, and DRUGS. By nature, these are destructive tests and usually are not the first choice, particularly if only small samples are available. The solvents used vary based on the material being tested. For drugs and explosives, testing water solubility can be useful. For example, if a drug such as COCAINE is diluted with flour, the flour will not dissolve in water. Solubility testing can thus be a prelude to SOLVENT EXTRACTION and separations that are performed prior to instrumental techniques. For paints, common solvents used include concentrated acids such as nitric and hydrochloric (HNO_3 and HCl), acetone, methyl ethyl ketone (MEK), and dimethylformamide (DMF). For fibers, solvents are generally organics such as DMF and nitromethane. Plastics, which are synthetic polymers with much in common with synthetic fibers, can also be tested with different solvents for classification and comparison purposes. Many of the solvent tests for paints and fibers include steps in which the sample is heated. Solubility testing can help establish similarities between two samples but cannot INDIVIDUALIZE them to a COMMON SOURCE. Thus, it is best considered as a means of determining CLASS CHARACTERISTICS.

solvent extraction A method frequently used in forensic CHEMISTRY and TOXICOLOGY to isolate compounds of interest from a complex sample matrix. The principle of solvent extraction is based on selective PARTITIONING between two different phases based on solubility. For example, if oil and water are placed in a jar and shaken, they will separate, forming two phases. If sugar is added and the mixture shaken again, the sugar will end up in the water layer since it dissolves in water but not in oil. This is an example of selective partitioning based on solubility.

In solvent extraction, a sample such as urine can be extracted with an organic solvent such as chloroform or methylene chloride (dichloromethane) to selectively isolate compounds in the organic layer. Other solvents commonly used are hexane, acetone, alcohols, and ethers. Acid/base extractions are also used to isolate drugs that have acidic and basic forms. COCAINE can exist as a white powder (a hydrochloride salt) or as a sticky brownish material, the freebase form. By dissolving the material in water and adjusting the pH, one can isolate the cocaine from other materials and compounds that might be present. By adjusting the pH, ionization of the drug molecules is minimized, ensuring solubility in the organic phase rather than in the water phase. Acid/base extractions using water (aqueous phase) and the organic phase are the first steps in screening procedures that target a large number of compounds. Drugs that are acidic include aspirin and BARBITURATES (based on barbituric acid), while AMPHETAMINE is basic. MORPHINE, a large molecule with several functional groups, can be either acidic or basic ("amphoteric") depending on the pH. Thus, by adjusting the pH of an aqueous solution and then extracting that solution, acidic drugs can be separated from basic ones. Extractions are usually performed using specialized glassware and equipment designed for continuous operation rather than the older method of manual shaking, a process not far removed from shaking oil and water in a jar. In recent years, SOLID PHASE EXTRACTION techniques have become more common since they can often yield a "cleaner" extract with fewer extraneous materials. This is particularly valuable for toxicology, in which the sample matrix itself, particularly blood, complicates solvent extraction.

See also LOG P; PKA.

Further Reading

Siek, T. J. "Sample Preparation." In *Principles of Forensic Toxicology*. 2nd edition. Edited by B. Levine. Washington, D.C.: American Association of Clinical Chemistry, 2003.

Southern blotting *See* DNA TYPING.

speciation The process of determining the species from which a body fluid stain originated. Blood, saliva, and semen are the body fluids most commonly tested for species. Prior to the widespread adoption of DNA TYPING using PCR amplification, immunological tests such as IMMUNODIFFUSION and crossed-over electrophoresis, an IMMUNOELECTROPHORESIS method, were employed for speciation. These tests involve the use of antisera created to react with blood and body fluids of species such as human, cat, and dog. FALSE POSITIVES can result from cross-reactivity that can occur between related species. For example, cross-reactivity would be expected between antihuman antiserum and stain extracts from human and chimpanzee blood. Because of this, the strength of the antisera used (also called the "titer") must be determined. In most cases, the central question relates to blood and whether it is human. Thus, cross-reactivity with chimpanzee serum is rarely a concern. However, in wildlife forensics, the question is usually not about human blood at all. Between the two extremes are cases in which animal blood would not be unexpected, say at a farm or in a home with pets. Regardless, when an unknown stain is encountered, the general approach is to employ PRESUMPTIVE TESTS to determine what the stain is (blood, semen, and so on) followed by confirmatory tests and speciation.

In DNA typing procedures using PCR, one step of the procedure requires determining how much human (higher primate) DNA is present in the recovered sample. In so doing, the test confirms the presence or absence of human DNA. Many forensic labs are adopting this as the speciation step, but immunodiffusion and crossed-over electrophoresis are still used. When very small stains are found, the analyst must decide whether presumptive and confirmatory tests can be performed in conjunction with or in addition to the typing.

HAIR evidence is also subject to speciation, although this is performed visually with a microscope rather than using chemical or biochemical analysis. The structure and appearance of human hair (its MORPHOLOGY) is distinct from that of animals, and as a result determining the species of a hair sample is usually relatively easy. Finally, in TOXICOLOGY and ENVIRONMENTAL FORENSICS, the term *speciation* may refer to how an atom is complexed with or bonded to other elements. Often this is dependent on the oxidation state of the element. For example, mercury can exist in its metallic state as a liquid, symbolized as Hg(l). Mercury in thermometers and barometers is in this state. This form of mercury is relatively nontoxic compared to methyl mercury (CH_3Hg^+) and dimethyl mercury (CH_3HgCH_3), both of which are much more toxic. All three forms represent different "species" of the same element.

See also ANTIGENS, ANTIBODIES, AND ANTISERA; FISH AND WILDLIFE SERVICE (WILDLIFE FORENSICS).

Further Reading

Shaler, R. C. "Modern Forensic Biology." In *Forensic Science Handbook*. Vol. 1. 2d ed. Edited by R. Saferstein. Upper Saddle River, N.J.: Prentice Hall, 2002.

Spalding, R. P. "Identification and Characterization of Blood and Bloodstains." In *Forensic Science: An Introduction to Scientific and Investigative Techniques*. 2nd edition. Edited by S. H. James and J. J. Nordby. Boca Raton, Fla.: CRC Press, 2005.

spectrofluorometer (spectrofluorometry) An instrumental technique that uses ELECTROMAGNETIC ENERGY (EMR or radiation) to promote an atom, compound, or molecule to an excited state and detect the radiation that is emitted when the species returns to the original state ("ground state"). As shown in the figure, incoming radiation is absorbed, and the energy is used to promote the species to an excited state. What constitutes an excited state varies depending on the type of spectrometry being used. In atomic absorption, for example, the energy of the incident (incoming) radiation is used to move an electron in an atom to a higher energy level. In UV/VIS SPECTROPHOTOMETRY, electrons are

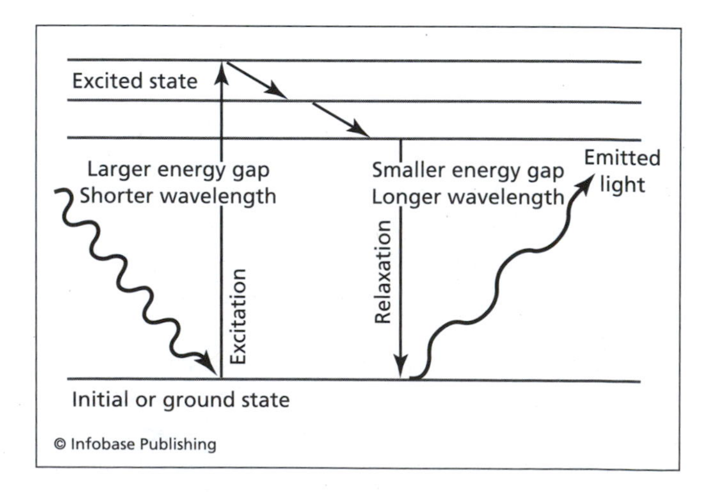

The process of fluorescence

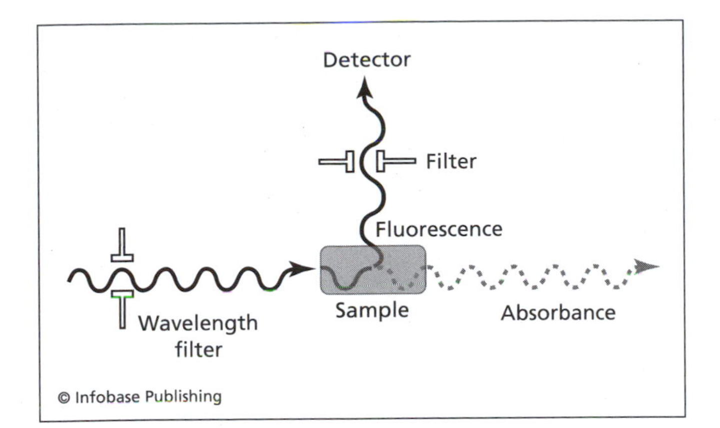

Box diagram of a spectrofluorometer

promoted in molecules. With infrared spectroscopy, a molecule absorbs energy and is excited to a different vibrational state; electrons are not moved at all. Regardless of the type of excitation, there are usually sublevels within the excited state, and small relaxation steps can be taken without the release of electromagnetic energy. However, when the target species takes the final step to relax to the ground state, electromagnetic energy is released. Since the energy gap is smaller than initially jumped, the energy of the emitted radiation is lower, and thus a longer wavelength is observed. This process is fluorescence, and it can be exploited to make very sensitive instrumental detectors that are widely used in forensic science.

The second figure shows a representation of a simple spectrofluorometer. In traditional SPECTROSCOPY, light interacts with the sample and is absorbed, decreasing the strength of the original incident beam. In spectrofluorometry, the detector is typically situated at a 90° angle to the light path, since fluorescence radiation is emitted in all directions. By placing the detector to the side, most interference is removed, resulting in a strong, clear signal.

Forensic applications of spectrofluorometry are widespread, and many use ultraviolet radiation (UV) to excite the target molecule so that it emits in the visible range. In forensic DRUG ANALYSIS and TOXICOLOGY, spectrofluorometers can be used to detect drugs and metabolites that naturally fluoresce, such as LSD. However, many target compounds do not naturally fluoresce and are thus treated to create derivatives that do. Fluorescent species, whether natural or derived, are referred to as fluorophores. Fluorescent detection is also used in automated DNA TYPING procedures in several areas. For example, with STR techniques, the amount of DNA extracted from a given sample is quantitated using fluorescence. In subsequent analyses, primers that are labeled with fluorescent constituents are detected automatically using a laser as the excitation source. Fluorescence techniques can also be applied to INORGANIC materials using X-ray fluorescence (XRF).

See also X-RAY TECHNIQUES.

spectrophotometry (spectroscopy, spectrophotometer) Methods and instruments of chemical analysis widely used in forensic science. At the broadest level, spectrophotometry is the study of the interaction of ELECTROMAGNETIC ENERGY (EMR) with matter. The interaction can be measured and used to determine what is present (QUALITATIVE ANALYSIS) and how much (QUANTITATIVE ANALYSIS). The simplest form of spectrophotometry and also the first developed was VISIBLE (VIS), also called COLORIMETRY, in which the visible portion of electromagnetic energy is used. Variations of VIS spectrophotometry is used forensically in the analysis of DRUGS, PAINT, and FIBERS, to name a few applications. All spectrometers consist of an energy (light) source, a mechanism or device to filter the source energy and select the wavelength(s) of interest, a device or method to hold the sample, and a detector system, which converts electromagnetic energy (light) to a measurable electrical current. Other kinds of spectrometry widely used in forensic science include INFRARED SPECTROPHOTOMETRY (IR), ATOMIC ABSORPTION (AA), and ULTRAVIOLET/VISIBLE (UV/VIS). MASS SPECTROMETRY (MS) is somewhat misnamed in that it is not based on absorbance of electromagnetic radiation or on detection of electromagnetic energy.

A number of spectrophotometric techniques are based not on the absorption of energy but on its emission. INDUCTIVELY COUPLED ATOMIC EMISSION SPECTROSCOPY (ICP-AES) relies on the emission of characteristic wavelengths of light for elemental analysis. Fluorescence methods exploit the absorption of energy, typically in the ultraviolet range, to produce emissions that reveal information about the element or compound. Examples include X-ray fluorescence (XRF) and SPECTROFLUOROMETRY. Finally, spectrophotometric detectors can be used coupled to separation instruments such as high pressure liquid chromatographs (HPLC) and CAPILLARY ELECTROPHORESIS SYSTEMS.

See also X-RAY TECHNIQUES.

Spilsbury, Sir Bernard (1877–1947) English *Physician* The 19th century saw enormous progress in the forensic investigation of death, but the discipline was still in its infancy. Within the medical community, it lacked the respect and prestige of many other specialties. Death investigators of this age were true generalists, and the best of them recognized the limitations of this approach and pushed other specialists such as microscopists, chemists, and biologists to bring their skills to bear on death investigation. These men were the forerunners and founders of the MEDICAL EXAMINER's systems (or the equivalent). The nature of the challenge and the work drew an interesting lot of characters, and as a result the late period of the 19th and early 20th century marks the beginning of forensic celebrity. Through their work, extensive public exposure, and involvement in high profile cases, these men jumped into the public and scientific limelight.

One of the first was Sir Bernard Spilsbury in England. His colorful career was the subject of several books including *The Scalpel of Scotland Yard: The Life of Sir Bernard Spilsbury*. His life and, ironic given his profession, his death had just the right touch of theater for the age. Spilsbury was born in 1877, and he died near his office in 1947, committing suicide using carbon monoxide gas. His public life was one of fame, his private later life one of tragedy. Two of his sons, also doctors, died during the World War II period. The first to die was Peter, who was killed in an air raid in 1940. This was also the year of Spilsbury's first stroke. The next year his sister died, and in 1945 his son Alan died

Sir Bernard Spilsbury at St. Bartholemew's hospital, London, in 1921 ***(Tropical Press Agency/Hulton Archive/Getty Images)***

of consumption. Before the 1940s signaled the beginning of the end, Spilsbury made an indelible mark as a forensic pathologist and death investigator. He performed thousands of AUTOPSIES, many of them on criminals he had helped send to the gallows. He used what he learned to study death by sudden blow and to recommend more humane methods of hanging.

Spilsbury began training in medicine in 1899 at St. Mary's hospital in London, one of the key facilities associated with legal medicine. St. Mary's became famous in 1928 because of the work of Alexander Fleming, who discovered penicillin while working there. Spilsbury did not discover a new substance while there, but rather his own incredible aptitude for meticulous investigation and analysis required for death investigation. He was able to study under some of the most famous forensic pathologists and toxicologists of the day, men who were themselves representative of the first and second generation of true forensic scientists who emerged after MATHIEU ORFILA. By 1909, Spilsbury was senior pathologist at St. Mary's and the Home Office pathologist.

His celebrity was launched with a murder case in 1910. The accused was a mild-mannered physician named Robert Crippen, whose wife had been reported missing in early January. The badly decomposed body was found in July, buried in their cellar. Spilsbury's work was hampered because the killer had removed the head, bones, limbs, and sexual organs from the body. All Spilsbury had to work with was a decomposed torso. Analysis of the tissue revealed the presence of a compound called hyoscine, better known as SCOPOLAMINE in the United States. This drug is a naturally occurring ALKALOID that is readily absorbed through the skin but that is not as toxic as other poisons common at the time. Spilsbury's testimony centered on the difficult task of identification, which he was able to do using a scar found on the abdomen. Crippen was convicted and executed.

The second case that established Spilsbury's reputation was the Seddon case in 1912. Frederick Seddon and his wife operated a rooming house that was home to Eliza Barrow, whose life ended soon after she changed her will and left all her property to the Seddons. She was buried but later exhumed and tested for ARSENIC by Spilsbury. The MARSH TEST was used, and the arsenic mirror was shown to the jury. Frederick Seddon was convicted and hanged for murder.

Spilsbury's most famous case came three years later in a case known as the "Brides of Bath." George Smith stood accused of murdering three young women. In sequence, he courted, married, and insured each, with their deaths following shortly after the last step. All three had died in the bathtub, but the deaths occurred in different places across England. Once suspicion was raised, all three bodies were exhumed. Spilsbury determined that all had died by drowning. By conducting experiments, Spilsbury showed that Smith had managed to pull the women by the feet while they lay reclined in the tubs. A strong, violent jerk, he demonstrated, could cause the victim to become immediately unconscious. They drowned without struggle. While doing the experiments, Spilsbury's willing subject, a nurse, passed out and had to be revived. The story added to Spilsbury's notoriety. In 1923, after several more high-profile cases and convictions, he became Sir Bernard Spilsbury.

One of the most ironic and interesting cases Spilsbury worked on occurred in 1924. Emily Kaye was killed inside a small house on the southeastern coast of England. Her body was dismembered by a saw, the parts further chopped, and then boiled. Parts, including the head, were never located. The bloody saw was found in a suitcase belonging to Patrick Mahon. Spilsbury arrived on the scene and managed to find a few bone fragments, which he concluded were from a large woman who was pregnant at the time of death. The remains also showed evidence of violent blows inflicted at the same time. Spilsbury's testimony was crucial in convicting Mahon, who was subsequently executed by hanging. Spilsbury performed the autopsy on Mahon. Based on his findings in this and subsequent autopsies of executed criminals, Spilsbury made several recommendations to make death by hanging more humane.

Spilsbury's prime as a medical detective spanned the 1920s and 1930s. In 1939, an important case in England involving Bernard Spilsbury and serologist Roche Lynch revolved around saliva evidence from a cigarette. The case involved a vicious rape-murder of an 11-year-old girl named Pamela Coventry. A few hours after disappearing on the way to school, her body was found in bushes. She was nearly naked, legs bound tight to her chest with wire. In the space between her legs and chest, a cigarette was found. Spilsbury collected it and gave it to Lynch. The serologist knew of secretor status and was aware of attempts to type body fluids. Since the techniques were relatively new, Lynch spent a great deal of time studying and practicing before turning his attention to the evidence.

Meanwhile, solid detective work had narrowed the suspect list. Unable to compel the one strong suspect to give a blood sample, detectives obtained a search warrant for the suspect's house, hoping to find dirty laundry (literally) that included used handkerchiefs. Lynch felt certain he could type the hankies using the techniques he had practiced and mastered for the cigarette. During the search, investigators found that and more, including a blood-stained raincoat, wire, and tape that were consistent with that used to bind the victim. Unfortunately, Lynch was unable to type the saliva on the cigarette since, as it turned out, the suspect was a nonsecretor. The suspect was eventually freed; however, an important forensic precedent had been set.

By the 1940s, stress began taking a toll on Spilsbury. In 1940, at the age of 63, he suffered the first of what would be several strokes and the death of one son, the first of several deaths of loved ones. His final case came in 1947, and shortly after it, he committed suicide. Speculation remains that his suicide was driven in part by mistakes he had made in earlier cases, amplified in the harsh light of his fame. Whatever the reason for his suicide, none since Spilsbury in the early 20th century have been able to claim the same degree of public recognition.

spoliation *See* DECOMPOSITION; MICROBIAL DEGRADATION.

spot tests *See* PRESUMPTIVE TESTS.

standardization The process of assuring that a method or instrument will produce accurate and reproducible results under specified operating conditions. Standardization is a vital part of QUALITY ASSURANCE/QUALITY CONTROL (QA/QC) and extends from simple things such as assuring that balances obtain accurate weights to the most complex DNA analysis. For example, when a police officer stops someone he or she suspects of driving under the influence of alcohol, a BREATH ALCOHOL test is usually performed using an instrument designed specifically for that purpose. It is crucial that the instrument be properly calibrated and tested using known standards of alcohol to ensure that the results the officer obtains are trustworthy. Similarly, analytical instruments such as GAS CHROMATOGRAPHS/MASS SPECTROMETERS (GC/MS) are routinely utilized to quantitate drugs and other substances. Standardization is used to link the instrument response to concentrations of the substance of interest.

standard methods Procedures for analysis of evidence that have been tested, validated, approved, and used routinely in the laboratory. The use of standard methods is critical to overall QUALITY ASSURANCE/QUALITY CONTROL (QA/QC) and ensures that results obtained are as accurate and reliable as possible. Standard methods are sometimes referred to as "SOPs," meaning standard operating procedures. Laboratories maintain files of standard methods and SOPs, which are used for training and for reference. The use of standard methods also facilitates comparison of results. For example, if Lab A obtains a given result for a DNA typing analysis, Lab B should obtain the same results; use of standard methods helps to eliminate procedural variation as the source of any disagreement in results.

starch A POLYMER of glucose, a simple sugar that has three principal roles in the forensic science context. First, starches, particularly cornstarch, which is inexpensive and widely available, is used as a CUTTING AGENT for drugs that are white powders such as COCAINE. Starch can also be a component of DUST and thus a form of TRACE EVIDENCE. In both cases, the identification of starch is made easier by its distinctive appearance under a POLARIZED LIGHT MICROSCOPE (PLM). Finally, starch-iodine is a common staining agent and has been used as part of the amylase test for the presence of SALIVA.

Starch Wars A term used to describe disagreements among forensic serologists in the 1980s concerning methods used for typing ISOENZYME systems. The use of ABO and starch gel electrophoresis dominated forensic SEROLOGY from the 1970s into the late 1980s and early 1990s. Although some labs could type as many as 15 different systems, four were commonly used: PGM, AK, EAP, and EsD. While powerful, the isoenzymes systems were often difficult to type in older stains. This period is also known informally in the FORENSIC SCIENCE community as "Starch Wars" due to the competing versions of substrates (gel versus acetates primarily) employed for isoenzymes typing. The conflict foreshadowed the contentious DNA WARS to come.

In most labs, serologists typed each system separately, each requiring a new sample, new gel, and additional time. The profusion of methods and media further added to the confusion. In response, the U.S. government commissioned a study that included Wraxall, B. Grunbaum and Robert Shaler from the United States. In 1978, the group issued a report and recommended techniques that could type more than one system with each electrophoretic run. This approach was referred to as multisystem in which the systems were broken into three groups. Group I included PGM, EsD, and GLO I; group II included ACP, AK, and C.E.A; and Group II permitted typing of Gc and Hp. After a period of review, the methods were well established in most U.S. and UK labs. The FBI and the Home Office establishment in the UK took the lead in researching IEF applications, adding them to the mix. The confusion led to the Starch Wars, a whimsical name for a period of intense and often bitter debate within the young forensic BIOLOGY community.

With the sudden abundance of available methods came the inevitable arguments about which were best and why. The Starch Wars and related issues, in a preview of what was to come with DNA, bled over into the courtroom with experts favoring one methodology over another squaring off in admissibility hearings. During the period from 1923 to 1993, the primary rule governing the admissibility of scientific evidence was the *Frye* rule (*see* FRYE DECISION). Under the *Frye* criteria, courts based their acceptance of a new scientific method or procedure on acceptance by the relevant scientific community.

Unlike the DNA battles to come, here it was easy to define the relevant scientific community as the forensic science community and the battles played out, for the most part, outside of public scrutiny and the public spotlight. Privacy did not make the battles any less vicious or painful for the scientists involved, and many of the wounds were just healing when DNA appeared.

Stas, Jean Servais (1813–1891) Belgian *Chemist and Toxicologist* The man who would make the crucial breakthrough for isolation and detection of ALKALOID poisons was Belgian Jean Servais Stas. Like many of his day, he had ventured to Paris to study TOXICOLOGY under the famous MATHIEU ORFILA. By 1850, Stas was back in Brussels serving as a professor of chemistry. The murder of a man named Gustav Fougnies was to alkaloid poisoning what the LaFarge case had been to those perpetrated by ARSENIC. Gustav had the unfortunate luck of being the brother of Lydie Fougnies. A man named Bocarme had married her for the money she supposedly had, but in fact did not. Undeterred by the lack of love and money, they enjoyed the wild life and generally living beyond their means.

By 1849, the hard-partying couple realized that their only hope of financial salvation was the death of Gustav and the inheritance that would provide. Gustav's bad luck continued; in 1849, he was recovering from an amputation and had not yet married. If he died as a bachelor, all his money would go to Lydie since this fortune originated from the death of their father. Lydie and her husband might have been willing to wait to see if the post-amputation trauma killed her dear brother, but Gustav sealed his fate by finding a suitable bride. The rumors of impending marriage surfaced in 1850, prompting a swift response. He died on the eve of announcing the wedding, his sister attributing his death to a stroke.

The initial postmortem immediately out ruled a stroke (*see* POSTMORTEM INTERVAL [PMI]). The investigators noted that Gustav's blackened throat and stomach and wondered if acid, another common poison, was involved. The examiners had the foresight to remove critical organs and place them in ALCOHOL, arrest Lydie and Bocarme, and request Stas's help.

Stas worked patiently from late 1850 and into early 1851. He employed alcoholic extractions of the tissues and evaporated off the alcohol, leaving thick syrup. During one of these evaporations, Stas detected the odor of vinegar and during another, the odor of coniine. This compound was one of two that had been identified by that time, the other being nicotine that killed in doses of as little as 100 milligrams. By way of comparison, a typical "baby" aspirin contains about 80 milligrams of the active ingredient.

Stas now believed that Gustav had alkaloid poison in his tissue, but he was not yet confident of which one. Meticulously, he refined his extractions, one of which yielded large quantities of nicotine, a fatal dose of which could easily be concealed in food. One of the confirmatory tests Stas used was the taste test; he noted that the typical burning of the nicotine lingered in his mouth for hours. While foolhardy in retrospect, early chemists relied on their senses for detection because it was often the only procedure available.

Stas compared the brownish material he extracted with known nicotine to confirm his identification. He also had an explanation for the vinegar odor he had detected. When an acid is mixed with a base, the result is neutralization of both. It was possible that Lydie and Bocarme had tried to neutralize the alkaloid poison using the vinegar to confound any chemical testing. Their cleverness betrayed them. By adding the acidic vinegar to the basic nicotine, the neutralized nicotine became water soluble and much easier to detect.

Stas reported the results to authorities who quickly determined that the suspects had extracted large amounts of tobacco in the days before the murder. Investigators located bodies of animals used in experiments on the property, and on the strength of the scientific evidence and the investigative information it provided, a court convicted Bocarme of murder and sent him to the guillotine. Lydie escaped, likely because the jury was squeamish about sending a woman to the same fate.

Stas's work, much as JAMES MARSH's and Mathieu Orfila's a decade earlier, led to a flurry of activity related to the detection and identification of alkaloid poisons. Many of the tests developed in the wake of Stas's breakthrough still are used in forensic labs, although their function has changed as the science has. What were definitive tests in the 19th century are now classified as presumptive tests, meaning that a positive test indicates that the suspected material is probably present, but not definitively present. The same holds for a negative result—the suspected material is probably, but not definitively, absent.

Stas's breakthrough had more to do with the chemistry of separation and extraction than with identification. With the separation methods now available, the way opened for other chemists to devise chemical tests to identify alkaloids present in the extracts. Finding caffeine in stomach contents is far less sinister than finding coniine or strychnine for example, even though all are alkaloids. Thus, it was important to be able to test extracts further to identify what alkaloids might be present. A series of tests appeared in the mid to late 1800s, most named after their German chemist inventors. The list includes the MARQUIS, Froedhe, MECKE, and MANDELIN tests, all of which are still used. These tests are called color tests because when the reagents are added to the alkaloid or an extract containing it, a distinctive color change is noted. The Marquis reagent turns a deep purple in the presence of HEROIN, MORPHINE, or related substances, but will turn orange in the presence of METHAMPHETAMINE (a synthetic alkaloid). The colored substances created are DYES, a subject of surprising importance in forensic science.

Stas collaborated with German chemist Friedrich Otto and improved upon Stas's method, allowing a larger number of alkaloids to be extracted and in higher purity. The Otto-Stas method was announced in 1856, heralding a marked acceleration in progress in forensic chemistry and toxicology. European chemists, particularly the English and Germans, led the way. In contrast, progress across the Atlantic was slower, partially due to the time it took news to cross the ocean. The United States also lacked the medico-legal infrastructure that had taken root across the Continent and in England.

statistics A key component underlying many forensic analyses. The most obvious role of statistics is in DNA TYPING, POPULATION GENETICS, and estimates of GENE FREQUENCIES. However, statistics play a role with many other kinds of physical evidence such as HAIRS, FIBERS, PAINT, and GLASS, to name a few. Simple statistical quantities such as the mean (average), standard deviation (average deviation from the mean), and the percent relative standard deviation (coefficient of variation) are often used when replicate measurements are made of the same sample. Finally, statistics are closely related to and integrated with many facets of QUALITY ASSURANCE/QUALITY CONTROL (QA/QC).

See also BAYESIAN STATISTICS; LIKELIHOOD RATIO.

stature estimations Considered primarily a task of forensic ANTHROPOLOGY, the determination of the likely height of an individual based on skeletal remains. This is accomplished by measurements of the long bones such as those found in the arm and the leg (femur, for example). Based on these measurements and databases of information, the anthropologist can estimate the height of the individual. Stature estimates are difficult if not impossible if only partial skeletal remains are recovered. However, this depends on what bones are recovered. For example, a partial femur or tibia can be sufficient to generate a reasonable estimate of stature.

stereoisomers Compounds that share the identical molecular formula and order of connection of atoms but that differ in the three dimensional arrangement of these atoms. If two stereoisomers are mirror images of each other, they are called ENANTIOMERS; otherwise, they are referred to as DIASTEREOISOMERS. Stereoisomers are a consideration principally in forensic DRUG ANALYSIS, particularly for COCAINE, AMPHETAMINE, and METHAMPHETAMINE. At one time, there were arguments raised in courts as to the importance of the identification of the different stereoisomers. However, this has been resolved by changing laws to include isomers of controlled substances under the definition of the illegal substance itself.

stereomicroscope *See* MICROSCOPY, MICROCHEMISTRY, AND MICROSPECTROPHOTOMETRY.

steroids *See* ANABOLIC STEROIDS.

Stielow case (Stielow, Charles) *See* FIREARMS.

stimulants *See* DRUG CLASSIFICATION.

stippling *See* BULLET WOUNDS.

stipulation Agreement reached voluntarily by the prosecution and the defense relating to some point in a case. When a forensic scientist completes an analysis and writes a report, it is not uncommon for the defense to stipulate to the report as fact rather than call the expert to testify. For example, in a drug case involving possession of a white powder that a forensic chemist has identified as COCAINE, by stipulation the defense might agree with the analysis and focus their case on other elements of the crime.

strangulation Death by ASPHYXIA that is brought about by compression of the neck. The cause of death is lack of blood flow to the brain. Strangulation can be performed manually with the hands ("throttling") or by use of a LIGATURE. HANGING is a form of ligature strangulation in which body weight is used to generate the compression. Manual strangulation is a common form of homicide used by men to kill women. Suicide by manual strangulation alone is not possible since as soon as the victim becomes unconscious, he or she relaxes, which releases the compression and restores blood flow to the brain. Ligature strangulation is seen in suicides, homicides, and accidental deaths. The time from the onset of compression to unconsciousness varies considerably depending on where the pressure is applied and how strong and steady it is. The time can range from just a few seconds to much longer if the grip is such that the airway and not the blood vessels are being constricted.

Manual strangulation can produce abrasions on the throat from the suspect's as well as from the victim's fingernails. Bruising may develop corresponding to the grip points used. If a victim struggles, the assailant may change grip several times, leading to widespread wounds. The victim's chin may also be scraped or bruised from lowering the chin to protect the throat, bringing it into contact with the assailant's hands. When a ligature such as a rope, cord, belt, or wire is used, distinctive markings on the throat are usually produced. However, if the ligature is soft, such as a towel or scarf, and is removed quickly after death, external markings may be difficult or impossible to detect on the neck area. In such cases, damage to internal structures will usually be seen during the AUTOPSY. In most strangulation deaths, pinpoint hemorrhages called petechiae, petechial hemorrhages, or Tardiue spots are found on the face and particularly in the eyes and on the eyelids. These hemorrhages are caused by capillary rupture brought on by pressure. However, finding petechiae by themselves is not sufficient to determine that the death was the result of strangulation.

Depending on the age of the victim, other injuries to the larynx area may be inflicted. The older the victim, the more rigid and calcified the structures in this area are, and the more likely they are to be broken or visibly damaged during strangulation. The hyoid bone, which is located near the base of the tongue, may be broken or fractured, but a broken hyoid alone is not sufficient to determine strangulation. In addition, it may not have formed completely and may appear broken. Similarly, cartilage that has become brittle can be damaged.

striations Marks that are made in a surface as the result of the motion of one surface across the other. Scratching, sliding, and scraping motions all can produce striations on one or both surfaces. In the case of metal-to-metal contact, the softer metal yields to the harder metal and will show striations. Forensically, striations are most often associated with FIREARM and TOOLMARK evidence such as the markings on BULLETS and CARTRIDGE CASINGS. Striations can also be observed in materials such as plastic bags and other materials in which the production process creates parallel markings in the material. Striations are observed and compared microscopically using a COMPARISON MICROSCOPE and can be critical in deriving a PHYSICAL MATCH between two pieces of evidence.

Striations from two bullets seen under a comparison microscope ***(Courtesy of Michael Bell, West Virginia University, Forensic and Investigative Sciences program)***

strychnine An extremely poisonous substance that can be extracted from the seeds of the plant *Strychnos nux vomica* L. or synthesized. It is a large molecule ($C_{21}H_{22}N_2O_2$) that can be found in the form of salts, many of which are bitter tasting. Since the substance is used in rat and rodent poisons, it is occasionally encountered in forensic cases of poisoning, including accidental, suicidal, and homicidal.

stylistics, forensic *See* LINGUISTICS, FORENSIC.

subpoena A document issued by a court that requires someone to appear at a set date and time to provide testimony in the matter at hand. The meaning in Latin is "under [sub] penalty [poena]," referring to penalties that will be applied if the person fails to appear. Forensic scientists routinely receive subpoenas to testify to their findings on evidence they have analyzed.

sugars (saccharides) In forensic science, the principal occurrence of sugars is in the area of DRUG ANALYSIS, where sugar is frequently used as a CUTTING AGENT. Sugars that are seen include table sugar (sucrose), fructose, xylose, glucose, galactose, lactose, and maltose. A presumptive test for sugars is Benedict's solution, which turns reddish brown in the presence of sugars known as "reducing sugars" that will reduce an oxidizing agent used in the test. Although not routinely performed, diluting sugars can be identified and quantified using HIGH-PERFORMANCE LIQUID CHROMATOGRAPHY (HPLC).

suicide Considered to be a MANNER OF DEATH, along with accidental, homicidal, and indeterminate. Forensic scientists, particularly medical examiners and coroners, are faced with determining if a suspicious death is suicide or other. The principal tool is the AUTOPSY, but it is certainly not the only one available to make this determination. As an example, if a person dies of a gunshot wound to the head and a gun is found at the scene, the determination of the manner of death might involve evidence gathered at autopsy as well as FIREARMS, FINGERPRINTS, and information developed by investigators through interviews of friends and family.

superimposition *See* ART, FORENSIC.

sweat An excretion that occasionally can have importance as evidence and as a source for DNA TYPING and, rarely, toxicological analysis. Sweat is also the fundamental component of FINGERPRINTS and acts as a carrier for the materials that react with compounds used to visualize them. Sweat is an excretion of the pores and originates from the eccrine sweat glands of the skin. These pores are found on all skin surfaces, while sebaceous glands (oil glands) are found in much more limited areas. Pore location is part of the minutiae of fingerprints, and the study of pore location as a tool for identification is referred to as poroscopy. Sweat is a complex mixture of inorganic and organic materials dissolved or suspended in water, which makes up about 98 percent of the volume. The amino acids found in sweat are the compounds that react with NINHYDRIN to produce a purple color and visualize a latent fingerprint. Dissolved salts, sugars, and ammonia are also components of sweat. As a medium for TOXICOLOGY, sweat can be used in some cases to detect drugs and metabolites and has the added advantage that it can be collected using noninvasive procedures.

T

Takayama test *See* HEMOCHROMOGEN TEST.

tape lifts A technique of evidence collection used to recover LATENT FINGERPRINTS, FIBERS, HAIRS, and other types of transfer and TRACE EVIDENCE. To remove fibers from a surface such as clothing, clear tape can be rolled into an inside-out (sticky side up) loop. This loop is then pressed against the surface where fibers and other clinging matter can be picked up. In the case of latent fingerprints, a tape lift technique can be used to preserve and transport the print to the examiner's laboratory. In the process of recovering a print, the first step is usually to photograph it in its original place, develop or visualize as necessary using fingerprint powder or other techniques, then photograph again. Lifting tape (similar to the familiar cellophane tape) is then carefully and firmly pressed over the print, causing the transfer. The final step is to place the tape on a card with an appropriate color that contrasts with the latent print.

taphonomy The systematic study of "death assemblages" and DECOMPOSITION applied to forensic science. Taphonomy is a subdiscipline of paleontology that was adopted by anthropology, archaeology, and now by forensic science. Strictly defined, taphonomy focuses on the process of death and the aftermath that ultimately led to the fossilization of remains, but in the forensic context, the time frame of interest is much shorter. By studying the scene of a death (or the scene at which a body was dumped), a forensic taphonomist attempts to determine what the CAUSE OF DEATH was, what the circumstances surrounding the death were, how long ago the person died (POSTMORTEM INTERVAL, or PMI), and what conditions at the scene are the result of natural processes versus human action.

Areas of interest in taphonomy include examination of bone weathering, scattering of remains by carnivores and scavengers, evaluation of marks on bone to determine their likely source, and transport of remains in and by water. The study of decomposition is also central to taphonomy, as are elements of forensic ENTOMOLOGY (insects). However, once remains have become skeletonized, determination of the PMI becomes more difficult, requiring analyses such as bone weathering to estimate it. Marks on bone can be created by natural processes such as chewing by scavengers or may reflect wounds received ante mortem or immediately after death. Taphonomy can be helpful in determining if disarticulated remains are a result of deliberate dismemberment by the killer or scattering by animals. These techniques are also valuable in reconstruction of the death event, the few moments before, and the events following it. Taphonomy is applied in attempts to determine if the location where a body is found is the actual death scene or a secondary scene used to dispose of the body. Although taphonomy is most often associated with remains found on land, the principles can be applied to bodies dumped in freshwater, such as lakes, streams, and rivers, to bodies in marine environments, and to bodies displaced by flooding.

Further Reading

Haglund, W. D., and M. H. Sorg, eds. *Forensic Taphonomy. The Postmortem Fate of Human Remains.* Boca Raton, Fla.: CRC Press, 1997.

TATP TATP stands for triacetone triperoxidase, an unstable EXPLOSIVE that is relatively easy to make and as such, is attractive to terrorists. It was used as part of the shoe bomb found on Richard Reid in 2001 and may have been an ingredient in the bombs used in the London Underground bombings in 2005. In 2006, transatlantic airline travel was disrupted by a threat originating in the United Kingdom. British police feared that terrorists were planning to smuggle the ingredients for making TATP onto aircraft disguised as ordinary liquids. It is questionable how feasible it would have been, but as a result, passengers traveling on U.S. airlines are no longer allowed to carry on liquids in large volumes.

Taylor, Albert Swaine (1806–1880) British *Toxicologist and Physician* Alfred Swaine Taylor established a reputation as one of the best of the early forensic toxicologists. As a young medical student, Taylor ventured to Paris to attend lectures by MATHIEU ORFILA, where on one memorable trip he studied gunshot wound patterns. These experiences solidified his interest in legal medicine. In 1831, Taylor assumed the position of lecturer in medical jurisprudence at Guy's Hospital in London; he was 25. He began to appear as an expert witness and studied the literature of the law with the same intensity he devoted to chemistry and medicine. Although he had a distinquished career, he was also involved in controversial cases that have marred his reputation.

Taylor recognized that chemical analysis of the time was not definitive but probabilistic in the same way PRESUMPTIVE TESTS are today. Taylor believed that the role of chemical findings, whatever they were, was to support medical and clinical findings. In his view, finding traces of ARSENIC in tissues alone was not enough to prove poisoning. Many other factors, including the limitations of the testing method, played into the overall conclusion. Equally or more important were clinical observations made by the physician while the victim still lived. Taylor also recognized that quantity of a poison was critical, echoing PARACELSUS, but he updated it to the forensic application. Arsenic was common in the environment of Taylor's time; finding it in small amounts in tissue alone was unremarkable. If arsenic was found in a suspected poisoning victim, Taylor integrated and weighed the merits and limits of clinical data with chemical findings to unravel the circumstances that led to finding the arsenic.

Taylor was forward-thinking in other ways as well. He noted the importance of the chain of custody and recommended that samples be stored in clean jars, tested immediately, and stored where there was no issue or question of tampering or adulteration. This practice amounted to tacit admission of the need for peer-review, independent confirmation, and the inevitable development of better methods to come. Taylor's insights sprung from scientific training, legal knowledge, and acumen that he passed along in his lectures. He wrote several books beginning in 1836, including the well-known *Manual of Medical Jurisprudence* and *On Poisons.*

Taylor was an early adopter of the MARSH TEST and used it extensively in casework. He also favored the REINSCH TEST despite the objections of others in the field who argued that the test was too sensitive. Without the proper controls, the argument went, such as testing reagents and apparatus to ensure they were not contaminated, results of the Reinsch test were problematic. Potential interferents included lead, copper, bismuth, and MERCURY, some of which were nearly as common in the 19th century world as was arsenic. Testing for all of these in addition to arsenic would be a chore in a modern lab, let alone a 19th-century one. Recognizing this, Taylor concluded that the complexity of eliminating every possible FALSE POSITIVE was much more likely to lead to false conclusions than a careful application of the test alone. Taylor also believed it was unlikely that this particular group of elements would innocently coexist with arsenic in tissues or other evidence associated with a suspected poisoning.

Taylor's reliance on the Reinsch test in spite of its limitations hurt him and his reputation during the Smethurst case, in which a physician was accused of murder. In his testimony, Taylor admitted that the arsenic he detected in the evidence could conceivably have come from his own reagents and techniques. Taylor was further misled by the medical findings that he valued so much; the clinical evidence strongly suggested that some type of irritating poison was involved. Taylor applied the Reinsch test to one sample and obtained a positive result, a finding that led to Smethurst's arrest. When he obtained other samples (more than 20) from other tissues, he found no arsenic and only traces of antimony. Some people thought that Smethurst, a physician himself and possibly knowing that Taylor would use the Reinsch test, had administered a chlorate substance to his victim to speed removal from the body. There was no evidence to support this conjecture. Even with the results of the chemical analyses thrown out, the court convicted Smethurst, although the conviction did not stand.

This was not Taylor's first baptism in controversy. In 1856, he had been involved in another murder case, another poisoning, and another conflict of chemistry and clinical findings. The suspect was William Palmer, the victim J. P. Cooke, and the method, suggested by predeath symptoms, was STRYCHNINE. Analysis of the stomach and intestines failed to reveal any. With modern hindsight, the lack of strychnine means little; the substance rapidly enters the bloodstream. Taylor and his contemporaries did not know this key bit of information, but again Taylor fell back to the clinical observations. Despite his inability to find strychnine, he believed Cooke died by murder with strychnine.

The defense countered that Cooke had died of tetanus and that the chemical evidence to the contrary was laughably inadequate. Taylor never wavered, to no avail. He regretted the outcome of the trial and was disheartened that courts and the public now often viewed chemical analysis as the weak link of a weak profession. The later Smethurst controversy added to the cloud that hovered over him and his profession in the later part of the 1800s. That cloud lingered until Sir BERNARD SPILSBURY established his reputation and returned a sense of public trust to forensic death investigation.

teeth and tooth marks *See* BITE MARKS.

Teichmann test *See* HEMATIN TEST.

tetrahydrocannabinol (THC) A member of a class of organic compounds called cannabinoids and the active ingredient in MARIJUANA and derivatives such as HASHISH. In marijuana, there are more than 50 cannabinoids, of which THC (a hallucinogen) is considered to be the most important active one. The structure of THC is shown in the figure, as are the two different numbering systems that are used (here, one underlined and the alternative circled). As a result of this, the principal active component in marijuana is referred to as either Δ^1-THC or Δ^9-THC. It is found in the oily resin, flowering tops, and leaves of the plant, with the highest concentrations found in the oil. Concentrations of THC in marijuana have been steadily increasing as growers improve the "quality" of their crop using selective breeding. THC is a thick, oily liquid that is actually the result of degradation, and so its concentration increases as the plant or extract ages. It was first isolated from marijuana in 1964.

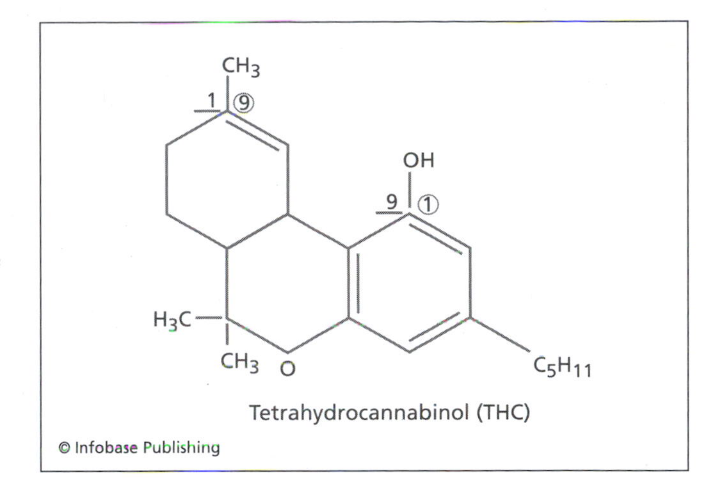

Chemical structure of tetrahydrocannabinol (THC)

Marijuana has a long history of medicinal use, and it was available until around 1940 for such applications. In 1996, two states (California and Arizona) passed laws that allowed doctors to prescribe marijuana for use in treating glaucoma and the nausea and vomiting associated with chemotherapy. A synthetic cannabinoid called Dronabinol (Marinol) is also available. When smoked, some of the THC is destroyed, but what remains enters the bloodstream almost immediately. Metabolites (*see* METABOLISM AND METABOLITES) are excreted primarily in the feces and urine, with the majority being removed within five days.

tetramethylbenzidine (TMB) A PRESUMPTIVE TEST for blood made by dissolving TMB in glacial (highly concentrated) acetic acid. Like other similar tests such as the BENZIDINE and KASTLE-MEYER (K-M) tests, TMB changes color in a reaction that is catalyzed by hemoglobin. The TMB test is an alternative to the older benzidine test, which has been abandoned due to the carcinogenic properties of that compound.

thermionic detector *See* NITROGEN PHOSPHORUS DETECTOR (NPD).

thin layer chromatography (TLC) A simple form of chromatography used extensively in forensic DRUG ANALYSIS, TOXICOLOGY, EXPLOSIVES analysis, and analysis of DYES and INKS. TLC separates mixtures by exploiting solvents traveling over a solid support phase and the resulting chemical interactions between the solvent, solid support, and molecules of interest. As shown in the figure, tiny spots of dissolved sample are placed in a line across the bottom of a plate or paper. The plate is coated with a thin layer of a silica or related powdery material. The line is called the "origin." The plate is then placed in a shallow solvent bath so the level of the solvent is below that of the origin. The solvent can be as simple as water (used for ink analysis), or it may consist of two or more organic solvents such as benzene, ethanol, or acetonitrile.

Capillary action draws solvent up the plate in the same way water is drawn up into a paper towel. As the solvent encounters the sample, some or all of it dissolves and begins moving along with the solvent "front" as it creeps up the plate. Some components in the mixture will interact with the silica material, causing it to fall behind components that interact less. Eventually, all separable components will be spread out in spots across the plate. Any components that are not soluble in the solvent will remain at the origin.

In the case of ink analysis, spots are clearly visible. However, in many cases a developer must be sprayed or otherwise applied to visualize the spots. CONTROL SAMPLES are routinely applied to the plate as well for comparative purposes. For example, TLC is a key component of the identification of MARIJUANA. After initial PRESUMPTIVE TESTS indicate that a sample of plant matter could be marijuana, a small amount is dissolved in a solvent such as petroleum ether. The extract is spotted on a thin layer plate along with extracts of known marijuana samples. After the runs are completed, the plates are sprayed with a developer such as "Fast Blue B," which reacts with the components to produce colored spots. If the sample is marijuana, the pattern of the questioned sample should be similar to that of the known marijuana.

throttling *See* STRANGULATION.

time of death *See* POSTMORTEM INTERVAL (PMI).

time since death *See* POSTMORTEM INTERVAL (PMI).

tire prints (tire impressions) A form of IMPRESSION EVIDENCE. A tire print in soil, mud, or snow is created by compression and is sometimes referred to as compression evidence or as an indentation. Tire prints can also be created without compression, such as when black tire marks are left on the concrete floor of a garage. When indented tire prints are discovered, they must be protected and preserved until documenta-

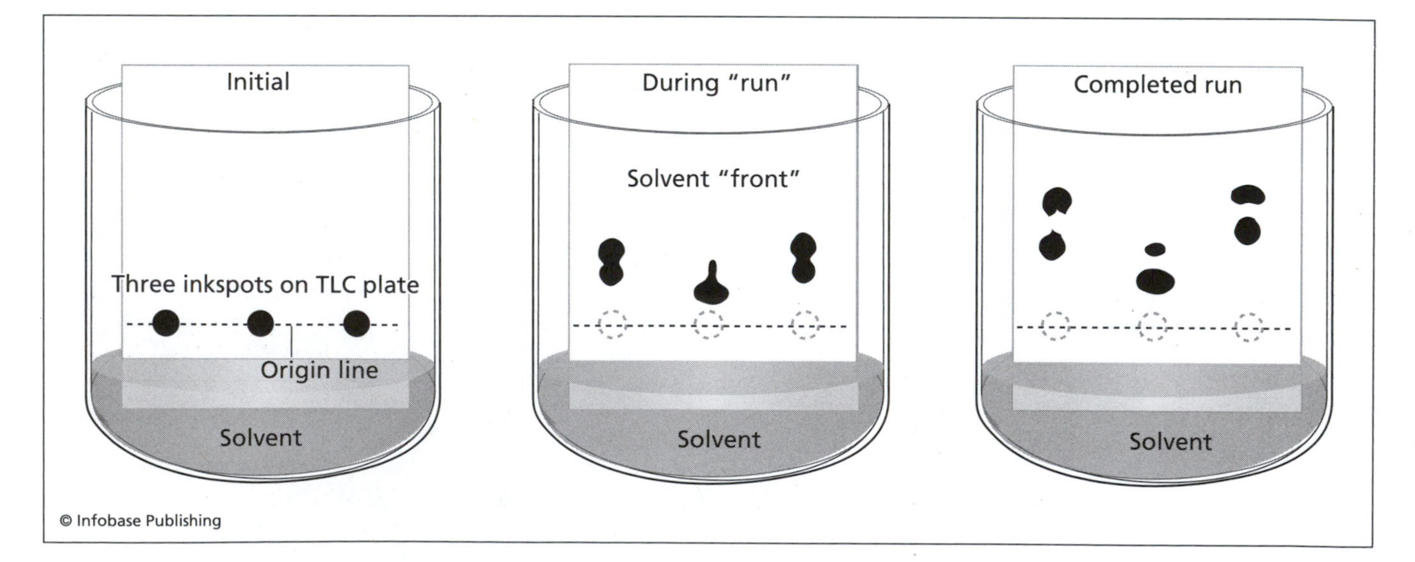

The process of separation in thin layer chromatography

tion and CASTING can be performed. Further analysis is conducted on the cast. The type of tread and pattern and dimensions such as width, depths of grooves, heights, and so on can be used to classify tire impressions. If the tracks of all wheels are visible, additional information such as wheelbase (distance between front and rear wheels) and front and rear track widths (distance between the center of the treads) may be sufficient to identify or narrow down the pool of vehicles that might have left the impressions. If a suspect tire is located, it may be possible to INDIVIDUALIZE the impression (assign a common source) based on comparing a collected suspect tire and the impression. Since tires are all used differently, they acquire WEAR PATTERNS that will be unique to the tire. Often the key to such individualizations depends on the quality of the impression, how well it was preserved when found, and the quality of the cast. To assist investigators and forensic scientists, the FBI maintains a special unit devoted to shoe and tire treads, patterns, and identification.

See also SHOE PRINTS.

Further Reading

Brodziak, W. J. "Forensic Tire Impression and Tire Track Evidence." In *Forensic Science: An Introduction to Scientific and Investigative Techniques.* 2nd edition. Edited by S. H. James and J. J. Nordby. Boca Raton, Fla.: CRC Press, 2005.

McDonald, P. *Tire Impression Evidence.* Boca Raton, Fla.: CRC Press, 1993.

TNT (2,4,6-trinitrotoluene) A secondary high EXPLOSIVE used extensively in World War II as a military explosive. As shown in the figure, it is prepared by treating toluene (a common solvent used as a paint thinner) with nitric and sulfuric acids (HNO_3 and H_2SO_4). A related compound, 1,3,5-trinitrotoluene, can also be prepared, which is often used as a standard in forensic analyses. A secondary high explosive, TNT must be detonated by a primary high explosive and thus is relatively shock insensitive. In addition, TNT is not affected by moisture, further increasing its utility as a military explosive. It can be found combined with other explosives such as RDX in products such as detonation cord. The forensic analysis of suspected TNT begins with the use of chemical screening tests (PRESUMPTIVE TESTS) followed by THIN LAYER CHROMATOGRAPHY (TLC), CRYSTAL TESTS, and instrumental tests such as INFRARED SPECTROSCOPY (IR) and HIGH-PERFORMANCE LIQUID CHROMATOGRAPHY (HPLC).

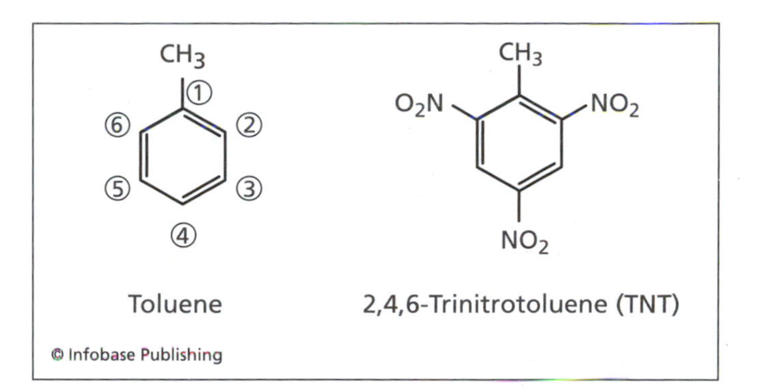

Toluene and 2,4,6 trinitrotoluene (TNT)

toners *See* QUESTIONED DOCUMENTS.

toolmarks A form of IMPRESSION EVIDENCE created when a tool comes in contact with softer materials such as WOOD or PAINT. In a forensic lab, the FIREARMS examiner usually analyzes toolmarks since the impressions made by guns on bullets and cartridge cases are specialized types of toolmarks. The most common source of toolmarks evidence is burglary cases. Tools can create indentation or compression imprints, scraping or STRIATION marks, or a combination of the two. Through two factors—grinding

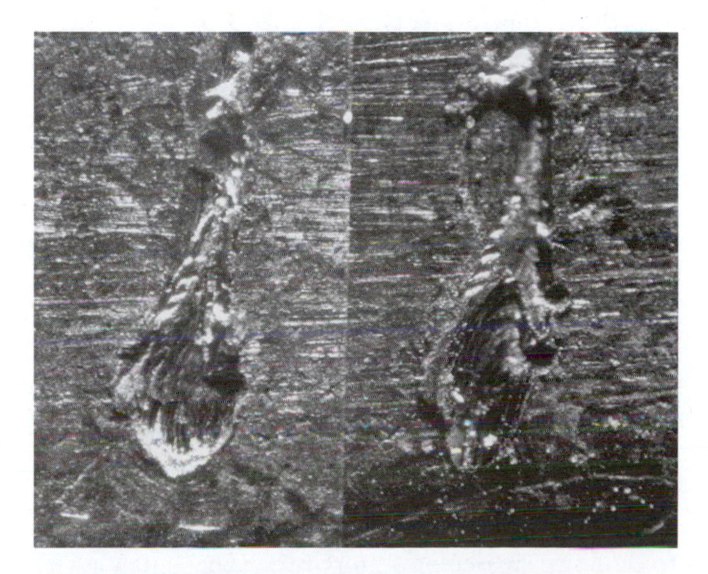

A cartridge case comparison ***(Courtesy of Dr. Keith Morris, West Virginia University, Forensic and Investigative Sciences program)***

Closeup and overlap of the previous comparison ***(Courtesy of Dr. Keith Morris, West Virginia University, Forensic and Investigative Sciences program)***

during manufacture and WEAR PATTERNS—toolmark impressions can often be linked to a specific tool if the quality of the impression is good enough and the suspect tool is recovered. The challenge in toolmark analysis is the difficulty of casting such impressions; often it is not possible to retain critical details needed for INDIVIDUALIZATION. The preferred method of preservation is to transfer the mark in place to the laboratory for analysis. Thus, if a lock is cut, rather than attempt to cast the impressions, the lock is studied directly. Toolmarks can be classified to some extent based on dimensions and patterns, but, for the most part, to be of value in an investigation, a tool must be obtained for comparison. To create test impressions, the examiner uses the tool on a soft material, such as lead, and compares the mark made with that found on the collected evidence. Variables that can influence the appearance of the marks include pressure applied and the angle at which the tool was held.

Further Reading

Rowe, W. F. "Firearm and Toolmark Examination." In *Forensic Science: An Introduction to Scientific and Investigative Techniques*. 2nd edition. Edited by S. H. James and J. J. Nordby. Boca Raton, Fla.: CRC Press, 2005.

toxicology (toxicology, forensic) The analysis of drugs and POISONS in blood and body fluids. Forensic toxicology uses many of the same techniques as does forensic DRUG ANALYSIS, the difference being the types of samples analyzed. In forensic drug analysis, the chemist works with physical evidence such as powders, plant material, paraphernalia, and equipment seized from CLANDESTINE LABS. The forensic toxicologist works with biological samples and must consider absorption of the material, distribution in body tissues, metabolic conversions (METABOLISM), movement through the system (PHARMACODYNAMICS and PHARMACOKINETICS), persistence of the drug, and modes of excretion. One of the largest sources of samples for forensic toxicologists is BLOOD ALCOHOL CONCENTRATION (BAC).

The history of toxicology dates back to the late 1700s and early 1800s, when incidents of ARSENIC poisoning were rampant. JAMES MARSH, an English chemist, developed a reliable test for arsenic in body tissues that was first used in a legal setting in 1840. In that case, MATHIEU ORFILA, credited with being the father of forensic toxicology, performed the analysis. Although in modern practice toxicology is more focused on drugs, detection of poisons is still an important component of the field. Toxicology officially entered the forensic system in the United States in 1915, when a toxicology lab was incorporated into the New York City medical examiner's office, which was replacing the CORONER system there. Currently, forensic toxicology can be divided into three areas: postmortem toxicology, drug testing (such as urine screening for employment or athletics), and human performance toxicology (including blood alcohol). A forensic toxicologist deals with tissue and organ samples, BLOOD, URINE, bile, VITREOUS FLUID, and, if gaseous poisons or drugs are suspected, lung tissue.

Toxicological analyses begin with extraction procedures that isolate the compounds of interest from the biological matrix. These extractions often exploit the natural acidity, basicity (alkalinity), or neutrality of the compounds. For example, BARBITURATES are acidic drugs and can be extracted into an acidic medium, whereas other drugs and poisons are basic or neutral. Samples are extracted under two sets of conditions. One recovers acidic and neutral components, while the other extracts basic components. The extracts are then subject to screening tests (PRESUMPTIVE TESTS) that will determine what further analyses are needed. Typical presumptive tests are THIN LAYER CHROMATOGRAPHY (TLC) and IMMUNOASSAY techniques. Once possible compounds of interest are identified, confirmatory analyses follow using instrumental techniques such as INFRARED SPECTROSCOPY (IR), GAS CHROMATOGRAPHY/MASS SPECTROMETRY (GC/MS), and HIGH-PERFORMANCE LIQUID CHROMATOGRAPHY/MASS SPECTROMETRY (HPLC-MS).

If the cause of a death is in question, the job of the toxicologist is critical and challenging. In addition to a drug overdose or interaction, illness or death can result from many common household items and materials if ingested in sufficient quantity. Any information concerning the deceased's behavior and symptoms prior to death can be helpful, as can a thorough search of the scene, where opened drug or other containers can provide clues.

See also PARACELSUS; PRODUCT TAMPERING.

Further Reading

Fenton, J. J. "Forensic Toxicology." In *Forensic Science: An Introduction to Scientific and Investigative Techniques*. 2nd edition. Edited by S. H. James and J. J. Nordby. Boca Raton, Fla.: CRC Press, 2005.

Goldsmith, R. H. "The Search for Arsenic." In *More Chemistry and Crime: From the Marsh Test to DNA Profile*. Edited by S. M. Gerber and R. Saferstein. Washington, D.C.: American Chemical Society, 1997.

Levine, B., ed. *Principles of Forensic Toxicology* second edition. Washington, D.C.: American Association for Clinical Chemistry, 2003.

trace evidence *See* TRANSFER EVIDENCE.

tracings *See* QUESTIONED DOCUMENTS.

traffic accident reconstruction A type of forensic engineering involving the study of automobile accidents and related accidents involving motorcycles, trucks, pedestrians, bicycles, and other vehicles. Reconstructions can be used in civil or criminal cases and can become crucial when there are no witnesses. For example, if a car crashes into a light pole in the middle of the night with no witnesses, the reconstruction will be the only method of assessing what might have happened. Points of investigation commonly include speed of the car(s), positions, directions of travel, braking, and points of impact. Information is derived from inspections of the car(s), evaluation of the scene, photographs and drawings, accident reports, police reports, interviews, calculations, and, in some cases, computer models. Central to any such investigation are basic principles of physics and engineering using quantities such as mass, velocity (speed), acceleration, momentum, and angular velocity.

One of the familiar signatures of many accidents is skid marks, which can be used to estimate speed. The length of the skid correlates to the speed at the time the brakes were applied and the coefficient of friction of the tires, which in turn depends on the road conditions. For example, the coefficient of friction is higher on dry asphalt than on wet asphalt, resulting in longer stopping distances on wet pavement. Other factors that come into play are tire conditions, slope of the road, and weight being pulled, such as if a trailer is being towed. Skid marks can also contain information about braking, although the use of antilock braking (ABS) systems can complicate interpretation. Related to accident reconstructions are failure analysis investigations in which the question revolves around the performance of equipment such as brakes and tires. The controversy over the safety of Firestone tires used on SUVs that arose in 2000–01 illustrates how reconstruction and failure analysis is crucial to identifying the root cause of accidents. Accident reconstruction is also used in cases of vehicle fires.

See also ENGINEERING, FORENSIC; HEADLAMPS AND HEADLIGHTS; TRANSPORTATION DISASTERS.

Further Reading

Brown, J. F., and K. S. Osborn. *Forensic Engineering. Reconstruction of Accidents*. Springfield, Ill.: Charles C. Thomas, 1990.

Carper, K. L., ed. *Forensic Engineering*. 2d ed. Boca Raton, Fla.: CRC Press, 2001.

Noon, R. K. "Vehicular Accident Reconstruction." In *Forensic Science: An Introduction to Scientific and Investigative Techniques*. 2nd edition. Edited by S. H. James and J. J. Nordby. Boca Raton, Fla.: CRC Press, 2005.

Noon, R. K. *Engineering Analysis of Vehicular Accidents*. Boca Raton, Fla.: CRC Press, 1994.

transfer evidence **(trace evidence)** Materials that are transferred from one person or place to another person or place. The evaluation of transfer evidence is the heart of traditional forensic science and often is conducted using MICROSCOPY. EDMOND LOCARD (1877–1966) is credited with developing what has become known as Locard's exchange principle, paraphrased as "every contact leaves a trace." Locard was particularly interested in everyday materials such as DUST that are among the most common forms of transfer evidence. Other ubiquitous types of materials found as transfer evidence are

HAIRS, FIBERS, GLASS, soil, and PAINT chips. To illustrate, consider a physical struggle between two people. Hairs from both may be exchanged, as may fibers of clothing. Once the two separate, each may bear trace evidence of their contact. However, the transfer evidence is not permanent. By definition, transfer evidence is easily exchanged, and, within a few hours, the original hairs and fibers transferred will likely be lost to SECONDARY TRANSFERS. The longer the time since the initial contact, the more evidence is lost.

Traditionally, transfer evidence is thought of as microscopic evidence, but the category is much broader than that. Large fragments of glass can be transferred from a car window to the victim of a hit and run. Back spatter from gunshot wounds can be transferred to the gun and the hand that holds it. By writing on a pad of paper, the tracing of the letters can be transferred to the pages below it, creating INDENTED WRITING. As in many forensic analyses, the key tests in the analysis of transfer evidence often involve comparing samples to determine if they had a COMMON SOURCE. For example, if a distinctive dust was found on the clothing of a suspected burglar, it would make sense to see if that dust could have come from the site of the burglary under investigation. In many ways, forensic science still revolves around Locard's ideas of transfer evidence, even if the scope has become much larger than that which he envisioned more than a century ago.

See also INDIVIDUALIZATION.

Further Reading

Kubic, T. A., and N. Petraco. "Microanalysis and the Examination of Trace Evidence." In *Forensic Science: An Introduction to Scientific and Investigative Techniques*. 2nd edition. Edited by S. H. James and J. J. Nordby. Boca Raton, Fla.: CRC Press, 2005.

transmission electron microscopy (**TEM**) A microscopic technique that uses electrons instead of photons (light) to create images. TEM uses equipment similar to that used for SCANNING ELECTRON MICROSCOPY (SEM). As in standard light MICROSCOPY, the incident radiation (in this case an electron beam) can interact with the sample by reflection or transmission. In light microscopy, this is referred to as transmitted light or reflected light. Similarly, in electron microscopy, reflected electrons (backscattered electrons, as well as others) are used to create an image in SEM, while transmitted electrons are used in TEM. Thus, for TEM studies, samples must be thin enough to allow for electron transmission. Currently, TEM has found limited forensic applications, as opposed to SEM, which is becoming commonplace.

transportation disasters Large-scale disasters, such as airline crashes and railroad mishaps, are investigated by the NATIONAL TRANSPORTATION SAFETY BOARD (NTSB). If criminal or terrorist actions are suspected, as in the crash of TWA Flight 800 (see essay), forensic scientists work closely with forensic engineers and other specialists involved in the investigations. In 1908, the first fatality in an airplane occurred outside Washington, D.C. Orville Wright was the pilot, and Lt. Thomas E. Selfridge was the passenger that day on a demonstration flight that ended in tragedy. Both men were injured, Selfridge fatally, when the plane crashed while attempting an emergency landing. As the airline industry grew along with other transportation systems, so did the need for thorough and impartial investigations of mishaps and crashes. The NTSB was established in 1966 as a part of the newly formed cabinet-level Department of Transportation. In 1974, the NTSB became a completely independent entity with investigative powers for air, rail, highway, marine, and pipeline accidents. In such investigations, the primary goals of the NTSB are to identify the probable causes and to recommend steps and actions that can help to avoid similar accidents in the future.

See also ENGINEERING, FORENSIC; MASS DISASTERS.

Further Reading

Carper, K. L., ed. *Forensic Engineering*. 2nd edition. Boca Raton, Fla.: CRC Press, 2001.

tricyclic antidepressants *See* BENZODIAZEPINES.

trier of fact The party responsible for making a decision or finding on a matter before a court. The trier of fact can include the jury and the judge or be just a judge if there is no jury present. However, if the question has to do with a point of law, such as admissibility of evidence or testimony, the trier of fact is the judge.

TWA flight 800 *See* "TWA Flight 800: The Interface of Criminal Investigation, Forensic Science, and Forensic Engineering" essay.

typewriter *See* QUESTIONED DOCUMENTS.

U

ultraviolet light and ultraviolet spectroscopy (UV/VIS) In the electromagnetic spectrum (ELECTROMAGNETIC RADIATION, EMR), ultraviolet energy lies on the high energy side of visible light. Since many of the analytical techniques that utilize the UV range also work in the visible range, the term *UV/VIS* is often used in the description. The UV range encompasses the wavelength range of 200 nm to 400 nm, at which point the energy can be detected visually as a violet light, hence the name "ultraviolet." The range is further subdivided into UVA (320–400 nm), UVB (280–320 nm), and UVC, the highest energy UV light, 200–280 nm. These designations are found on sunscreen products that are designed to prevent UV damage to skin.

Many organic compounds can absorb UV/VIS light, and this behavior is the basis of UV/VIS spectrophotometry. Spectrophotometry was introduced in the 1800s and utilized the eye (human vision) as the detection system (COLORIMETRY). By 1940 this capability had been extended to include the ultraviolet range. UV/VIS spectrophotometry is used primarily to study organic compounds that absorb radiation in patterns that can reveal information about the molecules' structure. However, UV spectra do not contain much detail, and, as a consequence, the technique is not used extensively in forensic applications. When employed, it is mostly as a screening technique, since UV spectra are not unique to a compound in the way INFRARED SPECTROPHOTOMETRY (IR) is. However, UV/VIS spectrophotometers are used as detectors in many types of instruments including HIGH-PERFORMANCE LIQUID CHROMATOGRAPHY (HPLC) and CAPILLARY ELECTROPHORESIS (CE), both of which are finding increasing use in forensic science.

UV illumination, known informally as "black light," does have many forensic applications in its own right. In QUESTIONED DOCUMENT analysis, UV light can be used to stimulate FLUORESCENCE and to reveal different inks or erasures and OBLITERATIONS. Similarly, drugs such as LSD fluoresce under UV light. Finally, microscopes can be designed to work with UV light, although images must be converted to the visible range to be recorded.

uncertainty A range, value, or other quantity that expresses the range of a value. For example, in a DRUG ANALYSIS the forensic chemist may perform a quantitative analysis for COCAINE in a white powder and return a result such as 25.0 percent cocaine, +/- 1.0 percent. The one percent value is the uncertainty associated with the result. Several methods can be used to determine the uncertainty including a technique called "propagation of error," in which the uncertainty of every step in an analysis is combined to determine the overall uncertainty. Other methods are empirical, meaning they rely on experiments and multiple trials, method VALIDATION, reliability studies, and other forms of QUALITY ASSURANCE/QUALITY CONTROL (QA/QC) to determine a realistic uncertainty value. A common measure is called the confidence interval, which is based on multiple runs of the same analysis. In the cocaine example above, the interval might be reported as "+/- 1.0 percent at the

95 percent confidence." This statement can be paraphrased as "repeated laboratory analysis coupled to simple statistical calculations shows that there is a 95 percent certainty the concentration of cocaine in this sample lies between 24.0 and 26.0 percent." However, in many types of forensic analysis, uncertainty is more difficult to specify. For example, if a FIREARMS analyst compares two bullets, they either match or they do not, and uncertainty is not reported. FINGERPRINTS, BITE MARKS, and other types of comparison and pattern evidence fall into the same category.

United States Postal Inspection Service The investigative branch of the Post Office that deals with mail fraud, letter and package bombs, mailing of controlled substances, trafficking in child pornography using the mail, counterfeit stamps, and related crimes. The roots of the postal inspection service can be traced back to the first postmaster, Benjamin Franklin, and, as early as 1853, nearly 20 special agents were working in the precursor to the inspection service. The first inspection service forensic laboratory was founded in 1940 and has grown to encompass a system of five labs around the country. Not surprisingly, a large contingent of their work lies in the area of QUESTIONED DOCUMENTS and FINGERPRINT analysis. Other units within the labs include physical evidence, chemistry, photography, digital evidence (forensic COMPUTING), and POLYGRAPH. The postal inspectors were deeply involved in the ANTHRAX mailing attacks in late 2001.

Further Reading

United States Post Office. "Chronology of U.S. Postal Inspection Service." Available online. URL: http://www.usps.com/postalinspectors. Downloaded May 5, 2007.

United States Post Office. "Forensic and Technical Services Division." Available online. URL: http://www.usps.com/postalinspectors/crimelab.htm. Downloaded May 5, 2007.

United States Post Office. "Jurisdictions and Laws." Available online. URL: http://www.usps.com/postalinspectors/jurislaw.htm. Downloaded May 5, 2007.

United States Post Office. "Who We Are." Available online. URL: http://www.usps.com/postalinspectors/missmore.htm. Downloaded May 5, 2007.

United States Secret Service *See* SECRET SERVICE.

urine (urinalysis) A body fluid frequently used in forensic TOXICOLOGY, occasionally as physical evidence, and as a source of cellular material for DNA TYPING. Urine contains urea, creatinine, uric acid, and a number of ionic materials such as chloride (Cl-) and phosphate (PO_4^{3-}). If a stain is encountered that might be urine, the simplest PRESUMPTIVE TEST is based on smell. Upon heating, urine gives off a characteristic and distinctive smell. There are also gel-based diffusion techniques that can be used. With the advent of DNA typing, it is sometimes possible to type the cellular material excreted in urine. However, the sample must be fairly concentrated for this to be an option.

By far the most important role of urine in forensic science is in the area of toxicology, in which it is a preferred medium for detection of many drugs and poisons. Traces of drugs and metabolites remain in the urine much longer than in blood, and it is generally easier to quickly screen a urine sample than it is to screen a blood sample. Urine is the principal medium for workplace and employment drug screening, which is often referred to as "urinalysis." Analysis of urine for drugs proceeds as does any other drug analysis, from presumptive and screening tests such as THIN LAYER CHROMATOGRAPHY (TLC) and IMMUNOASSAY to specific confirmatory analysis such as GAS CHROMATOGRAPHY/MASS SPECTROMETRY (GC/MS). In the case of employee drug testing or drug testing for athletes, the toxicologist must also check for purposeful adulteration of the sample to mask the use of drugs. Such adulteration can range from simple dilution to the addition of "masking compounds" either directly to the urine sample or indirectly by ingestion. Although urinalysis is effective and efficient, the levels of a drug or metabolite in urine cannot be directly related to a level of impairment. In addition, urine reflects only relatively recent ingestion of substances (within a week or so depending on the drug) and will not reveal drug use that occurred before then.

See also CREATINE AND CREATININE; IMMUNODIFFUSION; METABOLISM AND METABOLITES; PHARMACODYNAMICS; PHARMACOKINETICS.

Further Reading

Greenfield, A., and M. M. Sloan. "Identification and Biological Fluids and Stains." In *Forensic Science: An Introduction to Scientific and Investigative Techniques*. 2nd edition. Edited by S. H. James and J. J. Nordby. Boca Raton, Fla.: CRC Press, 2005.

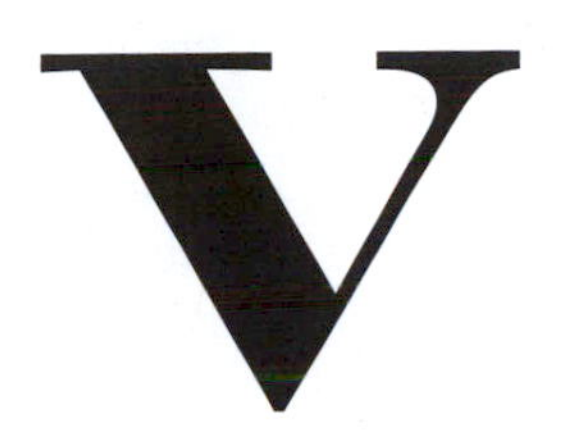

vacuum sweeping A method of collecting trace evidence from large areas such as rooms and cars. Sweepings have been employed since the early part of the 20th century using specialized attachments and systems designed to trap materials such as DUST, HAIRS, and FIBERS. While thorough, sweepings are indiscriminant and collect a large amount of evidence, most of which is usually irrelevant. For example, if a car interior is vacuumed, inevitably a large number of fibers will be recovered. However, unless some type of grid system is used, there will be no way to determine exactly where the fibers were found nor which had been in place for weeks, months, or years. Thus, vacuum collection is generally best used only after a detailed initial search for trace evidence.

vaginal acid phosphatase (VAP) *See* SEMEN.

vaginal swabs Swabs taken as part of a SEXUAL ASSAULT KIT. Portions of the swabs are smeared on microscope slides and allowed to dry, and both swabs and slides are turned in to the laboratory for analysis. On the slides, the analyst looks for sperm cells characteristic of SEMEN. However, a lack of visible sperm cells does not mean that intercourse has not occurred. Further tests are performed on the swabs to detect the presence of semen using the PSA (P30) test. The swabs can also be used as a source of DNA for DNA TYPING.

Further Reading

Greenfield, A., and M. M. Sloan. "Identification and Biological Fluids and Stains." In *Forensic Science: An Introduction to Scientific and Investigative Techniques*. 2nd edition. Edited by S. H. James and J. J. Nordby. Boca Raton, Fla.: CRC Press, 2005.

validation The process of determining the accuracy, precision, and limitations of a forensic analysis or procedure. Validation is an integral part of QUALITY ASSURANCE/QUALITY CONTROL (QA/QC) and takes place at many levels. Methods such as a DNA TYPING or the analysis of a drug are validated by repeat analysis of different types of samples (not casework) under controlled conditions. A method can be validated internally (within the lab) or among many different laboratories. Similarly, validation can be applied to a single analyst, who must show that he or she is capable of performing the test correctly and obtaining acceptable results. Even procedures such as documentation can be validated. The goal of validation is to ensure that reliable results are obtained when proper procedures are followed.

See also METHOD VALIDATION.

Valium *See* DIAZEPAM.

variable number of tandem repeat (VNTR) *See* DNA TYPING.

vegetable fibers *See* HAIRS; FIBERS.

vehicle paint *See* PAINT.

ViCAP (Violent Criminal Apprehension Program) A database of information maintained by the FEDERAL

Bureau of Investigation (FBI) that contains information on homicides and other violent crimes. The program is useful for linking seemingly unrelated crimes through similarities in crime scene patterns, behaviors, and physical evidence. Participation by police agencies is voluntary, and the FBI provides the software needed.

victimology *See* psychology, forensic, psychiatry, forensic, and profiling.

videography, forensic Techniques of video recording used at crime scenes to supplement (but not replace) photography and other traditional forms of documentation. Videotaping or digital video recordings offer the added advantage of sound, so the person recording can provide a narrative that will later be transcribed. The video can also record evidence in place and in context as well as record the process of photographing and collecting evidence. Like photographs, any video taken cannot be altered, and steps must be taken to ensure that what ends up in court has not been edited or enhanced.

Vinland Map Purported to be a map dating from the early 1400s, the Vinland Map controversy illustrates how forensic science and art forgery can intersect in a historical investigation. Many of the same tools used for forensic trace evidence and questioned document investigations have been applied to the map, and the central figure in the case was Dr. Walter McCrone.

The Vinland Map surfaced in 1957. It appeared to be a map showing the coastline of North America, but was purported to have predated Columbus's journey by 50 years. As with many objects of art and archaeology, the question of authenticity was immediately raised but any chemical analyses undertaken had to be nondestructive. If genuine, it would lend support to the

The Vinland Map ***(Courtesy of Yale University, Beinecke Rare Book and Manuscript Library)***

idea (now proven) that others, such as the Vikings, had been the first Europeans to reach the New World. More important, it would make an otherwise unremarkable parchment priceless. At the behest of Yale University, McCrone and others at the McCrone Research Institute studied the map and concluded that that it was a forgery, likely dating from the 1920s. The basis for the conclusion centered on the type of pigment found in the INK, a type not widely available until after 1916.

Additional microscopic analysis using Raman microspectroscopy supported McCrone's findings. Genuine medieval documents created with black inks often contained iron gallontannate, which slowly leached from the ink into the parchment underneath. Over time, the migrating iron causes the parchment to yellow and become brittle, leaving a faint yellow border around black inks. On the Vinland Map, the black ink was found to be carbon-based, and the parchment showed no evidence of the damage expected from an iron gallontannate ink. Rather, it appears that a forger placed a yellow line on the parchment before overwriting with the black ink in an attempt to simulate the appearance of a nearly 500-year-old document.

See also COLORANTS; MICROSCOPY; RAMAN SPECTROSCOPY.

Further Reading

Brown, K.L., and R. J. H. Clark. "Analysis of Pigmentary Materials on the Vinland Map and Tartar Relation by Raman Microprobe Spectroscopy." *Analytical Chemistry* 74 (2002): 3,658–3,661.

McCrone Research Institute. "Update for the Vinland Map and the Turin Shroud." Available online. URL: http://www.mcri.org/vm_shroud_update.html. Downloaded April 15, 2007.

visible spectroscopy and microspectrophotometry (MSP) *See* COLORIMETRY; MICROSCOPY, MICROCHEMISTRY, AND MICROSPECTROPHOTOMETRY; ULTRAVIOLET LIGHT AND ULTRAVIOLET SPECTROSCOPY (UV/VIS).

vitreous humor The fluid inside the eyeball that is frequently used for postmortem toxicological analyses. Since this fluid is contained and fairly isolated, it degrades more slowly than other fluids, such as blood and urine, and thus is usually preferred for cases in which any significant decomposition has begun.

voice recognition (voiceprint analysis) *See* PHONETICS, FORENSIC.

voir dire Literally, "to speak the truth," and sometimes incorrectly spelled *voie dire*. Prospective jurors as well as expert witnesses are subject to this process. The voir dire is a preliminary investigation by the court as to the qualifications, suitability, and, in the case of forensic scientists, expertise and competency in the matter at issue.

Vollmer, August (1876–1955) American *Police Officer* Vollmer was instrumental in linking university scientists to the analysis of evidence and in developing forensic science laboratory systems in the United States. He was a strong advocate of education of police officers, the use of scientific analysis, and proper evidence handling. In 1907, he recorded the first use of scientific analysis of evidence in a case in the United States, involving the analysis of BLOOD, SOIL, and FIBERS. He is best known for establishing the first forensic science laboratory in the United States while he served as chief of police in Los Angeles in 1923–24. In Berkeley in 1930, he established the first CRIMINOLOGY and CRIMINALISTICS program (police science) in the United States at the University of California, Berkeley, which became a separate school in 1950. The criminalistics program was headed by PAUL KIRK. Vollmer also laid the groundwork for the crime laboratory system in California, the most extensive state system in the United States.

Further Reading

Berkeley Police Department. "Our History." Available online. URL: http://www.ci.berkeley.ca.us/police/History/history.html. Downloaded May 5, 2007.

volume of distribution A quantity used in forensic TOXICOLOGY to describe how a drug or other substance is distributed among tissues in the body. The equation used is:

$$V_D = D/(C_p \times k)$$

where D is the dose; C_p is the plasma concentration once equilibrium has been reached in the plasma but before elimination has begun; and k is body weight in kg. Although the plasma concentration is used here, V_d can be calculated for any body fluid and should be specified.

The volume of distribution is not a physical volume, but rather a quantitative expression of how a drug is distributed in compartments of the body. For this reason, it is referred to as the apparent volume of distribution. In general, the higher the V_d value, the more lipophilic (fat-soluble) the drug. It provides a measure of the relative partitioning of the drug between the plasma and other tissues. If two compounds experience roughly the same degree of binding to plasma proteins in the blood, the compound with the greater affinity for association with the tissues will have a larger V_d relative to the compound with more affinity for the plasma. If the V_d of a drug is known, along with a person's body weight, toxicologists may employ the equation to estimate the dose of a drug.

See also ADME; METABOLISM.

vomitus (vomit) Evidence that is commonly found at death scenes. Vomit can be important for toxicological analysis and for the detection of drugs and poisons. Also, in some cases vomit can be a form of TRANSFER EVIDENCE, useful in tracking movements of a victim or a body.

Further Reading

Greenfield, A., and M. M. Sloan. "Identification and Biological Fluids and Stains." In *Forensic Science: An Introduction to Scientific and Investigative Techniques.* 2nd edition. Edited by S. H. James and J. J. Nordby. Boca Raton, Fla.: CRC Press, 2005.

Vucetich, Juan (Ivan) (1858–1925) *Argentinean Police Officer* Vucetich was a pioneer in the use of FINGERPRINTS in criminal investigation and is credited with the first case solved using them. In 1892, a woman murdered her two sons and, in an attempt to deflect suspicion, injured herself. However, Vucetich was able to locate her bloody fingerprint on a doorjamb, leading to a confession. Vucetich had read Sir FRANCIS GALTON's book (*Finger Prints*) and created a large collection of prints by 1891. This early collection still included BERTILLON measurements, although fingerprints would soon supplant them. By 1896, he had developed and instituted a fingerprint classification system that is still extensively used in Latin America.

wadding A plug made of plastic, cardboard, PAPER, or felt that is used to separate the propellant from the shot in shotgun AMMUNITION. An older term for wadding is "overpowder wadding." In newer ammunition, a one-piece plastic cup and wad combination is often used. The wadding material is blown out of the barrel when the ammunition is fired, and the position of the wadding at a shooting scene can sometimes be helpful in estimating distances and locations of the shooter and victim.

Wagenaar test *See* ACETONE CHLOR-HEMIN TEST.

Walker test A PRESUMPTIVE TEST for GUNSHOT RESIDUE (GSR) useful for clothing. The Walker test detects the nitrite ion (NO_2-) found in PROPELLANTS in AMMUNITION. The test works by transferring the residues to inactivated photographic paper that has been pretreated with sulfanilic acid and 2-napthylamine (or similar reagents). Development of an orange-red color is indicative of gunshot residues.

Washing Away of Wrongs, The An ancient Chinese manual describing methods of death investigation. Copies written in 1247 have been recovered, but the book probably had existed much earlier than that. The author is listed as Sung Tz'u, and the book is entitled *Hsi Yüan Lu,* translated as *The Writing of Wrongs* or *The Washing Away of Wrongs.* Incredibly, this book was used into the 20th century and included information on STRANGULATION, drowning, wound characteristics, poisons, and even a dash of forensic dentistry. Characteristic changes in the POSTMORTEM INTERVAL (PMI) were discussed to provide a means of estimating how long a person had been dead. Sung Tz'u's work was notable because it was a manual expressly for death investigators. A death investigator was charged with providing information that would be used during an inquest. (An inquest is an investigation of a questioned death held at the behest of government officials.) In English law, the procedure became known as a CORONER's inquest, and such investigations were common throughout recorded history. Sung Tz'u was the earliest known record of death investigation in the forensic tradition.

One oft-cited case recounted in the book involved the use of insects to assist in a death investigation. A sickle (a sharp tool used in agriculture) was apparently used in a murder. Several sickles were located from among a large collection of worker's tools, and all were placed in an open field. The object was to see which sickle drew blowflies. These tiny flies are the first to arrive at a death scene, sometimes within minutes of the death, and anecdotally, even just before death. Reportedly, the flies gathered around the sickle used in the killing, attracted by tiny flecks of blood.

See also CORONER; MEDICAL EXAMINER.

watermarks *See* PAPER; QUESTIONED DOCUMENTS.

wavelength dispersive spectroscopy (WDS) *See* SCANNING ELECTRON MICROSCOPY (SEM); X-RAY TECHNIQUES.

Wayne Williams case The case of a serial killer that was notable for its dependence on FIBERS and HAIRS as the principal form of physical evidence and statistics to assess significance. Starting in 1979, young black males were disappearing in Atlanta, Georgia, and their bodies were later being discovered dumped in wooded areas or near highways and streets. Most of the victims died of asphyxiation. A wealth of fiber evidence was found on the bodies, particularly a yellow-green fiber with an unusual cross section. Although hairs and fibers cannot be INDIVIDUALIZED, the evidence was crucial. SECONDARY TRANSFERS of fibers occur, and therefore fibers found on a person come from where they were most recently. In this case, the plethora of fiber evidence showed that the victims had all spent the time immediately before and after their deaths in the same environment. The fibers also offered convincing evidence that the same person was responsible for all the killings.

When a local paper published information relating to the fiber findings, the killer changed his modus operandi (MO) and began stripping his victims completely or down to their undershorts and dumping some into the Chattahoochee River. This change of pattern led to Williams's arrest in the early morning hours of May 22, 1981. A surveillance team was located under a bridge over the river and heard a loud splash. A car that was driving slowly over the bridge was stopped, and Wayne Williams was arrested. Two days later, another body was recovered from the river, the odd green-yellow fibers found entangled in his head hair. Subsequent searches of Williams's home and vehicles led to the discovery of the probable sources of the fibers, hair (a family dog), and, most important, the greenish carpet fiber with the unusual cross section. Additional research eventually led to the probable manufacturer of the carpet and allowed the forensic analysts to estimate that this type of carpet would be found in only one in approximately 8,000 homes in the Atlanta area. When the particular combinations of fibers were considered, the odds of a random match were even smaller. Williams was tried for two of the murders with evidence from 10 others introduced. However, suspicion remains that Williams may have killed others. In the absence of any other significant physical evidence, it was the totality of the fiber transfers and the combination of fibers found on the victims that could be linked to Williams that won the conviction. Williams was sentenced to life imprisonment. The case points to the importance of TRACE EVIDENCE even in an age of sophisticated chemical and biochemical analysis.

Further Reading

Deadman, H. "Fiber Evidence and the Wayne Williams Trial." In *Criminalistics: An Introduction to Forensic Science*. 7th ed. Edited by R. Saferstein. Upper Saddle River, N.J.: Prentice Hall, 2001.

wear patterns/wear characteristics Patterns acquired in or on an object as a result of normal usage. Wear patterns are found in objects ranging from clothing to tools and can be distinctive enough to allow INDIVIDUALIZATION. For example, something as common as boots are mass produced, and, for any given type or model, pairs manufactured one right after the other would be expected to be identical, or nearly so. However, if two pairs of identical boots are sold to two different people, through different wear the boots will become unique. One purchaser may wear the boots daily to work outdoors at a construction site, while another may wear them only once a month to work around the yard. Within a short time, the initially identical boots will be easily distinguished from one another. Uniqueness acquired through wear is often exploited in forensic science in areas such as IMPRESSION EVIDENCE (SHOE PRINTS and TIRE PRINTS, for example), TOOLMARKS, and FIREARMS.

See also ACCIDENTAL CHARACTERISTICS.

Wiener, Alexander (1907–1976) American *Biologist/Serologist* In the United States, the person generally credited with pushing ABO typing (see ABO BLOOD GROUP SYSTEM) into the realm of routine forensic use was Dr. Alexander S. Wiener, who worked in the New York City Medical Examiner's Office under Dr. THOMAS GONZALES. Appointed to his position, Wiener had worked with Dr. KARL LANDSTEINER, as had so many of the pioneering forensic serologists (see SEROLOGY, FORENSIC). Wiener was also the first to use BODY FLUIDS other than BLOOD for typing and, more important, to have the results lead to a confession and conviction. In March 1943, a woman was murdered in New York City. The key evidence was SWEAT-stained garments that Wiener was able to type as B, the same as the suspect. Faced with this evidence and other information, the male suspect confessed.

The "Top Ten" Cases in Forensic Science

There have been dozens of cases critical in charting the course of forensic science both in the courtroom and in the popular imagination. Some of these are new, some old, some are famous or infamous, and some have yet to go to trial. Many had (and continue to have) consequences far beyond forensic science. This list highlights some of these cases and articulates why they were pivotal.

1. **THE MARIE LAFARGE CASE (TOXICOLOGY), 1840:** This case was notable for the first legal testimony based on a chemical test (the MARSH TEST) and for the testimony of MATHIEU ORFILA, considered to be the father of forensic toxicology. The Marsh test was developed in 1836 and provided the first reliable method for the detection of arsenic in tissue. Given the near epidemic use of arsenic as a poison at that time, Marsh's test was pivotal not only in detection of poisonings but also in deterring future poisonings. The Lafarge case was the first in which Marsh test results were used to obtain a conviction and served notice that arsenic poisoning was no longer an easy crime.

   Marie's husband, Charles, a much older man, passed away under circumstances consistent with ARSENIC poisoning. Initial tests of Charles's stomach showed no arsenic, nor did the organs removed from the body. However, arsenic was detected in food and beverages that Charles had been given. The court charged Orfila, who at that time was a recognized expert in forensic TOXICOLOGY, to sort out the analytical results. Along with the other experts, he reviewed and reanalyzed the samples, finding small amounts of arsenic in the body. Orfila, a consummate chemist, also showed that the arsenic was not from the soil in which the body had been buried, nor had it come from contamination of reagents. To this day, such cognizance of QUALITY ASSURANCE/QUALITY CONTROL (QA/QC) and an awareness of possible contamination constitutes an integral part of forensic toxicology. Finally, Orfila was able to attribute the initial FALSE NEGATIVE results to analyst inexperience. Marie was convicted, forensic toxicology was recognized, and Orfila was acknowledged as its founder.
2. **THE STIELOW CASE (FIREARMS), 1917:** This case marked an early example of the dangers of self-professed experts and false convictions and of how competent science can be used to correct them. For Charles Stielow, the reversal saved him from the electric chair. He had been accused and convicted of shooting his landlord and his housekeeper. An expert at the trial stated that Stielow's pistol had fired the fatal shots, and Stielow was sentenced to die. Shortly thereafter, another man confessed to the crime, prompting a reexamination. No evidence of a conclusive match of the bullet to Stielow's pistol was found. Critical in the reversal was the work of Charles Waite, who proceeded to collect information about the rifling characteristics of firearms in the United States and Europe. In 1924, he started a private laboratory called the Bureau of Forensic Ballistics staffed by pioneers such as CALVIN GODDARD, John Fisher, and Philip Gravelle. Goddard went on to greater fame in cases such as the St. Valentine's Day Massacre.
3. **THE SACCO-VANZETTI CASE (FIREARMS), 1920–1927:** On April 20, 1920, a robbery took place in South Braintree, Massachusetts, in which five men robbed a paymaster, killing him and the guard. A month later, Sacco and Vanzetti were charged with the crime. The case took a political twist since both men were of foreign descent and were anarchists, a movement that was spreading across the United States and Europe at the time. Thus, the case became highly politicized and publicized, a precursor in many ways to the O. J. SIMPSON trial that was to follow nearly 70 years later. A .38 pistol was recovered from Vanzetti, but it could not be conclusively tied to any of the evidence recovered. From the bodies, .32 ACP bullets were recovered, and Sacco had a .32 pistol in his possession when arrested. Many experts testified about the FIREARMS evidence, while riots and protests raged outside the courtroom and around the world. The same expert who had been pivotal in convicting Charles Stielow, Dr. Albert Hamilton, was again involved and was caught trying to switch the barrel of one of the guns admitted into evidence. The judge witnessed the attempt and prevented him from completing the exchange. A review and reanalysis of the firearms evidence by Goddard linked one cartridge case and a fatal bullet to Sacco's gun. The executions took place seven years after the crime, on August 23, 1927. Although contentious at the time, later firearms examination supported Goddard's findings.
4. **THE LINDBERGH KIDNAPPING (QUESTIONED DOCUMENTS, PHYSICAL MATCHING, WOOD EVIDENCE) 1932:** From the forensic perspective, this case was noteworthy for the variety of physical evidence and the novel use of physical matching on the ladder used to abduct the child. It was the first famous American trial to bring forensic evidence into the public spotlight.

   Charles A. Lindbergh became a national and international hero in 1927 after flying the Atlantic alone in the *Spirit of St. Louis*. He later married Anne Morrow and their first child, a son, named Charles, Jr., was 20 months old when he was kidnapped around 9:30 P.M. on

(continues)

The "Top Ten" Cases in Forensic Science *(continued)*

March 1, 1932. The case took many bizarre turns and became the subject of a worldwide media frenzy. A ransom of $50,000 was paid, the majority of it in gold certificates, which were used as currency at the time. Serial numbers of the bills were recorded before the money was delivered. Sadly, the body of the child was found a month later in the woods close to the Lindbergh home. The kidnapper was never seen or heard from after the ransom was paid. The break came in September 1934, when an exchange bank received a $20 gold certificate from a man who matched the description of the kidnapper. More important, a license plate number had been written on the bill, which was quickly traced to a truck belonging to Bruno Richard Hauptmann, who lived in the Bronx. He was arrested, and a subsequent search of his garage uncovered a gun and several thousand dollars of the ransom money gold certificates.

During the time between the kidnapping and murder of the child and Hauptmann's arrest, forensic investigations were undertaken on the physical evidence, including the ransom notes (QUESTIONED DOCUMENTS), trace evidence, psychological and psychiatric aspects, and, perhaps most damning for Hauptmann, the ladder that had been left outside the window of Charles's second floor room. ALBERT S. OSBORN, a pioneer in the field of questioned documents, performed analysis of the handwriting found in the ransom notes. Arthur Koehler, a wood expert employed by the U.S. Forest Service, undertook a meticulous evaluation of the ladder, including the wood, construction techniques, and toolmarks found on the wood. Eventually, he was able to trace the lumber used to a lumberyard and mill located in the Bronx. Marks made by planers on the wood in the ladder matched a planer at the yard. A search of the attic above Hauptmann's apartment revealed a missing floorboard. Nail holes and tree ring patterns from a rail of the ladder lined up perfectly where the floorboard had been. Furthermore, Koehler was able to demonstrate this at the trial as well as to show how the planer marks from the ladder matched Hauptmann's planer. Hauptmann was convicted, and, after a series of appeals and reviews, including one by the Supreme Court, he was executed on April 3, 1936. Although conspiracy theories still emerge claiming to identify the "real" kidnapper, the forensic evidence against Hauptmann was overwhelming. The jury agreed. This decision would stand in sharp contrast to the next "trial of the century," in which the forensic evidence, almost equally overwhelming, was discounted in the O. J. Simpson case.

5. **THE SAM SHEPPARD CASE (BLOOD SPATTER AND CRIME SCENE EVIDENCE), 1954–ONGOING:** Sam Sheppard was an osteopathic physician living in a suburb of Cleveland with his wife, Marilyn, who was 31 and pregnant on the night she was murdered, July 4, 1954. Like the Lindbergh trial, it was a sensationalized case in which physical evidence played a key role. It evolved into a series of trials, one before the U.S. Supreme Court.

   Initially, forensic evidence was limited principally to the results of the AUTOPSY and testimony of the CORONER, all of which would become crucial. Sheppard was convicted December 21, 1954, and sentenced to life imprisonment. However, in 1966 the U.S. Supreme Court ordered Sheppard released on the basis that his trial was hopelessly prejudiced by publicity. A second trial began in October 1966, and, in the 12 years that had passed, additional forensic analysis was conducted. PAUL KIRK, an eminent forensic scientist in the 1950s, had reviewed the evidence, principally blood spatter, and concluded that Sheppard was likely innocent. However, in the days before DNA TYPING, the amount and type of information that could be extracted from BLOODSTAINS was limited to the ABO BLOOD GROUP SYSTEM and stain patterns. The second trial was also notable for the presence of F. Lee Bailey as the defense lawyer. The case was instrumental in making Bailey famous and, ironically, set the stage for his appearance in the O. J. SIMPSON trial decades later. Sheppard was found not guilty in November 1966. Despite being freed, his health quickly deteriorated due to drinking, and he died in 1970. Legal action and additional forensic study, including an exhumation of Marilyn's body, continued into 2002.

   Because of its length, advances in forensic science from the late 1950s to the present day, particularly in DNA typing, played a continuing role in the case. Everything from CRIME SCENE analysis to BITE MARKS, TOOLMARKS, bloodstains, profiling, BLOODSTAIN PATTERNS, forensic ANTHROPOLOGY, and forensic ODONTOLOGY has been applied to the evidence at one time or another, with a variety of findings, many contested. Unfortunately, as technology improved, the evidence deteriorated, memories dimmed, and principals died. As a result, it is doubtful that the truth will ever be known. Like the Lindbergh case before and the Simpson case to follow, the Sheppard case became a cottage industry producing articles, books, and speculation, all in their own small way adding to the public awareness of forensic science. The continuing controversy served as the inspiration for the 1960s television series *The Fugitive* and, later, the 1993 movie of the same name.

6. **THE ASSASSINATION OF PRESIDENT KENNEDY (FIREARMS, AUTOPSY), 1963:** In modern times, no single mur-

der has generated more speculation, conspiracy theories, conjecture, and fictionalization than the death of President John F. Kennedy on November 22, 1963, in Dallas, Texas. Kennedy was hit by two of three bullets fired from above and behind from the Texas School Book Depository building, and Lee Harvey Oswald was arrested within hours of the killing. Of the three shots fired, one missed; one dealt the fatal head shot to Kennedy; and one, the so-called magic bullet, struck Kennedy in the throat, exited, and passed through Governor John Connally of Texas, who was sitting in the front passenger seat of the open convertible. This bullet, recovered nearly intact on Connally's stretcher, was chemically analyzed and shown to be similar to minute fragments found in his wrist, but different from the fragments recovered from the president's brain tissue. Although as in any complex case or investigation, gaps and improbabilities crop up, an objective analysis of the physical evidence supports the theory that Oswald was the one and only gunman in Dallas. However, public skepticism was and remains much higher than exists for the Lindbergh and Sheppard cases, principally because the victim was the president of the United States. Aspects of the case, such as the magic bullet, continue to draw criticism even in the face of reasonable explanations and supporting chemical analysis. No doubt had Oswald lived to be tried, a guilty verdict would have been controversial, probably in much the same way the Sacco-Vanzetti verdict was. A public eager to accept conspiracy and cover-up would likely never have accepted the verdict even if a jury had. Thus, from a forensic point of view, the Kennedy case was important in illustrating how forensic evidence can take a back seat to other considerations if the circumstances are right.

7. **THE WAYNE WILLIAMS CASE (HAIR AND FIBERS), 1982:** Starting in 1979, young black men were disappearing in Atlanta, Georgia, and their bodies were later being discovered dumped in wooded areas or near highways and streets. Most of the victims died of asphyxiation. The bodies were found with a wealth of FIBER evidence that indicated that the victims had all spent the time immediately before and after their deaths in the same environment. When a local paper published information relating to the fiber findings, the killer began stripping his victims completely or down to their undershorts and dumped some of the bodies into the Chattahoochee River. This change of pattern led to Wayne Williams's arrest in the early morning hours of May 22, 1981, as his car passed slowly over a bridge being watched by a surveillance team. Subsequent searches of his home and vehicles led to the discovery of the probable sources of the fibers, hair (a family dog), and, most important, a greenish carpet fiber with an unusual cross section. Additional research eventually led to the probable manufacturer of the carpet and allowed the forensic analysts to estimate that this type of carpet could be found in only one in approximately 8,000 homes in the Atlanta area. When the particular combinations of fibers were considered, the odds of a random match were even smaller. Williams was tried for two of the murders with evidence from 10 others introduced. However, the suspicion remains that Williams killed others. In the absence of any other significant physical evidence, it was the totality of the fiber transfers and the combination of fibers found on the victims linked to Williams that won the conviction. The case confirms the importance of trace evidence and statistical assessments of significance, even in an age of sophisticated chemical and biochemical analysis.
8. **THE O. J. SIMPSON CASE (DNA TYPING, CRIME SCENE EVIDENCE), 1995:** The case was noteworthy, even notorious, for many reasons, including the forensic aspects, which focused around CRIME SCENE documentation, CHAIN OF CUSTODY, and DNA TYPING results. Many separate BLOODSTAINS were found at the scene, in Simpson's home, and in his car that had types consistent with one or both of the victims, and the odds that these were random matches were inconceivably high. The defense never attacked the results of forensic analysis directly; rather, their case rested on the supposition that members of the Los Angeles Police Department who harbored racial motivations framed Simpson. Their attacks centered on issues of potential contamination, deliberate alteration and planting of evidence, poor crime scene documentation, and poor evidence collection and analysis techniques. If any of these occurred, it was argued, a reasonable doubt as to Simpson's guilt had been raised. After months of courtroom presentation and endless media attention and analysis, the jury took a short time to acquit Simpson on October 3, 1995.

Implications for forensic science were significant and raised serious questions within the profession as to the conduct of the scientists involved, particularly defense experts. The trial also pointed to an increasingly important issue. As the complexity of analyses, particularly DNA analysis, increases, the procedures and results become harder to explain to a jury. Consequently, critical and unassailable physical evidence may have been ignored because it was not properly understood. The trial highlighted the fact that forensic analysis and evidence is not necessarily the key component of any jury decision. Many other factors, including jury selection, money available to the defense, socioeconomic factors, and racial issues, were raised

(continues)

The "Top Ten" Cases in Forensic Science *(continued)*

and exploited at the trial. In this sense, the Simpson trial set the stage for the 21st century, in which scientific analyses will become increasingly complex even as these other factors become more crucial in dispensing justice.

9. **SEPTEMBER 11, 2001 (ENGINEERING, MASS DISASTERS, CRIME SCENES), 2001–ONGOING:** The terrorist attacks on this fateful day created the largest CRIME SCENE and largest forensic science investigation to date. Sadly, it also ushered in a new era in forensic science in which the scope of investigations becomes enormous and the volume of evidence and information collected can be nearly overwhelming. Investigations crossed jurisdictional lines, involving local, state, and federal officials from multiple agencies. The last debris from the World Trade Center site was sorted and the last human remains collected in July 2002, but analysis of much of the evidence collected will continue for years. Engineering analysis of the collapse of the Twin Towers has already revealed information that can be used in future construction to help buildings withstand the nearly unimaginable forces involved when the jet liners, traveling several hundred miles an hour, smashed into the buildings. Thus, although the tasks are grim, the lessons learned from September 11 will help forensic science cope with the new realities of the 21st century.
10. **THE ANTHRAX MAILING (QUESTIONED DOCUMENTS, PROFILING), 2001–ONGOING:** On the heels of the September 11 attack came the anthrax mailings that killed postal workers and others along their paths through the postal system. In this case, physical evidence was seized, but normally routine forensic analysis was complicated by the deadly nature of the anthrax powder coating the envelopes. The crime scenes were spread thousands of miles in many directions as the envelopes moved from Trenton, New Jersey, to postal handling facilities along the east coast of the United States. Procedures had to be developed to make the evidence safe to handle without altering or destroying it, and genetic testing was used in an attempt to track the anthrax bacteria to an identifiable source. The FBI went public with a profile of their suspect, using the Internet to enlist the public in the investigation. In this case, the evidence was deadly, and the forensic scientists were as much at risk as anyone who unwittingly handled it before it became evidence. Like the September 11 attacks, the anthrax mailings and the subsequent changes and challenges will bring forensic science into the terrorist age.

Further Reading

Goldsmith, R. H. "The Search for Arsenic." In *More Chemistry and Crime: From the Marsh Test to DNA Profile.* Edited by S. M. Gerber and R. Saferstein. Washington, D.C.: American Chemical Society, 1997.

Inman, K., and N. Rudin. "Evolution of Forensic Science." In *Principles and Practices of Criminalistics: The Profession of Forensic Science.* Boca Raton, Fla.: CRC Press, 2001.

Rowe, W. F. "Firearms Identification." In *Forensic Science Handbook.* Vol. 2. Edited by R. Saferstein. Englewood Cliffs, N.J.: Regents/Prentice Hall, 1988.

Stajic, M. "Milestones in Forensic Toxicology." In *More Chemistry and Crime: From the Marsh Test to DNA Profile.* Edited by S. M. Gerber and R. Saferstein. Washington, D.C.: American Chemical Society, 1997.

wet chemistry A term used to describe older chemical analyses that do not rely on instrumentation. Examples of wet chemical analyses include gravimetric analysis and many types of volumetric titrations. In a gravimetric analysis, for example, the quantity of a given ion such as silver (Ag^+) in a solution is determined by combining it with a solution that results in the formation of an insoluble material that precipitates out of solution. The solid is then dried and weighed, and the concentration of silver in the original solution is determined by calculation. No instrumentation other than a simple balance is required. In forensic science, wet chemical methods are used primarily as PRESUMPTIVE TESTS such as screening tests for drugs or EXPLOSIVES.

Widmark, Erik (1889–1945) Swedish *Chemist* Dr. Erik Widmark developed the first reliable methods for determining ALCOHOL concentration in BLOOD in 1922 (*see* BLOOD ALCOHOL CONCENTRATION (BAC)). Widmark, a Swede, worked in the University of Lund. His methods remained the standard worldwide into the 1970s. Widmark's studies extended to the metabolic pathways ethanol followed, once ingested.

Widmark received an M.D. degree in 1917, but he gravitated toward research early in his career. His thesis centered on measurement of acetone in blood and other tissues and fluids, a common occurrence in people with diseases such as diabetes. This work set the stage for his research with alcohol, a compound that behaved similarly to acetone. Unlike most drugs, acetone and alcohol (specifically ethanol) are appreciably volatile. Consequently, ethanol is detectable in exhaled breath, and its concentration depends directly on the blood concentration responsible for intoxication. Widmark's research allowed work to move from concept to applications such as the modern Breathalyzer.

Upon graduation, Widmark accepted a position at the University of Lund in 1918, and he competed successfully for a full professorship there in 1921. His main competitor for the position was Dr. L. Michaelis, who later became famous for his work relating to the role of enzymes in biochemical reactions. A fundamental equation related to enzyme reactions, the Michaelis-Menton equation bears his name in testament to his accomplishments. This equation remains central in describing how quickly enzymes in the body work to eliminate ethanol from the system.

Alcohol intoxication has been an issue throughout human history, but it became much more of a problem with the introduction of the automobile. Operating a horse while drunk is a nuisance; operating several tons of metal while drunk endangers the public. The same applied to other modern machinery such as airplanes and boats. Being able to detect and quantify the degree of intoxication became a medicolegal question. Law enforcement agencies and the courts needed reliable methods to gauge the severity of the intoxication, the degree of public threat, and ultimately the penalty or treatment.

By the late 1920s, several trends were clear. First, automobiles were soon to become the primary means of transport for most people. More cars meant more drivers, more drunk drivers, and more accidents. Given poor roads and lack of even rudimentary safety engineering, even seemingly minor accidents could cause injury and death. The judicial system needed a quantitative chemical test of blood alcohol concentration that could prove intoxication. Widmark was the right man in the right place at the right time.

The first tests he devised for blood alcohol were crude but effective. For each, he collected five small tubes of blood collected from the earlobe. He then transferred a carefully weighed portion to a small cup suspended over a solution containing an oxidizer of sulfuric acid and dichromate ion. As the ethanol evaporated, it diffused into the oxidizing liquid below. The resulting chemical reaction consumed the oxidizer in proportion to the amount of ethanol, and by calculating the concentration of the remaining oxidizer, the concentration of ethanol in the blood could be determined. Testing five samples instead of one increased the reliability of the results. With a valid analytical method in place, Sweden was among the first countries to establish a DUI law with the intoxication limit set at 0.80 milligram per gram (0.08 percent) in 1941. As methods improved, the limit fell to 0.05 percent in 1957 and 0.02 percent in 1990, which remains one of the lowest limits worldwide. By comparison, the U.S. federal standard is 0.08 percent.

Another of Widmark's key contributions was in deciphering the process and speed of ethanol elimination from the body. Beginning in 1924, he published a series of papers describing mathematical equations relating blood alcohol concentration to the time and amount of ethanol ingested. Alcohol is rapidly absorbed into the bloodstream in the stomach and upper gastrointestinal tract, leading to a spike in blood concentration. Assuming an empty stomach and no further ingestion, metabolic processes remove alcohol at a steady and predictable rate. Variables include sex (women on average metabolize and remove ethanol faster than men) and genetic variations. Widmark's equation remains the standard for determining how much alcohol a person drank and when he or she started from the blood alcohol concentration (BAC).

Widmark's blood test and metabolic understanding was critical to forensic TOXICOLOGY, but the problem of how to determine intoxication in the field remained. A police officer cannot draw blood from each person he or she stops; nor will everyone stopped on suspicion of intoxication actually be so. Law enforcement needed a field test that could provide both probable cause of intoxication or reliable evidence of its absence. In the United States, the National Safety Council became involved in issues of alcohol intoxication, and they appointed a commission to study the issue. This work led to the first discussions of using breath as a means of estimating intoxication.

Ethanol, with two carbons, is a small molecule by biological standards. It is soluble in water and water-like systems such as blood, but it also appreciably volatile. The warmer the liquid, the greater the tendency of ethanol to evaporate out of it. In the human body, blood comes in direct contact with the atmosphere deep in the lungs. There, some ethanol will evaporate out of the blood and into the exhaled air. Because human body temperature is nearly constant for all people, the concentration of ethanol in exhaled breath is directly proportional to the concentration of ethanol in the blood, and that proportion can be calculated using simple equations. These equations are the basis of breath alcohol testing used today.

wildlife forensics *See* FISH AND WILDLIFE SERVICE.

Williams, Wayne *See* WAYNE WILLIAMS CASE.

wood A form of evidence that can be encountered in many forms such as wooden objects or BUILDING MATERIALS. Analysis of wood evidence is usually considered to be part of TRACE EVIDENCE or forensic BOTANY, depending on the type of evidence and where it was recovered. For example, wood found clinging to a body could provide useful information about where the victim had been recently. The use of wood evidence, specifically the analysis of a wooden ladder, was crucial in the LINDBERGH KIDNAPPING case and trial in 1932.

wool *See* FIBERS.

wound ballistics The study of wound patterns and forms, principally by forensic PATHOLOGISTS, to determine such things as what type of weapon was used, what movement took place during the struggle, when the wounds were inflicted, and whether wounds might have been defensive or self-inflicted. Types of wounds studied include gunshot wounds, stabbing and cutting wounds inflicted by a sharp object, blunt force injuries such as are incurred by beatings with weapons such as clubs, and wounds caused by accidents such as airplane and automobile crashes. Wound patterns are studied during AUTOPSY, both by external examination and dissection, as well as by experimentation. Wound ballistics is also useful in accident reconstruction, MASS DISASTERS, and forensic ENGINEERING. As an example, if a bomb were suspected in the crash of an airliner, the types of injuries inflicted by the blast, particularly those near the detonation, would be different from the types of injuries associated with other possible causes such as an in-flight breakup.

Firearms Wounds (Gunshot Wounds)

Several issues are considered when injuries are caused by FIREARMS, such as the type of weapon (handgun, rifle, or shotgun), caliber, distance between the shooter and the victim, sequence of shots, and whether a wound is an entrance or exit wound. Fundamentally, the damage inflicted by a fired projectile is related to its mass and velocity when it strikes the target. This energy (kinetic energy, or energy of motion) is defined mathematically as: $KE = \frac{1}{2}mv^2$, where m is the mass and v is the velocity. The squared term is critical and explains why a rifle bullet that is smaller than a handgun bullet (for example, a .223 CALIBER rifle bullet versus a .45 pistol bullet nearly twice the size) can cause much more massive damage. The rifle bullet is fired at a much higher velocity, and, since that term is squared, the damage increases proportionately. Additional and detailed information on BULLET WOUNDS is covered in that entry.

Blunt Force Wounds

Blunt force injuries include scrapes (abrasions), bruises (contusions), broken bones or other bone injuries, and cuts (lacerations). Lacerations caused by blunt force injury are not the same as stabbing or cutting wounds, although the initial appearance may be similar. A laceration is a tear in the skin (and underlying tissues) that is caused by a blunt force impact, whereas a sharp object causes a cut or stab. Similarly, blunt force impact can cause ruptures of arteries and organs beneath the surface of the skin even when the skin itself does not show significant damage.

Blows to the head ("bludgeoning"), common in homicides, can fall into this category, and the injury can result in a fatal hemorrhage and subsequent swelling that compresses the brain or in other injuries inside the brain itself. The location of injury can also give clues as to how the head was moving when the blow was incurred. If the head is moving and strikes a stationary surface, such as when someone falls to the floor, the hemorrhage inside the skull will be most pronounced on the side opposite the impact. This is referred to as the "countercoup." If, on the other hand, the head is stationary and is hit by a weapon such as a baseball bat, the impact site (the "coup") will show greater damage.

In blunt force injury, bruises sometimes form in the pattern of the object that caused the injury, such as the bumper of a car or the striking surface of a weapon, even a hand. Bruises (also called contusions) are usually formed from blows delivered before death, but some postmortem injuries can form bruises. Bruises can also form as a result of defensive actions such as attempting to block an attack using a hand or forearm.

Stabbing and Cutting Wounds

Characteristics of a stab wound can be used to determine what type of weapon was used, whether it was sharp on one side or two, how long it was, and how many wounds were inflicted. Aside from knives, other stabbing weapons include scissors, razors, picks, screwdrivers, and anything that can be forcefully thrust into or through the skin. However, the appearance of a stab wound is not always simple to interpret and can be complicated by such factors as angles, force applied, movement of the victim, twisting of the blade, and other variables. Thus, it is not always possible to deduce everything about a weapon from the wound pattern alone. As with blunt force trauma, the external appearance of the wound does not necessarily reflect the depth, force, or severity of injury inflicted. Stab wounds are often clustered, but not always. In cases in which suicide might have been the cause of death, "hesitation marks" are frequently observed. These are usually superficial or shallow cuts that run parallel to the track of the fatal wound, and there may be several such marks. Suicidal stabbings of the chest and slitting of the throat are not unheard of, but they are less common than slitting of the wrists. Thus, a study of the wound patterns, particularly in this type of situation, is essential to determine if the wounds could have been self-inflicted. Stabbing attacks also frequently result in defensive wounds on the victim, usually on the back of the hand or forearm or on the fingers. Severe lacerations of the fingers are observed when victims grasp the blade in an attempt to stop the attack.

See also BIOMECHANICS, FORENSIC; BULLET WOUNDS; DISTANCE DETERMINATION; PATHOLOGY, FORENSIC.

Further Reading

DiMaio, D. J., and V. J. M. DiMaio. *Forensic Pathology.* Boca Raton, Fla.: CRC Press, 1993.

Dix, J., and R. Calaluce. *Guide to Forensic Pathology.* Boca Raton, Fla.: CRC Press, 1999.

Missliwetz, J., W. Denk, and I. Weiser. "Study on the Wound Ballistics of Fragmentation Protective Vests Following Penetration by Handgun and Assault Rifle Bullets." *Journal of Forensic Sciences* 40, no. 4 (1995): 582.

Moenssens, A. A., J. E. Starrs, C. E. Henderson, and F. E. Inbau. *Scientific Evidence in Civil and Criminal Cases.* Westbury, N.Y.: The Foundation Press, 1995.

X–Z

xanthine alkaloids *See* ALKALOIDS.

xenobiotic *See* ADME.

X-ray techniques A family of instrumental techniques that exploits the interaction of X-rays with matter. Broadly defined, these techniques include forensic RADIOLOGY, fluoroscopy, and analytical techniques such as X-ray fluorescence (XRF) and X-ray diffraction (XRD). Many related techniques exist, but XRF and XRD are the most widely used in forensic applications. X-rays are a highly energetic form of ELECTROMAGNETIC ENERGY (EMR), characterized by high frequencies and short wavelengths. The different ways in which X-rays interact with matter are shown in the first figure.

X-rays, like any form of electromagnetic radiation, can be partially absorbed by the sample. This is the basis of diagnostic medical and dental X-rays that are produced by exposing the body to a short burst of energy. In general, the heavier the element (the larger its atomic number), the more X-ray radiation it will absorb. Thus bones, which contain large amounts of elements such as calcium (atomic number 20), absorb more radiation than do soft tissues containing large amounts of lighter elements such as hydrogen (atomic number 1) and oxygen (16). Fluoroscopy, such as that used in airport luggage screeners, also depends on X-ray absorption. In this technique, emerging X-rays are directed to a screen that fluoresces, giving off visible light that can be focused and viewed on a screen. In

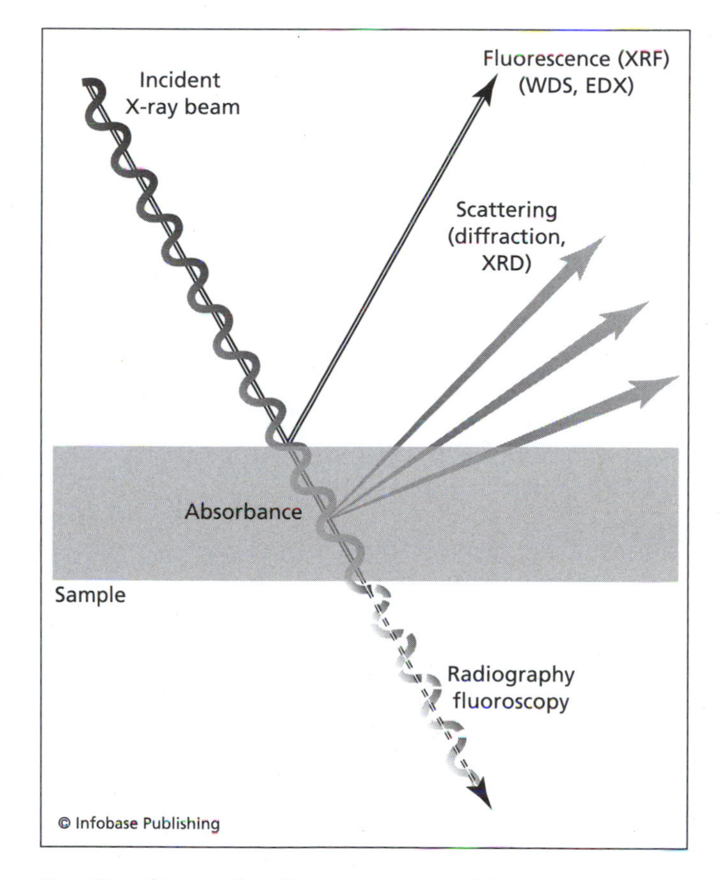

Sampling of interactions between matter and X-rays

addition to absorption, X-rays can cause elements to emit radiation of longer wavelengths (lower energy); this effect is exploited in XRF. The X-rays can also be scattered and bent (diffracted). X-ray diffraction

(XRD) is used to study the crystal structure of materials. Diffraction techniques were used to help decipher the double helix structure of DNA.

X-Ray Fluorescence Methods
When X-rays interact with the electrons surrounding an atomic nucleus, they can cause inner shell electrons to be ejected completely from the atom. This creates instability that is remedied when an electron from an outer shell moves in to fill the vacancy. When an electron "falls" into an empty orbital, it must release energy and does so in the form of an X-ray. The emitted X-ray is of a longer wavelength (lower energy) than was the original incident X-ray. If a filling electron comes from one shell away, the process is referred to as an α (alpha) transition; from two shells away, a β (beta) transition; and from three shells, a γ (gamma) transition. Because more than one electron can be ejected and more than one electron can change shells during the refilling process, each atom can give off more than one wavelength of radiation. The pattern of emitted X-rays is characteristic of a given element. For example, for the element lead (Pb), which is part of GUNSHOT RESIDUE (GSR), electron transitions are labeled $K_{\alpha 1}$, $K_{\alpha 2}$, $L_{\beta 2}$, $L_{\alpha 1}$, and $L_{\beta 1}$. The letter corresponds to the shell where the falling electron originated. Within each shell, additional sublevels exist, leading to the numerical designation. XRF is a QUALITATIVE technique, in that it can identify the elements present, and a QUANTITATIVE technique, since the intensity of emitted radiation will be proportional to the amount of the element present. It is not, however, considered to be the ideal choice for the identification of trace elements present at levels below 0.001 percent.

There are two principal methods for implementation of XRF, namely, energy dispersive and wavelength dispersive. These techniques are sometimes referred to as energy dispersive spectroscopy (EDS) and wavelength dispersive spectroscopy (WDS). In EDS, the detection system differentiates between X-rays emitted from elements based on the different energies produced. In WDS, this differentiation is achieved by separating and detecting the various wavelengths emitted from the sample. Another variant of XRF is called the electron microprobe. The fluorescence is produced not by X-rays but by bombarding the sample surface with a beam of focused electrons, similar to what occurs in SCANNING ELECTRON MICROSCOPY (SEM).

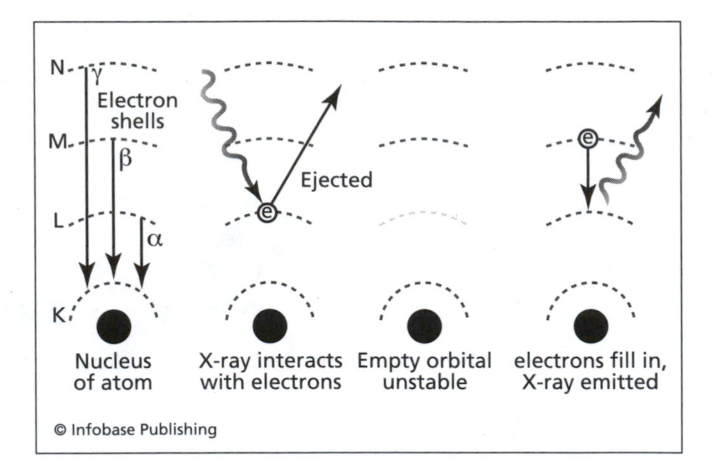

Process of X-ray emission

In forensic science, XRF is used for elemental analysis of evidence such as gunshot residue (GSR) and paints and occasionally for soils, DRUGS, FIBERS, and other types of trace evidence.

X-Ray Diffraction Methods
While XRF is used for identification of individual elements, XRD is used to identify solid crystalline compounds such as potassium nitrate (KNO_3). Crystals are characterized by a regular and orderly structure with distances separating the atoms on the order of size of the wavelengths of X-ray radiation. As a result, X-rays can interact with the atoms in the crystals, resulting in the bending (diffraction) of the radiation in ways that are characteristic of the crystal. The principle of diffraction is based on constructive and destructive interference patterns created by reflection and scattering of the radiation within the crystal. XRD is particularly valuable for the analysis of GSR and explosives, which often contain compounds such as potassium nitrate (KNO_3) and potassium chlorate (K_2ClO_4). The technique can also be used to study inorganic pigments in paint, drugs, minerals, and compounds in soils.

Further Reading
Dean, J. A. *Analytical Chemistry Handbook.* New York: McGraw Hill, 1995.

Jenkins, R. "X-ray Fluorescence Spectrometry." Vol. 99. In *Chemical Analysis: A Series of Monographs on Analytical Chemistry and Its Applications.* New York: John Wiley and Sons, 1988.

Zwikker test *See* DILLE-KOPPANYI TEST.

APPENDIX I

BIBLIOGRAPHIES AND WEB RESOURCES

PRINT

Organized by forensic discipline

Accounting

Bologna, G. J., and R. J. Lindquist. *Fraud Accounting and Forensic Accounting New Tools and Techniques.* New York: John Wiley and Sons, 1995.

Analytical Chemistry

Buffington, R., and M. K. Wilson. *Detectors for Gas Chromatography: A Practical Primer.* Avondale, Pa.: Hewlett-Packard Inc, 1987.

Dean, J. A. *Analytical Chemistry Handbook.* New York: McGraw Hill, 1995.

Eiceman, G. A., and Z. Karpus. *Ion Mobility Spectrometry.* Boca Raton, Fla.: CRC Press, 1995.

Fritz, J. S. *Analytical Solid-Phase Extraction.* New York: Wiley-VCH, 1999.

Harris, D. C. *Quantitative Chemical Analysis.* 4th ed. New York: W. H. Freeman and Company, 1995.

Jenkins, R. "X-ray Fluorescence Spectrometry." Vol. 99. In *Chemical Analysis: A Series of Monographs on Analytical Chemistry and Its Applications.* New York: John Wiley and Sons, 1988.

Skoog, D. A., D. M. West, F. Holler, F. James, and S. R. Crouch. *Analytical Chemistry, An Introduction.* 7th ed. Orlando, Fla.: Harcourt College Publishers, 2000.

Yinon, J., ed. *Forensic Applications of Mass Spectrometry.* Boca Raton, Fla.: CRC Press, 1995.

Anthropology and Archaeology

Fairgrieve, S. I., ed. *Forensic Osteological Analysis: A Book of Case Studies.* Springfield, Ill.: Charles C. Thomas, 1999.

Maples, W. R., and M. Browning. *Dead Men Do Tell Tales, The Strange and Fascinating Cases of a Forensic Anthropologist.* New York: Doubleday, 1994.

Morse, D., J. Duncan, and J. Stoutamire, eds. *Handbook of Forensic Archaeology and Anthropology.* Tallahassee, Fla.: Rose Printing Co., 1983.

Nafte, M. *Flesh and Bone: An Introduction to Forensic Anthropology.* Durham, N.C.: Carolina Academic Press, 2000.

Pickering, R. B., and D. C. Bachman. *The Use of Forensic Anthropology.* Boca Raton, Fla.: CRC Press, 1997.

Reichs, K. J., ed. *Forensic Osteology.* 2d ed. Springfield, Ill.: Charles C. Thomas, 1998.

Ubelaker, D., and H. Scammell. *Bones: A Forensic Detective's Casebook.* New York: HarperCollins, 1992.

Arson and Related Topics

Redsicker, D. R., and J. J O'Connor. *Practical Fire and Arson Investigation.* 2d ed. Boca Raton, Fla.: CRC Press, 1997.

Art

Taylor, K. T. *Forensic Art and Illustration.* Boca Raton, Fla.: CRC Press, 2001.

Blood Evidence and Blood Spatter Patterns

James, S. H., and W. G. Eckert, eds. *Interpretation of Bloodstain Evidence at Crime Scenes.* 2d ed. Boca Raton, Fla.: CRC Press, 1999.

Bloodstain Pattern Analysis

Bevel, T., and R. M. Gardner. *Bloodstain Pattern Analysis with an Introduction to Crime Scene Reconstruction.* Boca Raton, Fla.: CRC Press, 1997.

James, Stuart H., ed. *Scientific and Legal Applications of Bloodstain Pattern Interpretation.* Boca Raton, Fla.: CRC Press, 1999.

Wonder, A. Y. *Blood Dynamics.* San Diego, Calif.: Academic Press, 2001.

Chemistry

Bell, S. *Forensic Chemistry.* Upper Saddle River, N.J.: Prentice Hall, 2006.

Computer Forensics

Casey, E. *Digital Evidence and Computer Crime: Forensic Science, Computers, and the Internet.* London: Cambridge University Press, 2000.

Crime Scenes

Fisher, B. A. J. *Techniques of Crime Scene Investigation.* 7th ed. Boca Raton, Fla.: CRC Press, 2003.

Lee, H. C., T. Palmbach, and M. T. Miller. *Henry Lee's Crime Scene Handbook.* San Diego, Calif.: Academic Press, 2001.

Deception Analysis

Hall, H. V., and D. A. Pritchard. *Detecting Malingering and Deception: Forensic Deception Analysis.* Delray Beach, Fla.: St. Lucie Press, 1996

Turvey, B. E. *Criminal Profiling: An Introduction to Behavioral Evidence Analysis.* San Diego, Calif.: Academic Press, 1999.

DNA Typing

Inman, K., and N. Rudin. *An Introduction to Forensic DNA Typing.* 2d ed. Boca Raton, Fla.: CRC Press, 2002.

Engineering

Brown, J. F., and K. S. Osborn. *Forensic Engineering: Reconstruction of Accidents.* Springfield, Ill.: Charles C. Thomas, 1990.

Carper, K. L., ed. *Forensic Engineering.* 2d ed. Boca Raton, Fla.: CRC Press, 2001.

Milton, P. *In the Blink of an Eye, The FBI Investigation of TWA Flight 800.* New York: Random House, 1999.

Noon, R. K. *Engineering Analysis of Vehicular Accidents.* Boca Raton, Fla.: CRC Press, 1994.

Shepherd, R., and J. D. Frost, eds. *Failures in Civil Engineering: Structural, Foundation, and Geoenvironmental Case Studies.* Reston, Va.: American Society of Civil Engineers, 1995.

Entomology (Insects)

Byrd, J. H., and J. L Caster, eds. *Forensic Entomology: The Utility of Arthropods in Legal Investigations.* Boca Raton, Fla.: CRC Press, 2001.

Fingerprints

Lee, H. C., and R. E. Gaensslen. *Advances in Fingerprint Technology.* Boca Raton, Fla.: CRC Press, 1994.

Lee, H. C., and R. E. Gaensslen. *Advances in Fingerprint Technology.* 2d ed. Boca Raton, Fla.: CRC Press, 2001.

Firearms and Gunshot Residue

Schwoeble, A. J., and D. L. Exline. *Current Methods in Forensic Gunshot Residue Analysis.* Boca Raton, Fla.: CRC Press, 2000.

Footwear, Shoeprints, and Tire Impression Evidence

Bodziak, W. J. *Footwear Impression Evidence.* New York: Elsevier, 1990.

McDonald, P. *Tire Imprint Evidence.* Boca Raton, Fla.: CRC Press, 1993.

General Textbooks and References

De Forest, P. R., R. E. Gaensslen, and H. C. Lee. *Forensic Science: An Introduction to Criminalistics.* New York: McGraw Hill, 1983.

Eckert, W. E., ed. *Introduction to Forensic Sciences.* 2d ed. Boca Raton, Fla.: CRC Press, 1992.

Houck, M. M., and J. Siegel. *Fundamentals of Forensic Science.* Burlington, Mass.: Academic Press, 2006.

James, S. H., and J. J. Nordby, eds. *Forensic Science: An Introduction to Scientific and Investigative Techniques.* 2nd ed. Boca Raton, Fla.: CRC Press, 2005.

Lane, B. *The Encyclopedia of Forensic Science.* London: Headline Book Publishing, 1992.

Nickell, J. N., and J. F. Fisher. *Crime Science: Methods of Forensic Detection.* Lexington: University Press of Kentucky, 1998.

Nordby, J. J. *Dead Reckoning—The Art of Forensic Detection.* Boca Raton, Fla.: CRC Press, 2000.

Geology

Murray, R. C., and J. C. F. Tedrow. *Forensic Geology.* Englewood Cliffs, N.J.: Prentice Hall, 1992.

Law

Becker, R. F. *Scientific Evidence and Expert Testimony Handbook.* Springfield, Ill.: Charles C. Thomas, 1997.

Black's Law Dictionary. 6th ed. St. Paul, Minn.: West Publishing Company, 1990.

Moenssens, A. A., J. E. Starrs, C. E. Henderson, and F. E. Inbau. *Scientific Evidence in Civil and Criminal Cases.* 4th ed. Westbury N.Y.: The Foundation Press, Inc., 1995.

Miscellaneous

Gerber, S. M., ed. *Chemistry and Crime: From Sherlock Holmes to Today's Courtroom.* Washington, D.C.: American Chemical Society, 1983.

Kaye, B. H. *Science and the Detective, Selected Reading in Forensic Science.* Weinheim, Germany: VCH Press, 1995.

Kurland, M. *A Gallery of Rogues, Portraits in True Crime.* New York: Prentice Hall General Reference, 1994.

Lee, J. A. *The Scientific Endeavor: A Primer on Scientific Principles and Practice.* San Francisco: Addison Wesley Longman, 2000.

Saferstein, R., and S. M. Gerber, eds. *More Chemistry and Crime: From Marsh Arsenic Test to DNA Profile.* Washington, D.C.: American Chemical Society, 1997.

Pathology and Taphonomy

Brogdon, B. D. *Forensic Radiology.* Boca Raton, Fla.: CRC Press, 1998.

DiMaio, D. J., and V. J. M. DiMaio. *Forensic Pathology.* Boca Raton, Fla.: CRC Press, 1993.

Dix, J., and R. Calaluce. *Guide to Forensic Pathology.* Boca Raton, Fla.: CRC Press, 1999.

Dix, J., and M. Graham. *Time of Death, Decomposition, and Identification, An Atlas.* Boca Raton, Fla.: CRC Press, 2000.

Haglund, W. D., and M. H. Sorg, eds. *Forensic Taphonomy. The Postmortem Fate of Human Remains.* Boca Raton, Fla.: CRC Press, 1997.

Knight, B. *Simpson's Forensic Medicine.* 10th ed. London: Edward Arnold, 1991.

Spitz, W. U., ed. *Spitz and Fisher's Medicolegal Investigation of Death.* 3d ed. Springfield, Ill.: Charles C. Thomas, 1993.

Psychology, Psychiatry, Profiling, Etc.; Behavioral Evidence

Sweet, J. J., ed. *Forensic Neuropsychology Fundamentals and Practice.* Lisse, Netherlands: Swets and Zeitlinger Publishers, 1999.

Turvey, B. E. *Criminal Profiling: An Introduction to Behavioral Evidence Analysis.* San Diego, Calif.: Academic Press, 1999.

Questioned Documents

Ellen, D. *The Scientific Examination of Documents, Methods and Techniques.* 2d ed. London: Taylor and Francis, 1997.

Hilton, O. *Scientific Examination of Questioned Documents.* Rev. ed. Boca Raton, Fla.: CRC Press, 1993.

Koppenhaver, K. M. *Attorney's Guide to Document Examination.* Westport, Conn.: Quorum Books, 2002.

Toxicology

Levine, B. *Principles of Forensic Toxicology.* 2nd ed. Washington, D.C.: American Association of Clinical Chemistry, 2003.

Trace Evidence/Hairs/Fibers

Houck, M. *Mute Witness: Trace Evidence Analysis.* San Diego, Calif.: Academic Press, 2001.

Robertson, J., ed. *Forensic Examination of Fibers.* London: Ellis Horwood, 1992.

WEB RESOURCES

General Information

Armed Forces Institute of Pathology. www.afip.org

Zeno's Forensic Page. www.forensic.to/forensic.html

Reddy's Forensic Page. www.forensicpage/com

Nikon's Microscopy University. www.microscopyu.com

Professional Organizations and Societies

American Academy of Forensic Psychology (AAFP, www.abfp.com)

American Academy of Forensic Sciences (AAFS, www.aafs.org)

American Academy of Psychiatry and the Law (AAPL, www.emory.edu/AAPL)

American Association of Physical Anthropologists (AAPA, www.physanth.org)

American Board of Criminalistics (ABC, www. criminalistics.com/)

American Board of Forensic Anthropology (ABFA, www.csuchico.edu/anth/ABFA/)

American Board of Forensic Document Examiners (ABFDE, www.abfde.org/)

American Board of Forensic Entomology (ABFE, www.research.missouri.edu/entomology/)

American Board of Forensic Odontologists (ABFO, www.abfo.org)

American Board of Forensic Toxicology (ABFT, www.abft.org)

American Board of Medicolegal Death Investigators (ABMDI, www.slu.edu/organizations/abmdi)

American Board of Questioned Document Examiners (ABQDE, www.asqde.org)

American Chemical Society (ACS, www.chemistry.org)

American Society for Testing Materials (ASTM, www.astm.org)

American Society of Crime Laboratory Directors (ASCLD, www.ascld.org)

Association for Environmental Health and Science (AEHS, www.aehs.com)

Association of Certified Fraud Examiners (ACFE, www.acfe.com)

Association of Firearm and Toolmark Examiners (AFTE, www.afte.org)

California Association of Criminalists (CAC, www.cacnews.org)

Canadian Association of Forensic Sciences (CAFS, www.csfs)

College of American Pathologists (CAP, www.cap.org)

Forensic Sciences Society (FSS, www.forensic-science-society.org.uk)

International Association of Bloodstain Pattern Analysis (IABPA, www.iabpa.org)

International Association of Forensic Nurses (IAFN, www.iafn.org)

International Association of Forensic Toxicologists (IAFT, www.tiaft.org)

International Association of Identification (IAI, www.theiai.org)

National Association of Forensic Engineers (NAFE, www.nafe.org)

National Association of Medical Examiners (NAME, www.thename.org)

National Association of Traffic Accident Reconstructionists and Engineers (NATARI, www.natari.org)

Society for Forensic Toxicologists (SOFT, www.soft-tox.org)

APPENDIX II

COMMON ABBREVIATIONS AND ACRONYMS

The following table contains abbreviations commonly used in forensic science. Except for a few instances, chemical names are not included.

Abbreviation	Meaning
AA or AAS	Atomic absorption spectroscopy
AAFS	American Academy of Forensic Sciences
AAR	Amino acid racemization
Ab	Antibody
ABC	American Board of Criminalistics
ABFA	American Board of Forensic Anthropology
ABFE	American Board of Forensic Entomologists
ABFO	American Board of Forensic Odontologists
ABO	ABO blood group system
ACE-V	analysis, comparison, evaluation, and verification
ACFE	Association of Certified Fraud Examiners
ACP	Acid phosphatase
ADA	Adenosine deaminase
ADME	absorption, distribution, metabolism, and excretion
AEHS	Association for Environmental Health and Science
AFIS	Automated fingerprint identification system
AFLP	Armed Forces Laboratory of Pathology
AFLP	Amplified fragment length polymorphism
AFTE	Association of Firearm and Toolmark Examiners
Ag	Antigen
AK	Adenylate kinase
ALS	Alternate light source
AP	Acid phosphatase
ARF	Anthropological Research Facility ("Body Farm")
As	Arsenic
ASCLD	American Society of Crime Laboratory Directors
ASQDE	American Society of Questioned Document Examiners
ASTM	American Society for Testing Materials
ATF	Bureau of Alcohol, Tobacco, and Firearms (federal)
ATR	Attenuated total reflectance
BAC	Blood alcohol concentration
BSA	Bovine serum albumin
BZ	Benzoylecgonine
CAB	Civil Aeronautics Board (federal)
CAC	California Association of Criminalists
CA II	Carbonic anhydrase
CAT	Computerized axial tomography
CCSC	Criminalistics certification study committee
CDC	Centers for Disease Control (federal)
CE	Capillary electrophoresis
CFC	Chlorofluorocarbon
CFN	Clinical forensic nursing
CGE	Capillary gel electrophoresis
CI	Chemical ionization
CIEF	Capillary isoelectric focusing
CIL	Central Identification Laboratory (US Army)
CNS	Central nervous system
CO	Carbon monoxide
CODIS	Combined DNA Indexing System
CPA	Certified public accountant
CSA	Controlled Substances Act
CSI	Crime scene investigation

CZE	Capillary zone electrophoresis
DA	District attorney
DE	Diatomaceous earth
DEA	Drug Enforcement Agency (federal)
DHS	Department of Homeland Security
DNA	Deoxyribonucleic acid
DOJ	Department of Justice (federal)
DOT	Department of Transportation (federal)
DRE	Drug recognition expert
DRIFTS	Diffuse reflectance infrared Fourier transformation spectroscopy
EAP	Enzyme acid phosphatase
EAP	Erythrocyte acid phosphatase
EDS	Energy dispersive spectroscopy
EDTA	Ethylenediamine tetraacetic acid
EDXRF	Electron diffraction X-ray fluorescence spectroscopy
EI	Electron impact ionization
ELISA	Enzyme linked immunoassay
EMIT	Enzyme multiplied immunoassay
EMP	Electron microprobe
EMR	Electromagnetic radiation
EP	Electrophoresis
EPA	Environmental Protection Agency (federal)
ESD	Esterase D
ESDA	Electrostatic detection apparatus
FAA or FAS	Flame atomic absorption spectroscopy
FBI	Federal Bureau of Investigation
FDA	Food and Drug Administration (federal)
FEPAC	Forensic Education Program Accreditation Commission
FID	Flame ionization detector for gas chromatography
FISH	Forensic information system for handwriting
FISH	Fluorescent in-situ hybridization sex determination
FPIA	Fluorescent polarization immunoassay
FSS	Forensic science service (UK)
FTC	Federal Trade Commission
FTIR	Fourier transform infrared spectroscopy
GAO	General Accounting Office (federal)
GC	Gas chromatography
Gc	Group specific component
GC/MS or GC-MS	Gas chromatography-mass spectrometry
GKE	General Knowledge Examination
GLO I	Glyoxalase I
GLP	Good laboratory practice
GPR	Ground penetrating radar
GSR	Gunshot residue
Hb	Hemoglobin
HLA	Human leukocyte antigen
Hp	Haptoglobin
HPD	Heavy petroleum distillates
HPLC	High performance liquid chromatography
IABPA	International Association of Bloodstain Pattern Analysis
IAFIS	Integrated automatic fingerprint identification system
IAFN	International Association of Forensic Nurses
IAI	International Association of Identification
IBIS	Integrated ballistics identification system
IC	Ion chromatography
ICP	Inductively coupled plasma
ICP-AES	Inductively coupled plasma-atomic emission spectroscopy
ICP-MS	Inductively coupled plasma-mass spectroscopy
IED	Improvised explosive device
IEF	Isoelectric focusing
IMS	Ion mobility spectrometry
IOC	International Olympic Committee
IR	Infrared region of the electromagnetic spectrum
IRS	Internal Revenue Service (federal)
ISO	International Organization for Standardization
LD_{50}	Lethal dose 50
LPD	Light petroleum distillates
LSD	Lysergic acid diethylamide
MALDI	Matrix assisted laser desorption/ionization
ME	Medical examiner
MECC or MEKC	Micellular electrokinetic capillary chromatography
MO	Modus operandi
MPD	Medium petroleum distillates
MRI	Magnetic resonance imaging
MS	Mass spectrometry
MSP	Microspectrophotometry
MtDNA or mDNA	Mitochondrial DNA
NAA	Neutron activation analysis
NAFE	National Association of Forensic Engineers
NASA	National Aeronautics and Space Administration
NASH	Natural, accidental, suicidal, homicidal (cause of death)
NC	Nitrocellulose
NCAVC	National Center for the Analysis of Violent Crime
NCFS	National Center for Forensic Sciences
NCIC	National Crime Information Center
NFPA	National Fire Protection Association
NG	Nitroglycerin
NHTSA	National Highway Transportation Safety Administration
NIJ	National Institute of Justice (federal)
NMR	Nuclear Magnetic Resonance

NPD Nitrogen phosphorus detector for gas chromatography
NTSB National Transportation Safety Board (federal)
OTC Over the counter; drugs and medicines available without a prescription
P/M Parent to metabolite ratio
P2P Phenyl-2-propanone
PCP Phencyclidine
PCR Polymerase chain reaction
Pd Power of discrimination
PD Physical developer
PDA Personal digital assistant
PDQ Paint data query
PDR Physician's Desk Reference
PGM Phosphoglucomutase
PLM Polarizing light microscopy
PMI Post mortem interval
PSA Prostate specific antigen
QA/QC Quality assurance/quality control
RBC Red blood cell
RCMP Royal Canadian Mounted Police
RFLP Restricted fragment length polymorphisms
RI Refractive index
RIA Radioimmunoassay
RMNE Random man not excluded
SANE Sexual assault nurse examiner
SAP Seminal acid phosphatase
SEC Securities and Exchange Commission (federal)
SEC Size exclusion chromatography
SEM Scanning electron microscopy
SIM Selected ion monitoring
SNP Single nucleotide polymorphism
SPF Spectrofluorometry
SPME Solid phase microextraction
STR Short tandem repeats
SWG Scientific Working Groups
TEM Transmission electron microscopy
Tf Transferrin
THC Tetrahydrocannabinol
TLC Thin layer chromatography
TNT 2,4,6-trinitrotoluene
TOF Time of flight
USPS United States Postal Service
USSS United States Secret Service
UV Ultraviolet region of the electromagnetic spectrum
V/V Volume to volume ratio
VAP Vaginal acid phosphatase
ViCAP Violent criminal apprehension program
VIS Visible region of the electromagnetic spectrum
VNTR Variable number of tandem repeats
W/V Weight to volume ratio
WBC White blood cell
WDS Wavelength dispersive spectroscopy
WTC World Trade Center
XRD X-ray diffraction spectroscopy
XRF X-ray fluorescence spectroscopy

APPENDIX III

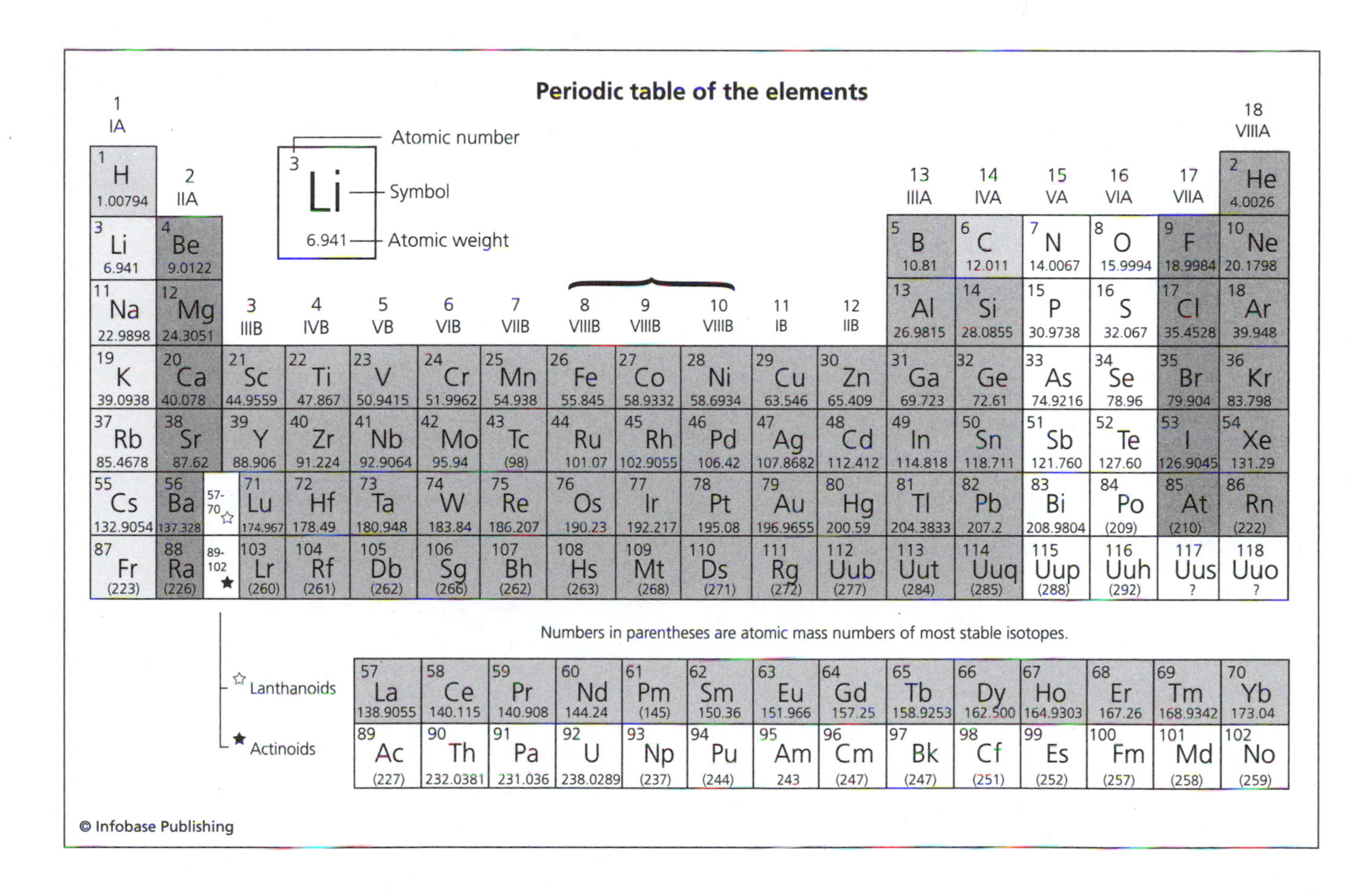

The chemical elements

(g) none
(c) nonmetallics

element	symbol	a.n.
carbon	C	6
hydrogen	H	1

(g) chalcogen
(c) nonmetallics

element	symbol	a.n.
oxygen	O	8
polonium	Po	84
selenium	Se	34
sulfur	S	16
tellurium	Te	52
ununhexium	Uuh	116

(g) alkali metal
(c) metallics

element	symbol	a.n.
cesium	Cs	55
francium	Fr	87
lithium	Li	3
potassium	K	19
rubidium	Rb	37
sodium	Na	11

(g) alkaline earth metal
(c) metallics

element	symbol	a.n.
barium	Ba	56
beryllium	Be	4
calcium	Ca	20
magnesium	Mg	12
radium	Ra	88
strontium	Sr	38

(g) none (c) metallics

element	symbol	a.n.	element	symbol	a.n.
aluminum	Al	13	scandium	Sc	21
bohrium	Bh	107	seaborgium	Sg	106
cadmium	Cd	48	silver	Ag***	47
chromium	Cr	24	tantalum	Ta	73
cobalt	Co	27	technetium	Tc	43
copper	Cu***	29	thallium	Tl	81
darmstadium	Ds	110	titanium	Ti	22
dubnium	Db	105	tin	Sn	50
gallium	Ga	31	tungsten	W	74
gold	Au***	79	ununbium	Uub	112
hafnium	Hf	72	ununtrium	Uut	113
hassium	Hs	108	ununquadium	Uuq	114
indium	In	49	vanadium	V	23
iridium	Ir ****	77	yttrium	Y	39
iron	Fe	26	zinc	Zn	30
lawrencium	Lr	103	zirconium	Zr	40
lead	Pb	82			
lutetium	Lu	71			
manganese	Mn	25			
meitnerium	Mt	109			
mercury	Hg	80			
molybdenum	Mo	42			
nickel	Ni	28			
niobium	Nb	41			
osmium	Os****	76			
palladium	Pd****	46			
platinum	Pt ****	78			
rhenium	Re	75			
rodium	Rh****	45			
roentgenium	Rg	111			
ruthenium	Ru****	44			
rutherfordium	Rf	104			

(g) pnictogen (c) metallics

element	symbol	a.n.
arsenic	As*	33
antimony	Sb*	51
bismuth	Bi	83
nitrogen	N	7
phosophorus	P**	15
ununpentium	Uup	115

(g) none (c) semimetallics

element	symbol	a.n.
boron	B	5
germanium	Ge	32
silicon	Si	14

(g) actinoid (c) metallics

element	symbol	a.n.
actinium	Ac	89
americium	Am	95
berkelium	Bk	97
californium	Cf	98
curium	Cm	96
einsteinium	Es	99
fermium	Fm	100
mendelevium	Md	101
neptunium	Np	93
nobelium	No	102
plutonium	Pu	94
protactinium	Pa	91
thorium	Th	90
uranium	U	92

(g) halogens (c) nonmetallics

element	symbol	a.n.
astatine	At*	85
bromine	Br	35
chlorine	Cl	17
fluorine	F	9
iodine	I	53
ununseptium	Uus*	117

(g) lanthanoid (c) metallics

element	symbol	a.n.
cerium	Ce	58
dysprosium	Dy	66
erbium	Er	68
europium	Eu	63
gadolinium	Gd	64
holmium	Ho	67
lanthanum	La	57
neodymium	Nd	60
praseodymium	Pr	59
promethium	Pm	61
samarium	Sm	62
terbium	Tb	65
thulium	Tm	69
ytterbium	Yb	70

(g) noble gases (c) nonmetallics

element	symbol	a.n.
argon	Ar	18
helium	He	2
krypton	Kr	36
neon	Ne	10
radon	Rn	86
xenon	Xe	54
ununoctium	Uuo	118

a.n. = atomic number
(g) = group
(c) = classification

* = semimetallics (c)
** = nonmetallics (c)
*** = coinage metal (g)
**** = precious metal (g)

INDEX

Note: Page numbers in **boldface** indicate main entries; *italic* page numbers indicate photographs and illustrations.

G

J

K